Statics and Mechanics of Structures

Steen Krenk • Jan Høgsberg

Statics and Mechanics of Structures

Springer

Prof. Steen Krenk
Department of Mechanical Engineering
Technical University of Denmark
Kongens Lyngby, Denmark
sk@mek.dtu.dk

Prof. Jan Høgsberg
Department of Mechanical Engineering
Technical University of Denmark
Kongens Lyngby, Denmark
jhg@mek.dtu.dk

ISBN 978-94-007-6112-4 ISBN 978-94-007-6113-1 (eBook)
DOI 10.1007/978-94-007-6113-1
Springer Dordrecht Heidelberg New York London

Library of Congress Control Number: 2013933869

Cover photo: Golden Terraces Shopping Mall, Warsaw. Designed by the Jerde Partnership, completed 2007. Photo by Steen Krenk

Printed on acid-free paper

Springer is part of Springer Science+Business Media (www.springer.com)

Preface

The theory of statics of structures has developed from intuition via gradual refinement to its current state, where the basic principles are put into a systematic framework that enables precise analysis. Although the basic laws governing statics of structures have been known for several centuries, the methods of analysis have developed considerably over the last decades. At the current state of this development an introductory book on statics should aim at the dual goal of providing sufficient background for developing an intuitive understanding of structures, and at the same time lay a solid foundation for modern analysis, typically made by computational techniques. In this vein the present book makes extensive use of simple but realistic examples to develop familiarity and understanding of how structures carry and distribute the loads through the structural members to the supports. This is then supplemented by a few simple computer programs that illustrate, how the theories for trusses and frames are implemented, and open up to a more general approach to computational mechanics as a natural extension of the present book.

The book is organized as follows. The first five chapters build up a basic understanding of the statics of structures. It starts with force systems and reactions in Chapter 1, then proceeding to the intuitively very accessible theory of trusses, first analyzed by hand calculation procedures and then reformulated as a small systematic finite element program MiniTruss in Chapter 2. Chapter 3 develops the statics of beams and introduces the concept of internal forces. The internal forces are then related to deformation mechanisms of curvature, shear and extension in Chapter 4, and the principle of virtual work is developed in a concise form and used for calculation of specific displacements. The introductory part is rounded off in Chapter 5 on the analysis of columns, describing instability as a bifurcation problem, solved by eigenvalue analysis, and design principles based on the existence of a characteristic imperfection. This part of the book covers material suitable for an introductory one-semester course on basic statics of structures.

The remaining six chapters treat various extensions, that are typically included in one form or another in a second semester course. The Chapters 6 and 7 deal with analysis of statically indeterminate frame structures. The first of these chapters gives a systematic development of the force method and describes how simple structures can conveniently be analyzed by hand. The following chapter then develops the deformation method in which the displacements of individual nodes play the key role. This then serves to introduce the idea of the finite element formulation of frame structures. This development is supported by the small program MiniFrame for internal forces and displacements, and an extension MiniFrameS for linearized stability analysis. The Chapters 8 and 9 introduce three-dimensional states of stress and strain, and present the theory of linear elasticity and some common failure conditions. This material provides the background for the Chapters 10 and 11, in which the simple two-dimensional beam theory used in the previous chapters is extended to flexure and torsion of non-symmetric beams, and the associated shear stress distributions.

The three small computer programs are coded in Matlab. The syntax and input structure are described in connection with the corresponding theory in the text, and the code is available from the authors via e-mail.

The authors are grateful for the permission to include photographs provided by the following companies: Chapter 7, Rafsanjan Bridge, Waagner-Biro AG, Vienna, Austria; Chapter 8, Test of wind turbine blade, LM Wind Power, Kolding, Denmark; Chapter 10, Wind turbine, Siemens Wind Power, Brande, Denmark.

Kgs. Lyngby *Steen Krenk*
September 2012 *Jan Høgsberg*

Contents

1 Equilibrium and Reactions

Statics of structures deals with structures that are exposed to loads and develop reactions and internal forces that leave the structure stationary. The present book deals with buildings and civil engineering structures that are supported to prevent motion, as opposed to space structures, trains etc. where motion is an integral part of the behavior. A fundamental tool of statics is the concept of equilibrium. In order to remain stationary the total effect of the loads and the reactions provided by the supports must be in equilibrium. This applies to the full structure and also to its different parts. In this chapter the equilibrium conditions for the full structure are used to identify requirements for the supports and to determine the reactions provided by the supports. The concept of equilibrium is developed further in the following chapters to deal with hypothetical parts of the structure, and thereby obtain knowledge of the distribution of the forces inside the structure.

First the notion of a force is introduced in Section 1.1. A force is specified by its magnitude and its line of action, and is closely related to the mathematical concept of a vector. If two forces have intersecting lines of action they combine as vectors, and act at the point of intersection. However, this concept of intersecting forces is too limited, and it is necessary to introduce the notion of a moment, as described in Section 1.2. When considering forces and moments together, the concept of equilibrium takes a precise mathematical form, discussed in Section 1.3. The direct form of the equilibrium conditions constitutes two vector equations for the total force and the total moment, respectively. It is explained in detail, how these equations can be combined

S. Krenk, J. Høgsberg, *Statics and Mechanics of Structures*,
DOI 10.1007/978-94-007-6113-1_1,

into a single scalar equation of virtual work – that is the work that the forces and moments would perform, if subjected to an arbitrary small virtual displacement. The virtual work is here introduced in its basic form, but appears in a more advanced form later in connection with deformation of beams and frames. The concept of virtual work plays a central role in the modern formulation of theories for structures and solid bodies, e.g. in connection with the formulation of numerical methods. The two last sections of the chapter deal with the support conditions and the reactions developed in the supports.

1.1 Forces

The notion of a force is fundamental to the theory of structures. A force is associated with a magnitude, a direction, and a point of action. In the analysis of forces it is convenient first to focus on the direction and magnitude, combined in the boldface vector symbol $\mathbf{P}$. The magnitude is represented by the length of the vector and is denoted by $P = |\mathbf{P}|$. In practice a force often has a specific point of action, but it is often convenient to consider the force as acting in a line of action, defined as the line obtained by extending the force vector in space. This notion permits the force to be translated along its line of action, and leads to a fairly intuitive formulation of the theory of equilibrium of a set of one or more forces.

1.1.1 The parallelogram rule

It is an important property of a force $\mathbf{P}$ that it can be resolved into components according to the parallelogram rule, known from elementary vector analysis, see e.g. Strang (2001). The parallelogram rule is illustrated in Fig. 1.1 showing the force $\mathbf{P}$ and two directions intersecting the line of action of $\mathbf{P}$. For convenience and clarity the point of intersection is shown as the point of action of the force $\mathbf{P}$ in the figure. If this is not the case for the initial location of $\mathbf{P}$, it is translated to the point of intersection along the line of action.

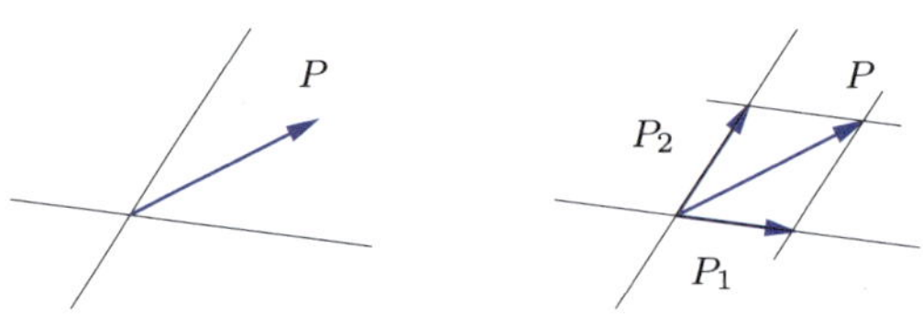

Fig. 1.1: Decomposition of force $\mathbf{P}$ in given direction.

The parallelogram rule for resolving a force $\mathbf{P}$ into components $\mathbf{P}_1$ and $\mathbf{P}_2$ with given directions consists in forming a parallelogram with $\mathbf{P}$ along a diagonal and the components $\mathbf{P}_1$ and $\mathbf{P}_2$ along adjoining sides as shown in Fig. 1.1b.

The parallelogram rule can also be used to form the resulting force from two given forces $\mathbf{P}_1$ and $\mathbf{P}_2$, when these forces have intersecting lines of action. The construction of the parallelogram follows from sliding the forces $\mathbf{P}_1$ and $\mathbf{P}_2$ to the point of intersection. They then form the sides of a parallelogram with the resultant $\mathbf{P}$ along the diagonal as shown in Fig. 1.1b.

Example 1.1. Force on a string. Figure 1.2 shows a simple example of the force parallelogram rule. A vertical force is acting on a string, which is stretched, forming two linear parts. These parts carry constant forces $\mathbf{T}_1$ and $\mathbf{T}_2$, in the direction of the respective strings. □

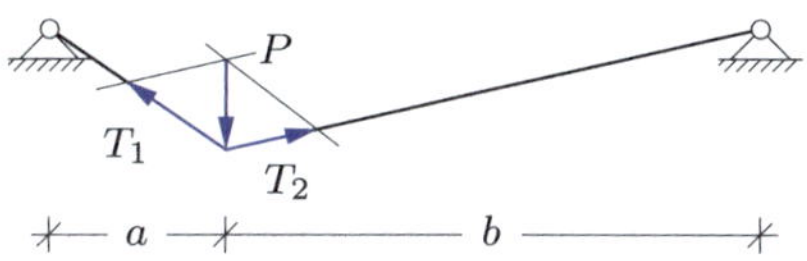

Fig. 1.2: Cable carries $\mathbf{P}$ via the forces $\mathbf{T}_1$ and $\mathbf{T}_2$.

In a static analysis of simple structures the forces may be referred directly to directions in the structure, e.g. as along or transverse to a beam. However, in larger analyzes, and when using a computer for the numerical computations, it is often convenient to represent forces by their components in a Cartesian coordinate system. In this case two forces $\mathbf{P}_1$ and $\mathbf{P}_2$ are represented by their xyz-components as

$$\mathbf{P}_1 = \begin{bmatrix} P_x^1 \\ P_y^1 \\ P_z^1 \end{bmatrix}, \qquad \mathbf{P}_2 = \begin{bmatrix} P_x^2 \\ P_y^2 \\ P_z^2 \end{bmatrix}, \tag{1.1}$$

and it then follows from the parallelogram composition rule that the resultant force $\mathbf{P}$ has the components

$$\begin{bmatrix} P_x \\ P_y \\ P_z \end{bmatrix} = \begin{bmatrix} P_x^1 \\ P_y^1 \\ P_z^1 \end{bmatrix} + \begin{bmatrix} P_x^2 \\ P_y^2 \\ P_z^2 \end{bmatrix}. \tag{1.2}$$

It should be noted that this standard addition rule of vector components must be accompanied by an account of the resultant's line of action.

1.1.2 Parallel forces

In the case of two parallel forces their lines of action do not intersect, and thus the parallelogram rule needs an extension. The problem is illustrated in Fig. 1.3 showing two parallel forces with distance a and magnitude P_1 and P_2, respectively. In principle the magnitude and location of the resulting force $\mathbf{P}$ can be obtained as a limit of the two forces, if inclined slightly with their original common direction. However, it is more direct to obtain the result by introducing two auxiliary forces as demonstrated here.

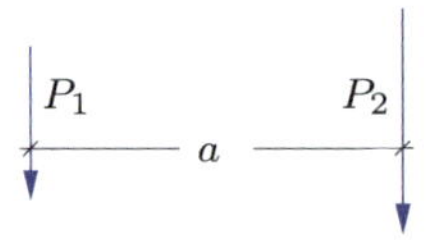

Fig. 1.3: Parallel forces with distance a.

In order to increase the clarity of the geometric construction the two forces $\mathbf{P}_1$ and $\mathbf{P}_2$ are first translated along their respective lines of action, until their points of application A_1 and A_2 lie on a line orthogonal to the lines of action as shown in Fig. 1.4a. Two forces of equal magnitude Q but opposite direction along the connecting line are now added as shown in the figure. As these forces are opposite with the same line of action they have the sum zero, and therefore do not change the resulting force of the system. There is now a force $\mathbf{P}_1 + \mathbf{Q}$ acting at A_1 and a force $\mathbf{P}_2 - \mathbf{Q}$ acting at A_2. These forces are not parallel, and they can therefore be combined by the parallelogram rule, whereby the resulting force $\mathbf{P} = \mathbf{P}_1 + \mathbf{P}_2$ passes through the point of intersection C of the lines of action.

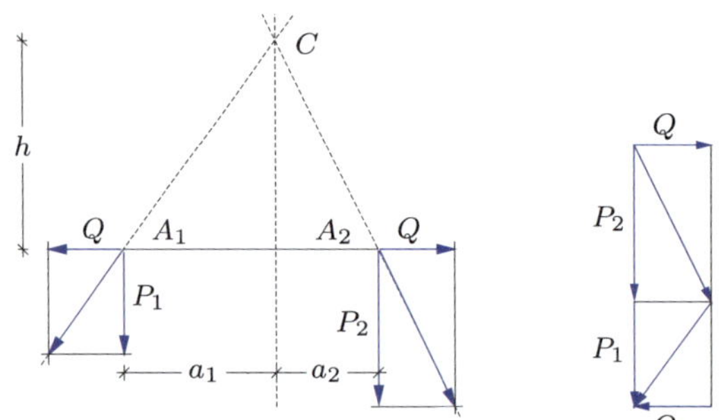

Fig. 1.4: Composition of parallel forces.

The line of action of the resulting force $\mathbf{P}$ is characterized by the distance a_1 from P_1 and the distance a_2 from P_2. The figure contains similar triangles formed by the force components and geometric distances. Thus, the force triangle with sides P_1 and Q is similar to the geometric triangle with sides h and a_1. An equivalent relation holds for P_2 and a_2. From this the following two relations are obtained:

$$\frac{P_1}{Q} = \frac{h}{a_1}, \qquad \frac{P_2}{Q} = \frac{h}{a_2}. \tag{1.3}$$

Elimination of the product hQ between these equations then gives the relation

$$a_1 P_1 = a_2 P_2 . \tag{1.4}$$

This is the lever rule, used since Antiquity for scales where two weights are placed on a lever at different distance from a point of fixture.

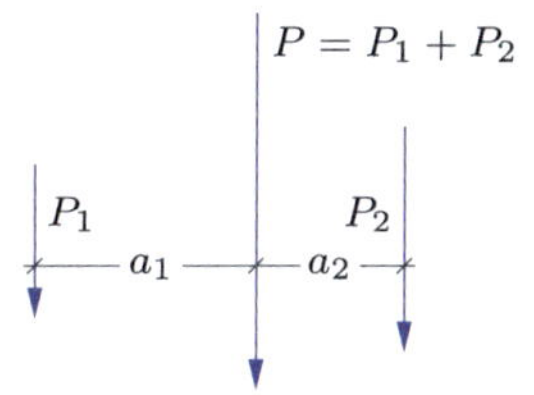

Fig. 1.5: Composition or resolution of parallel forces.

The magnitude of the resulting force is illustrated in Fig. 1.4b, showing the resolution of the forces $\mathbf{P}_2$ and $-\mathbf{Q}$ on top of the resolution of the forces $\mathbf{P}_1$ and $\mathbf{Q}$. The double occurrence of $\pm\mathbf{Q}$ implies that the resulting force retains the original direction and has the magnitude

$$P = P_1 + P_2 . \tag{1.5}$$

Thus, the vector components can also be found by direct summation as already indicated by the component summation formula (1.2).

1.2 Moments

The geometric construction of the line of action of the sum of two parallel forces demonstrates that the two forces are translated in such a way that the product $a_j P_j$ is equal for the two forces. This is a special instance of a moment. Moments play a central role in the mechanics of structures.

The basic form of a moment generated by a force is illustrated in Fig. 1.6. Let C be a point and $\mathbf{P}$ a force in space. The moment of the force $\mathbf{P}$ about the point C is defined by its magnitude and its direction, and can therefore be represented by a vector. The magnitude is hP, where h is the distance of

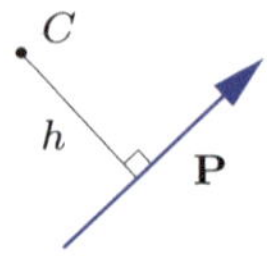

Fig. 1.6: Moment of force **P** about the point C.

C from the line of action of the force. The direction is orthogonal to the plane defined by the line of action and the point C. The concept of a moment is considered in the following subsections: first in the plane containing the line of action and the point C, then by extending the concept to three-dimensional vector form, and finally the relation between force couples and moments.

1.2.1 Moment from forces in a plane

Many problems within structural analysis can be resolved into one or more planar problems. It is therefore convenient first to consider the moment of a force **P** about a point C as a planar problem. The problem is illustrated in Fig. 1.7a, showing the force **P** located in the xy-plane of a Cartesian coordinate system with origin C. The components of the force in this coordinate system are $[P_x, P_y]$. The moment of the force **P** about the point C can be calculated as the sum of the moment of each of the force components P_x and P_y.

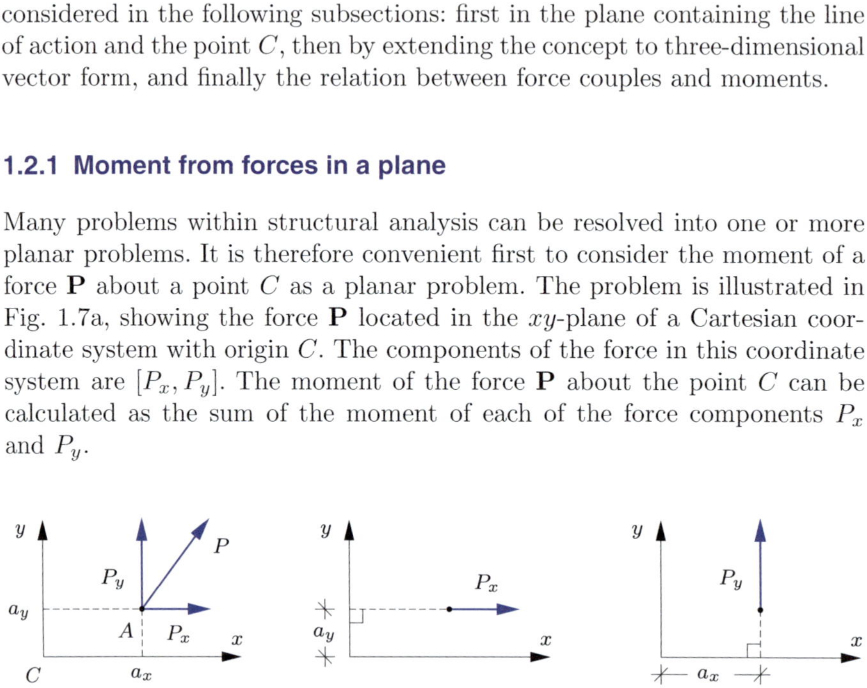

Fig. 1.7: Moment calculation via components.

The procedure is illustrated in Fig. 1.7, where the point A with components $[a_x, a_y]$ indicates the point of application of the force. The moment M_C is considered positive when acting in a counter clockwise direction about C. Thus, the force component P_x gives the moment $-a_yP_x$, while the force component P_y gives the moment a_xP_y. As a result the moment of the force **P** about the point C is

$$M = a_x P_y - a_y P_x . \qquad (1.6)$$

This procedure, where the moment is calculated from suitable components instead of using lengths and angles of the corresponding vectors, is often the most convenient, as illustrated extensively throughout the book.

The moment was here defined in terms of the components of a vector $[a_x, a_y]$ from the origin C to the point A of application of the force, here taken as the origin of the force vector. It is easily demonstrated that the moment remains unaffected if the point A is replaced by any point A' on the line of action of the force. Any such point may be represented in terms of the original point A by a parameter representation of the form

$$\begin{bmatrix} a'_x \\ a'_y \end{bmatrix} = \begin{bmatrix} a_x \\ a_y \end{bmatrix} + \alpha \begin{bmatrix} P_x \\ P_y \end{bmatrix}, \tag{1.7}$$

where α is a scalar parameter defining the location of the point A'. When substituting these vector components into the moment definition (1.6) it is seen that the result is independent of the parameter α. Thus, any point on the line of action leads to the same moment.

Example 1.2. Moment of a force. This example illustrates two alternative methods of computing the moment of a force in a plane: a) straightforward calculation using given components, and b) translating the force along its line of action before computing the moment.

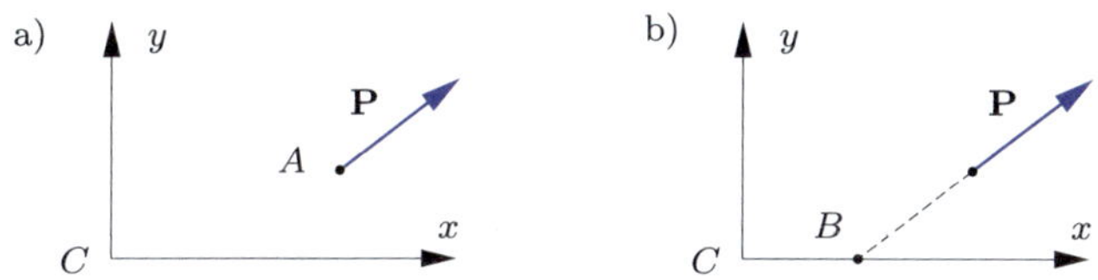

Fig. 1.8: Moment by horizontal and vertical force components.

Figure 1.8 shows a force $\mathbf{P}$ acting at the point A located in the xy-plane with components

$$[P_x, P_y] = [4, 3], \qquad [a_x, a_y] = [6, 3],$$

with dimension of force and length, respectively. Straightforward application of (1.6) gives the moment about the origin C as

$$M = a_x P_y - a_y P_x = 6 \cdot 3 - 3 \cdot 4 = 6.$$

It is observed that if the force was translated along its line of action to the intersection with one of the coordinate axes, then the resulting expression would contain only one product. In the present case the force can be translated to the point B on the x-axis with coordinates $[b_x, b_y] = [2, 0]$. The moment would then be obtained as

$$M = b_x P_y = 2 \cdot 3 = 6,$$

where it has been used that $b_y = 0$. In the analysis of e.g. truss structures it is quite common to simplify the calculation of moments by sliding the forces along their line of action, as illustrated in Chapter 2. □

In the form presented here the moment M appears as a scalar, i.e. a quantity associated with a numerical value. However, in a more general context the moment M calculated here is the moment about an axis normal to the plane through C. Thus, the moment is actually the z-component M_z of a moment vector, acting as a moment about the z-axis of the xyz-coordinate system. While this may be of less importance in planar problems, it is important to recognize the present problem as a special case applicable to forces located in the xy-plane. The next subsection gives the extension to the fully three-dimensional case.

1.2.2 Moment from forces in space

The three-dimensional problem of the moment of a force $\mathbf{P}$ about a point C is illustrated in Fig. 1.9. The force vector $\mathbf{P}$ and the point C define a plane, and the moment is defined by a vector $\mathbf{M}$ through C orthogonal to that plane, with magnitude hP. This coincides with the definition of the vector product between a vector $\mathbf{a}$ from the point C to any point on the line of action of the force and the force vector $\mathbf{P}$,

$$\mathbf{M} = \mathbf{a} \times \mathbf{P}. \tag{1.8}$$

This relation defines the moment $\mathbf{M}$ about C as a vector.

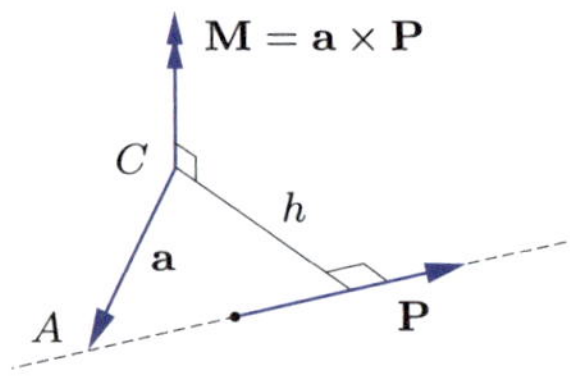

Fig. 1.9: Moment vector $\mathbf{M}$ from force $\mathbf{P}$.

The components of the moment vector $\mathbf{M}$ defined in (1.8) are given by the determinant relation

$$\begin{bmatrix} M_x \\ M_y \\ M_z \end{bmatrix} = \begin{bmatrix} * & * & * \\ a_x & a_y & a_z \\ P_x & P_y & P_z \end{bmatrix}. \tag{1.9}$$

The individual moment components are the sub-determinants defined by this relation,

$$\begin{bmatrix} M_x \\ M_y \\ M_z \end{bmatrix} = \begin{bmatrix} a_y P_z - a_z P_y \\ a_z P_x - a_x P_z \\ a_x P_y - a_y P_x \end{bmatrix} . \tag{1.10}$$

It is seen that the moment component M_z corresponds to the planar case considered in (1.6).

1.2.3 Force couples

The construction from Section 1.1.2 for addition of parallel forces gives finite distances, if the resultant force has a magnitude different from zero. This leaves a special case, where one force is **P** while the other is −**P** with a parallel line of action. This is called a force couple with two forces of equal magnitude but opposite direction acting along parallel lines as illustrated in Fig. 1.10. Let C be a point in the plane, and let $\mathbf{a}_1$ and $\mathbf{a}_2$ be vectors connecting C to a point on the line of action of the force **P** and −**P**, respectively. This is shown in Fig. 1.10a. Applying the vector product formula (1.8), the total moment of the two forces about the point C then is

$$\mathbf{M} = \mathbf{a}_1 \times \mathbf{P} + \mathbf{a}_2 \times (-\mathbf{P}) = (\mathbf{a}_1 - \mathbf{a}_2) \times \mathbf{P} = \mathbf{a} \times \mathbf{P} , \tag{1.11}$$

where $\mathbf{a} = \mathbf{a}_1 - \mathbf{a}_2$ is a vector connecting the two lines of action. Thus, the moment **M** of the force couple is independent of the initial reference point C. It follows from (1.10) that the magnitude of the moment component orthogonal to the plane of the forces is $M = hP$, where h is the distance between the lines of action and P is the length of the force vectors, see Fig. 1.10b.

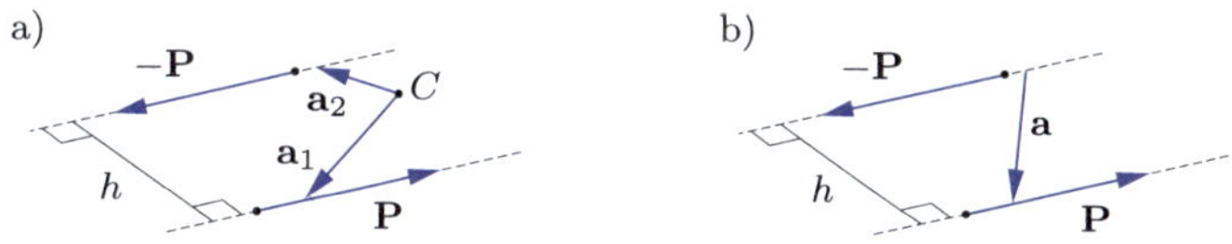

Fig. 1.10: Force couple with **P** and −**P** along parallel lines.

The force couple plays a central role, when a force is translated to a new line of action. This problem is illustrated in Fig. 1.11a, showing the force P in its original position in blue. This force is translated into a new line of action by placing two new forces **P** and −**P** in the new line of action. As they act in the same line and are of equal but opposite magnitude they are equivalent to zero force. The original force **P** and the new force −**P** constitute a force couple with moment $M = hP$, and thus the total effect of the three forces is a translated force **P** and the moment M illustrated in Fig. 1.11b.

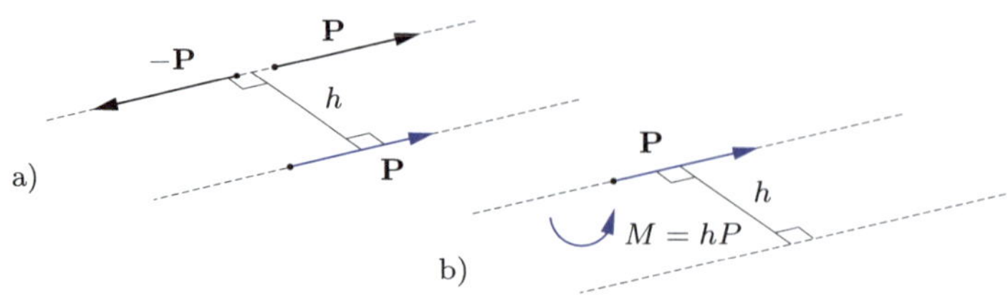

Fig. 1.11: Offset of force by introduction of moment.

1.3 Equilibrium

Statics is the theory of equilibrium of force systems and the use of this theory to study the equilibrium of solids and fluids. In particular statics of structures deals with the equilibrium of structures under time-independent loads.

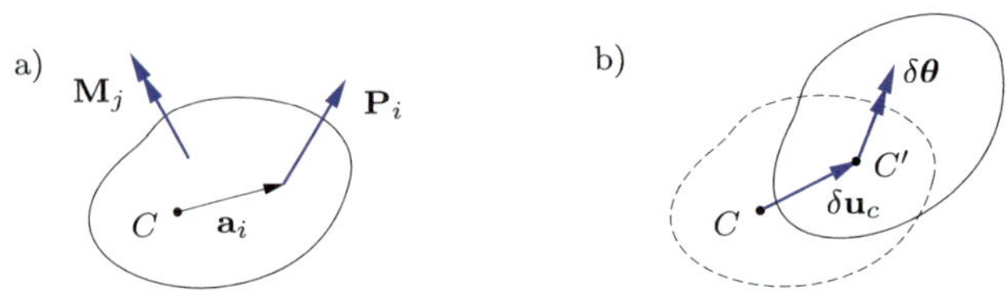

Fig. 1.12: Statics and kinematics of a rigid body.

A basic concept of statics is the notion of equilibrium. Figure 1.12a illustrates a rigid body acted upon by forces $\mathbf{P}_1, \mathbf{P}_2, \cdots$ at points located relative to a common point C by the vectors $\mathbf{a}_1, \mathbf{a}_2, \cdots$ and by moments $\mathbf{M}_1, \mathbf{M}_2, \cdots$. In order for the system of forces and moments, and thereby the rigid body, to be in equilibrium the resulting total force and moment vectors must vanish. The resulting force is found by direct summation of the individual forces as vectors. The resulting moment consists of the vector sum of the applied moments plus the sum of the moments of the forces about a point, e.g. the point C. Thus, equilibrium of a rigid body is expressed by the two vector equations

$$\mathbf{P} = \sum_i \mathbf{P}_i = \mathbf{0}, \qquad \mathbf{M}_C = \sum_j \mathbf{M}_j + \sum_i \mathbf{a}_i \times \mathbf{P}_i = \mathbf{0}. \tag{1.12}$$

The general equilibrium conditions, stated in vector form, are often used for numerical computations, where generality is important. On the other hand most calculations by hand are made by isolating one or more of the components and treating the equilibrium conditions sequentially. Efficient procedures for handling this problem is the main issue in the remaining examples and the exercises in this chapter.

1.3.1 Virtual work of rigid bodies

The equations of equilibrium are closely related to the conditions needed to prevent motion of the body. In essence the three force components must vanish in order not to produce accelerated motion in any of the directions of three-dimensional space. Similarly, the three moment conditions prevent the occurrence of rotation with angular acceleration about any of three independent directions in space. The theory of motion of bodies described in terms of translations and rotations is termed kinematics, while equilibrium described in terms of forces and moments is within the field of statics. There is an interesting and important relation between the kinematic quantities describing a possible motion of a body and the forces and moments acting on this body. This relation is called the equation of virtual work. It is called virtual work, because it deals with a hypothetical situation in which a body is acted upon by actual forces $\mathbf{P}_i$ and moments $\mathbf{M}_j$, and then a small hypothetical motion in terms of displacements $\delta\mathbf{u}$ and rotations $\delta\boldsymbol{\theta}$ is introduced. The symbol δ is used on the displacement parameters to indicate that these are infinitesimal quantities, similar in nature to the notation 'd' used for differentials, e.g. as $\mathrm{d}x$ and $\mathrm{d}y$.

A virtual displacement of a rigid body can be described by the virtual translation $\delta\mathbf{u}_C$ of a point C and the virtual rotation $\delta\boldsymbol{\theta}$ about this point as illustrated in Fig. 1.12b. The virtual rotation is infinitesimal, and the resulting displacement at a point described by the position $\mathbf{a}_i$ relative to the point C can then be described as

$$\delta\mathbf{u}_i = \delta\mathbf{u}_C + \delta\boldsymbol{\theta} \times \mathbf{a}_i \,. \tag{1.13}$$

The idea is now to consider the work that would result if the forces $\mathbf{P}_i$, located at $\mathbf{a}_i$, were translated via the virtual displacements $\delta\mathbf{u}_i$, and the moments $\mathbf{M}_j$ were rotated by the common virtual rotation $\delta\boldsymbol{\theta}$. Because the displacements are virtual this work is called the virtual work and denoted δV. The virtual work of the forces $\mathbf{P}_i$ and the moments $\mathbf{M}_j$ through the virtual displacements $\delta\mathbf{u}_i$ and the common virtual rotation $\delta\boldsymbol{\theta}$ is

$$\delta V = \sum_i \delta\mathbf{u}_i \cdot \mathbf{P}_i + \sum_j \delta\boldsymbol{\theta} \cdot \mathbf{M}_j \,, \tag{1.14}$$

where a dot denotes the scalar product of two vectors. It follows from the representation (1.13) that an infinitesimal virtual rigid body motion can be described in terms of the virtual translation $\delta\mathbf{u}_C$ of the point C and the virtual rotation $\delta\boldsymbol{\theta}$ about this point. When the representation (1.13) is substituted, the expression for the virtual work becomes

$$\delta V = \sum_i \delta\mathbf{u}_C \cdot \mathbf{P}_i + \sum_i (\delta\boldsymbol{\theta} \times \mathbf{a}_i) \cdot \mathbf{P}_i + \sum_j \delta\boldsymbol{\theta} \cdot \mathbf{M}_j \,. \tag{1.15}$$

The middle term is a summation of so-called triple vector products. A vector triple product consists of a vector product of two vectors, multiplied via a scalar product with a third vector – here in the form of the vector product $\delta\boldsymbol{\theta}\times\mathbf{a}_i$ multiplied by $\mathbf{P}_i$. The triple product has a simple geometric interpretation as the volume of the parallelepiped spanned by the three vectors. This volume does not depend on which of the two vectors are used to form the vector product, and thus the factors can be interchanged according to the relation

$$(\delta\boldsymbol{\theta}\times\mathbf{a}_i)\cdot\mathbf{P}_i = \delta\boldsymbol{\theta}\cdot(\mathbf{a}_i\times\mathbf{P}_i)\,. \tag{1.16}$$

When this expression is substituted into (1.15) the virtual work takes the form

$$\delta V = \delta\mathbf{u}_C\cdot\Big(\sum_i\mathbf{P}_i\Big) + \delta\boldsymbol{\theta}\cdot\Big(\sum_i(\mathbf{a}_i\times\mathbf{P}_i)+\sum_j\mathbf{M}_j\Big)\,. \tag{1.17}$$

The terms in the large parentheses have already been identified in (1.12) as the resulting force $\mathbf{P}$ and the total moment $\mathbf{M}_C$ about the point C. Furthermore it was found that equilibrium corresponds to the conditions $\mathbf{P}=\mathbf{0}$ and $\mathbf{M}_C=\mathbf{0}$ for an arbitrary point C. With these observations the virtual work of a rigid body in equilibrium can be expressed as

$$\delta V = \delta\mathbf{u}_C\cdot\mathbf{P} + \delta\boldsymbol{\theta}\cdot\mathbf{M}_C = 0\,. \tag{1.18}$$

The first equality defines the virtual work as the work of the resulting force $\mathbf{P}$ through the virtual displacement $\delta\mathbf{u}_C$ at some reference point C plus the work of the resulting moment $\mathbf{M}_C$ about C through the virtual rotation $\delta\boldsymbol{\theta}$ of the body.

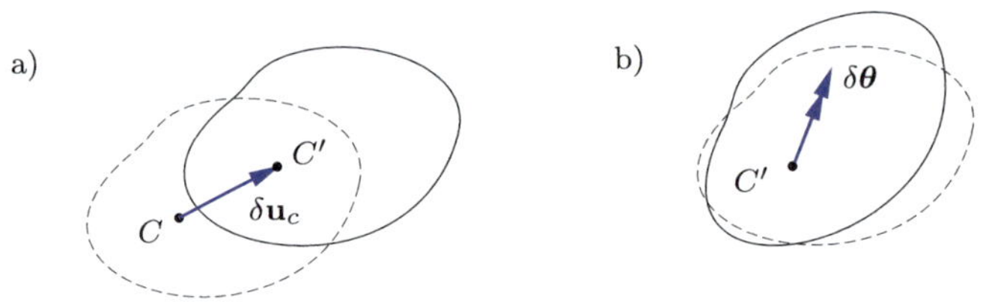

Fig. 1.13: Virtual translation $\delta\mathbf{u}_C$ and rotation $\delta\boldsymbol{\theta}$ of a rigid body.

The second equality states that the virtual work, formed through six virtual displacement components, must vanish. This holds for any choice of the virtual displacement components, and thus the six equilibrium equations are equivalent to the statement that the virtual work must vanish for *any* choice of the virtual displacement components. This is called the principle of virtual work. It is a fundamental tool in statics and mechanics of structures. The principle of virtual work establishes a close relationship between the statics and kinematics of a structure. Thus, the force equilibrium equation $\mathbf{P} = \mathbf{0}$ can be stated as: the work done by the forces by *any* virtual rigid body translation $\delta\mathbf{u}_C$ must vanish. This is illustrated in Fig. 1.13a. Similarly, moment equilibrium about a point C originally expressed as $\mathbf{M}_C = \mathbf{0}$ can be restated in the form: the work done by forces and moments via *any* rotation about C must vanish. This relation is illustrated in Fig. 1.13b.

The principle of virtual work is important in understanding and analyzing structures. It will appear in the discussion of support conditions in Section 1.4 and calculation of reactions in Section 1.6. It is extended to deformable bodies in the following chapters.

1.3.2 Equilibrium in a plane

Many structures can be decomposed into planar parts, i.e. parts that can be described in a plane and are loaded in this plane. For these structures only a reduced set of equilibrium equations are needed. The basic form of these conditions consists of the projection of the force equilibrium condition into the plane, supplemented by the component of the moment equation along a normal to the plane. This corresponds to a special case of the principle of virtual work including only motion in the plane, e.g. in the form of two translation components and one rotation component.

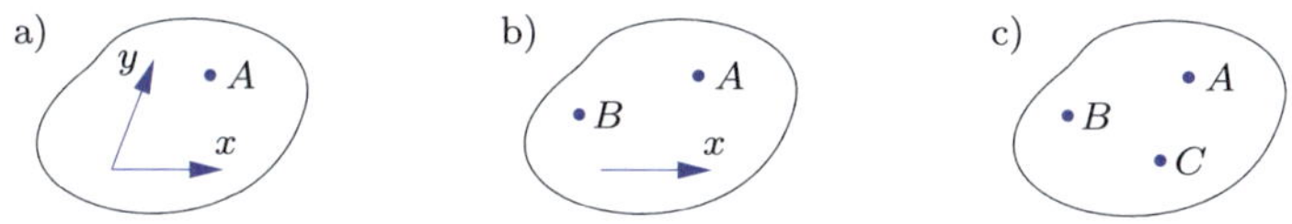

Fig. 1.14: In-plane equilibrium by a) 2 force components and moment about 1 point, b) 1 force component and moment about 2 points, c) moment about 3 points.

It is often useful to modify the initial formulation of the equilibrium equations in terms of two projections and a moment equation by replacing one or both projection equations with moment equations. This results in the following three possibilities, illustrated in Fig. 1.14,

a) Two force projection equations and one moment equation, e.g. $P_x = 0$, $P_y = 0$ and $M_A = 0$.

b) One force projection equation and moment equations for two points, e.g. $P_x = 0$, $M_A = 0$ and $M_B = 0$. The line connecting A and B must not be orthogonal to the line used in the force projection.

c) Moment equations for three points, $M_A = 0$, $M_B = 0$ and $M_C = 0$. The points A, B and C must not lie on a common line.

The examples and exercises of this chapter illustrate the use of each of these sets of equilibrium conditions for calculation of reactions: a) for a clamped structure, b) for a structure with parallel reactions, and c) for structures with non-parallel reactions.

1.3.3 Distributed load

Structures may be exposed to loads that are distributed over the length, surface or volume of the structure. These distributed loads typically arise from e.g. the weight of the structure itself, load from wind or fluid pressure, or earth pressure on foundations. Distributed loads are characterized by their intensity p, denoting the force per unit volume, surface area or per unit length. In plane problems loads are distributed along e.g. the length of a beam, and will have the unit N/m, Newton per meter.

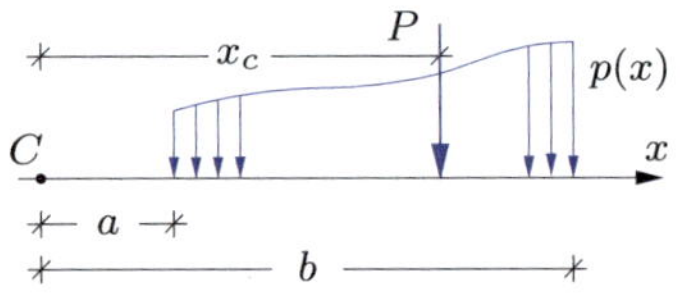

Fig. 1.15: Distributed load $p(x)$ and equivalent load P.

In the analysis of structures with distributed loads it is often convenient to represent the load by an equivalent load in the form of one or more concentrated forces, or a concentrated force and a moment. In order to be equivalent with the original distributed load the concentrated forces (and moments) must have the same resulting force and resulting moment. The situation is illustrated in Fig. 1.15, showing a load distributed along a line with intensity p. The line is taken as x-axis with origin C, and the load extends from a to b. The load is to be represented by an equivalent force P, acting at the distance x_c from C. The magnitude of the distributed load and its moment about C are determined by the integrals

$$P = \int_a^b p(x)\,\mathrm{d}x\,, \qquad M_C = \int_a^b p(x)\,x\,\mathrm{d}x\,. \tag{1.19}$$

The force P is located at the distance x_C from C that produces the same moment M_C as the distributed load, and thus

$$x_c P = M_C \qquad \Rightarrow \qquad x_c = \frac{M_C}{P}\,. \tag{1.20}$$

The calculation is seen to be similar to that of calculating the location of the 'center of gravity' of the load distribution.

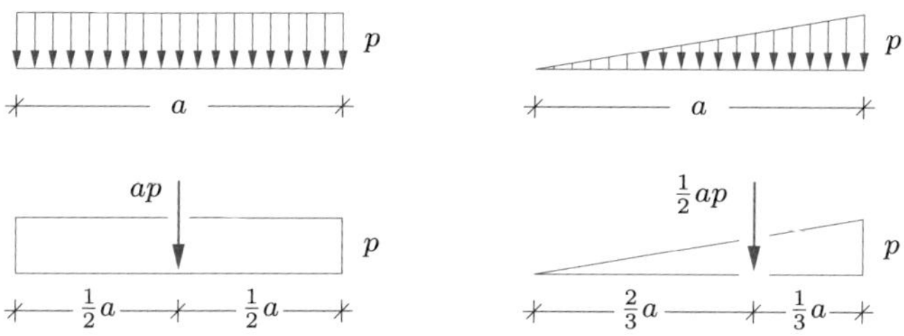

Fig. 1.16: Distributed load and equivalent load.

The two most common load distributions are the constant intensity and a linear variation of the intensity, illustrated in Fig. 1.16. For the uniform intensity load distributed over the interval $[0, a]$ the force and the moment about the left end point follow from the integrals (1.19) as

$$P = \int_0^a p \, \mathrm{d}x = ap \,, \qquad M = \int_0^a p\, x \, \mathrm{d}x = \tfrac{1}{2}a^2 p \,.$$

This gives the distance x_c of the force from the left end as

$$x_c = \frac{M}{P} = \frac{\frac{1}{2}a^2 p}{ap} = \tfrac{1}{2}a \,.$$

The result is illustrated in the left half of Fig. 1.16.

The linear load intensity variation can be represented via triangular distributions as illustrated in the right side of Fig. 1.16. With origin C located at the left end, the load distribution is px/a, where p is used to denote the maximum intensity as shown in the figure. In this case the force and moment integrals are

$$P = \int_0^a px/a \, \mathrm{d}x = \tfrac{1}{2}ap \,, \qquad M = \int_0^a (px/a)\, x \, \mathrm{d}x = \tfrac{1}{3}a^2 p \,.$$

Thus, in this case the distance x_c of the force from the left end is

$$x_c = \frac{M}{P} = \frac{\frac{1}{3}a^2 p}{\frac{1}{2}ap} = \tfrac{2}{3}a \,.$$

This result is illustrated in the right half of Fig. 1.16.

A common load distribution is a linear variation between p_1 at the left end and p_2 at the right end, illustrated in Fig. 1.17. In the representation of the distributed load it is convenient to associate the load intensities p_1 and p_2 with each their triangular distributions. Each of these distributions are then associated with a concentrated force at distance $\frac{1}{3}a$ from the corresponding

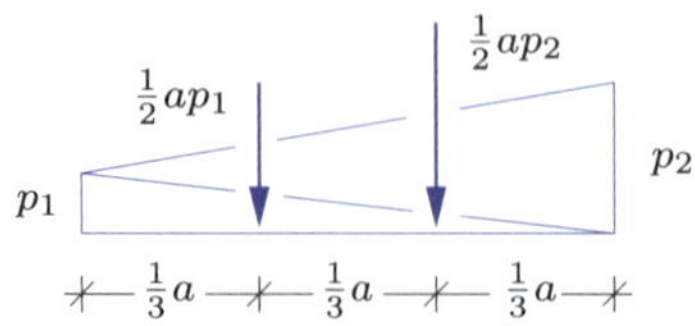

Fig. 1.17: Distributed and equivalent load.

end of the load distribution. The equivalent load then consists of the two concentrated forces $\frac{1}{2}ap_1$ and $\frac{1}{2}ap_2$, located as shown in the figure.

1.4 Support conditions

Structures are typically connected to the surroundings via local supports. A simple example is shown in Fig. 1.18a, where the rotor of a wind turbine is exposed to a distributed horizontal load from the wind. The wind turbine is placed on a foundation that is sufficiently strong and large to secure the wind turbine against motion. As discussed in the previous section equilibrium is associated with the balance of forces and moments. Thus, in plane problems the reaction forces transmitted to the foundation, and from there to the soil, can be represented by two orthogonal force components, typically a horizontal and a vertical force, and a single moment. The wind turbine must be secured against vertical motion due to its own weight, against horizontal motion due to the wind loading P and against rotation in A due to the overturning moment Ph. This requires a vertical reaction force R_A, and horizontal reaction force R'_A and a reaction moment M_A, which are shown in the so-called free body diagram in Fig. 1.18b, where the foundation and the rotor have been replaced by the representative loading and reaction forces acting on the main structure, which in this case is the wind turbine tower.

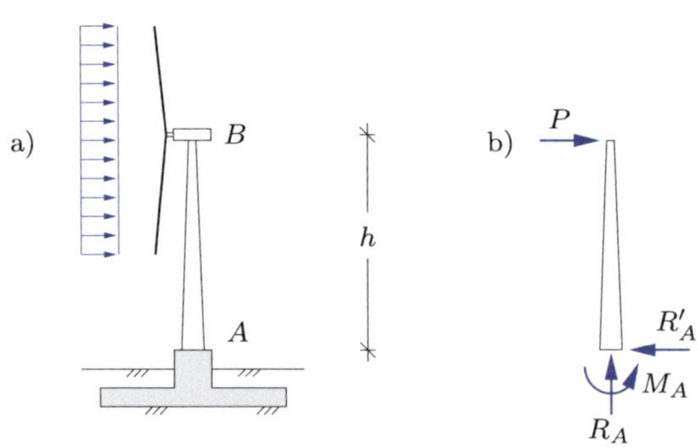

Fig. 1.18: Reaction forces and reaction moment on wind turbine structure.

In the case of the wind turbine all three reactions are required. If for instance the horizontal reaction is removed the structure is free to move in the horizontal direction, thereby violating the concept of equilibrium. The number of support conditions matches the number of reactions necessary for general equilibrium, and the wind turbine structure is therefore *statically determinate.* As demonstrated in the following this means that the unknown reactions can be determined directly from the available equilibrium equations.

Figure 1.19a shows a frame structure placed on two small supporting foundations in A and B. The load on the frame is transmitted to the surrounding soil by the foundation, which is assumed sufficiently large to actually do this for the reaction forces needed in the present problem. The frame structure is supported at two separated points A and B. This separation secures the frame structure against rotation and overturning. Thus, for the frame structure reaction moments in A and B are not required, whereby the local foundations are fairly small. The lack of significant local moment capacity implies that a conservative supporting system only assumes vertical and horizontal reaction forces, i.e. no reaction moments, as indicated by the free body diagram in Fig. 1.19b. This leads to a frame structure with four reactions. As discussed in connection with Fig. 1.14 equilibrium in the plane requires three equilibrium conditions composed of forces and moments. In Fig. 1.19b it is shown that the frame structure has four reactions, which is one more than the number of available equilibrium equations in plane problems. Thus, the frame structure, with the present support conditions, is *statically indeterminate.* This is also illustrated by the fact that one reaction force can be removed without allowing any rigid body motion or mechanisms. For instance, the horizontal reaction R'_A can be removed, and the remaining horizontal reaction R'_B will still secure the structure against horizontal motion. The analytical and numerical analysis of statically indeterminate structures is presented in Chapters 6–7.

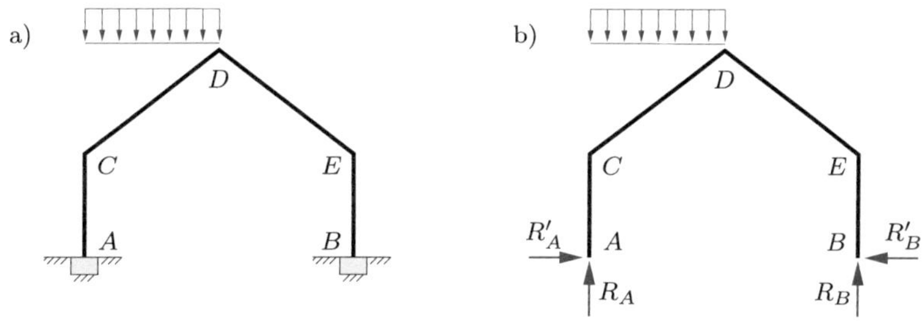

Fig. 1.19: Plane frame: a) Structure with supports, b) Free body diagram with reactions.

In the analysis of structures a number of specific symbols are commonly used to indicate the various support conditions. For basic static analysis the most important symbols are presented in Fig. 1.20, illustrated for a beam type structure. In Fig. 1.20a the circle indicates a hinge that permits free rotation.

Thus, a hinge has no moment capacity. This is also called a simple support, and the reactions are the vertical reaction force R and the horizontal reaction force R'. If supports are allowed to translate in a particular direction, this is indicated by rollers. For the simple support this is shown in Figs. 1.20b,c for unconstrained motion in the horizontal and vertical direction, respectively. Note, that in these situations the support imposes no reaction force in the specific direction of the roller, leaving only a reaction force in the orthogonal direction.

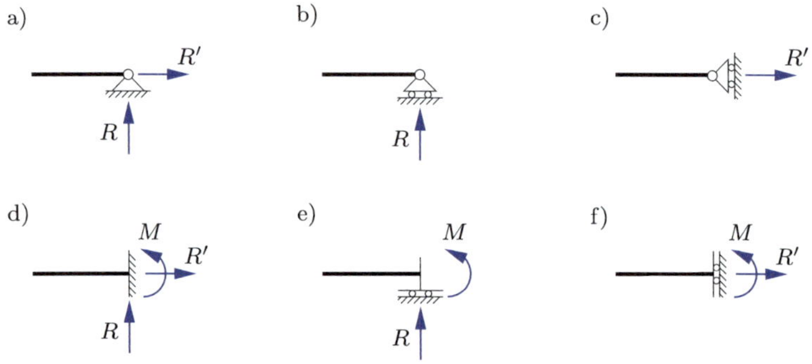

Fig. 1.20: Symbols for various support conditions.

If the support is assumed to be rigidly fixed, it constrains all displacement and rotation components. The symbol for the fixed support is shown in Fig. 1.20d, where all three reaction components are in general non-vanishing. Thus, the wind turbine in Fig. 1.18 has a fixed support. If the moment capacity of a fixed support is removed, the simple support in Fig. 1.20a is recovered. If the fixed support can translate this is indicated by rollers showing the direction of translation as shown in Figs. 1.20e,f. The reaction force is normal to the direction of translation. Most support conditions can be described by the combination of a fixed support, a hinge and/or rollers. In the case of flexible supports these may be represented via spring connections.

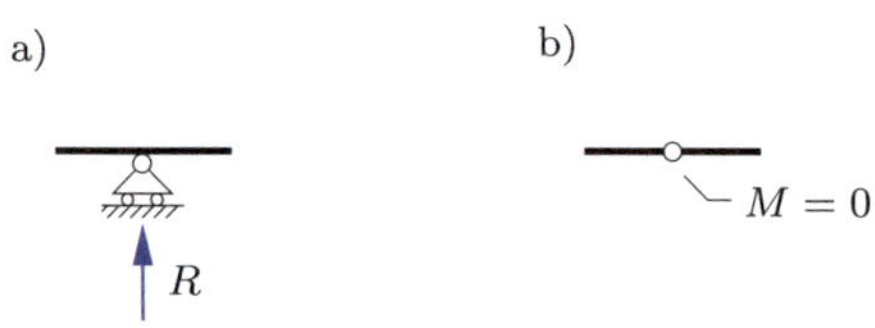

Fig. 1.21: Intermediate and internal support conditions.

Structures often have intermediate supports, such as pylons carrying cables and bridge decks, or columns carrying roof structures. These intermediate support conditions can often also be represented by the symbols presented above. Figure 1.21a shows a simple support with horizontal rollers supporting a beam structure. In this case the only non-vanishing reaction force is the vertical reaction R. It is important to note that the beam in Fig. 1.21a is resting on top of the intermediate simple support, whereby the internal moment capacity of the beam is not changed by the support. Figure 1.21b shows a hinge in the beam, whereby the structure is free to rotate at the hinge. This implies that no internal moment can be transmitted through the hinge, and thus $M = 0$ in the beam at the location of the hinge. This particular type of internal support is common in structural engineering, because structural parts are often connected by hinge type connections with negligible moment capacity. The notion of internal section forces is considered in more detail in Chapter 2 for trusses and in Chapter 3 for beams and frames.

1.5 Reactions by equilibrium equations

If the reactions of a structure can be determined by equilibrium conditions alone, the structure is classified as statically determinate with respect to reactions. A necessary condition for static determinacy of a rigidly connected structure then is that the number of reaction components corresponds to the number of equilibrium conditions. Thus, a rigid structure in space will have six reaction components, while a rigid plane structure will have three reaction components – relating to the plane.

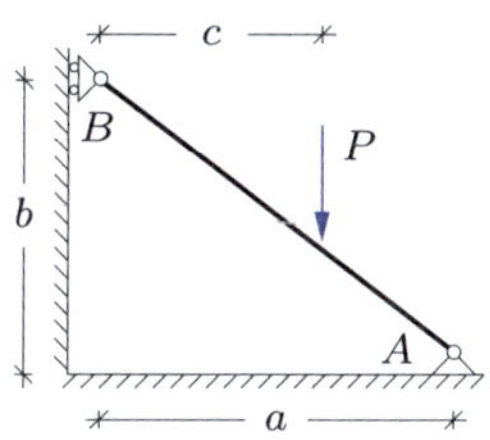

Fig. 1.22: Inclined beam with concentrated load.

It is often convenient to organize the description of the structure and the following analysis of the reactions in a systematic way. The procedure is illustrated by the simple inclined beam shown in Fig. 1.22, and supplemented by a number of examples on the reactions of beams and frames in the later part of this section.

The first step in the analysis of the beam in Fig. 1.22 is to make a description of the structure, including support conditions and loading. This step is illustrated in Fig. 1.23. It consists of making a representative sketch of the static system including reactions and loads as follows:

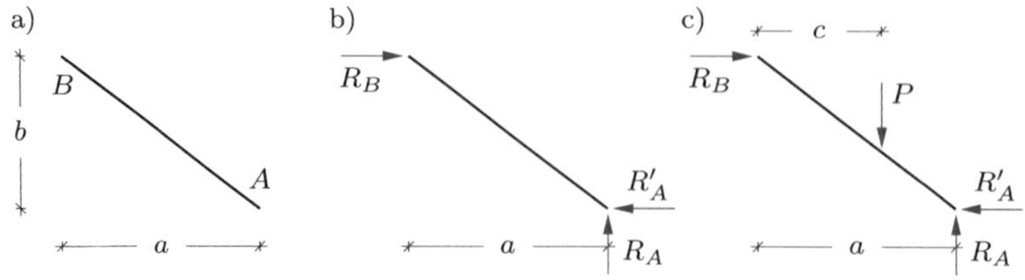

Fig. 1.23: Description of structure with system lines and supports.

a) Make a sketch of the static system with internal joints (as appropriate) and supports. The structure is represented by its system lines, and sufficient dimensions to define the geometry of the structure are included.

b) Indicate all *possible* reactions. It is important that all reactions, that can be generated by the supports are included to ensure that the structure has sufficient support to prevent motion. Compare the number of reaction components with the available number of equilibrium conditions. In the present case the reactions consist of the three components R_A, R_B and R'_A shown in Fig. 1.23.

c) Introduce the loads. This is the last point, as many structures are analyzed for several load cases, and the previous points are unaffected by the specific load case to be analyzed.

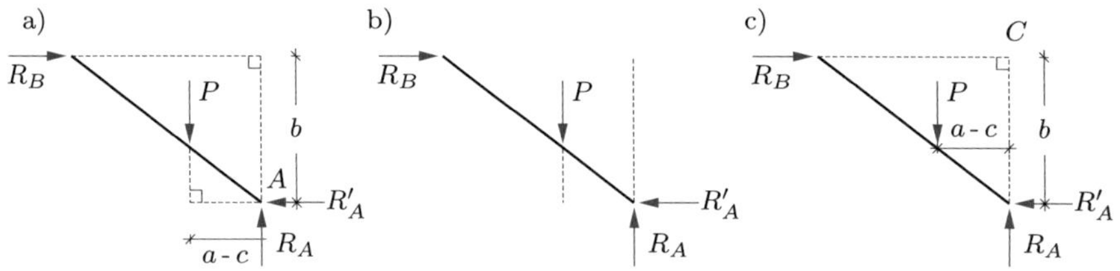

Fig. 1.24: Sketches for static analysis with loads and reactions.

The second step consists in the calculation of the reaction components. It is often preferable to compute each reaction component by a relation that only involves this component, and this can often be accomplished by a suitable choice of the equilibrium condition to be used.

a) There are two reaction components R_A and R'_A at A. Thus R_B can be determined via moment about A. This equation is independent of the components R_A and R'_A,

$$\overset{\frown}{A} \qquad b\,R_B - (a-c)P = 0 \qquad \Rightarrow \qquad R_B = \frac{a-c}{b}\,P\,.$$

b) The reactions R'_A and R_B are both horizontal. Thus, R_A can be determined independently by a vertical projection,

$$\uparrow \qquad R_A - P = 0 \qquad \Rightarrow \qquad R_A = P\,.$$

c) The line of action of the reaction components R_A and R_B intersect at a point C, and R'_A is therefore determined by moment equilibrium about this point.

$$\overset{\frown}{C} \qquad b\,R'_A - (a-c)P = 0 \qquad \Rightarrow \qquad R'_A = \frac{a-c}{b}\,P\,.$$

Note the notation, in which the symbol $\overset{\frown}{A}$ identifies the point as well as the direction of positive moments, while arrow symbols $\uparrow$ and $\rightarrow$ are used for vertical and horizontal projection, respectively.

1.5.1 Plane beams

The beam plays an important role in many structures. The beam is a structural element which is relatively long compared to a characteristic cross-section. Typical examples are wooden or concrete beams with rectangular cross-section and steel beams with I or H-section. Beams are often used in contexts where the statics can be analyzed as one or more two-dimensional problems. A typical first step in a static analysis is determination of the reactions associated with a certain load on the structural element. In the following a number of two-dimensional examples involving beams and frames are considered, and the reactions are determined. The flow of the load *through* beams and frames is considered later in Chapter 3, and the ability of the beams to withstand the effects of the load is discussed in Chapters 8–11.

Example 1.3. Simply supported beam. The horizontal beam shown in Fig. 1.25 is a common structural element. The beam has a fixed simple support at the left end A and a moving simple support with horizontal motion at the right end B. The beam carries a vertical downward concentrated force P acting at the distance a from the left end A. Thus, the support at A permits a vertical reaction R_A and a horizontal reaction R'_A, while the support at B permits a vertical reaction R_B. The reactions are shown in the figure as they would act *on* the beam, if positive. The supports and the associated reactions are precisely sufficient to prevent motion of the beam, and thus the beam is statically determinate.

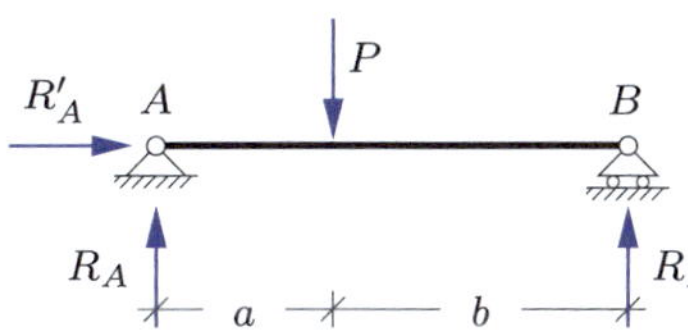

Fig. 1.25: Simply supported beam with concentrated force.

The three reactions are determined from the equilibrium conditions, expressed in the form of one force projection equation and two moment equations as illustrated in Fig. 1.14b. It is advantageous to use the equilibrium equations in a form, where each new condition determines a new reaction. In the present case a simple straightforward procedure consists in using a horizontal projection, followed by moment about A, and finally moment about B.

The only force with a horizontal component is the reaction R'_A, and thus horizontal force equilibrium directly gives $R'_A = 0$. When taking moment about A the two reactions through this point do not contribute, and the resulting clockwise moment about A then takes the form

$$\overset{\frown}{A} \qquad a\,P - (a+b)R_B = 0 \qquad \Rightarrow \qquad R_B = \frac{a}{a+b}\,P\,.$$

Similarly the moment equation about B does not involve the reactions pointing through B, and thus the counterclockwise moment equation is

$$\overset{\frown}{B} \qquad b\,P - (a+b)R_A = 0 \qquad \Rightarrow \qquad R_A = \frac{b}{a+b}\,P\,.$$

It is noted that each of the three reactions have been determined independently, i.e. without using reactions already determined.

It is often desirable to carry out a simple check of the reactions. In the present case a simple check is obtained by calculating the sum of the vertical reactions,

$$\uparrow \qquad R_A + R_B = \frac{b}{a+b}\,P + \frac{a}{a+b}\,P = P\,.$$

Thus, the sum of the vertical reactions is equal to the imposed load P. It is noted that the force is distributed to the supports according to the 'lever rule' illustrated in Fig. 1.5. □

Example 1.4. Cantilever beam. A beam supported only at one end is called a cantilever. An example is illustrated in Fig. 1.26. The cantilever is supported by a fixed rigid support at A, while the other end B is free. The fixed rigid support prevents vertical and horizontal motion via the reaction forces R_A and R'_A, while rotation is prevented by the reaction moment M_A. The reactions and their positive directions are indicated in the figure.

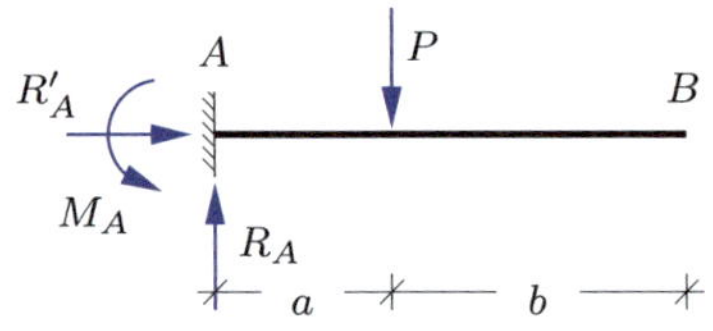

Fig. 1.26: Cantilever beam with concentrated force.

Equilibrium of the horizontal projection of all loads and reactions directly determine the horizontal reaction $R'_A = 0$. Similarly, the vertical projection of the load and the reactions determine the vertical reaction,

$$\downarrow \qquad P - R_A = 0 \qquad \Rightarrow \qquad R_A = P\,.$$

It is seen that the reaction forces R'_A and R_A, determined by projection, are independent of the location of the force a.

The reaction moment M_A is determined by taking the clockwise moment about A,

$$\overset{\frown}{A} \qquad a\,P - M_A = 0 \qquad \Rightarrow \qquad M_A = aP\,.$$

The moment M_A is seen to increase linearly with a, confirming the intuitive feeling that it requires more to support a load at a large distance.

Note, that the load P and the vertical reaction R_A constitute a force couple aP, balanced by the reaction moment M_A. □

Example 1.5. Beams with hinge. Figure 1.27 shows two beams AD and DB, joined at D by a hinge. The French term 'charnier' for a hinge is often used in connection with structures. The total system consisting of the two beams must be in equilibrium, thus providing three equilibrium conditions. In addition, the moment at the hinge vanishes, and thus the moment of all loads and reactions on each side of the hinge must vanish. The hinge introduces an extra degree of freedom in the motion of the structure, and thereby a need for an extra reaction, bringing the number of reactions of the present structure to four. There are three simple supports, of which the support at C is fixed, while the supports at A and B permit horizontal motion. Thus, there is one horizontal reaction R'_C and three vertical reactions R_A, R_B and R_C, as indicated in the figure. In the present example the load consists of a concentrated force P acting at the center of the beam DB.

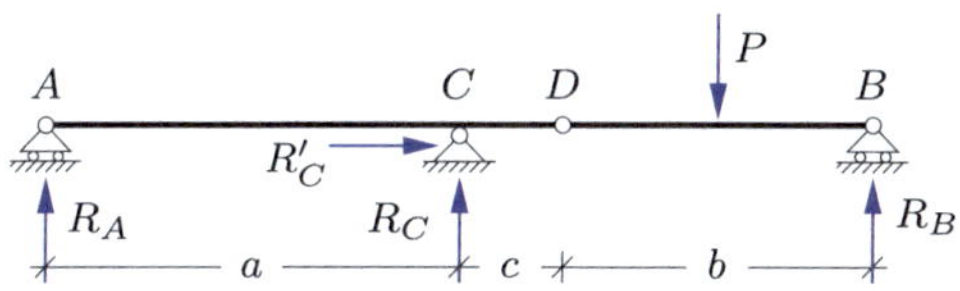

Fig. 1.27: Hinged beams with concentrated force.

When determining the reactions on structures with hinges, the forces transmitted through the hinges are often included in the analysis. This is illustrated in Fig. 1.28, showing the parts slightly separated to make room for the force components R_D and R'_D acting on the beam DB through the hinge.

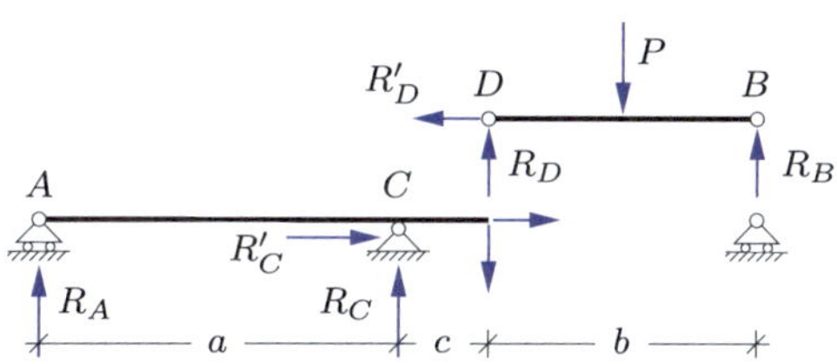

Fig. 1.28: Reactions and hinge forces.

First the reactions on the beam DB are determined. It follows from horizontal force equilibrium that $R'_D = 0$. Thus, the hinge at D only transmits a vertical force R_D. The beam DB now appears as simply supported, carrying a concentrated force P at its center. Symmetry implies that the two reactions R_D and R_B are equal, and it then follows from vertical force projection that $R_D = R_B = \frac{1}{2}P$. These reactions could also have been determined by moments about the end points of the beam as in Example 1.3.

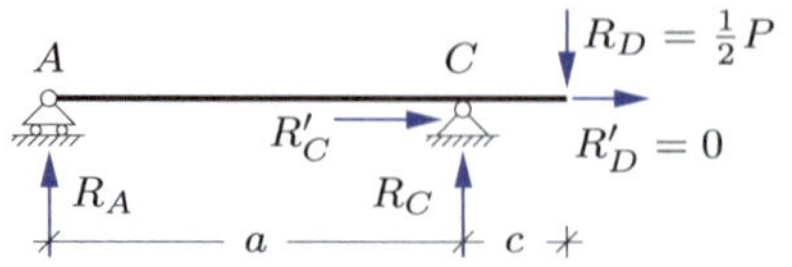

Fig. 1.29: Reactions on left beam.

Finally, the remaining reactions are determined by considering the beam AD as simply supported and loaded by R_D and R'_D as shown in Fig. 1.29. Horizontal force equilibrium gives

$$\rightarrow \qquad R'_C + R'_D = 0 \qquad \Rightarrow \qquad R'_C = 0\,.$$

Counterclockwise moment about A gives

$$\overset{\curvearrowleft}{A} \qquad a\,R_C - (a+c)\,\tfrac{1}{2}P = 0 \qquad \Rightarrow \qquad R_C = \frac{a+c}{2a}\,P\,,$$

while clockwise moment about C gives

$$\overset{\curvearrowright}{C} \qquad a\,R_A + c\,\tfrac{1}{2}P = 0 \qquad \Rightarrow \qquad R_A = -\frac{c}{2a}\,P\,.$$

Note, that the beam AD carries half the load, and thus $R_A + R_C = \frac{1}{2}P$. However, as R_A is negative, the reaction R_C is greater than the load $R_D = \frac{1}{2}P$ actually carried by this part of the structure. □

1.5.2 Simple frames

Frames offer some additional features in the determination of reactions. The following two examples illustrate how to deal with supports placed at different heights and supports with rollers at inclined angles. As in the case of beams the main point is to arrange the analysis in such a way that the individual reaction components can be computed sequentially.

Example 1.6. Portal frame. This example deals with inclined load and a frame with supports at different height. Figure 1.30a shows a rigid frame with a fixed simple support at A and a simple support on horizontal rollers at B. The frame is loaded by a concentrated force of magnitude $\sqrt{2}P$ with inclination $-45°$ at C.

The load, decomposed in horizontal and vertical components of magnitude P, is shown in Fig. 1.30b together with the three reaction components R'_A, R_A and R_B.

The reactions are determined sequentially by the following procedure. First the component R'_A is determined by horizontal projection of all load and reaction force components,

$$\rightarrow \qquad R'_A + P = 0 \qquad \Rightarrow \qquad R'_A = -P\,.$$

Counterclockwise moment about A then gives

$$\overset{\curvearrowleft}{A} \qquad 2a\,R_B - 2a\,P = 0 \qquad \Rightarrow \qquad R_B = P\,.$$

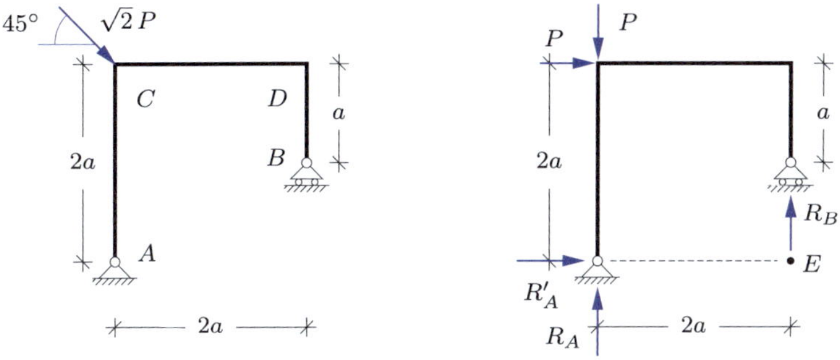

Fig. 1.30: Portal frame with inclined load.

The last reaction R_A can be determined by moment about the support point B. However, this would imply that both reaction components R_A and R'_A would appear in the equation. Although the horizontal component R'_A has already been determined it may be advantageous to find an equilibrium condition in which only the unknown reaction component R_A appears together with the given loads. This condition is found by taking moment about the point of intersection E of the two other reaction components R'_A and R_B. At this point these two reaction forces do not contribute to the moment. Clockwise moment about E gives

$$\overset{\frown}{E} \quad 2a\,R_A + 2a\,P - 2a\,P = 0 \qquad \Rightarrow \qquad R_A = 0\,.$$

The results are controlled by translating the reactions R'_A and R_B to the point of intersection E, and observing that they combine to a force along the diagonal EC. □

Example 1.7. Angle frame. The distribution of the load to the supports depends on the support conditions as illustrated in this example. The issue is illustrated by comparing the angle frame in Fig. 1.31a with vertical reaction at B with the similarly loaded frame in Fig. 1.32a with a reaction at B inclined by 45°. The example also illustrates the use of the equivalent load in the form of an equivalent concentrated force in the calculation of the reactions.

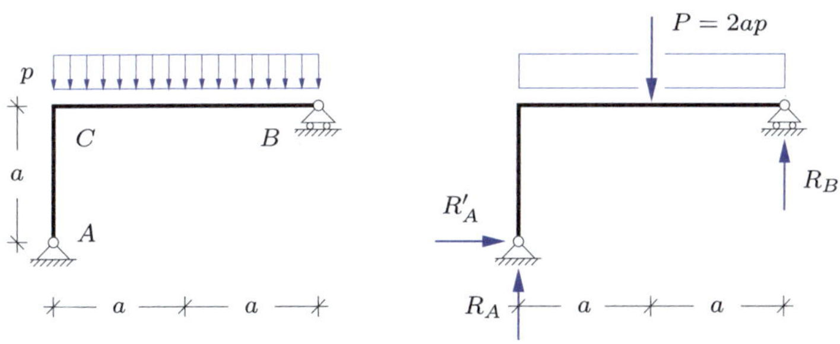

Fig. 1.31: Angle frame with distributed load.

The distributed load with intensity p is equivalent to a concentrated force $P = 2ap$ acting at the center of the beam CB. This is independent of the support condition, and thus applies to both cases as shown in Fig. 1.31b and Fig. 1.32b.

In the case of the frame with vertical reaction in B shown in Fig. 1.31 the calculation proceeds in a manner quite similar to that of the simply supported beam in Example 1.3. It follows immediately from horizontal projection that $R'_A = 0$, and the vertical reaction components then follow from moment about B and A, respectively.

$$\overset{\frown}{B} \qquad 2a\,R_A - a\,(2ap) = 0 \qquad \Rightarrow \qquad R_A = ap\,,$$

$$\overset{\frown}{A} \qquad 2a\,R_B - a\,(2ap) = 0 \qquad \Rightarrow \qquad R_B = ap\,.$$

Vertical equilibrium is verified by projection, giving the sum of the reactions $R_A + R_B = 2ap$, matching the vertical load.

In the case of the frame with inclined support at B the calculation of reactions is arranged in a slightly different way. In this case there are no parallel reaction components, and each reaction component can therefore be determined independently by taking moment about the intersection point of the two other reaction components. By this procedure the calculation of each reaction component is independent of the order in which they are computed.

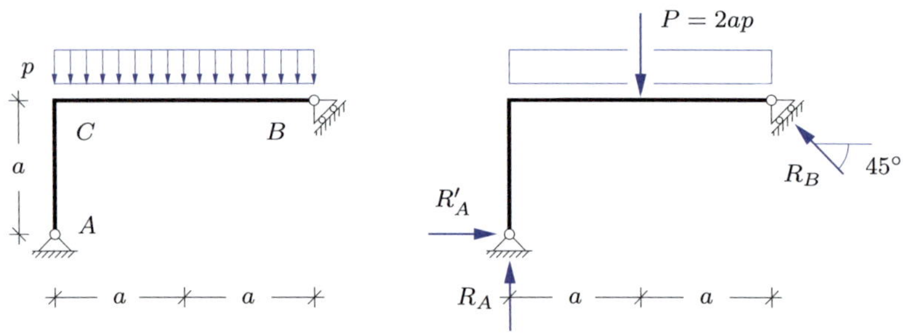

Fig. 1.32: Angle frame with inclined support.

First the horizontal reaction R'_A is calculated by taking moment about the point D defined by the intersection of the lines of action of R_A and R_B. The point D is located vertically above C at distance $2a$. Thus, the moment equation is

$$\overset{\frown}{D} \qquad 3a\,R'_A - a\,(2ap) = 0 \qquad \Rightarrow \qquad R'_A = \tfrac{2}{3}ap\,.$$

The intersection E of the lines of action of R'_A and R_B is located at the same height as A at a distance $3a$ to the right. This gives the moment equation

$$\overset{\frown}{E} \qquad 3a\,R_A - 2a\,(2ap) = 0 \qquad \Rightarrow \qquad R_A = \tfrac{4}{3}ap\,.$$

The final equilibrium equation is a moment equation about A. The distance from the point A to the line of action of R_B is $(\frac{3}{2}a)\sqrt{2}$, and thus the moment equation is

$$\overset{\frown}{A} \qquad \tfrac{3}{2}\sqrt{2}a\,R_B - a\,(2ap) = 0 \qquad \Rightarrow \qquad R_B = \tfrac{2}{3}\sqrt{2}ap\,.$$

Alternatively, sliding the reaction R_B along its line of action to the point E at the same level as A and using its vertical component $\frac{1}{2}\sqrt{2}R_B$ gives the same equation. Finally, a check is obtained by horizontal and vertical projection of the reaction forces,

$$\rightarrow \quad R'_A - \tfrac{1}{2}\sqrt{2}\,R_B = 0\,, \qquad \uparrow \quad R_A + \tfrac{1}{2}\sqrt{2}\,R_B = 2ap\,,$$

corresponding to zero horizontal load and a vertical load of $P = 2ap$.

An alternative procedure consists in determining the inclined reaction R_B first, and then obtaining R_A and R'_A by vertical and horizontal projection, respectively. However, in that procedure the determination of the components are not independent. □

1.5.3 Three-hinge frame

A classic form of a statically determinate frame is the three-hinge frame, sometimes called a three-charnier frame, shown in Fig. 1.33. The frame has two fixed supports with hinges, and has a third hinge in the frame structure. As a result of the hinge within the frame it now has two reaction components at each of the supports. The shape of the frame and the location of the internal hinge may vary without changing the basic principles of analysis of the frame. The task here is to determine the reactions at A and B.

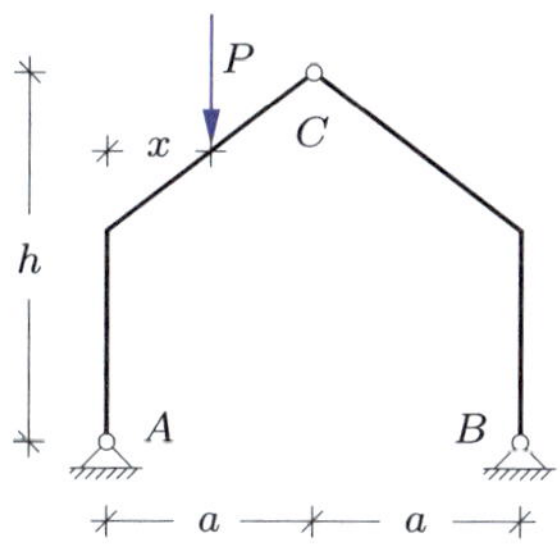

Fig. 1.33: Three-charnier frame.

The reactions may be determined by purely graphical means using boldface notation for vectors, as illustrated in Fig. 1.34 for a concentrated force $\mathbf{P}$ acting on the left part of the frame at the distance $x \leq a$ from the left side. The reaction force $\mathbf{R}_B$ at B is the only force to act on the right half of the frame, apart from any force transmitted through the hinge at C. Therefore its line of action must pass through C in order not to produce any moment about C. Thus, the line of action of $\mathbf{R}_B$ goes through BC as shown in Fig. 1.34. By extending the lines of action of the load $\mathbf{P}$ and the reaction $\mathbf{R}_B$ the intersection point is identified. When the load is translated to this point, it can be resolved into a force $-\mathbf{R}_B$ through the support point B and another force $-\mathbf{R}_A$ through the support point A by use of the force parallelogram construction introduced in connection with Fig. 1.1 in Section 1.1.1. This purely geometric construction gives the reaction forces $\mathbf{R}_A$ and $\mathbf{R}_B$ as

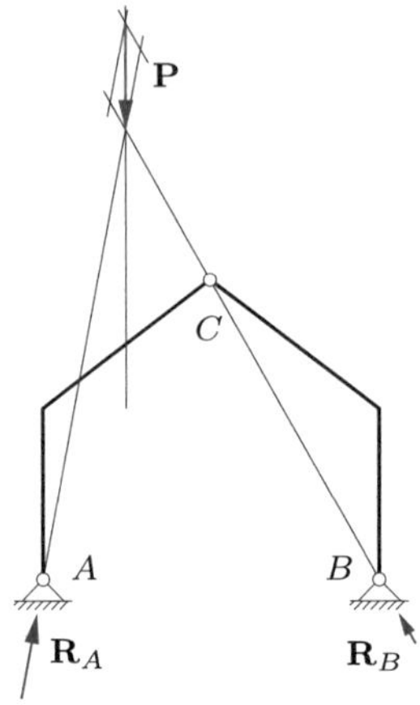

Fig. 1.34: Force resolution for three-hinge frame.

vectors. While the graphic procedure establishes a direct relation between the geometry of the frame and the resulting reactions, and thereby a good intuitive understanding of the statics of the frame, it is less convenient for numerical computations.

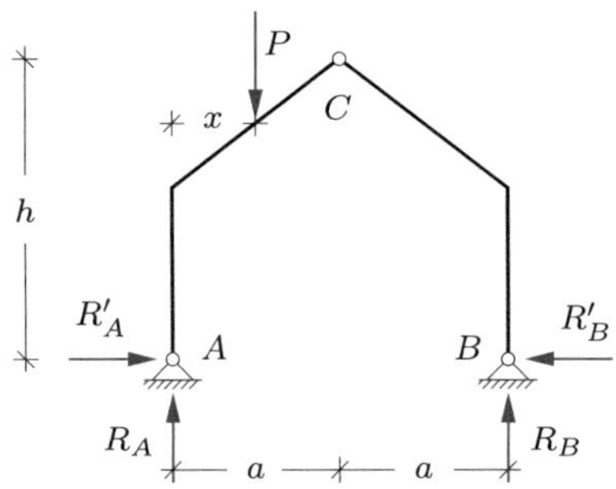

Fig. 1.35: Reactions of three-hinge frame.

A straightforward computation of the reactions can be performed by use of the principles already used in the previous examples. First the reactions are represented in terms of their vertical components R_A, R_B and their horizontal components R'_A, R'_B, shown in Fig. 1.35. These components can then be determined sequentially by moment and force projection equations as follows.

Example 1.8. Reactions of the three-hinge frame. First the vertical reaction R_B is determined by moment equilibrium of the total frame about A,

$$\overset{\frown}{A} \qquad x\,P - 2a\,R_B = 0 \qquad \Rightarrow \qquad R_B = \frac{x}{2a}\,P\,.$$

In a similar way the vertical reaction R_A is determined by moment equilibrium of the total frame about B,

$$\overset{\frown}{B} \qquad (2a - x)\,P - 2a\,R_A = 0 \qquad \Rightarrow \qquad R_A = \frac{2a - x}{2a}\,P\,.$$

The hinge at C imposes a condition of zero moment of the forces on *the right half* of the frame about C,

$$\overset{\frown}{C} \qquad h\,R'_B - a\,R_B = 0 \qquad \Rightarrow \qquad R'_B = \frac{x}{2h}\,P\,.$$

By using the forces on the right half of the frame the moment relation only involves the two reaction components R_B and R'_B. If the left half were used, the moment equation would involve the similar reaction components at A and in addition the contribution form the load P. The determination of the reactions is completed by using a horizontal projection of all forces acting on the frame,

$$\rightarrow \qquad R'_A = R'_B \qquad \Rightarrow \qquad R'_A = \frac{x}{2h}\,P\,.$$

It is noted that the vertical reactions are independent of the height h of the frame, while the horizontal components are proportional to $x/2h$, $(x < a)$. Thus, increasing the height of the location of the central hinge will reduce the horizontal reactions correspondingly. □

1.5.4 Space structures

Although many structures can be represented via planar problems, some must be analyzed in three dimensions. In three-dimensional space the number of equilibrium equations increases to six, e.g. projection in three directions and moment about three lines. The following example demonstrates the procedure for a simple space frame.

Example 1.9. Reactions of simple crane. Figure 1.36a shows a space structure representing a simple crane, carrying a tip load P. The structure is simply supported in A, B and C, permitting horizontal displacement in the x-direction in A and C and in the y-direction in B. Thus, the structure is statically determinate. Figure 1.36b shows a top view of the crane structure. The vertical forces are indicated by circles with a bullet or a cross inside, indicating positive upward or downward direction, respectively. First the horizontal reactions are determined. Equilibrium in the x-direction directly gives zero horizontal reaction $R'_B = 0$ in B. Moment about the vertical line through B only receives a contribution from the horizontal reaction in A, and thus $R'_A = 0$. The horizontal reaction in C can be found by equilibrium in the y-direction, or by moment about the vertical line through A, leading to $R'_C = 0$. Thus, all horizontal reactions vanish.

The vertical reaction in A is obtained directly by moment about BC:

$$\overset{\frown}{BC} \qquad a\,R_A + a\,P = 0 \qquad \Rightarrow \qquad R_A = -\,P\,.$$

Symmetry implies that $R_B = R_C$. This can also be verified by moment about the line DE, where the only non-vanishing reactions are R_B and R_C. The magnitude of these identical

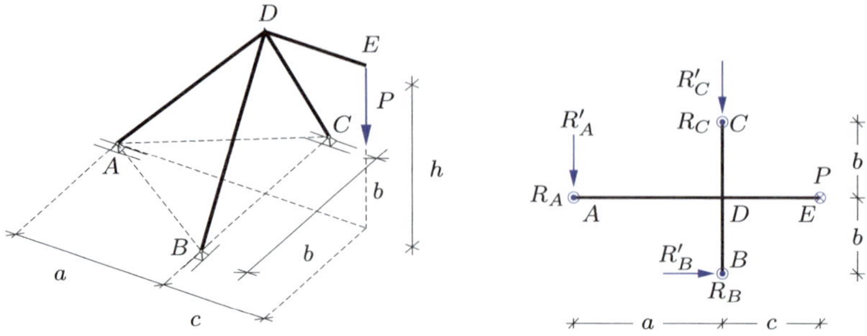

Fig. 1.36: a) Crane carrying vertical tip load P. b) Top view with forces.

reactions can then be found by vertical equilibrium:

$$\downarrow \quad P - R_A - R_B - R_C = 0 \qquad \Rightarrow \qquad R_B = R_C = P\,.$$

This shows that the three reactions have the same magnitude, because the support at A has the same distance from the central plane BCD as the load P. □

1.6 Reactions by virtual work

The determination of statically determinate reactions on a structure was discussed in Section 1.5 based on direct use of equilibrium conditions. While this is always possible for statically determinate reactions, it is sometimes convenient to use the principle of virtual work instead. The advantage of this is partly the ability to concentrate the calculation on individual reaction components, even for compound structures with internal hinges, and partly the visual display of displacement mechanisms that determine the contributions from loads applied anywhere on the structure.

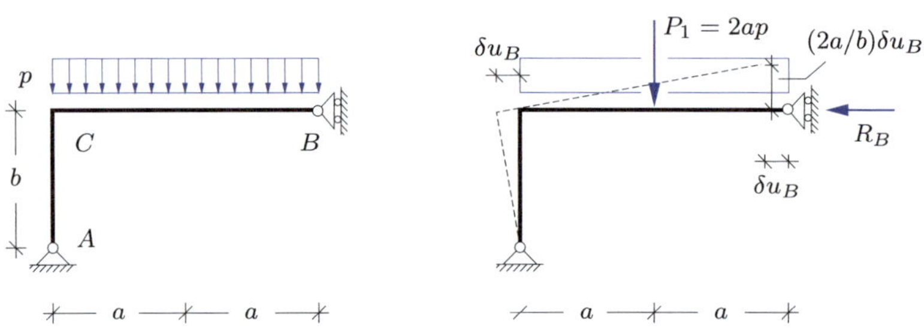

Fig. 1.37: Angle frame with distributed load.

The basic idea is illustrated in Fig. 1.37 showing a simply supported angle frame. The frame has a simple support permitting vertical motion of B. Thus, there is one horizontal reaction component R_B at the support B. The idea behind the use of the principle of virtual work for determination of reactions is to release the support, and to impose a virtual displacement δu_B of the support. When the structure is statically determinate the release of one support condition creates precisely one mechanism. The mechanism created by releasing the horizontal support in B is a rotation about the fixed support A, illustrated by the dashed line in the figure. The displacements associated with the mechanism are considered infinitesimally small, and the magnitudes of the displacements are then proportional to the imposed virtual displacement δu_B. For concentrated loads the corresponding virtual work equation takes the form

$$\delta V = \delta u_B R_B + \sum_j \delta u_j P_j = 0, \tag{1.21}$$

with loads P_j and corresponding virtual displacements δu_j, defined by the mechanism corresponding to releasing the support with reaction R_B. The particular mechanism shown in the figure corresponds to an infinitesimal counter clockwise rotation of magnitude $\delta u_B/b$, and thus the virtual displacements of the force is $\delta u_1 = -(a/b)\delta u_B$. Hereby the equation of virtual work takes the form

$$\delta V = \delta u_B R_B - \frac{a}{b}\delta u_B P_1 = 0 \qquad \Rightarrow \qquad R_B = \frac{a}{b} P_1 . \tag{1.22}$$

The displacement field from the virtual rotation about A is here proportional to δu_B. In some cases it appears more direct to represent the displacement field in terms of a virtual rotation – in this case the rotation about A, given by $\delta\theta_A = \delta u_B/b$.

Example 1.10. Simply supported beam. This example illustrates the use of the principle of virtual work to determine the reactions of the simply supported beam AB of length ℓ with a concentrated force P located at the distance a from the left support A shown in Fig. 1.38a. The reactions have already been determined by use of force projection and moments in Example 1.3, and thus the present example mainly illustrates the idea of the virtual work equation, that will find more general applications later.

Each of the three equilibrium equations for the beam are replaced by an equivalent virtual motion of the beam, found by releasing a support component and imposing an infinitesimal positive virtual displacement in its place. The three virtual motions are illustrated in Fig. 1.38b. The first is a virtual horizontal translation $\delta u'_A$ of the beam. The vertical force P does not contribute to the virtual work produced by a horizontal motion, and thus the corresponding virtual work equation is

$$\delta V = \delta u'_A R'_A = 0 \qquad \Rightarrow \qquad R'_A = 0 .$$

Clearly, this condition is equivalent to a horizontal force projection.

The second virtual motion consists in lifting the support at A by the virtual displacement δu_A, whereby the beam rotates about B. The corresponding virtual work equation is

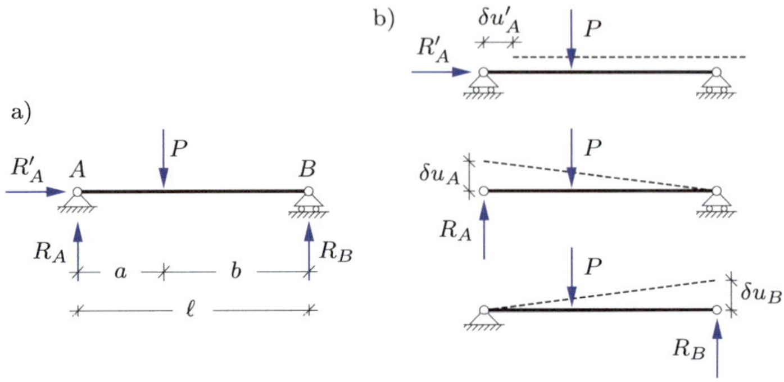

Fig. 1.38: a) Simply supported beam. b) Virtual displacements.

$$\delta V = \delta u_A \, R_A - \frac{b}{\ell} \, \delta u_A \, P = 0 \qquad \Rightarrow \qquad R_A = \frac{b}{\ell} \, P \, ,$$

where $b = \ell - a$ is the distance of the force from the right support B. It is seen that by rotating the beam the contribution of the forces to the virtual work is proportional to the distance from the center of rotation. This is quite analogous to the contributions in a moment equation about the center of rotation.

Finally, the last reaction R_B is found via the virtual work equation by lifting the support at B by the virtual displacement δu_B, whereby the beam rotates counter clockwise about B,

$$\delta V = \delta u_B \, R_B - \frac{a}{\ell} \, \delta u_B \, P = 0 \qquad \Rightarrow \qquad R_B = \frac{a}{\ell} \, P \, .$$

This relation is equivalent to a moment equation about A. □

Example 1.11. Beam with distributed load. In the case of distributed loads the determination of reactions proceeds in the same way via mechanisms obtained by releasing one support at a time. Figure 1.39 shows a uniformly loaded beam ABC with a fixed simple support at A and a simple support permitting horizontal motion at B. After releasing the horizontal support component a virtual horizontal translation immediately gives $R'_A = 0$.

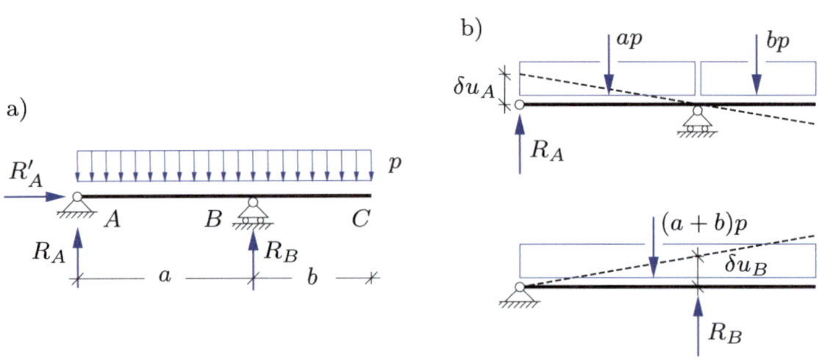

Fig. 1.39: Simply supported beam with distributed load.

The reaction R_A is determined via a vertical virtual displacement δu_A of point A. This corresponds to a clockwise rotation of the beam with angle $\delta u_A / a$ about B. The corresponding virtual work equation is determined by considering the load on AB as an equivalent force of magnitude ap acting at distance $\frac{1}{2}a$ to the left of B, and similarly the load on BC as an equivalent force of magnitude bp acting at the distance $\frac{1}{2}b$ to the right of B. The corresponding virtual work equation is

$$\delta V = \delta u_A R_A + \frac{\delta u_A}{a}\left[\tfrac{1}{2}b(bp) - \tfrac{1}{2}a(ap)\right] = 0 \quad \Rightarrow \quad R_A = \frac{a^2 - b^2}{2a}\,p\,.$$

The reaction R_B is determined from the virtual displacement corresponding to lifting the support B vertically by $\delta u_B = 1$. Hereby the beam rotates the angle $\delta u_B/a$ about A, and the virtual work equation becomes

$$\delta V = \delta u_B R_B - \frac{\delta u_B}{a}\,\tfrac{1}{2}(a+b)^2 p = 0 \quad \Rightarrow \quad R_B = \frac{(a+b)^2}{2a}\,p\,.$$

The sum of the vertical reactions is $R_A + R_B = (a+b)p$. It is observed that the procedure is nearly identical to the use of the similar moment equations. However, this changes when the principle of virtual work is applied to composite structures connected by hinges as illustrated in the following example. □

Example 1.12. Beams with hinge. The problem of a composite structure was illustrated in Fig. 1.27, showing two beams connected by a hinge. The reactions of the composite structure were determined sequentially in Example 1.5 by use of the equilibrium conditions. The principle of virtual work offers an alternative, in which each reaction component of the composite structure can be determined independently of the others. The procedure is illustrated by application to the vertical reaction R_C at the support C. A mechanism is created by releasing the support C and then imposing a virtual vertical displacement δu_C as shown in Fig. 1.40.

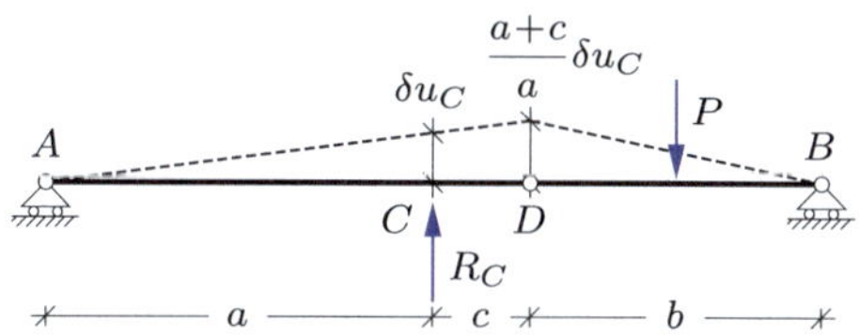

Fig. 1.40: Simply supported beam with distributed load.

There is no moment at the hinge at D, and therefore the different rotations at the two sides of the hinge do not lead to any contribution to the virtual work. It is seen directly from the figure that the vertical motion of D is $\delta u_D = \delta u_C(a+c)/a$, and thus the virtual work equation is

$$\delta V = R_C\,\delta u_C - \tfrac{1}{2}\delta u_D\,P = 0 \qquad \Rightarrow \qquad R_C = \frac{a+c}{2a}\,P\,.$$

The other reactions can be determined similarly from mechanisms created by releasing that particular support and imposing a virtual displacement in the direction of the reaction. □

1.7 Exercises

Exercise 1.1. The figure shows two planar forces $P_1 = [-1, 2]$ and $P_2 = [1, 2]$, both acting at the point $A = [3, 2]$.

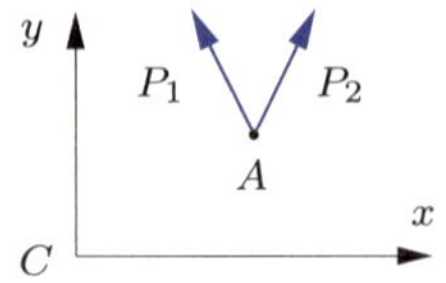

a) Determine the moment around C for each of the two forces P_1 and P_2 individually and for the two forces combined.

Exercise 1.2. The figure shows a force $P_\alpha = [2\cos\alpha, 2\sin\alpha]$, acting at the point $A = [3, 2]$.

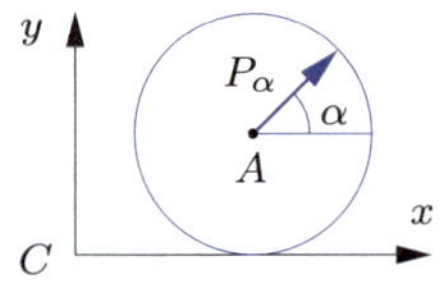

a) Determine the angle α such that the moment about C vanishes.

b) Determine the angle α where the moment about C attains its largest magnitude.

Exercise 1.3. The top figure shows two parallel planar forces $P_1 = -P_2 = [0, 2]$ acting at $A_1 = [3, 2]$ and $A_2 = [-3, 2]$, respectively.

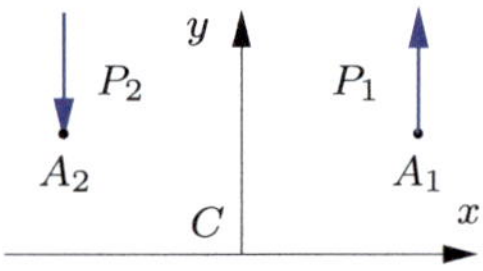

a) Determine the moment around C for the force couple P_1 and P_2.

In the lower figure the forces are now inclined similar to Exercise 1.1a such that $P_1 = [-1, 2]$ and $P_2 = [-1, -2]$.

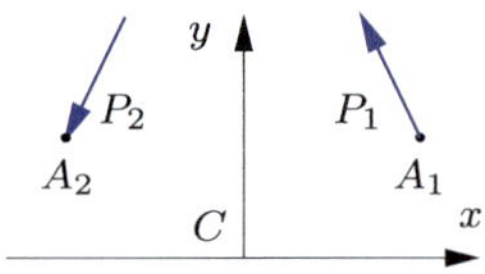

b) Determine the moment around C for the force couple P_1 and P_2.

c) Locate the points where the moment vanishes.

Exercise 1.4. The figure shows a three-dimensional space with a force $P = [-1, 2, 1]$ acting at the point $A = [3, 1, 2]$.

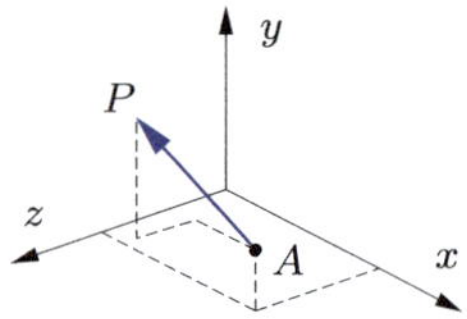

a) Sketch the projection on each of the coordinate planes and calculate the moments M_x, M_y and M_z as in Fig. 1.7.

b) Determine the moment vector $[M_x, M_y, M_z]$ about the origin of the coordinate system by use of (1.10) and compare with a).

Exercise 1.5. Show that the 'lever rule' (1.4) for two parallel forces P_1 and P_2 implies

$$\frac{a_1}{P_2} = \frac{a_2}{P_1} = \frac{a}{P}$$

where $a = a_1 + a_2$ and $P = P_1 + P_2$.

Exercise 1.6. The figure shows a simply supported beam loaded by two concentrated forces.

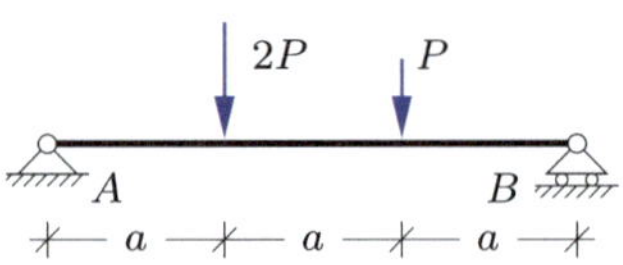

a) Show all possible reaction components with indication of the sign convention used.

b) Determine the magnitude of the reactions.

Exercise 1.7. The figure shows a cantilevered simply supported beam loaded by two concentrated forces.

a) Show all possible reaction components with indication of the sign convention used.

b) Determine the magnitude of the reactions.

c) Interchange the two forces and determine the magnitude of the reactions.

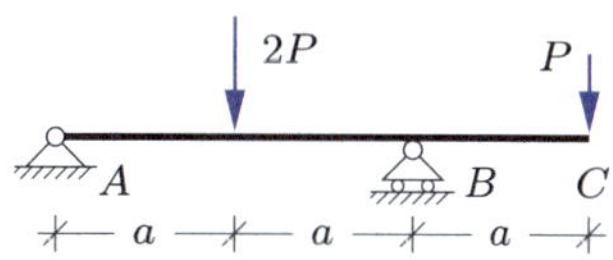

Exercise 1.8. The figure shows a cantilevered simply supported beam loaded by a moment M at the free end.

a) Show all possible reaction components with indication of the sign convention used.

b) Determine the magnitude of the reactions.

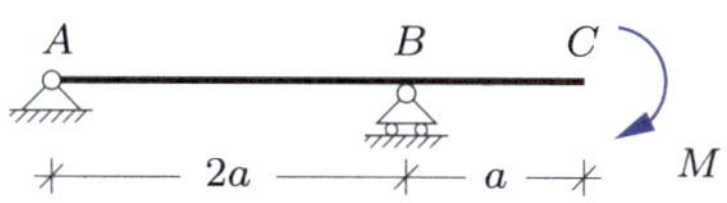

Exercise 1.9. The figure shows a simply supported beam with a vertical distributed load with intensity p acting only on part of the beam.

a) Show all possible reaction components with indication of the sign convention used.

b) Determine and show the location and magnitude of the force P that is equivalent to the distributed load p.

c) Determine the magnitude of the reactions.

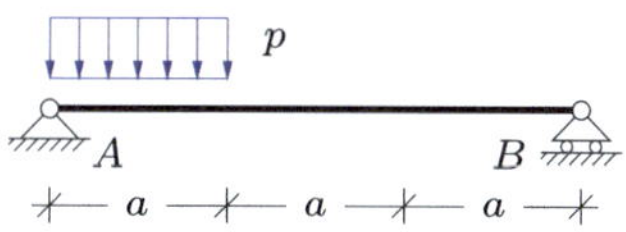

Exercise 1.10. The figure shows a beam with a fixed simple support at the left end and a simple support at the right end permitting motion at an angle of 45°. The beam is loaded by a uniform distributed load with intensity p.

a) Show all possible reaction components with indication of the sign convention used.

b) Determine and show the location and magnitude of the force P that is equivalent to the distributed load p.

c) Determine the magnitude of the reactions.

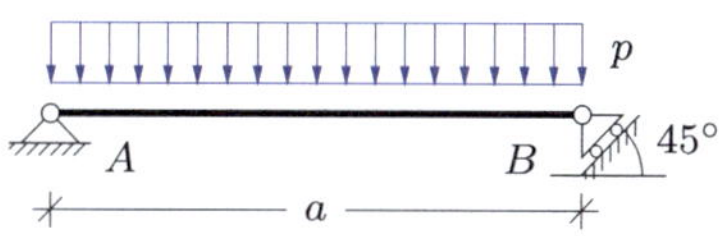

Exercise 1.11. The figure shows three different load cases for a cantilever beam of length a: i) linearly increasing load, ii) constant intensity load, iii) linearly decreasing load. The following questions should be answered for each of the three load cases.

a) Show all possible reaction components with indication of the sign convention used.

b) Determine and show the location and magnitude of the force P that is equivalent to the distributed load.

c) Determine the magnitude of the reactions.

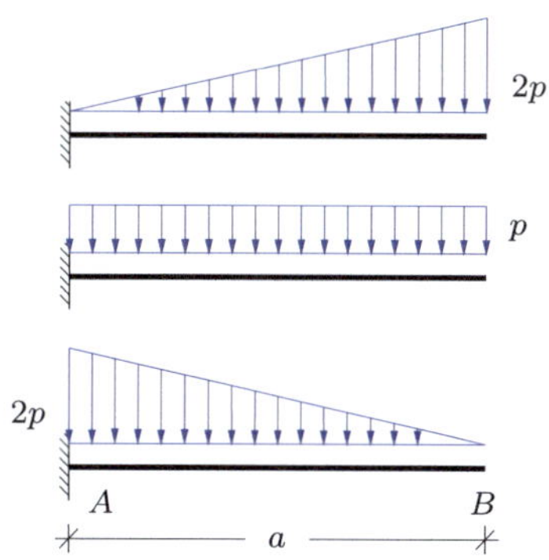

Exercise 1.12. The figure shows two simply supported beams connected by a hinge at C, loaded by a uniformly distributed load with intensity p acting over the full length of the two beams.

a) Show all possible reaction components and interaction force components at C with indication of the sign convention used.

b) Determine and show the forces P_1 and P_2 that are equivalent to the distributed load p on the two beams.

c) Determine the magnitude of the reactions and the interaction forces at C.

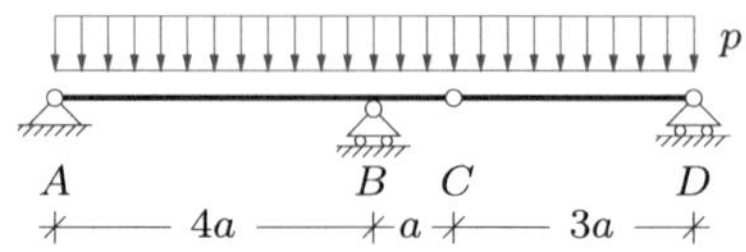

Exercise 1.13. The figure shows a three-hinge frame with the internal hinge located at the center of the span. The frame is loaded by a uniformly distributed load with intensity p along the top of the frame.

a) Show all possible reaction components and interaction force components at E with indication of the sign convention used.

b) Determine and show the forces P_1 and P_2 that are equivalent with the distributed load p on BE and EC, respectively.

c) Determine the magnitude of the reactions and the interaction forces.

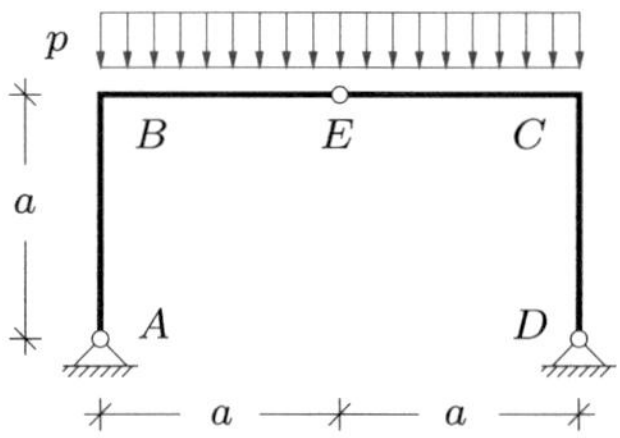

Exercise 1.14. The figure shows a simply supported angle frame, with a uniformly distributed load along CB. The frame has a fixed simple support at B and a simple support permitting motion at an angle of 45° at A.

a) Show all possible reaction components with indication of the sign convention used.

b) Determine and show the force P that is equivalent to the distributed load p.

c) Determine the magnitude of the reactions.

d) Use the force parallelogram rule to illustrate, how the load represented by the equivalent force P is decomposed into the reaction forces.

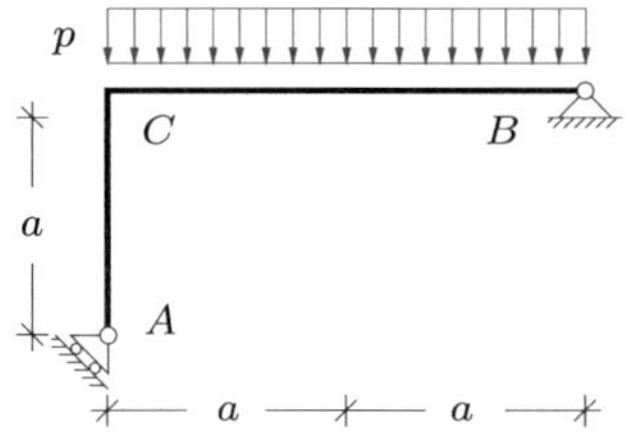

Exercise 1.15. The figure shows a three-hinge frame with the internal hinge at the top point C of a straight inclined roof beam. The roof is loaded by a vertical load of intensity p per unit *horizontal* length.

a) Show all possible reaction components and interaction force components at C with indication of the sign convention used.

b) Determine and show the force P that is equivalent to the distributed load p.

c) Determine the magnitude of the reactions and the interaction forces at C.

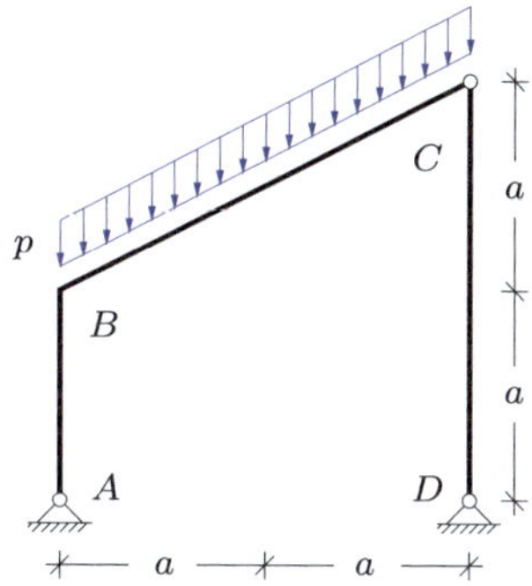

Exercise 1.16. The figure shows an angle frame ABC located in the x, y-plane. The frame has a fixed simple support in A, and simple supports permitting motion in the y-direction and in both directions in C and B, respectively. The length of AB is a, while the length of BC is B. At the center of BC a transverse force P is acting in the downward out-of-plane direction.

a) Define the reaction forces in a figure, and explain why the structure is statically determinate.

b) Show that the three in-plane reaction forces are zero.

c) Determine the remaining three out-of-plane reaction forces.

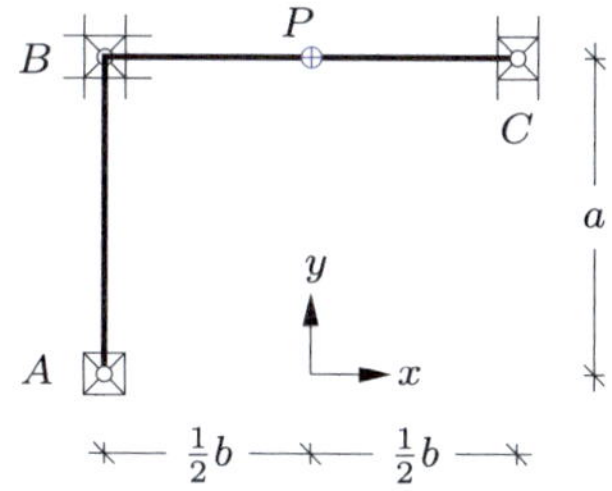

Exercise 1.17. Three beams of equal length AD, BE and CF are placed in a horizontal plane as shown in the figure. Each beam has a simple support at one end, while the other end rests on the mid-point of one of the other beams. The beam AD is loaded by a downward force of magnitude 14 N, located in the middle between A and B.

a) Determine the vertical reaction forces at the supports D, E and F as well as the forces transferred between the beams at A, B and C.

It is convenient to formulate the equilibrium conditions for each individual beam.

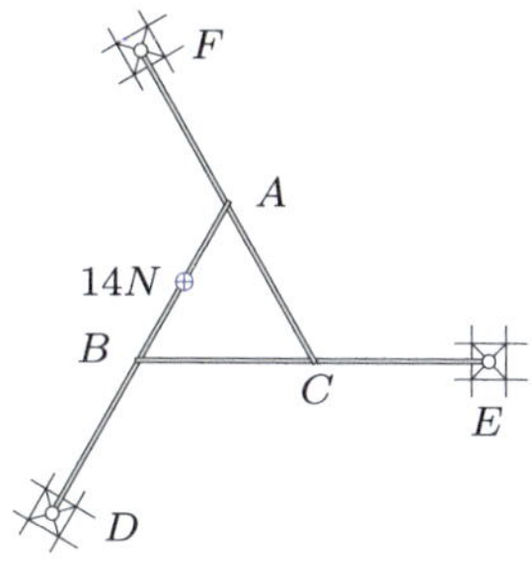

Exercise 1.18. The figure shows a simply supported girder $ABCD$ with a hinge at C. The girder is loaded by a uniform distribution of vertical forces of intensity p, similar to Exercise 1.12.

a) Determine all reaction components R_A, R'_A, R_B and R_D by use of the principle of virtual work.

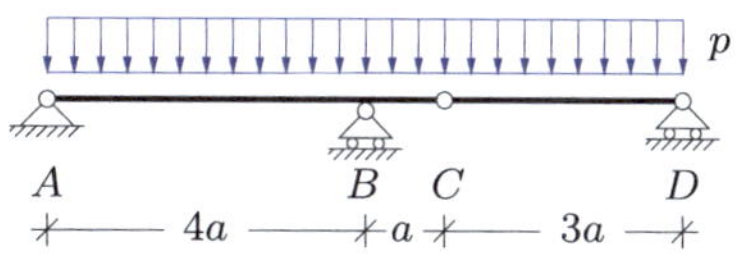

Exercise 1.19. The girder $ACEFDB$ is simply supported at A, B, C and D. The central part is connected via hinges at E and F. The vertical reactions at A, B, C and D are to be determined by the principle of virtual work for the following load cases:

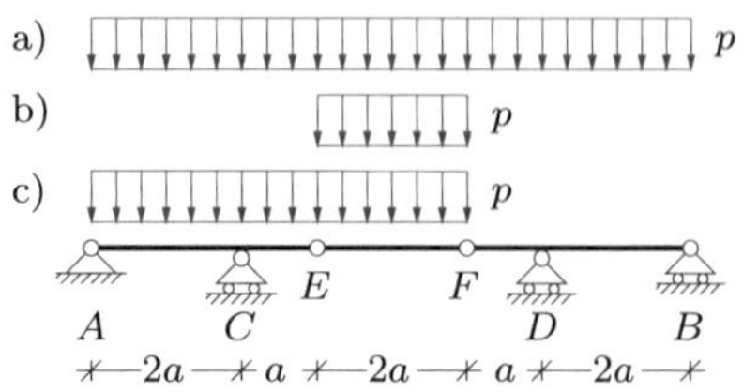

a) Uniform vertical load p over the full length AB.

b) Uniform vertical load p over the central part EF.

c) Uniform vertical load p over $ACEF$.

Exercise 1.20. Determine the remaining reactions in Example 1.12 by the principle of virtual work.

Truss Structures | 2

Truss structures constitute a special class of structures in which individual straight members are connected at joints. The members are assumed to be connected to the joints in a manner that permit rotation, and thereby it follows from equilibrium considerations, to be detailed in the following, that the individual structural members act as bars, i.e. structural members that can only carry an axial force in either tension or compression. Often the joints do not really permit free rotation, and the assumption of a truss structure then is an approximation. Even if this is the case the layout of a truss structure implies that it can carry its loads under the assumption that the individual members act as bars supporting only an axial force. This greatly simplifies the analysis of the forces in the structure by hand calculation and undoubtedly contributed to their popularity e.g. for bridges, towers, pavilions etc. up to the middle of the twentieth century. The layout of the structural members in the form of a truss structure also finds use with rigid or semi-rigid joints, e.g. space truss roofs, girders for suspension bridges, or steel offshore structures. The rigid joints introduce bending effects in the structural members, but these effects are easily included by use of numerically based computational methods.

S. Krenk, J. Høgsberg, *Statics and Mechanics of Structures*,
DOI 10.1007/978-94-007-6113-1_2,

In a statically determinate truss all the bar forces can be determined by the equilibrium equations, applied to the bars and joints of the truss. There are several strategies for carrying out the corresponding calculations, and three of these will be described in this chapter. The first and conceptually simplest method consists in considering each joint as an isolated body, for which the equilibrium equations must be satisfied. As there are no moment equations due to the hinge property of the joint this gives two equilibrium equations for a joint in a planar truss and three equilibrium equations for a joint in a space truss. The calculation of the bar forces proceeds by considering the individual joints sequentially as explained in Section 2.2. Alternatively, the bar forces can be calculated by using sections to separate larger parts of the structure and then applying suitable equilibrium equations for these larger parts. This method is dealt with in Section 2.3.

It is characteristic of the classic methods of joints and of sections, that they are arranged to determine the bar forces sequentially, and thus are convenient for calculation of the bar forces or a subset of these by hand. However, in their basic form these methods are limited to statically determinate trusses, and even for this class of structures the calculations may become quite elaborate for space trusses and larger planar trusses. Alternatively, a general systematic method can be developed for elastic trusses, irrespective of whether they are statically determinate or indeterminate. The method consists of setting up the equilibrium equations of all joints in a systematic way, using the elastic property of the bars. This method is a special case of the Finite Element Method, used in its general form for a wide range of problems within structural mechanics, see e.g. Cook et al. (2002) and Zienkiewicz and Taylor (2000). The formulation of the Finite Element Method takes on a particularly systematic form, when using the principle of virtual work, already mentioned for rigid bodies in Section 1.6. The extension of the principle of virtual work to truss structures is described in Section 2.4.3, and is used to formulate the Finite Element Method for elastic truss structures in Section 2.5.

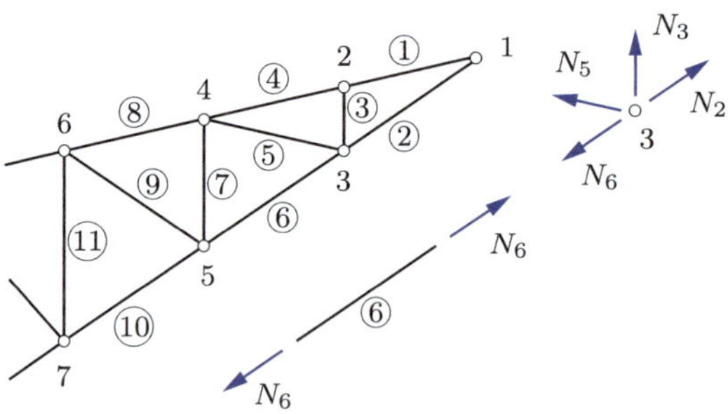

Fig. 2.1: Plane truss with joints and bars.

2.1 Basic principles

A truss structure consists of a number of joints, connected by bars. This is illustrated in Fig. 2.1 showing a truss structure consisting of the joints numbered as $1, 2, \cdots, 7$, connected by bars indicated by numbers in a circle. The joints are assumed to act like hinges, permitting free rotation of the bars around the joint. It is furthermore assumed that the truss structure is only loaded by concentrated forces acting at the joints. As a consequcnce of the assumption of hinges the bar elements can only support an axial force. This is easily demonstrated by considering Fig. 2.2. Due to the hinge there can be no moment at the ends of the bar element. Furthermore, if the force N in the bar were not aligned along the direction of the bar, there would be a non-zero moment about the hinge at the other end of the bar. Thus, the bar can only support a force of magnitude N, aligned along the direction of the bar. By convention the force in the bars of a truss structure are defined as positive when corresponding to tension, and they are then negative when representing compression.

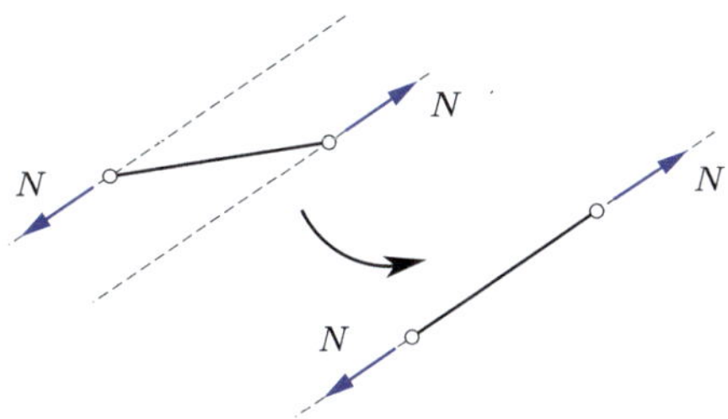

Fig. 2.2: Force in a bar along the bar axis.

A basic principle in the analysis of structures is the section. A section is used to represent a hypothetical separation of a part of the structure from the rest. This hypothetical separation enables a concise discussion of the forces exchanged between the parts on the two sides of the section. The situation is illustrated in simple form in Fig. 2.1. The figure shows the hypothetical situation in which the bar number 6 is separated from the structure by a section right next to the joints at the ends of the bar. The bar will be acted on by a force along the bar axis, and it follows from equilibrium, that the force at the sections at the ends of the bar must be of equal magnitude but opposite direction. The magnitude is denoted N_6 and is shown in the figure as positive corresponding to tension.

The figure also shows the effect of separating the joint 3 from the structure. In order to maintain the same state as in the structure the joint is acted upon by forces from each of the connected bars. The bar forces are defined as positive in tension, and positive bar forces therefore appear as forces N_2, N_3,

N_5 and N_6 pointing away from the joint. It is important to note that when the bar 6 has a tension force N_6, the bar is acted on by a force of magnitude N_6 pointing away from the bar towards the connecting joint. The connecting joints will similarly be acted on by a force of magnitude N_6 pointing away from the node. By this sign convention positive forces will point away from the member – joint or bar – on which they act. When making a sketch of a structural part, i.e. a joint or a bar, the forces will always be shown corresponding to their positive direction, i.e. as tension forces. If a bar force is determined to be compressive, this corresponds to a negative value of the magnitude N, and the sign of the arrow in the figure will be retained in the direction corresponding to tension.

2.1.1 Building with triangles

The triangle plays an important role in the geometric layout of truss structures. The reason for this is illustrated by the three planar trusses shown in Fig. 2.3. To be specific they can be envisaged to have a simple fixed support at the left end, and a simple support permitting horizontal motion at the right end. At first glance they may look as 'plausible' candidates for a truss, but are they satisfactory structures?

The truss shown in Fig. 2.3a consists of two triangles, connected by a quadrilateral at the center. The original structure is shown in full line, while the dotted line shows a possible deformation mechanism, by which the central quadrilateral changes shape without need for changing the length of any of the bars in the structure. Thus, even for perfectly rigid bar members the structure has a deformation mechanism. The existence of one or more free deformation mechanisms within a structure is termed kinematic indeterminacy. In the present case the implication is that the structure can not be used with the prescribed support conditions, but will need an extra support preventing the mechanism.

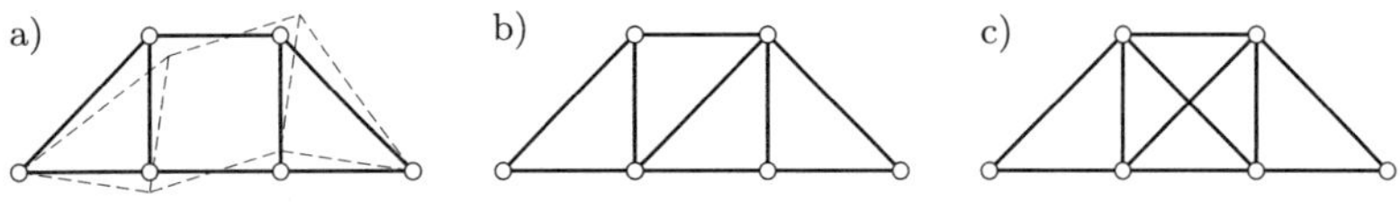

Fig. 2.3: From kinematic to static indeterminacy.

The mechanism can be locked by introducing a diagonal bar in the center quadrilateral as shown in Fig. 2.3b. It is seen that this prevents the free deformation mode, and also leads to a structure formed by triangles. The structure thereby becomes kinematically determinate. It is demonstrated below that this structure with supports providing three reaction components permits determination of all bar forces by use of the equilibrium conditions only. This property is termed static determinacy.

The equilibrium conditions imply that the force at the two ends of each bar must be of identical magnitude but opposite direction. The remaining equilibrium conditions then express force equilibrium at each joint, illustrated e.g. as equilibrium of the four forces acting on the joint 3 of the truss in Fig. 2.1. It follows from this principle that introduction of an extra bar in a truss, as shown in Fig. 2.3c, will introduce a new undetermined force in this bar. However, for a statically determinate truss the equilibrium equations are precisely sufficient to determine the forces in all bars, and consequently the introduction of an extra bar will leave the number of equilibrium conditions one short. A truss structure, in which the number of equilibrium equations is insufficient to determine all bar forces, is termed statically indeterminate. In contrast to structures with deformation mechanisms, that are generally unsuitable, statical indeterminacy does not constitute a limitation of the potential usefulness of the structure. It just implies that the specific distribution of the forces between the bars, or some of the bars, depends on the deformation properties of these bars. In the present example the two crossing diagonals in Fig. 2.3c share in preventing the deformation mechanism of the quadrilateral of the original structure. However, due to the static indeterminacy the precise ratio in which they share depends on their relative stiffness. Thus, the analysis of statically indeterminate structures require additional information about the stiffness of the structural members. In this chapter hand calculation type methods are developed for statically determinate trusses, while statically indeterminate trusses are left as part of the Finite Element formulation developed in Section 2.5.

2.1.2 Counting joints and bars

Some typical planar trusses are shown in Figure 2.4. It is seen that they are formed by triangles, and this suggests that they are statically determinate, when supported appropriately by three independent reaction components. It is now demonstrated by a common method for planar trusses, that they are indeed statically determinate. The method leads to a necessary relation between the number of joints and the number of bars. However, and probably equally important, it identifies a rational way of thinking about a truss structure, in which a process is constructed by which the structure is extended joint by joint, simulating an actual construction of the truss from bar elements connected by joints.

For a planar truss the hypothetical construction process starts from a simple triangle, and in order to be specific this triangle is supported by a fixed and a movable support as shown in Fig. 2.5a. Equilibrium of the nodes can be established by two projection equations for the unsupported node, and

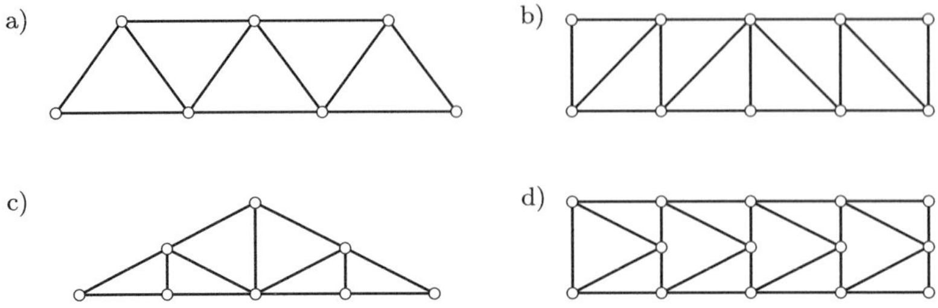

Fig. 2.4: a) V-truss, b) N-truss, c) Roof truss, d) K-truss.

a vertical projection equation of the forces on the node with the moveable support. This gives three equations, corresponding to the three bar forces to be determined. Thus, the initial triangle is statically determinate.

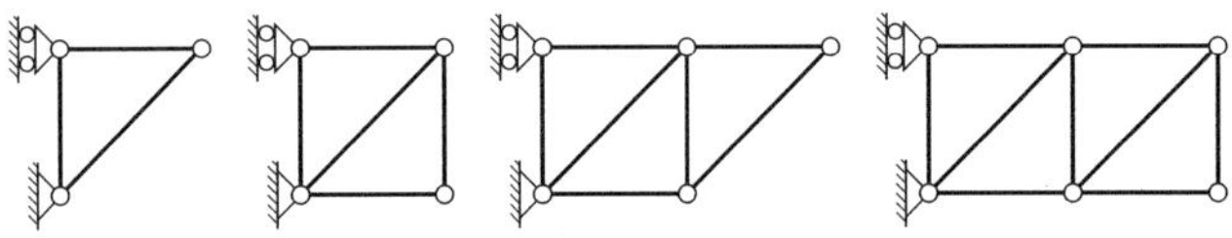

Fig. 2.5: Construction of plane truss girders by triangles.

The process is continued by attaching a new joint by two new bars as illustrated in Fig. 2.5. If the bars are not parallel, they will uniquely determine the position of the new joint, and two projection equations for the forces on the new joint will determine the bar forces. This step, in which a new joint is added and fastened by two new bars, can be continued as illustrated in the figure. The process defines a simple relation between the number of bars b and the number of joints j in a statically determinate planar truss:

$$b = 2j - 3 . \tag{2.1}$$

This relation is easily verified by observing that it is correct for the original triangle with $b = 3$ and $j = 3$, and that inclusion of one new joint leads to two additional bars. The relation between the number of bars and the number of joints is necessary, but clearly not sufficient. This becomes obvious e.g. by considering removing one of the diagonals and rejoining it as a diagonal crossing the remaining diagonal. Hereby part of the truss becomes kinematically indeterminate, implying a mechanism, while the additional bar in the remaining structure makes this part statically indeterminate. Thus, the process, in which a gradual construction of the truss by statically determinate steps is imagined, is probably more valuable than the formula, if left alone.

Fig. 2.6: Change of supports after 'constructing' the truss.

At first sight it may appear that the process is dependent on the supports being applied to the initial triangle. However, this is not the case. The result is independent of the specific support conditions as long as they provide three independent reaction components. After completing the truss structure, the supports can be moved as illustrated in Fig. 2.6.

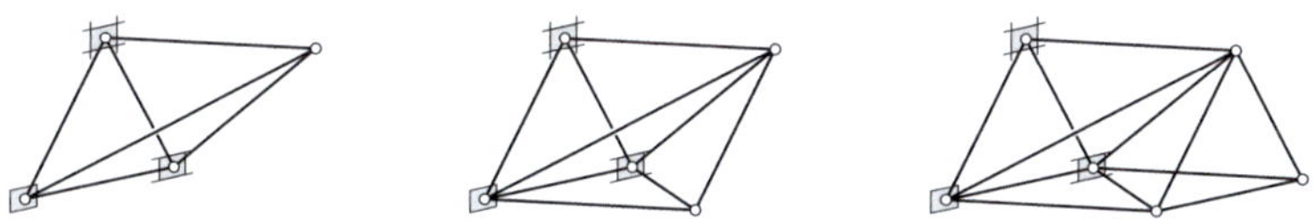

Fig. 2.7: Construction of space truss by addition of tetrahedra.

The results for planar trusses are easily extended to space trusses as illustrated in Fig. 2.7. The starting point is a tetrahedron (pyramid), formed by 4 joints and 6 connecting bars. The tetrahedron is supported by 6 independent reaction components. This leaves $4 \cdot 3 - 6 = 6$ equilibrium conditions from the 4 nodes for determination of the 6 bar forces. The process is continued in steps consisting in the addition of 1 new joint connected by 3 new bars. The three bars keep the joint fixed in space, and the three force projection equations associated with equilibrium of the new joint determine the three new bar forces. The figure shows the two first steps in this process leading to a truss girder of a type typically used for building cranes. This leads to the following relation between the number of bars b and the number of joints j of a statically determinate space truss:

$$b = 3j - 6\,. \tag{2.2}$$

Also in this case the relation is necessary but not sufficient, and the imaginary process of constructing the space truss constitutes an important part.

2.1.3 Qualitative tension-compression considerations

It is often possible to identify whether a bar member in a truss is loaded in tension or compression by a simple qualitative argument involving an estimate of the actual deformation of the loaded truss, or by constructing the mechanism that would result if the member were removed from the truss. Figure 2.8a shows a simply supported N-truss girder consisting of the 'head', the 'foot', the 'verticals' and the 'diagonals'. It is observed, that with $j = 10$ joints and $b = 17$ bars the truss satisfies the condition (2.1) for a statically determinate truss. Figure 2.8b shows a sketch of the deformed girder after loading by distributed downward forces. It is clearly seen that the bars in the foot are extended, indicating tension, while the bars in the head become shorter, indicating compression. However, it is more difficult to identify elongation or shortening of the verticals and the diagonals.

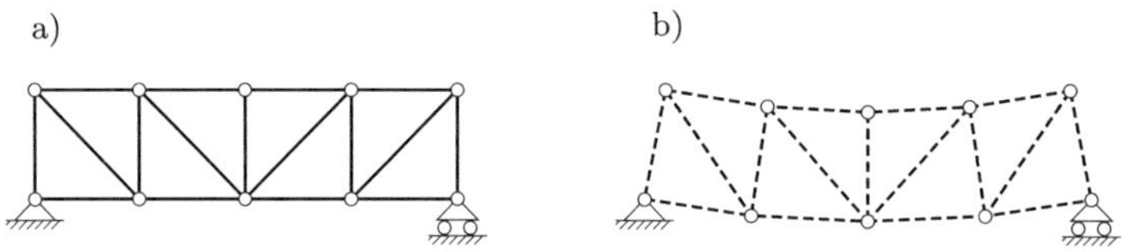

Fig. 2.8: Tension and compression members in truss girder.

A somewhat different and more precise way of estimating whether a bar is in tension or compression consists in imagining that the bar were removed from the truss. For a statically determinate structure this would create a mechanism. Figure 2.9a illustrates the mechanism generated by removing the second bar in the head, while Fig. 2.9b illustrates the mechanism associated with removal of the third bar in the foot. The mechanisms are shown corresponding to a downward load. It is clearly seen that the distance between the two joints constituting the end points of the removed bar approach each other in the case of the bar in the head, while they become further separated in the case of the bar in the foot. Thus, the bar in the head will experience compression, while the bar in the foot will experience tension, when the structure carries a downward load.

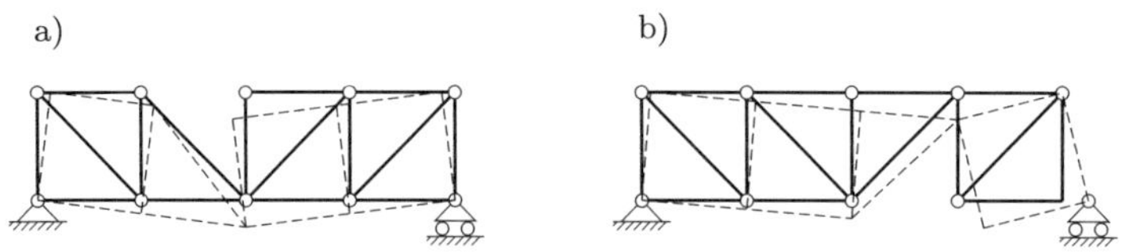

Fig. 2.9: Mechanisms by removing a bar in the head or in the foot.

A similar geometric argument can be used to identify the sign of the force in the diagonals and the verticals. Figure 2.10a illustrates the deformation mechanism generated when removing the second diagonal from the left. It is seen that diagonal would be extended by the illustrated mechanism. Thus there will be tension in the diagonal when the loads perform positive work through the mechanism. this would be the case for a downward load at the cental or right nodes of the head or the foot. However, a downward load in the first set of node to the right of the left support would create negative work and therefore contribute a compressive force. Figure 2.10b illustrates the mechanism generated by removing the second vertical from the left. A vertical downward load at any of the three inner nodes of the head would lead to compression in this vertical.

a) b)

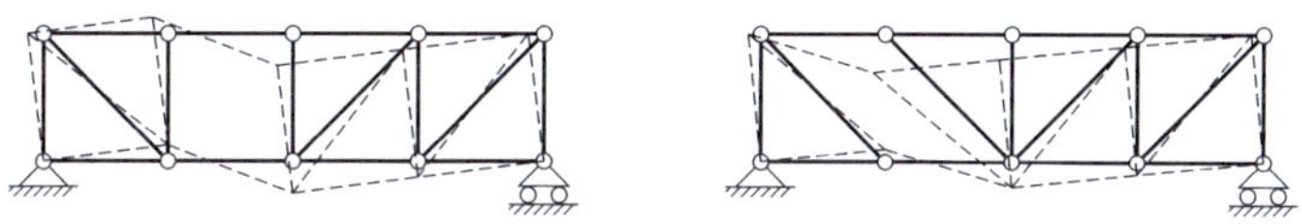

Fig. 2.10: Mechanisms by removing a diagonal or a vertical bar.

The qualitative arguments used to explain the implication of the mechanisms generated by removing a single bar from the truss can be made precise if the geometry of the infinitesimal motion of the mechanism is described exactly and used within the context of virtual work, described in Section 2.4.

2.2 Method of joints

The magnitude of the forces in the bars of a statically determinate truss structure can be determined by the method of joints. The idea of the method of joints is to consider each joint as separated from the rest of the truss structure by the introduction of a virtual section. The parts on the two sides of the section will exchange identical but opposite forces, and by introducing the section and identifying these forces explicitly, they can be analyzed by the equilibrium equations. The principle is illustrated in its simplest form in Fig. 2.11. The left part of the figure shows a joint C in a planar truss loaded by the vertical force P and connected to the rest of the truss by the two bars AC and BC. A section is now introduced, separating the joint from the rest of the structure. The forces N_{AC} and N_{BC}, by which the bars act on the joint, are indicated as acting on the joint together with the load P. Thus, the joint C is acted on by three forces. The forces in the bars are considered as positive, when representing tension in the bar. Thus, the effect on the joint is a force directed away from the joint. By the law of action and reaction equal but opposite forces act on the bars. As seen, these forces represent tension in the bars. It is noted that the forces N_{AC} and N_{BC} are uniquely defined as being positive in tension. A representation in terms of vectors is less direct, as it would require identification of the part on which the force acts.

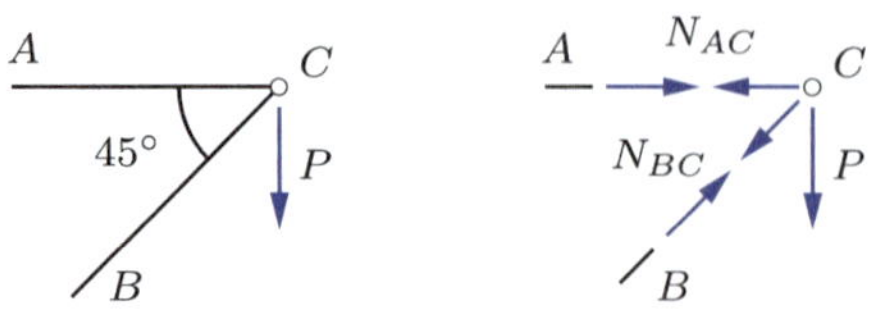

Fig. 2.11: Node C with load P and bar forces N_{AC} and N_{BC}.

Equilibrium of the joint C requires that two force projection equations are satisfied. Vertical projection gives

$$\downarrow \qquad N_{BC}\sin 45^\circ + P = 0 \quad \Rightarrow \quad N_{BC} = \frac{-P}{\sin 45^\circ} = -\sqrt{2}\,P\,.$$

By taking a vertical projection, the force N_{AC} in the horizontal bar AC does not contribute to the equilibrium equation.

The remaining bar force N_{AC} can be determined by projection on the direction orthogonal to BC. The present case is simple due to the angle 45°, and gives $N_{AC} = P$ directly. In many cases it will be more convenient to use a horizontal projection, whereby

$$\leftarrow \qquad N_{AC} + N_{BC}\cos 45^\circ = 0 \quad \Rightarrow \quad N_{AC} = -N_{BC}\cos 45^\circ = P\,.$$

Thus, there is compression in the inclined bar BC, while the horizontal bar AC is in tension to ensure horizontal equilibrium.

In this simple illustration there were only two bar forces, and thus they could be determined directly by the two equilibrium equations available for the planar joint C. Most joints in truss structures are connected by more bars than there are equilibrium equations available for the particular joint. The bar forces can therefore only be determined sequentially, if the joints are considered in a certain order. This is illustrated in the following examples.

2.2.1 Planar truss structures

Many truss structures can conceptually be broken down into planar parts, and this section illustrates the calculation of bar forces for some simple planar trusses.

Example 2.1. Double triangle. Figure 2.12 shows a planar truss consisting of two triangles. There are 4 joints, providing $4 \times 2 = 8$ equilibrium equations, that determine the three reaction components R_A, R_D and R'_D, plus the forces in the five bars in the truss.

In principle the analysis could be carried out completely on a joint by joint basis, starting from C and then proceeding through B, A and D. At each node there would be two

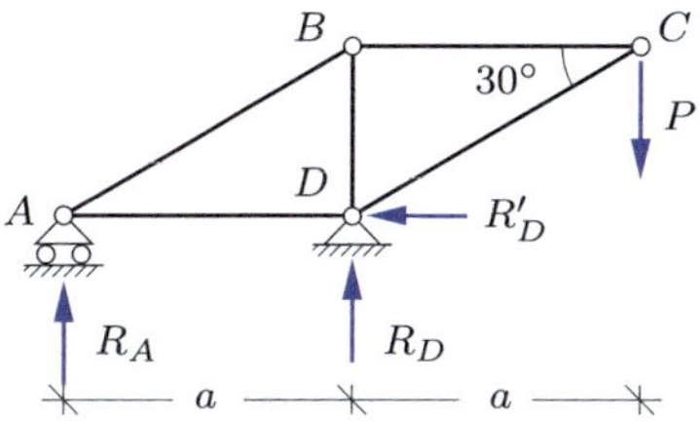

Fig. 2.12: Double triangle truss with load P.

unknown forces, and at the end all bar forces and reactions will be determined. However, for many truss structures it is not possible to start from a node with only two unknown forces, unless appropriate reaction components are determined first. Therefore, the analysis of a statically determinate truss usually starts with determination of the reactions, using equilibrium of the full truss or parts of the truss as discussed in Section 1.5 on reactions. In the present case the reaction components R'_D, R_D and R_A are determined by horizontal projection and moment about A and D, respectively:

$$R'_D = 0\,, \qquad R_D = 2P\,, \qquad R_A = -P\,.$$

Note the negative reaction in A, indicating downward direction of the force.

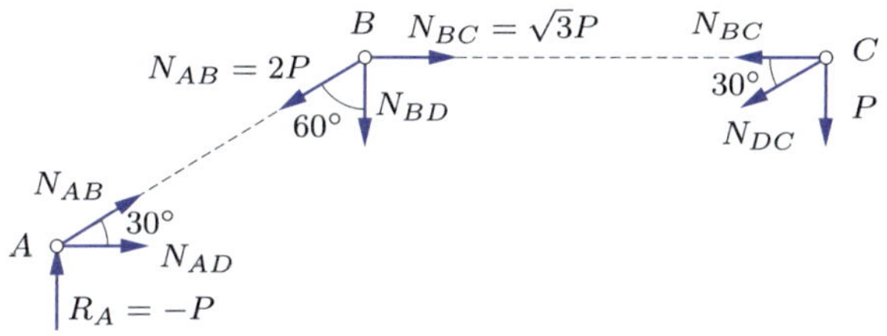

Fig. 2.13: Joints A, B and C of dounle triangle truss.

Figure 2.13 shows all the nodes with the forces indicated in their positive direction as tension. If a compression force is found in the analysis, this shows up as a negative force. The bar forces are determined from equilibrium of the forces acting on the individual nodes. Thus, vertical and horizontal projection of the forces acting on joint C lead to

$$\downarrow \qquad N_{DC}\cos 60^\circ + P = 0 \qquad \Rightarrow \qquad N_{DC} = -2P\,.$$

$$\leftarrow \qquad N_{BC} + N_{DC}\cos 30^\circ = 0 \qquad \Rightarrow \qquad N_{BC} = \sqrt{3}P\,.$$

At joint A vertical projection gives

$$\uparrow \qquad N_{AB}\cos 60^\circ + R_A = 0 \qquad \Rightarrow \qquad N_{AB} = 2P\,,$$

while horizontal projection then leads to

$$\rightarrow \qquad N_{AD} + N_{AB}\cos 30^\circ = 0 \qquad \Rightarrow \qquad N_{AD} = -\sqrt{3}P\,.$$

Finally, vertical projection of the forces acting on joint B gives

$$\downarrow \qquad N_{AB}\cos 60^\circ + N_{BD} = 0 \qquad \Rightarrow \qquad N_{BD} = -P\,.$$

This completes the calculation of the bar forces. There are still an unused projection equation at joint B and two projection equations for joint D. This corresponds to the three reaction components that were determined initially from total equilibrium. □

Example 2.2. V-truss by the method of joints. Figure 2.14 shows a simply supported V-truss, loaded by a concentrated force at B. The support conditions permit the vertical reaction components R_A and R_C plus the horizontal reaction R'_A. They are determined from the equilibrium conditions for the total truss structures as discussed in detail in Chapter 1. The reactions follow from horizontal projection, moment about C and moment about A as

$$R'_A = 0\,, \qquad R_A = \tfrac{1}{2}P\,, \qquad R_C = \tfrac{1}{2}P\,.$$

Control by vertical projection gives $R_A + R_C = P$, corresponding to the load P.

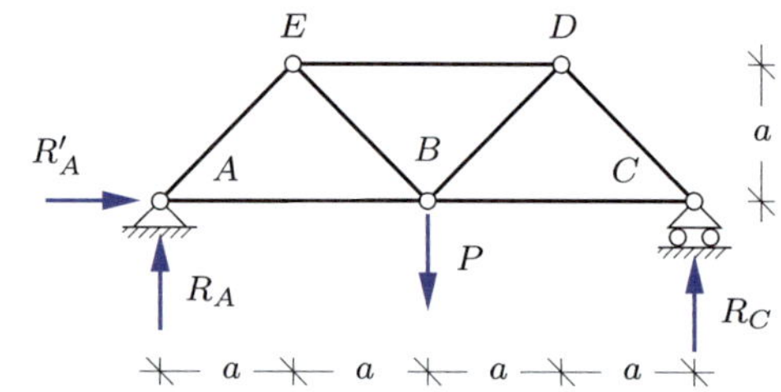

Fig. 2.14: Simply supported V-truss with load P.

The loading is symmetric, and because the horizontal reaction component R'_A vanishes, so are the reactions. Thus, the structure and its bar forces are symmetric with respect to a vertical line through B, and only the right half with the nodes B, C and D need to be considered to determine all the bar forces. These nodes and the corresponding forces from loads, reactions and bars are shown in Fig. 2.15.

It is seen from the figure that joint C only contains two unknown forces. Once they are determined, the joint D only contains two unknown forces. These forces, and their symmetric counterparts, determine equilibrium at joint B which can be used to check the previous calculations.

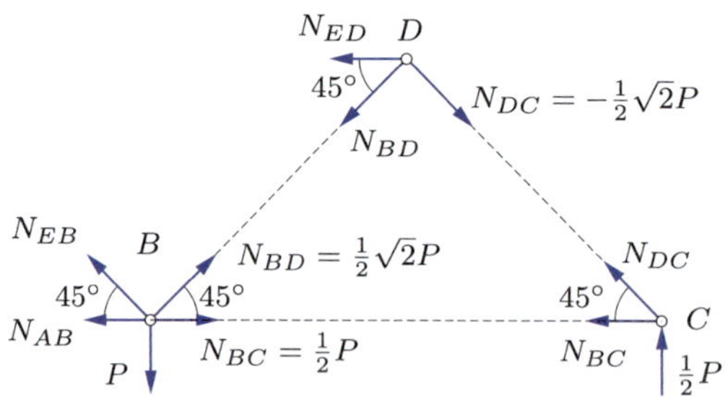

Fig. 2.15: Joints B, C and D of V-truss.

Equilibrium of joint C determines the forces N_{DC} and N_{BC} by vertical and horizontal projection, respectively:

$$\uparrow \qquad N_{DC}\sin 45^\circ + \tfrac{1}{2}P = 0 \qquad \Rightarrow \qquad N_{DC} = -\tfrac{1}{2}\sqrt{2}\,P\,.$$

With N_{DC} known, horizontal projection then gives

$$\leftarrow \qquad N_{BC} + N_{DC}\cos 45^\circ = 0 \qquad \Rightarrow \qquad N_{BC} = \tfrac{1}{2}P\,.$$

At joint D vertical projection gives

$$\downarrow \quad N_{BD}\sin 45^\circ + N_{DC}\sin 45^\circ = 0 \qquad \Rightarrow \qquad N_{BD} = \tfrac{1}{2}\sqrt{2}\,P\,.$$

The remaining bar force N_{ED} then follows from horizontal projection:

$$\leftarrow \quad N_{ED} + N_{BD}\cos 45^\circ - N_{DC}\cos 45^\circ = 0 \quad \Rightarrow \quad N_{ED} = -P\,.$$

This completes the calculation of bar forces, because the remaining bar forces now follow from their symmetric counterparts, e.g. $N_{EB} = N_{BD}$ and $N_{AB} = N_{BC}$. Using these forces from the left side of the structure, the equilibrium conditions for joint B may be used as control of the calculations, with vertical projection

$$\uparrow \qquad N_{EB}\sin 45^\circ + N_{BD}\sin 45^\circ - P = 0$$

and horizontal projection

$$\leftarrow \qquad N_{AB} - N_{BC} + (N_{EB} - N_{BD})\cos 45^\circ = 0\,,$$

demonstrating equilibrium. □

Example 2.3. Roof truss by method of joints. For some loads one or more of the bars in a truss structure may have zero force, thus being essentially inactive in this load case. The identification of these bars is illustrated in this example by considering the roof truss in Fig. 2.16.

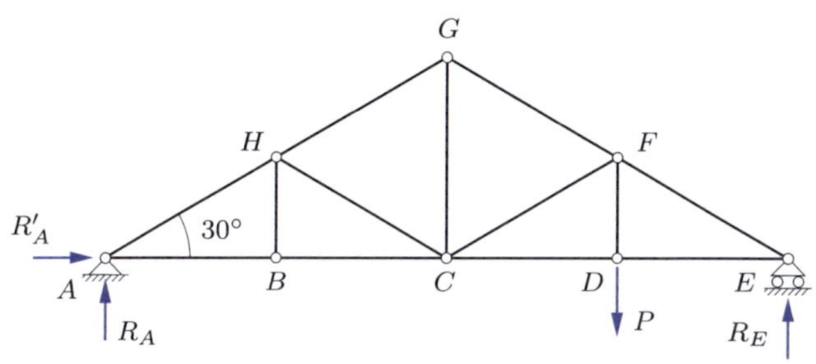

Fig. 2.16: Roof truss with load P at truss foot.

Bars with zero force may occur, when two bars at an unloaded joint are aligned. A transverse projection of the forces at this node will then not contain these forces. Thus, if there is only a single transverse bar, the force in this bar must vanish if there is no transverse load. This applies to the joint B as shown in Fig. 2.17a. The two bars AB and BC are horizontal, and thus only the force N_{BH} in the vertical bar BH contributes to vertical equilibrium. When there is no load at the node, this implies that $N_{BH} = 0$. In plain words, the argument is that when two bars share a common joint and follow the same direction, then they can not support a transverse force.

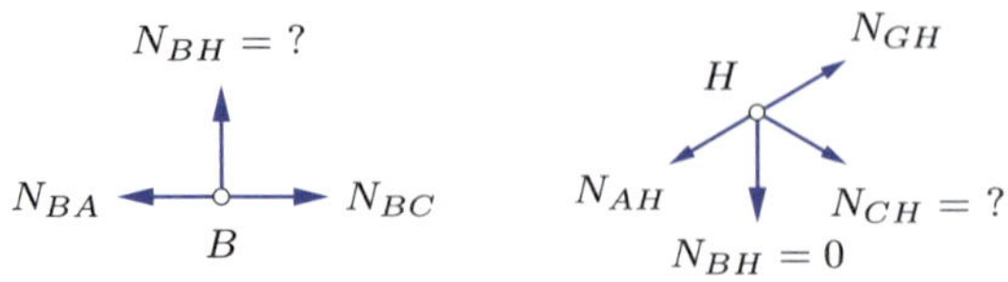

Fig. 2.17: Identification of zero force bars in roof truss.

This argument can be continued by now considering the joint H shown in Fig. 2.17b. There are two bar forces N_{BH} and N_{HC} that potentially could contribute to the transverse equilibrium of the joint H. However, as $N_{BH} = 0$ according to the previous calculation, the remaining transverse force must also vanish, whereby $N_{CH} = 0$. It should be noted that the argument does not depend on the angle between the two aligned bars and the bar providing the transverse component, as the latter is the only bar contributing to the transverse equilibrium. The presence of zero force bars simplify the computation of the remaining bar forces as illustrated below.

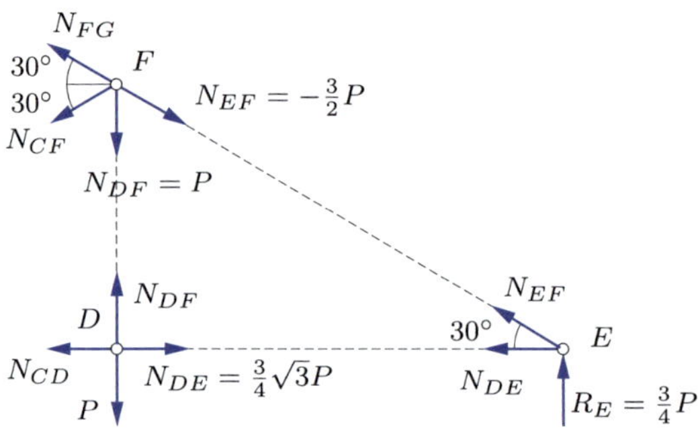

Fig. 2.18: Joints D, E and F of roof truss.

The computation of the non-zero bar forces conveniently starts with determination of the reactions from equilibrium of the complete truss structure as in the previous two examples. Horizontal projection and moments about E and A give

$$R'_A = 0\,, \qquad R_A = \tfrac{1}{4}P\,, \qquad R_E = \tfrac{3}{4}P\,.$$

The bar forces in the right half of the truss can then be calculated by considering equilibrium of the joints E, D and F, as illustrated in Fig. 2.18. For the joint E vertical equilibrium gives

$$\uparrow \qquad N_{EF}\cos 60^\circ + R_E = 0 \qquad \Rightarrow \qquad N_{EF} = -\tfrac{3}{2}P\,,$$

while horizontal equilibrium gives

$$\leftarrow \qquad N_{DE} + N_{EF}\cos 30^\circ = 0 \qquad \Rightarrow \qquad N_{DE} = \tfrac{3}{4}\sqrt{3}P\,.$$

It is now advantageous to move on to joint D due to the particularly simple form of the vertical and horizontal equilibrium conditions, each consisting of two opposing forces of equal magnitude:

$$N_{DF} = P\,, \qquad N_{CD} = N_{DE} = \tfrac{3}{4}\sqrt{3}P\,.$$

At node F projection on the direction orthogonal to the truss head gives

$$N_{CF}\cos 30^\circ + N_{DF}\cos 30^\circ = 0 \quad \Rightarrow \quad N_{CF} = -N_{DF} = -P\,,$$

while N_{FG} follows from horizontal projection with the common angle 30°:

$$\leftarrow \qquad N_{FG}\cos 30^\circ + N_{CF}\cos 30^\circ - N_{EF}\cos 30^\circ = 0$$

$$\Rightarrow \qquad N_{FG} = N_{EF} - N_{CF} = -\tfrac{1}{2}P\,.$$

The remaining bar forces are left as Exercise 2.10. □

2.2.2 Space trusses

In the case of space trusses hand calculation methods typically make use of special features of the truss geometry, and rapidly become fairly impractical for larger structures. Most space structures are therefore analyzed by the numerical Finite Element Method described in Section 2.5. A glimpse of the hand calculation procedure is provided by the following example.

Example 2.4. Simple space truss. Figure 2.19 shows a simple cantilever space truss carrying a tip load P. The dimensions of the structure are given in terms of a, b and h as shown in the figure. The length of the inclined bars then is

$$\ell = \sqrt{a^2 + b^2 + h^2}\,.$$

The bar forces are determined by the method of joints, placing a section around the tip D. The truss and the load are symmetric about a vertical plane through AD, and thus the internal forces in the inclined bars are equal, $N_{BD} = N_{CD}$. Vertical projection for node D then gives

$$\downarrow \quad \frac{h}{\ell}N_{BD} + \frac{h}{\ell}N_{CD} + P = 0 \qquad \Rightarrow \qquad N_{BD} = N_{CD} = -\frac{\ell}{2h}P\,.$$

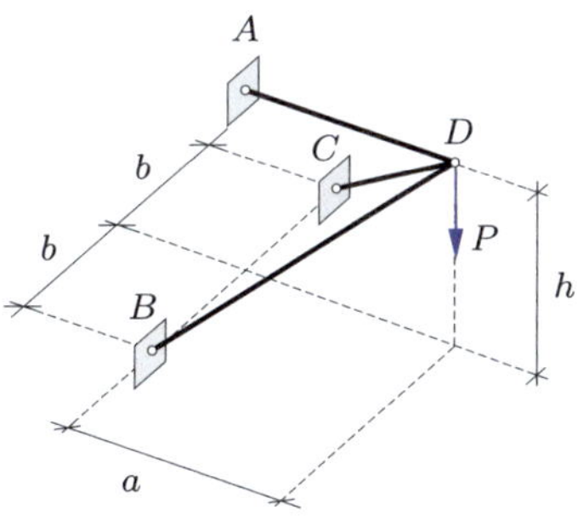

Fig. 2.19: Simple space truss carrying vertical tip load P.

The remaining bar force N_{AD} is obtained by horizontal projection, in the direction of AD,

$$\leftarrow \qquad 2\,N_{BD}\,\frac{a}{\ell} + N_{AD} = 0 \qquad \Rightarrow \qquad N_{AD} = \frac{a}{h}P\,.$$

Note, that all bar forces increase for smaller depth h of the truss. Also observe that the force N_{AD} is independent of the width $2b$ of the truss, and therefore is the same if the inclined bars were collapsed in the vertical plane containing the horizontal bar AD. □

2.3 Method of sections

The idea of introducing a section, whereby a structure is separated into two parts dates several hundred years back. The method of joints is a special case, in which the section is introduced in such a way that it separates precisely one joint. Hereby, all forces identified via the section pass through the released joint, and therefore are governed by force equilibrium of the joint. The idea of a section to identify the interaction of two parts of a structure is much more general and plays a central role in the theory of structures including beams and frames, but also in the general theory of a continuous bodies as discussed in Chapter 8.

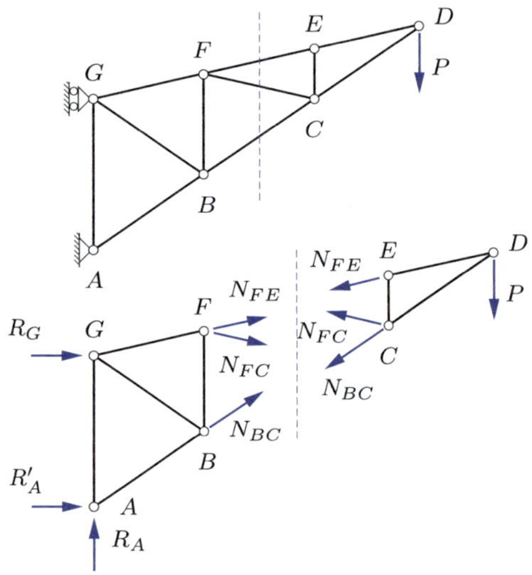

Fig. 2.20: Truss divided by vertical section.

A first step in the generalization of the idea of the section is to introduce a section that divides the truss structure into two parts. This is illustrated for the case of a planar truss in Fig. 2.20a. A section is introduced that intersects the bars BC, FC and FE and divides the structure into two parts shown in Fig. 2.20b. The section identifies three bar forces, acting on each part of the structure with different direction. Thus, three new forces N_{BC}, N_{FC} and N_{FE} have appeared. In the original structure three equilibrium equations were available for determining the reactions as discussed in Chapter 1. After separating the original structure into two parts, each part must satisfy three equilibrium equations, and thus three new equations are now available for calculation of the bar forces N_{BC}, N_{FC} and N_{FE}. In the truss illustrated in Fig. 2.20 the three bar forces may be obtained from equilibrium of the right part of the truss, and the three reactions may then be determined from equilibrium of the left part.

2.3.1 Bar forces via the method of sections

The use of the method of sections to determine the bar forces in a planar truss is first illustrated by a simple example, and then summarized in concise form.

Example 2.5. V-truss by method of sections. Figure 2.21 shows a planar V-truss that has already been analyzed by the method of joints in Example 2.2. It is here analyzed by the method of sections to demonstrate the principles involved. First the reactions are determined by equilibrium of the full truss:

$$R'_A = 0\,, \qquad R_A = \tfrac{1}{2}P\,, \qquad R_C = \tfrac{1}{2}P\,.$$

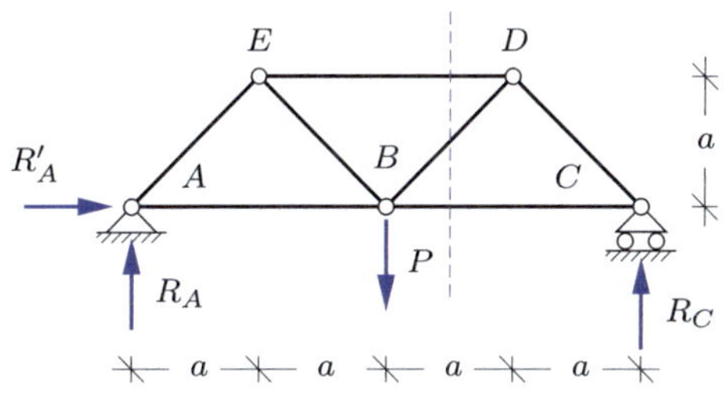

Fig. 2.21: Simply supported V-truss with load P.

In the method of joints the analysis would start from a node with two unknown bar forces – in the present case either of the joints A and C. This can also be used in the method of sections by introducing a vertical section, isolating the supported node. However, the method of sections also permits direct determination of the forces in the central bars. To illustrate the general procedure in the method of sections, a vertical section is introduced just to the right of the joint B as shown in the figure. This section intersects the bars BC, BD and DE and is used for calculating the corresponding bar forces N_{BC}, N_{BD} and N_{DE}.

Equilibrium of either the left or the right part of the structure is now used to determine the three bar forces N_{BC}, N_{BD} and N_{DE}. Note, that in contrast to the case of a single node the two parts have finite extent, and equilibrium therefore involves three equilibrium equations, and not just the two force projection equations associated with a single node. In the present problem equilibrium of the right part is the simpler, because it involves only one reaction component and not the load. The right part is shown in Fig. 2.22 together with all the forces acting on this part.

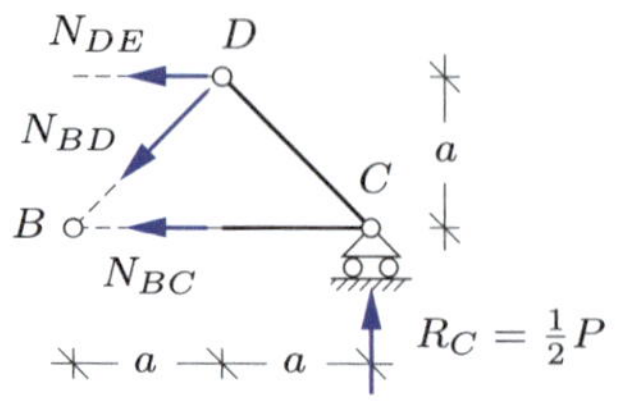

Fig. 2.22: Equilibrium in section.

The calculation of the bar forces proceeds in a systematic fashion by the following steps. First it is observed that two of the bar forces to be determined, namely N_{DE} and N_{BC}, are parallel. The remaining force N_{BD} in the inclined diagonal can then be determined by use of vertical equilibrium:

$$\downarrow \qquad N_{BD}\cos 45^\circ - R_C = 0 \quad \Rightarrow \quad N_{BD} = \sqrt{2}R_C = \tfrac{1}{2}\sqrt{2}\,P\,.$$

This force intersects the two still unknown bar forces N_{BC} and N_{DE} in B and D, respectively. Thus, a moment equation about any of these two points will involve only one unknown bar force. The bar force N_{BC} is determined by moment about D:

$$\overset{\curvearrowright}{D} \qquad a\,N_{BC} - a\,R_C = 0 \quad \Rightarrow \quad N_{BC} = R_C = \tfrac{1}{2}P\,.$$

Finally, the bar force N_{DE} is determined by moment about B:

$$\overset{\curvearrowleft}{B} \qquad a\,N_{DE} + 2a\,R_C = 0 \quad \Rightarrow \quad N_{DE} = -2R_C = -P\,.$$

It is seen that each of the three forces N_{BD}, N_{BC} and N_{DE} has been calculated from an equilibrium equation, that does not involve any of the other two forces. □

The method of sections for planar trusses can be formalized by the following procedure.

i) Determine the reactions on the truss structure.

ii) Divide the truss structure into two parts by a section, intersecting two or three bars.

iii) Consider each of the bar forces in turn and determine the bar force by: moment about the point of intersection of the other two forces, or projection on the transverse direction, if they are parallel.

It follows from the independence of the calculation of each bar force associated with a given section, that the order of the calculations can be changed, and indeed any of the bar forces can be calculated without calculating the others. As a consequence the method of sections can often be used to calculate isolated bar forces of a truss structure, without need for calculating the forces in adjoining bars. In addition to its computational simplicity, the method of sections often provides direct insight into the systematic variation of the forces in e.g. diagonals or verticals of regular truss structures.

2.3.2 Special types of planar trusses

Planar trusses appear in many contexts and often in the form of truss girders e.g. in bridges and cranes. Typical examples were illustrated in Fig. 2.4. The following examples illustrate the analysis for four types of trusses – three typical truss girders and a roof truss. The truss girder examples illustrate the method of analysis by introducing a typical section and calculating the bar forces associated with that section. The full analysis requires a sequence of similar sections, and omitting the repetitions associated with the typical section, the full results are summarized to illustrate how the girder layout determines the distribution of bar forces in the girder. The roof girder is more an individual type, where modification of the inner bars leads to modification of the analysis.

Example 2.6. N-truss girder. N-truss girders have constant or moderately changing height, filled with interchanging vertical and inclined bars. They find application in bridges, both traditional steel truss bridges and more recently in girders carrying the load on suspension bridges and cable stayed bridges. They also find use as supporting structure for roofs in industrial buildings where an important load component is a distributed vertical load. A simple illustration is shown as Fig. 2.23, where equal vertical forces P are applied to the 7 joints in the foot of a simply supported N-girder. For simplicity of analysis the horizontal spacing of the joints is taken equal to the height of the girder. While simplifying the expressions appearing in the analysis this has no principal impact on the procedure. The effect of the height of a regular truss girder under distributed load is discussed in the following example.

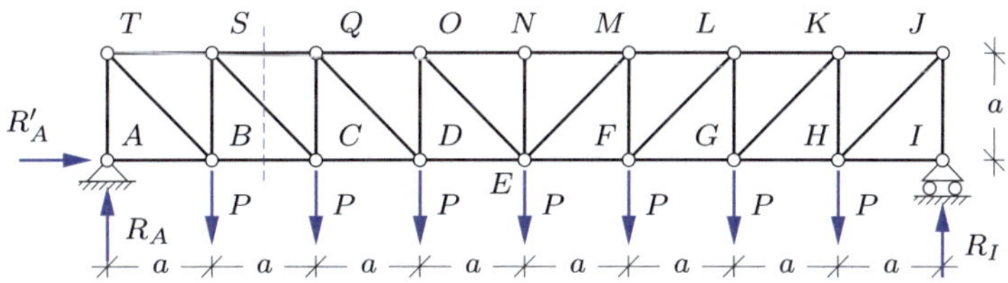

Fig. 2.23: Simply supported N-truss with distributed loads P.

The reactions are determined by horizontal projection and moments about nodes I and A:

$$R'_A = 0\,, \qquad R_A = R_I = \tfrac{7}{2}P\,.$$

The forces in the bars are then determined by introducing vertical and inclined sections as illustrated in Fig. 2.24.

The vertical section shown in the figure intersects the diagonal SC and the corresponding bars SQ and BC in the head and the foot, respectively. The bar forces in the head and the foot are parallel, and the force N_{SC} in the diagonal is therefore determined by vertical projection of all forces on the left part of the truss girder:

$$\downarrow \qquad N_{SC}\cos 45^\circ - R_A + P = 0 \quad \Rightarrow \quad N_{SC} = \tfrac{5}{2}\sqrt{2}P\,.$$

The force N_{BC} then follows from moment about S:

$$\overset{\curvearrowleft}{S} \qquad a\,N_{BC} - a\,R_A = 0 \quad \Rightarrow \quad N_{BC} = R_A = \tfrac{7}{2}P\,.$$

Finally, the force N_{SQ} in the head follows via moment about C:

$$\overset{\curvearrowright}{C} \qquad a\,N_{SQ} + 2a\,R_A - aP = 0 \quad \Rightarrow \quad N_{SQ} = -6P\,.$$

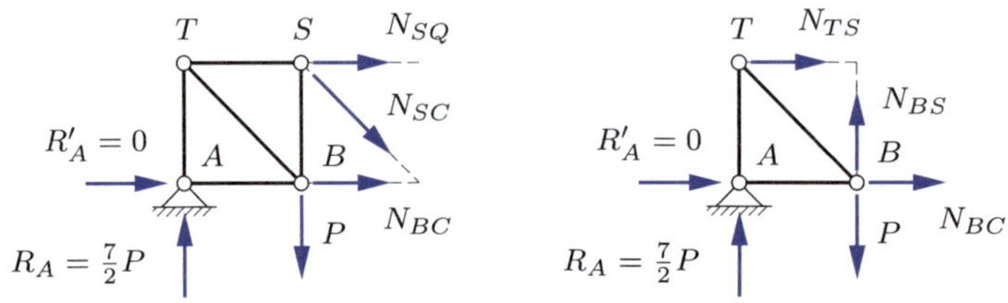

Fig. 2.24: Equilibrium at section.

The forces in the verticals are determined by use of inclined sections as shown in Fig. 2.24b. Vertical equilibrium determines the force N_{BS} in the vertical via

$$\uparrow \qquad N_{BS} + R_A - P = 0 \quad \Rightarrow \quad N_{BS} = -\tfrac{5}{2}P\,.$$

The force N_{TS} in the head can be determined via moment about B, while the force N_{BC} in the foot has already been determined above by the vertical section. The procedure used for the four bar forces here is repeated along the left half of the girder, and the remaining bar forces then follow by symmetry. The resulting bar forces are shown in Fig. 2.25.

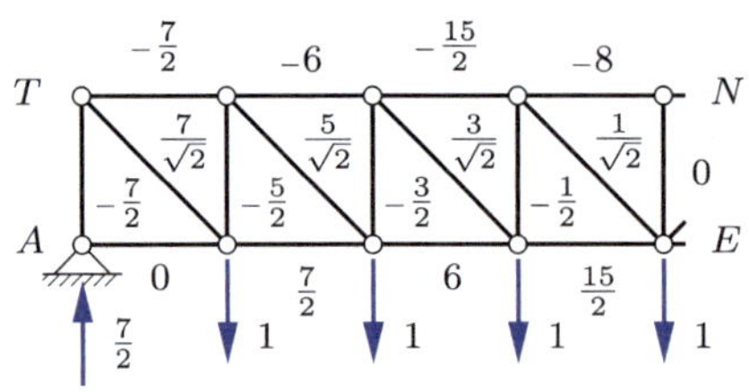

Fig. 2.25: Bar forces in N-truss.

A clear pattern can be seen in the magnitude of the forces in the verticals and diagonals of the truss. When starting at the left support A the vertical carries a compressive force $N_{AT} = -\frac{7}{2}P$. The force N_{BS} in the next vertical is P less due to the load P acting at B. This pattern continues towards the center, where the force in the vertical vanishes due to symmetry. The forces in the diagonals are closely related to those in the corresponding verticals. This is perhaps most easily seen by considering one of the unloaded joints in the head, e.g. S. Vertical equilibrium of this joint requires $N_{SC} = -\sqrt{2}N_{BS}$, and thus the pattern from the verticals is repeated in the diagonals, but with opposite sign. It is interesting to observe, that if the diagonals were turned the other way, the forces would retain their magnitude but become compression instead of tension. By the argument concerning equilibrium of the joints in the head of the girder it follows that the forces in the verticals would then also change sign. The forces in head and foot increase towards the center. The pattern of this variation will become clear in connection with the analysis of beams in Chapter 3. □

Example 2.7. V-truss. In a V-truss the diagonals are inclined to the right and to the left as illustrated in Fig. 2.26. The angle α with horizontal is given in terms of the truss height h and the length a of the horizontal bars by

$$\sin\alpha = \frac{2h}{\sqrt{4h^2 + a^2}}.$$

The truss of the present example is simply supported at nodes A and G, and loaded by vertical forces of magnitude P at all the nodes of the girder foot as shown in the figure.

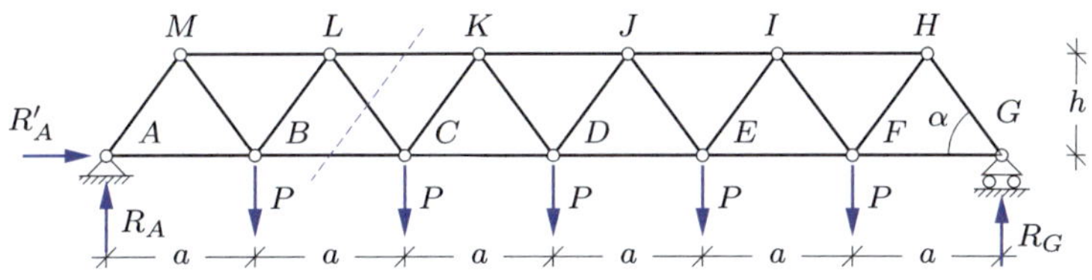

Fig. 2.26: Simply supported V-truss girder with distributed loads P.

The section shown in the figure intersects the diagonal CL. The analysis considers equilibrium of the left part of the structure as shown in Fig. 2.27a. The forces in the bars in the head and the foot are parallel and horizontal. Thus, the force N_{CL} in the diagonal is determined by vertical projection:

$$\downarrow \qquad N_{CL}\sin\alpha - R_A + P = 0 \quad \Rightarrow \quad N_{CL} = \frac{3\,P}{2\sin\alpha}.$$

The force N_{BC} in the foot is determined via moment about L:

$$\overset{\frown}{L} \qquad hN_{BC} - \tfrac{3}{2}aR_A + \tfrac{1}{2}aP = 0 \quad \Rightarrow \quad N_{BC} = \frac{13}{4}\frac{a}{h}\,P.$$

Finally, the force N_{KL} in the head is found by moment about C:

$$\overset{\frown}{C} \qquad hN_{KL} + 2aR_A - aP = 0 \quad \Rightarrow \quad N_{KL} = -4\frac{a}{h}\,P.$$

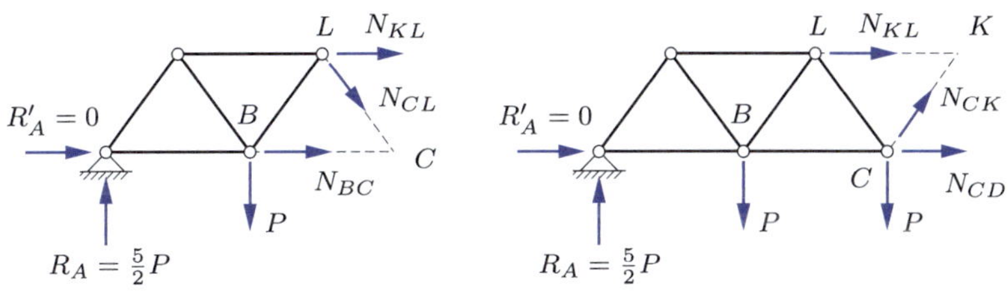

Fig. 2.27: Section equilibrium.

The calculation proceeds along the girder by next considering the section that intersects the diagonal CK as illustrated in Fig. 2.27b. The forces in the intersected bars now follow from equilibrium of the left part. The diagonal force N_{CK} is found by vertical projection:

$$\uparrow \qquad N_{CK}\sin\alpha + R_A - 2P = 0 \qquad \Rightarrow \qquad N_{CK} = -\frac{P}{2\sin\alpha}\,.$$

The force N_{CD} in the foot follows from moment about the node K in the head:

$$\overset{\curvearrowleft}{K} \qquad hN_{CD} - \tfrac{5}{2}aR_A + \tfrac{3}{2}aP + \tfrac{1}{2}aP = 0 \quad \Rightarrow \quad N_{CD} = \frac{17}{4}\frac{a}{h}\,P\,.$$

Finally, the force N_{KL} in the head follows from moment about the node C in the foot:

$$\overset{\curvearrowright}{C} \qquad hN_{KL} + 2a\,R_A - aP = 0 \qquad \Rightarrow \qquad N_{KL} = -4\frac{a}{h}\,P\,.$$

It should be noted that this force has already been calculated by the previous section.

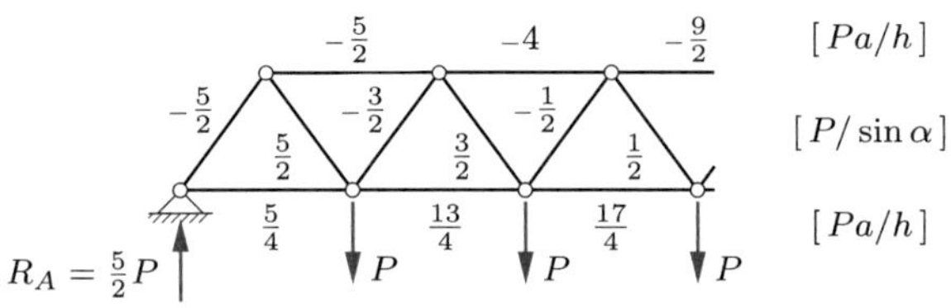

Fig. 2.28: Bar forces in V-truss girder.

By use of similar sections along the V-truss girder the member forces shown in Fig. 2.28 are found. It is noted that the forces in the diagonals are proportional with $P/\sin\alpha$, while the forces in the head and the foot are proportional with aP/h, i.e. smaller for larger girder height. □

Example 2.8. K-truss. The so-called K-truss finds application e.g. in towers, masts, and in the legs of offshore jackup platforms. In spite of the fact that the K-truss is typically used in vertical orientation the following analysis will address the horizontal orientation and use the names head and foot for the two sides of the truss. In a K-truss the connection between head and foot is established by a combination of transverse and inclined bars, meeting at the center of the transversal bars as shown in Fig. 2.29. The angle α between the diagonals and direction of the truss girder is determined by

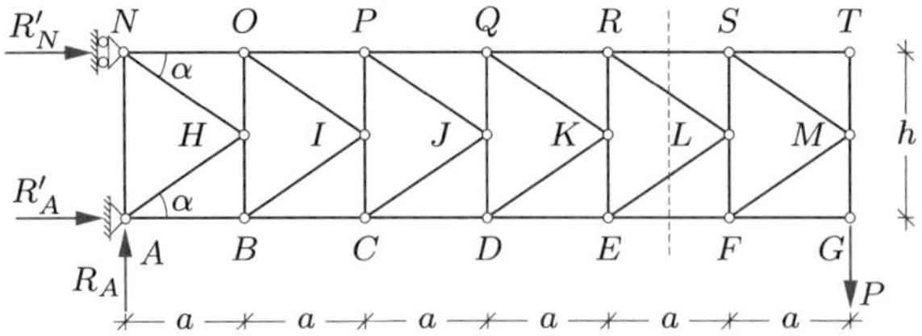

Fig. 2.29: Cantilever K-truss girder loaded by P at foot. ***

$$\sin\alpha = \frac{h}{\sqrt{h^2+4a^2}}\,, \qquad \tan\alpha = \frac{h}{2a}\,,$$

where h is the transverse girder dimension and a is the length of a single K-section as indicated in the figure.

Due to the double diagonals in the K-truss a general section will intersect four bars, and thus the bar forces can not be computed by use of a single section as in the case of N-trusses and V-trusses. However, this problem is easily overcome by considering equilibrium of the inner nodes of the K-truss. A typical inner node is shown in Fig. 2.30. The node is connected to two transverse bars with forces N_{LS} and N_{LF}, and to two diagonals with forces N_{LR} and N_{LE}. Horizontal projection only involves the forces in the diagonals, and thus

$$N_{LR}\cos\alpha + N_{LE}\cos\alpha = 0 \quad \Rightarrow \quad N_{LR} = -N_{LE}\,.$$

Thus, the resulting force from the two diagonals is a downward force of magnitude $2N_{LE}\sin\alpha$.

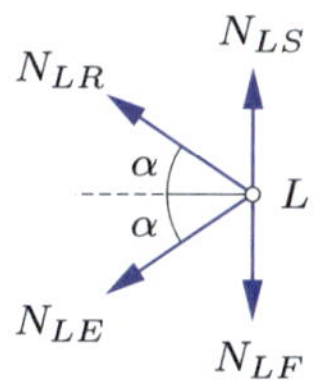

Fig. 2.30: Equilibrium of center node L.

The calculation now proceeds from a transverse section, here selected as shown in Fig. 2.31b. The two diagonal forces combine to a vertical downward force of magnitude $2N_{LE}\sin\alpha$, and thus vertical equilibrium of the right part of the structure gives

$$\downarrow \qquad 2N_{LE}\sin\alpha + P = 0 \quad \Rightarrow \quad N_{LR} = -N_{LE} = \frac{P}{2\sin\alpha}\,.$$

Note, that this is tension in the upper diagonal and compression in the lower diagonal. The force N_{EF} in the foot is determined by moment about the node S in the head:

$$\overset{\frown}{S} \qquad hN_{EF} + aP = 0 \quad \Rightarrow \quad N_{EF} = -\frac{a}{h}\,P\,.$$

This corresponds to compression in the foot. Finally, the force N_{SR} follows from moment about node F:

$$\overset{\frown}{F} \qquad hN_{SR} - aP = 0 \quad \Rightarrow \quad N_{SR} = \frac{a}{h}\,P\,.$$

corresponding to tension in the head.

The force in the transverse bars can be determined by equilibrium of the node in the foot or head, once the corresponding diagonal bar force has been determined. Thus, as shown in Fig. 2.31a, equilibrium of node E in the transverse direction gives

$$\uparrow \qquad N_{EK} + N_{LE}\sin\alpha = 0 \quad \Rightarrow \quad N_{EK} = -N_{LE}\sin\alpha = \tfrac{1}{2}P\,.$$

The similar argument for node R in the head gives

$$\downarrow \qquad N_{RK} + N_{LR}\sin\alpha = 0 \quad \Rightarrow \quad N_{RK} = -N_{LR}\sin\alpha = -\tfrac{1}{2}P\,.$$

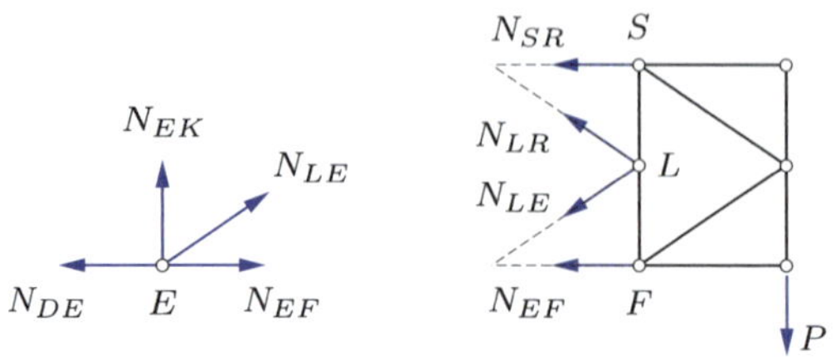

Fig. 2.31: Equilibrium of section.

The forces in the diagonals are found by projection, and in the present problem this involves only the end load P. Thus, all upper diagonals have the same tension force $\frac{1}{2}P/\sin\alpha$, while all lower diagonals have the compression force $-\frac{1}{2}P/\sin\alpha$. Similarly all upper transverse bars have the compression force $-\frac{1}{2}P$, while the lower transverse bars have the tension force $\frac{1}{2}P$.

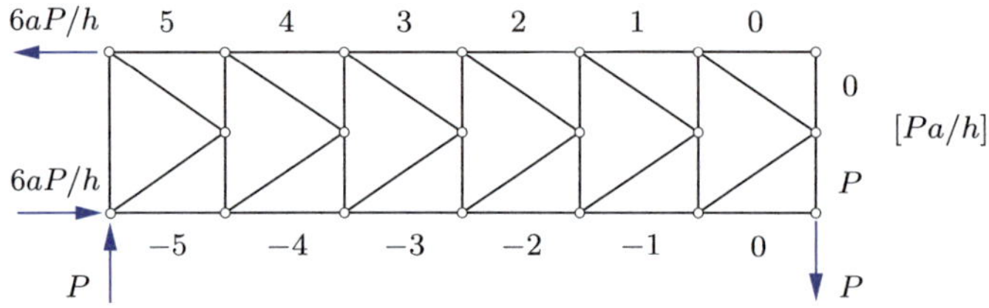

Fig. 2.32: Bar forces in K-truss.

The forces in the girder head and foot are determined by moment equilibrium about a node at distance h. The moment arm of the load increases by a when moving one step towards the support. The bar forces in the full K-truss girder are shown in Fig. 2.32. They are seen to follow a simple pattern with identical forces in similarly placed transverse and diagonal bars, while the forces in head and foot increase towards the support. This corresponds closely to the distribution of shear force and moment in a cantilever beam treated in the following chapter. □

Example 2.9. Roof truss. Roofs of houses are often supported by truss structures, and the particular W-type shown Fig. 2.33 is quite common. The geometry is determined by the width $12a$ and the height $2h$ of the truss, together with the information that the inner bars are connected to the mid-points of the head, and divides the foot into three equal parts as shown in the figure. It follows from this definition of the geometry that all four inner bars form the same angle β with horizontal. This can be seen by observing that the longer internal bars have vertical projection $2h$ and horizontal projection $2a$, while the shorter internal bars have vertical projection h and horizontal projection a. The two angles α and β are then defined from the dimensions a and h by

$$\sin\alpha = \frac{h}{\sqrt{h^2 + (3a)^2}}\,, \qquad \sin\beta = \frac{h}{\sqrt{h^2 + a^2}}\,.$$

In the present case the load consists of a single vertical force P acting at node E. The reactions are determined by horizontal projection, moment about D, and moment about A:

$$R'_A = 0\,, \qquad R_A = \tfrac{1}{4}P\,, \qquad R_D = \tfrac{3}{4}P\,.$$

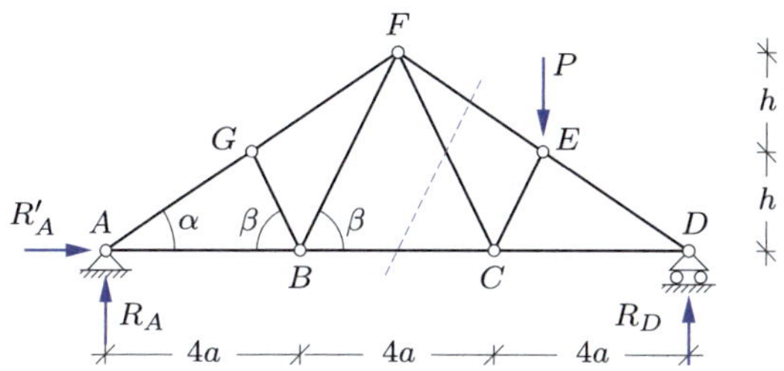

Fig. 2.33: W roof truss.

First, a vertical section is made to the right of the top node F as shown in Fig. 2.33. The right part of the structure and the exposed bar forces are shown in Fig. 2.34a. None of the exposed bar forces are parallel, and thus they are determined by a sequence of three independent moment equations. The first equation is moment about the supported node D. The load P and the bar force N_{CF} contribute to moment equilibrium. The contribution from the inclined force N_{CF} is most conveniently found by resolving it into a horizontal and a vertical component through node C. Of these only the vertical component $N_{CF}\sin\beta$ contributes to moment equilibrium, expressed by

$$\overset{\frown}{D} \qquad 4a\,N_{CF}\sin\beta \;-3a\,P \;=\;0 \quad\Rightarrow\quad N_{CF}=\frac{3\,P}{4\sin\beta}\,.$$

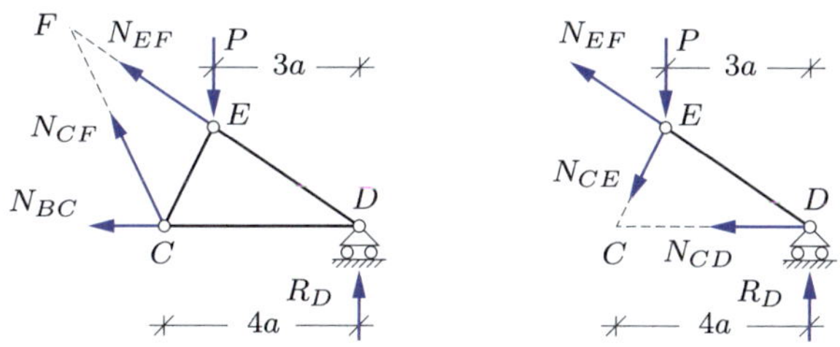

Fig. 2.34: Section forces in W-truss.

The second equation is moment equilibrium about node C. The contribution from the inclined force N_{EF} is most conveniently found by sliding the force along its line of action until it has origin in node D. It is then resolved into a horizontal component, and a vertical component $N_{EF}\sin\alpha$. Only the vertical component contributes to the moment equation, which takes the form

$$\overset{\frown}{C} \quad 4aN_{EF}\sin\alpha - aP + 4aR_D \;=\;0 \quad\Rightarrow\quad N_{EF}=-\frac{P}{2\sin\alpha}\,.$$

Node F is the intersection point of the two forces just determined, and thus moment about node F gives

$$\overset{\frown}{F} \qquad 2h\,N_{BC} \;+3a\,P - 6a\,R_D \;=\;0 \quad\Rightarrow\quad N_{BC}=\frac{3a}{4h}\,P\,.$$

This completes the computation of the three bar forces from the first section.

The section is now moved to the left of node C, whereby the right part of the structure is as shown in Fig. 2.34b. Two of the forces pass through node D, and thus only the load and the bar force N_{CE} contribute to moment equilibrium about D. By sliding the force N_{CE} along its line of action to node C, and then resolving it in a horizontal component and the vertical component $N_{CE}\sin\beta$, the following moment equation is obtained:

$$\overset{\curvearrowleft}{D} \qquad 4a\,N_{CE}\sin\beta + 3a\,P = 0 \quad \Rightarrow \quad N_{CE} = \frac{-3\,P}{4\sin\beta}\,.$$

Moment about node E determines the bar force N_{CD}:

$$\overset{\curvearrowright}{E} \qquad h\,N_{CD} - 3a\,R_D = 0 \quad \Rightarrow \quad N_{CD} = \frac{9a}{4h}\,P\,.$$

Finally, a vertical section through ED isolates the supported node D, and vertical equilibrium gives

$$\uparrow \qquad N_{DE}\sin\alpha + R_D = 0 \quad \Rightarrow \quad N_{DE} = -\frac{3P}{4\sin\alpha}\,.$$

This completes the analysis of the right half of the W-truss. With the present loading equilibrium of nodes G and B leads to the conclusion that the forces in the inner bars GB and BF of the left half of the truss vanish. Thus, the remaining forces are found by the two projection equations for the supported corner node A. □

2.4 Stiffness and deformation of truss structures

In most structures strength and stiffness play important roles, even if it implies just having 'enough strength' and 'sufficient stiffness'. Basically the concepts of strength and stiffness are material properties, and their effect in a structure depends on how the materials are used to form the structure. The concepts of strength and stiffness will be introduced gradually, when needed. Thus, the present section is devoted to stiffness of bars – the so-called uniaxial stiffness – while a general description of material stiffness and strength is given in Chapter 8.

2.4.1 Axial stress and strain

The stiffness of a bar relates the elongation u to the axial force N in the bar. The problem of relating these properties of the bar to material properties was discussed by Galileo Galilei in 1638. The essential part of this discussion is given here in a more modern form with reference to Fig. 2.35a. The figure shows a homogeneous bar of length ℓ and cross-section area A. The bar is loaded by application of an axial force of magnitude N, which is considered positive in tension. This force leads to an elongation of the bar of magnitude u.

Now, a thought experiment is conducted, in which the bar is split lengthwise into two parts, each of area $\frac{1}{2}A$ as shown in Fig. 2.35b. Each of these parts support half the load, while maintaining the original elongation. Therefore

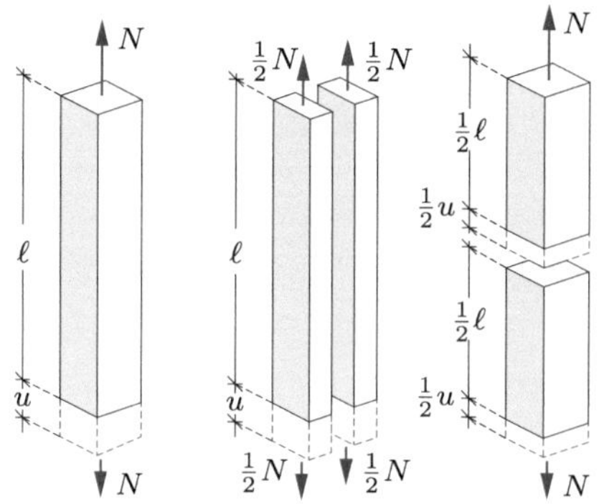

Fig. 2.35: Bar of length ℓ and cross-section area A.

the elongation must depend on the force normalized by the cross-section area. This normalized force is called the stress and is expressed as

$$\sigma = \frac{N}{A} = \frac{\text{Force}}{\text{Area}}, \qquad \left[\frac{\text{N}}{\text{m}^2}\right] = [\text{Pa}]. \tag{2.3}$$

In this formula the units are shown in square brackets. When the force is expressed in Newton [N] and the area in square meters [m^2] the resulting unit for the stress is Pascal [Pa]. The stress is an expression of the magnitude of the loading of the material.

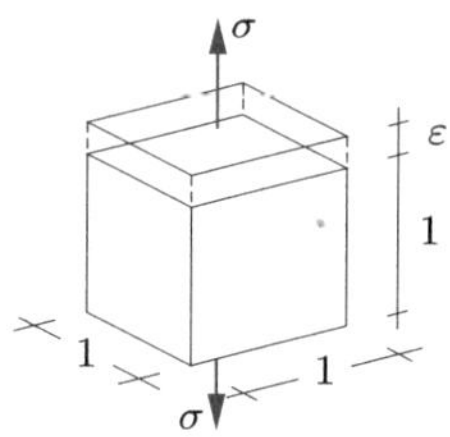

Fig. 2.36: Axial stress σ and axial strain ε.

A suitable measure of deformation is identified by cutting the original bar through the mid cross-section as shown in Fig. 2.35c. The length of each part is $\frac{1}{2}\ell$, and for a homogeneous bar each part contributes half of the full elongation. Thus, the extension experienced at the material level is the relative elongation. This is called the strain, defined by

$$\varepsilon = \frac{u}{\ell} = \frac{\text{Elongation}}{\text{Length}}, \qquad \left[\frac{\text{m}}{\text{m}}\right] = [\text{-}]. \tag{2.4}$$

It follows from this definition of strain as the relative elongation that strain is non-dimensional.

The concept of stress and strain introduced here is just a special case, often denoted as axial stress σ and axial strain ε, illustrated in Fig. 2.36. A general discussion of stress and strain is given in Chapter 8. When considered as part of a general state of stress and strain, the stress and strain defined here are called normal stress and normal strain, because they act normal to the surface area defined by the section.

2.4.2 Linear elastic bars

The operating stress level of a structure is normally considerably below the stress level that would lead to irreversible processes and failure. For many materials used in structures this implies that there is proportionality between the stress σ and the strain ε in any part of the structure. This behavior is called linear elasticity, and is described by the relation

$$\sigma = E\varepsilon\,, \qquad E\,\big[\,\mathrm{Pa}\,\big]. \tag{2.5}$$

This relation is often called Hooke's law after ROBERT HOOKE (1635–1703), who proposed it in 1675 and demonstrated it experimentally for several mechanical systems in 1678. The parameter E is called the modulus of elasticity. It is the factor of proportionality between the axial stress and axial strain in an experiment, where the loading is purely axial. Generally such an experiment leads to transverse contraction in addition to the axial elongation. The transverse contraction is not central to the present use in connection with trusses and will be dealt with in Chapter 8 in connection with the general discussion of elastic materials. The value of the elastic modulus varies between different materials as illustrated in Table 2.1 – from Gordon (2003).

Table 2.1: Typical elastic constants.

Material	MPa
Rubber	7
Nylon	1400
Plywood	7000
Wood	14000
Concrete	30000
Aluminum	70000
Steel	210000

The elastic relation of a bar follows by multiplication of the stress-strain relation with the cross-section area A of the bar, whereby

$$N = A\sigma = AE\varepsilon\,, \qquad AE\,\big[\,\mathrm{N}\,\big]. \tag{2.6}$$

It is seen that the elastic stiffness of the bar, AE, is the product of the material parameter E and the area A of the structural member. Thus, there are two contributing factors to the stiffness of a bar: its material stiffness, represented by E, and a geometric parameter of the structural element, here represented by the area A. This product form for the stiffness is general for beams and frames and plays an important role e.g. in design against column instability, discussed in Chapter 5.

2.4.3 Virtual work for truss structures

It was demonstrated in Section 1.3.1 that the force and moment equilibrium equations of a rigid body can be expressed in terms of the so-called virtual work. The idea of virtual work is that the structure in question is subjected to an infinitesimally small virtual displacement. The name 'virtual motion' indicates that it is a motion associated with a thought experiment, and the virtual motion need not be related to any real motion of the body. The principle of virtual work was used in Section 1.6 to calculate the reactions of statically determinate beam structures. A virtual motion was constructed where the constraint corresponding to the reaction in question was released, while all other support constraints were maintained. This produces a balance equation between the virtual work of the reaction and the virtual work of all external loads. The virtual displacement field was defined by letting the structure – or its individual parts – move as rigid bodies. It is of great importance in modern structural engineering that this simple form of the principle of virtual work can be extended to deformable bodies. The extension of the principle of virtual work to deformable bodies will here be illustrated for the fairly simple case of truss structures. It is demonstrated, how this principle can be used to determine the displacement of individual nodes in an elastic truss structure, and a general numerically oriented computational procedure in terms of finite elements is developed for truss structures in Section 2.5. The method of virtual work is extended to beam and frame structures in Chapter 4, and is used to formulate the finite element method for beams and frames in Chapter 7.

Vector algebra

The analysis of elastic truss structures and the stiffness relations for truss elements with general orientation is conveniently performed by use of vector analysis. Vectors will generally be described by boldface letters $\mathbf{x}$, $\mathbf{a}$ etc. The default Cartesian component representation has the individual components arranged in column format. The following analysis requires that component arrays can be written either in column format $\mathbf{a}$ or in row format indicated by the transpose $\mathbf{a}^T$,

$$\mathbf{a} = \begin{bmatrix} a_x \\ a_y \\ a_z \end{bmatrix}, \qquad \mathbf{a}^T = [\, a_x, a_y, a_z \,]. \tag{2.7}$$

The component arrays are considered as matrices. The scalar product of two vectors **a** and **b**, previously denoted by a dot as $\mathbf{a}\cdot\mathbf{b}$, can then be written as a matrix product,

$$\mathbf{a}\cdot\mathbf{b} = \mathbf{a}^T\mathbf{b} = [\, a_x, a_y, a_z \,] \begin{bmatrix} b_x \\ b_y \\ b_z \end{bmatrix} = a_x b_x + a_y b_y + a_z b_z \,. \tag{2.8}$$

In the matrix product the components of the rows of the first factor are multiplied by the components of the columns of the second factor, and the terms are then added. In the present case of the scalar product this leads to the indicated summation, of which the result is a scalar, i.e. a number. A special case is the scalar product of a vector with itself,

$$|\mathbf{a}|^2 = \mathbf{a}^T\mathbf{a} = [\, a_x, a_y, a_z \,] \begin{bmatrix} a_x \\ a_y \\ a_z \end{bmatrix} = a_x^2 + a_y^2 + a_z^2 \,. \tag{2.9}$$

The result of this operation is the square of the length of the vector, indicated as $|\mathbf{a}|$.

In matrix products the order of the factors is typically important. Thus, $\mathbf{a}^T\mathbf{a}$ is the scalar product, while $\mathbf{a}\,\mathbf{a}^T$ is the matrix

$$\mathbf{a}\,\mathbf{a}^T = \begin{bmatrix} a_x \\ a_y \\ a_z \end{bmatrix} [\, a_x, a_y, a_z \,] = \begin{bmatrix} a_x a_x & a_x a_y & a_x a_z \\ a_y a_x & a_y a_y & a_y a_z \\ a_z a_x & a_z a_y & a_z a_z \end{bmatrix}, \tag{2.10}$$

formed by products of the original vector components. Both the products $\mathbf{a}^T\mathbf{a}$ and $\mathbf{a}\,\mathbf{a}^T$ find application in the following theory of the elastic bar element.

Virtual work for a bar

Figure 2.37 shows a bar with end points A and B. The bar is described by the vector $\mathbf{a} = \overrightarrow{AB}$ with length $a = |\mathbf{a}|$. The bar supports the internal force N which is defined as positive for tension. Thus, the external forces $\mathbf{f}_A$ and $\mathbf{f}_B$ at the nodes A and B of the bar are given by

$$\mathbf{f}_B = -\mathbf{f}_A = \frac{1}{a}\mathbf{a}\,N \,. \tag{2.11}$$

This form secures equilibrium of the bar.

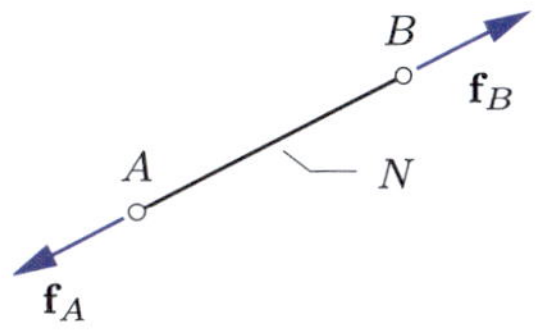

Fig. 2.37: Bar element with nodal loads.

Now, let the nodes A and B move by the virtual displacements $\delta\mathbf{u}_A$ and $\delta\mathbf{u}_B$, respectively. The nodal forces then perform the external virtual work

$$\delta V_{\mathrm{ex}} = \delta\mathbf{u}_A^T\,\mathbf{f}_A + \delta\mathbf{u}_B^T\,\mathbf{f}_B\,. \tag{2.12}$$

Note, that the external virtual work is formed by the work of the external forces by the virtual displacement of the bar. The idea now is to demonstrate that this external virtual work can be reformulated to an internal virtual work, expressed in terms of the internal force N and a virtual strain of the bar. The key component in this reformulation is the equilibrium equation (2.11). When substituting the nodal forces $\mathbf{f}_A$ and $\mathbf{f}_B$ from the equilibrium equation the external virtual work takes the form

$$\delta V_{\mathrm{ex}} = (\delta\mathbf{u}_B^T - \delta\mathbf{u}_A^T)(\mathbf{a}/a)N\,. \tag{2.13}$$

The task now is to show that the first factor can be expressed in the form of the virtual strain $\delta\varepsilon$, i.e. the strain that would arise in the bar, if the nodes were given the virtual displacements $\delta\mathbf{u}_A$ and $\delta\mathbf{u}_B$. First it is observed that the length of the bar can be expressed by the reltion

$$a^2 = |\,\mathbf{a}\,|^2 = \mathbf{a}^T\mathbf{a}\,. \tag{2.14}$$

Differentiation of this relation gives the incremental relation

$$2a\,\delta a = 2\mathbf{a}^T\delta\mathbf{a}\,, \tag{2.15}$$

where the factor 2 enters because both factors contribute to the differentiation. From this relation the virtual elongation of the bar follows as

$$\delta a = \frac{1}{a}\,\mathbf{a}^T\,\delta\mathbf{a} = \frac{1}{a}\,\mathbf{a}^T(\delta\mathbf{u}_B - \delta\mathbf{u}_A)\,. \tag{2.16}$$

The order of the two factors in a scalar vector product can be interchanged, and it is then seen that the factor to the internal force N in (2.13) is precisely the elongation of the bar as given in (2.16). Thus, the internal virtual work δV_{in} can be defined by

$$\delta V_{\mathrm{in}} = \delta a\,N = a\,\delta\varepsilon\,N\,. \tag{2.17}$$

The last form uses the definition of the virtual strain $\delta\varepsilon = \delta a/a$ corresponding to the virtual elongation δa. This definition of the internal virtual work as the work of the internal force N through the virtual strain $\delta\varepsilon$ then gives the equality between external and internal virtual work,

$$\delta V_{\rm ex} = \delta V_{\rm in} \,. \tag{2.18}$$

The result, that the virtual work of the external loads is equal to the work of the internal forces through the strain is of general validity in structural mechanics. It is generalized to truss structures below, and to beams and frames in Chapter 4.

Virtual work for a truss structure

The equality between external and internal virtual work for a bar only requires equilibrium of the bar. It must therefore apply to all bars of a truss structure, and therefore also to the sum of the contributions from each bar,

$$\sum_{\rm bars} \delta V_{\rm ex} = \sum_{\rm bars} \delta V_{\rm in} \,. \tag{2.19}$$

The external virtual work is now rewritten in terms of the external loads on the nodes. The basic principle is illustrated in Fig. 2.38 with reference to a two-dimensional truss structure. However, the principles are general and apply to three-dimensional trusses as well as to other structures.

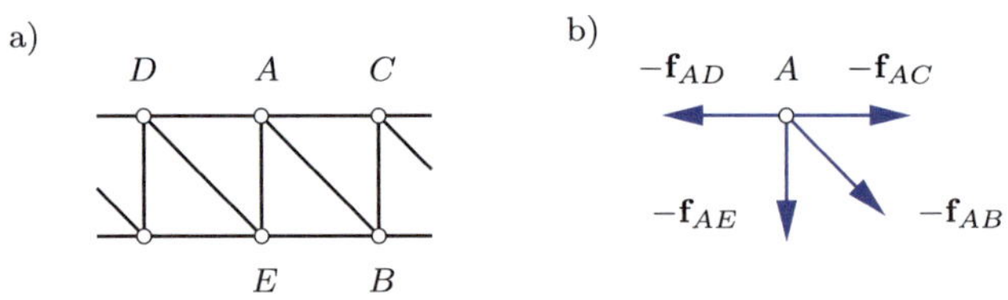

Fig. 2.38: Bar forces $\mathbf{f}_{A*}$ acting on node A.

Node A is acted on by all forces $\mathbf{f}_{AB}, \mathbf{f}_{AC}, \cdots$ from bars attached to this node. Thus, by the principle of action and reaction the similar forces but in opposite direction, $-\mathbf{f}_{AB}, -\mathbf{f}_{AC}, \cdots$, are the forces acting on the node. In addition to these internal forces the node may also be acted on by an external force $\mathbf{P}_A$ corresponding to a load. Equilibrium of the node requires the vector sum of all forces on the node to vanish. When arranging internal forces on the left side of the equation and the external force on the right, the equilibrium condition reads

$$\sum \mathbf{f}_{A*} = \mathbf{P}_A \,, \tag{2.20}$$

where $\mathbf{f}_{A*}$ is the force in the bar $A*$, with $*$ denoting an arbitrary node connected to A by a bar – in the present example the nodes B, C, D, E.

This gives the virtual work equation for trusses in the form

$$\sum_{\text{nodes}} \delta\mathbf{u}_j^T \mathbf{P}_j = \sum_{\text{bars}} a_i \, \delta\varepsilon_i \, N_i \, , \tag{2.21}$$

where $\mathbf{P}_j$ are the actual external forces acting at the nodes $j = 1, 2, \cdots$ and N_i are the actual internal forces in the bars $i = 1, 2, \cdots$. In contrast, the nodal displacements $\delta\mathbf{u}_j$ are virtual and define the virtual strains $\delta\varepsilon_i$ in the bars. The virtual displacements can be selected in a special way that enables explicit computation of the actual displacements of any node of an elastic truss structure as discussed in the next section.

2.4.4 Displacements of elastic truss structures

The procedure for calculation of the displacement of a node of an elastic truss structure is illustrated in Fig. 2.39. The top figure shows the structure with the actual loads – here consisting of vertical concentrated forces of magnitude P at all nodes in the girder foot. The bar forces corresponding to this load are calculated and denoted $N_1^0, \cdots, N_i^0$, where the superscript 0 indicates that these are the actual forces in the bars. The node displacements corresponding to the actual load are similarly denoted $\mathbf{u}_1^0, \cdots, \mathbf{u}_j^0$.

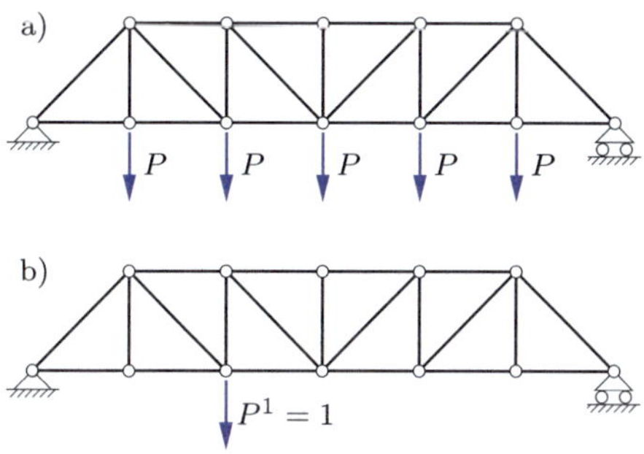

Fig. 2.39: Truss girder: a) Actual load, b) Unit test force.

The lower figure shows the same truss, but now loaded with only a single force $P^1 = 1$. This is an assumed load, used to determine the displacement component corresponding to the force P^1. The assumed concentrated load generates the bar forces $N_1^1, \cdots, N_i^1$.

If all bars in the truss are elastic the displacement component corresponding to the assumed load P^1 can now be determined by the principle of virtual work. The principle of virtual work is an equality between the external and internal virtual work calculated as the work of a set of loads and corresponding internal forces, when exposed to a virtual displacement field. In the present case the roles are interchanged, such that the structure with the actual loads in Fig. 2.39a provides the displacement field, while the static forces are taken from the assumed load in Fig. 2.39b. The virtual equation (2.21) then takes the form

$$u^0 P^1 = \sum_{\text{bars}} a_i \, \varepsilon_i^0 N_i^1 \, . \tag{2.22}$$

Due to the assumed load consisting of a single force, the external work consists of a single term $u^0 P^1$, where u^0 is the displacement in the direction of the assumed load P^1, when the structure is loaded by the actual load.

When the truss is elastic, the strains ε_i^0 in the bars can be expressed in terms of the corresponding bar forces N_i^0 according to the elastic relation (2.6),

$$\varepsilon_i^0 = \frac{N_i^0}{(EA)_i} \, , \tag{2.23}$$

where $(EA)_i$ is the elastic stiffness of bar i. Substitution of this into the virtual work equation (2.22) together with the condition that the test load is of unit magnitude, $P^1 = 1$, gives the final result in the form

$$u^0 = u^0 \, P^1 = \sum_{\text{bars}} a_i \, \varepsilon_i^0 N_i^1 = \sum_{\text{bars}} \frac{a_i}{(EA)_i} N_i^0 N_i^1 \, . \tag{2.24}$$

This is an explicit formula for the displacement u^0 in the direction of a unit test force in terms of a summation over all bars of the structure of the product of the bar force for the actual and the test load, respectively. It is notable, that the geometry of the truss does not appear directly in the formula. It has already been accounted for in the calculation of the bar forces.

Example 2.10. Displacement of node in a truss. The calculation of node displacements in elastic trusses is illustrated by considering the simple cantilever truss in Fig. 2.40a, supporting a single vertical force P at node C. All bars are assumed to have identical elastic stiffness parameter EA. In this example it is desired to calculate both the vertical and the horizontal displacement components of node C. This is done by considering two independent load cases shown in Fig. 2.40b: a vertical unit test force P^1 and a horizontal unit test force P^2, both acting at node C.

The total computation consists of calculating three sets of bar forces: N_i^0 for the actual load, and N_i^1 and N_i^2 for the two test load cases. It is convenient to collect the bar lengths and forces in a table as illustrated by Table 2.2. The bar lengths are denoted by the symbol ℓ here, because a has been used for a specific dimension of the truss. If the bars had different elastic stiffness, the values $(EA)_i$ should also be included in the table.

The vertical displacement of node C is found from (2.24) as

$$u_C^1 = \sum_i \frac{\ell_i}{EA} N_i^0 N_i^1 \, ,$$

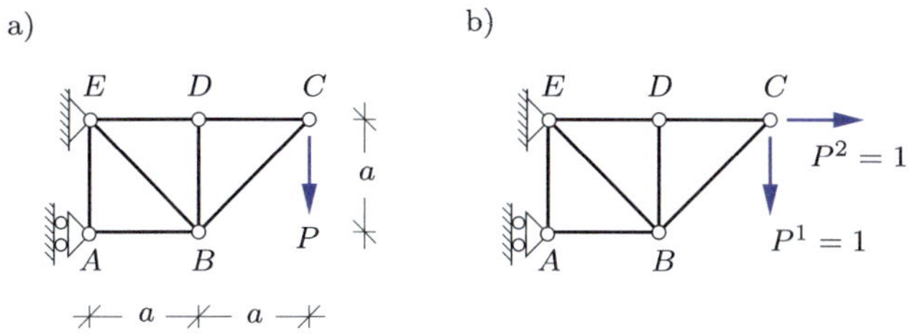

Fig. 2.40: Cantilever girder. a) actual load P, b) unit test loads P^1, P^2.

Table 2.2: Tabular calculation of virtual work.

	ℓ	N^0	N^1	N^2
AB	a	$-2P$	-2	0
BC	$\sqrt{2}a$	$-\sqrt{2}P$	$-\sqrt{2}$	0
BD	a	0	0	0
CD	a	P	1	1
BE	$\sqrt{2}a$	$\sqrt{2}P$	$\sqrt{2}$	0
DE	a	P	1	1
AE	a	0	0	0

and substitution of the numerical values from the table gives

$$u_C^1 = \frac{aP}{EA}(4 + 2\sqrt{2} + 1 + 2\sqrt{2} + 1) = (6 + 4\sqrt{2})\frac{aP}{EA} = 11.66\,\frac{aP}{EA}\,.$$

Similarly the horizontal displacement of node C is

$$u_C^2 = \sum_i \frac{\ell_i}{EA} N_i^0 N_i^2 = \frac{aP}{EA}(1+1) = 2\,\frac{aP}{EA}\,.$$

The vertical displacement associated with bending of the truss girder is much larger than the axial displacement associated directly with the elongation of the bars ED and DC. This behavior will also be seen in beams, where most of the displacement is usually associated with bending. □

2.5 Finite element analysis of trusses

The analysis methods developed so far in this chapter for trusses have mainly been based on statics, i.e. use of equilibrium conditions for the full truss and the individual bars. This approach works well for smaller structures and analysis carried out by hand. For larger truss structures and analysis carried out by computer a systematic approach in which the individual bar elements and nodes are treated in a repetitive way is desirable. In order to isolate an individual bar element from the rest of the structure it is desirable to consider the structure as flexible and to use the displacements of the nodes as the primary variables in the analysis. This represents a change in the point of view relative to the previous methods of nodes and sections, where the forces in the bars were the primary variables.

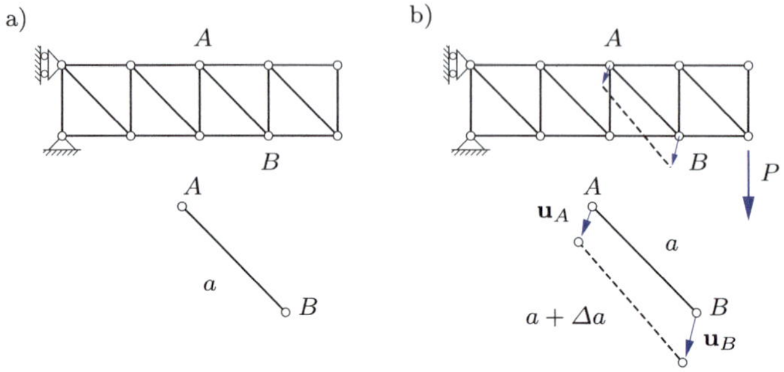

Fig. 2.41: Displacement of nodes A and B leads to elongation of the bar AB.

The basic idea of using the displacement of each of the nodes is illustrated in Fig. 2.41. Consider a flexible bar AB connecting the nodes A and B. Loading of the structure introduces a displacement $\mathbf{u}_A$ of node A and $\mathbf{u}_B$ of node B. These displacements may introduce a change of length of the bar AB from a to $a + \Delta a$. This change of length corresponds to a force N_{AB} in the bar.

An efficient analysis method, particularly suited for computer implementation, can be developed by expressing all bar forces in terms of the displacements $\mathbf{u}_A, \mathbf{u}_B, \cdots$ of the nodes, and then formulating and solving the equilibrium conditions for all the nodes of the structure. This task requires:

i) A constitutive relation between the elongation of each bar and the developed bar force.

ii) A general formulation for the elongation of a bar with arbitrary orientation in space, expressed in terms of the displacement of the two nodes of the bar.

iii) A systematic formulation of the equilibrium conditions for each node in terms of the relevant bar forces.

iv) Introduction of suitable support conditions.

These tasks are described in the following subsections, leading to the development of a small computer program MiniTruss.

2.5.1 Elastic bar element

The derivation of the elastic bar element consists in first determining the strain in the element in terms of the displacements of the element nodes, and then expressing the forces in the nodes in terms of this strain.

Strain in bar element

The properties of a bar element are conveniently formulated by use of vector algebra as illustrated in Fig. 2.42.

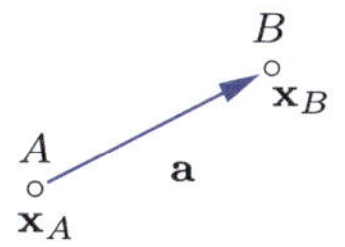

Fig. 2.42: Bar element represented as vector **a**.

The bar element AB is described by the initial position of the nodes A and B with Cartesian coordinates

$$\mathbf{x}_A = [x_A, y_A, z_A]^T, \qquad \mathbf{x}_B = [x_B, y_B, z_B]^T. \tag{2.25}$$

The bar element is given by the vector $\mathbf{a}$, represented in terms of the node coordinates as

$$\mathbf{a} = [x_B - x_A, y_B - y_A, z_B - z_A]^T. \tag{2.26}$$

The length a of the bar element in its initial position is therefore given as $a^2 = |\mathbf{a}|^2 = \mathbf{a}^T\mathbf{a}$.

Fig. 2.43: Elongation of bar element by projection of node displacements.

When the structure is loaded, the nodes A and B move as described by the displacement vectors $\mathbf{u}_A$ and $\mathbf{u}_B$, shown in Fig. 2.43. The nodes A and B are then located at $\mathbf{x}_A + \mathbf{u}_A$ and $\mathbf{x}_B + \mathbf{u}_B$, respectively. When the displacements of the nodes are small, e.g. relative to the length of the element, the elongation of the element Δa, and thereby the strain, can be calculated via projections as indicated in Fig. 2.43b. The projection of the displacement is obtained by scalar multiplication with the unit vector $a^{-1}\mathbf{a}$, and thus the elongation of the bar element is approximated by

$$\Delta a \simeq a^{-1}\mathbf{a}^T\mathbf{u}_B - a^{-1}\mathbf{a}^T\mathbf{u}_A. \tag{2.27}$$

The strain then follows from division by the bar length a, whereby

$$\varepsilon = \frac{1}{a^2}\mathbf{a}^T(\mathbf{u}_B - \mathbf{u}_A). \tag{2.28}$$

Note, that while use of projections to evaluate the elongation generally involves an approximation, the special case of a translation with identical displacements of the two nodes $\mathbf{u}_B = \mathbf{u}_A$ leads to zero strain, $\varepsilon = 0$.

Stiffness matrix of elastic bar element

Equilibrium of the bar element requires that the forces $\mathbf{f}_A$ and $\mathbf{f}_B$ at the nodes of the bar element are equal in magnitude but in opposite directions. The direction is described by the unit vector $a^{-1}\mathbf{a}$, and the magnitude of the force is denoted N with tension as positive. The force vectors in A and B are then similar to those previously given in (2.11),

$$\mathbf{f}_B = -\mathbf{f}_A = \frac{1}{a}\mathbf{a}N\,. \tag{2.29}$$

For an elastic bar with normal force $N = AE\,\varepsilon$ the force vectors $\mathbf{f}_A$ and $\mathbf{f}_B$ can then be expressed in terms of the displacements of the nodes via substitution of the expression (2.28) for the strain, whereby

$$\mathbf{f}_B = -\mathbf{f}_A = \frac{AE}{a^3}\mathbf{a}\,\mathbf{a}^T(\mathbf{u}_B - \mathbf{u}_A)\,. \tag{2.30}$$

Note the occurrence of the product $\mathbf{a}\,\mathbf{a}^T$, forming the matrix shown in (2.10).

When assembling all the nodal forces of a truss structure into a model for the full structure it is advantageous to combine the two vector equations from (2.30) in an explicit block matrix format. For this purpose the forces and displacements at the two element nodes A and B are arranged in an expanded vector format of double size,

$$[\,\mathbf{f}_A^T, \mathbf{f}_B^T\,] = [\,f_x^A, f_y^A, f_z^A\,,\, f_x^B, f_y^B, f_z^B\,]\,, \tag{2.31}$$

and in the same way for the displacements $[\,\mathbf{u}_A^T, \mathbf{u}_B^T\,]$. With this notation the expressions (2.30) for the nodal forces can be expressed in the following block matrix format:

$$\begin{bmatrix}\mathbf{f}_A\\ \mathbf{f}_B\end{bmatrix} = \underbrace{\frac{AE}{a^3}\begin{bmatrix}\mathbf{a}\,\mathbf{a}^T & -\mathbf{a}\,\mathbf{a}^T\\ -\mathbf{a}\,\mathbf{a}^T & \mathbf{a}\,\mathbf{a}^T\end{bmatrix}}_{\mathbf{K}_{\text{bar}}}\begin{bmatrix}\mathbf{u}_A\\ \mathbf{u}_B\end{bmatrix}. \tag{2.32}$$

The matrix $\mathbf{K}_{\text{bar}}$ in this relation is called the element stiffness matrix of the bar. The derivations have been illustrated for the three-dimensional case, where the individual vectors have 3 components, leading to a 6 by 6 element stiffness matrix. In the case of a plane truss the vector dimension is 2, and the element stiffness matrix has the dimension 4 by 4.

The forces acting on the bar element AB are conveniently expressed in the generic block matrix format

$$\begin{bmatrix} \mathbf{f}_A \\ \mathbf{f}_B \end{bmatrix} = \underbrace{\begin{bmatrix} \mathbf{K}_{AA} & \mathbf{K}_{AB} \\ \mathbf{K}_{BA} & \mathbf{K}_{BB} \end{bmatrix}}_{\mathbf{K}_{\text{bar}}} \begin{bmatrix} \mathbf{u}_A \\ \mathbf{u}_B \end{bmatrix}. \tag{2.33}$$

In this format the block matrices $\mathbf{K}_{AA}$ and $\mathbf{K}_{AB}$ represent the force at node A from the displacement $\mathbf{u}_A$ of node A and the displacement $\mathbf{u}_B$ of node B, respectively.

2.5.2 Finite Element Method for trusses

The next step is to use the information about the individual bar elements to set up conditions for all the nodes of the truss structure. The principle was illustrated in Fig. 2.38, where it was demonstrated that equilibrium of a node A requires the sum of the forces $\mathbf{f}_{A*}$ from all connecting bars to balance the external load $\mathbf{f}_A^{\text{ex}}$ at node A,

$$\sum \mathbf{f}_{A*} = \mathbf{f}_A^{\text{ex}}. \tag{2.34}$$

The forces in the individual bar elements are available from a element matrix relation of the form (2.32), and a central point in the formulation of the finite element method is the procedure used to assemble the contributions from the individual elements into a model for the structure.

Assembling the global stiffness matrix

The structure of the element stiffness matrix (2.33), where the force contribution at the element nodes is given in terms of the displacements of the nodes via a block matrix, leads to the following simple procedure to create a model of the complete truss structure.

i) Identify all nodes of the structure by numbers $1, \cdots, n$. Denote the corresponding coordinate set of the nodes by $\mathbf{x}_1, \cdots, \mathbf{x}_n$.
ii) Associate each bar element with two nodes, e.g. bar element AB with the nodes i and j of the structure, as illustrated in Fig. 2.44. This association between the element nodes A, B and the global structural nodes i, j is called the topology of the model.

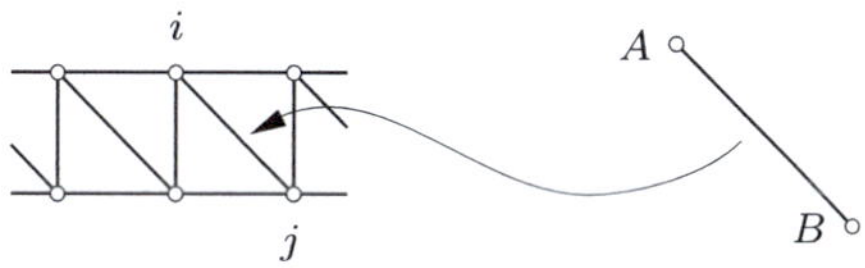

Fig. 2.44: Bar member AB as element ij in truss structure.

iii) The contribution of the forces from the individual elements can now be obtained by placing the submatrices from the element stiffness relation (2.33) in the global format as shown here,

$$\begin{bmatrix} \vdots \\ \mathbf{f}_i \\ \cdot \\ \mathbf{f}_j \\ \vdots \end{bmatrix}_{AB} = \begin{bmatrix} & \vdots & & \vdots & \\ \cdots & \mathbf{K}_{AA} & \cdot & \mathbf{K}_{AB} & \cdots \\ & \cdot & & \cdot & \\ \cdots & \mathbf{K}_{BA} & \cdot & \mathbf{K}_{BB} & \cdots \\ & \vdots & & \vdots & \end{bmatrix}_{AB} \begin{bmatrix} \vdots \\ \mathbf{u}_i \\ \cdot \\ \mathbf{u}_j \\ \vdots \end{bmatrix} \begin{matrix} \\ i \\ \\ j \\ \\ \end{matrix} . \tag{2.35}$$

When placed in this global format the displacements $\mathbf{u}_i$ and $\mathbf{u}_j$ contribute in the correct way to the internal forces $\mathbf{f}_i$ and $\mathbf{f}_j$ at nodes i and j. Adding the contributions from all elements to form the global stiffness matrix of the structure is seen to correspond to adding the internal forces at each of the nodes as prescribed in (2.33).

Support conditions

The model must also include provisions for supports, typically in the form of constraints on the displacements of certain nodes. Constraint of a node can typically be introduced by imposing one or more relations between the displacement components $\mathbf{u}_i = [\,u_x, u_y, u_z\,]_i^T$ at the corresponding node. The introduction of such a constraint reduces the number of unknown displacement components in the model. Before presenting the implementation of general constraints two simple alternative methods of implementing the support conditions are discussed.

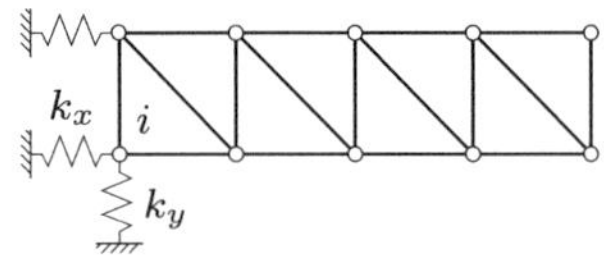

Fig. 2.45: Support springs attached to node i.

A simple method is to constrain the supported nodes by introducing stiff springs as illustrated in Fig. 2.45. The springs connect the node to a rigid support. They act essentially as bar elements, but because the 'other end' of the spring is fully constrained, the corresponding stiffness matrix to be included in the model is just a block matrix $\mathbf{K}_s$ appearing in the diagonal position corresponding to the supported node as illustrated by the corresponding global force stiffness matrix contribution

$$\begin{bmatrix} \vdots \\ \mathbf{f}_i \\ \vdots \end{bmatrix}_s = \begin{bmatrix} & \vdots & \\ \cdot & \mathbf{K}_s & \cdots \\ & \vdots & \end{bmatrix}_s \begin{bmatrix} \vdots \\ \mathbf{u}_i \\ \vdots \end{bmatrix} \quad i . \tag{2.36}$$

A support consisting of springs with stiffness constants k_x, k_y, k_z in the coordinate directions correspond to the diagonal block stiffness matrix

$$\mathbf{K}_s = \begin{bmatrix} k_x & & \\ & k_y & \\ & & k_z \end{bmatrix} . \tag{2.37}$$

This format permits some of the springs to have vanishing stiffness. The stiffness of an inclined spring with spring constant k_n along a direction described by the unit vector $\mathbf{n} = [n_x, n_y, n_z]$ follows from the bar element stiffness matrix (2.32) as

$$\mathbf{K}_s = k_n \, \mathbf{n}\mathbf{n}^T = k_n \begin{bmatrix} n_x n_x & n_x n_y & n_x n_z \\ n_y n_x & n_y n_y & n_y n_z \\ n_z n_x & n_z n_y & n_z n_z \end{bmatrix} . \tag{2.38}$$

Several springs can be applied to a node, simply be adding their stiffness contributions.

The use of stiff springs to represent constraints involves a compromise. Ideally the springs should be infinitely stiff, relative to the stiffness components of the structure itself. This would lead to ill-conditioning of the equations, and numerical roundoff errors in the solution of the equation system sets a limit on the magnitude of the spring constants that can be used without compromising the accuracy of the solution procedure. A simple modification of the idea of springs can be used when the involved degrees of freedom are constrained to zero. In that case the rows and columns corresponding to the constrained degrees of freedom can be set to zero, except for the diagonal element that is retained. When removing any loads associated with these degrees of freedom, the equations for the constrained degrees of freedom are uncoupled from the unconstrained degrees of freedom, and the equations can be solved directly, retaining the original size and organization of the matrix and displacement components.

A third more general alternative exists, that has a particularly elegant implementation in MATLAB. The first step is to separate the displacement vector into two parts: a vector $\mathbf{u}_c$ containing the constrained displacement components, and a vector $\mathbf{u}_u$ containing the remaining unconstrained displacement components. Rearrangement of the equilibrium equations gives the block matrix equation format

$$\begin{bmatrix} \mathbf{K}_{uu} & \mathbf{K}_{uc} \\ \mathbf{K}_{cu} & \mathbf{K}_{cc} \end{bmatrix} \begin{bmatrix} \mathbf{u}_u \\ \mathbf{u}_c \end{bmatrix} = \begin{bmatrix} \mathbf{f}_u \\ \mathbf{f}_c + \mathbf{r} \end{bmatrix}. \tag{2.39}$$

The stiffness sub-matrices follow from rearrangement of the original stiffness matrix $\mathbf{K}$. At the constrained degrees of freedom the total force consists of any imposed load $\mathbf{f}_c$ plus the reaction force components $\mathbf{r}$ produced by the support. In this format $\mathbf{u}_c$ represents imposed displacements, that may be non-zero. The solution proceeds in two steps. First the unconstrained displacements $\mathbf{u}_u$ are obtained from the top part of the equations,

$$\mathbf{K}_{uu}\mathbf{u}_u = \mathbf{f}_u - \mathbf{K}_{uc}\mathbf{u}_c, \tag{2.40}$$

and then the reaction forces $\mathbf{r}$ follow from the lower part as

$$\mathbf{r} = \mathbf{K}_{cu}\mathbf{u}_u + \mathbf{K}_{cc}\mathbf{u}_c - \mathbf{f}_c. \tag{2.41}$$

In classic programming this procedure would imply the formation of the corresponding sub-matrices. However, in high-level programming languages like MATLAB the operations can be implemented via the corresponding index sets without rearranging the data as explained in connection with the MINITRUSS program in the next section.

2.5.3 The *MiniTruss* program

The principles described in the previous sections have been implemented in a small Finite Element program MINITRUSS using the high level programming language MATLAB. However, other high-level programming languages permit similar implementations. The main structure of the program and its data structure are explained in relation to the specific roof truss analysis described in Example 2.9. Figure 2.46 shows the roof truss with the nodes numbered from 1 to 7 and the bars from 1 to 11. The program receives information about the nodes via their coordinates $[x_1, y_1], \cdots, [x_7, y_7]$, while the attachment of the elements to the nodes is defined by a topology matrix $\mathbf{T}$.

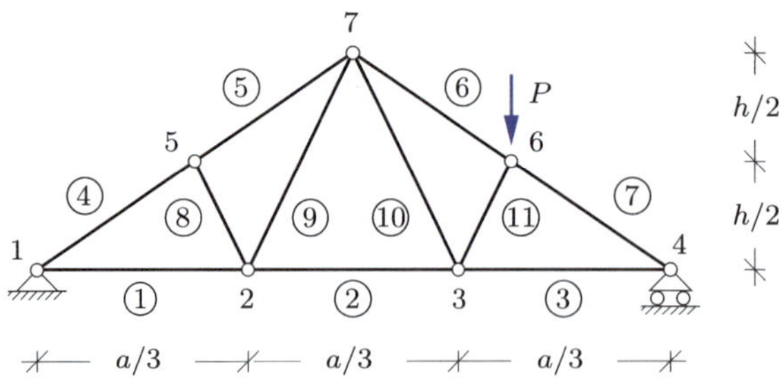

Fig. 2.46: Roof truss with concentrated vertical force at node 6.

The program is built as a script file `MiniTruss.m` that serves as a driver that reads a data file and activates subroutines that set up the model, form the global stiffness matrix, introduce the support conditions, apply the load, and finally solves for the displacement of all nodes and calculates the forces in all bars. The data of the present example is contained in the data file `W_Truss.m`. The structure and content of this file is described in the following.

Node data. The W-truss is described in an xy-coordinate system with origo at the center of the truss foot, and the y-axis vertical upwards. The width a and height h of the truss are given in parametric form in terms of variables `a` and `h`. The node coordinates are given in the form of an array `X`, with each node corresponding to one row. The first part of the data file then is

```
% Width 'a' and height 'h' of truss
a = 12.0; h =  4.0;

% Coordinates of nodes X = [x y (z)],
X = [  -a/2   0.00
       -a/6   0.00
        a/6   0.00
        a/2   0.00
       -a/4   h/2
        a/4   h/2
        0.00  h    ];
```

The node coordinates [`x,y`] are given in the order of the node number, starting with node 1. Thus, the node number is not given explicitly, but implied by the row number in the node coordinate matrix `X`.

The program identifies the truss structure as being 2 or 3-dimensional by counting the number of columns in the node coordinate matrix `X`. If the truss is 3-dimensional the displacement vector for a node has 3 components, while a 2-dimensional structure has 2 displacement components per node.

Element data. The truss elements are defined in the topology matrix `T`. Each row of this matrix defines an element, by listing its two nodes by their node number, and by giving a third number identifying a set of element properties, area A and elasticity modulus E, given in a material property matrix `H`.

```
% Topology matrix T = [node1 node2 propno],
T = [ 1  2   1
      2  3   1
      3  4   1
      1  5   2
      5  7   2
      6  7   2
      4  6   2
      2  5   3
      2  7   3
      3  7   3
      3  6   3 ];

% Element property matrix H = [ A E ],
H = [ 1.0  100.0
      1.0  100.0
      0.8  100.0 ];
```

In this example there are three element types: type 1 for the foot, type 2 for the head, and type 3 for the diagonals.

Loads. The loads are specified in the load matrix P. This matrix contains a row for each loaded node. The data line specifies the node number and the force components. In the present example node 6 is loaded by a force with components $[f_x, f_y] = [0.000, -1.000]$.

```
% Prescribed loads P = [node fx fy (fz)]
P = [ 6   0.000 -1.000 ];
```

Support conditions. The support conditions are given in the constraint matrix C. The constraint matrix contains a row for each constrained displacement component. In the present example there are 3 constrained displacement components u_1 and v_1 at node 1 and v_4 at node 4.

```
% Constraints C = [node 'dof' (uc)]
C = [ 1  1
      1  2
      4  2 ];
```

The optional parameter uc indicates the magnitude of a prescribed displacement component. If not included as a third column in C, constrained displacement components are set to zero.

Graphics. The MiniTruss program produces two plots of the structure: a plot of the initial undeformed geometry including node numbers, and another plot of the deformed structure after application of the load. For most real structures it is necessary to scale the displacements to be able to see the deformation. The coordinate window used for the plots is controlled via definition of the plot axes, specified in the array

```
% Axes used for geometry plots [Xmin Xmax Ymin Ymax]
PlotAxes = [-0.55*a 0.55*a  -0.5*h 1.2*h];
```

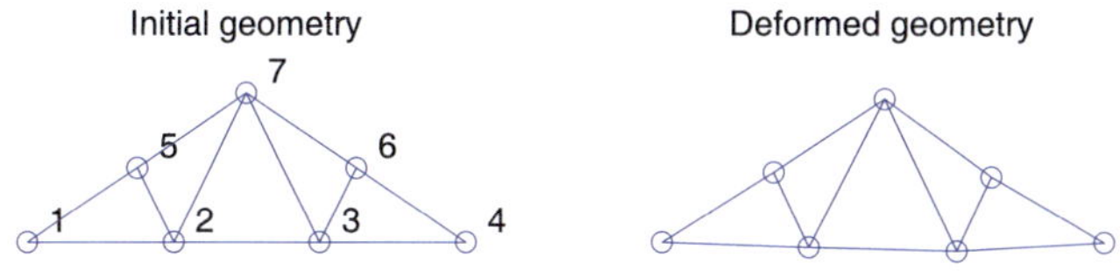

Fig. 2.47: Plots of initial and deformed geometry of W-truss.

The default in the program MiniTruss is to use the two top subplots in a 2×2 plot layout. For long trusses, as e.g. bridges etc., a 2×1 plot format can be introduced by changing to `subplot(2,1)`.

Analysis process. The first step in an analysis with the MiniTruss program is to read the appropriate data file into memory. This is done either by writing `W_Truss` in the command window, or by uploading `W_Truss` to the MATLAB editor and pressing the `F5` key from the editor. The data is now available in active memory and the analysis is carried out by activating the script file `MiniTruss.m`, either by writing `MiniTruss` in the command window or by pressing the `F5` key with `MiniTruss.m` in the editor.

```
% Nodal loads into load vector
if exist('P','var')
    f = loadnode(f,P,dof);
end

% Global stiffness matrix
K = kbar(K,T,X,H);

% Solve stiffness equation
[u,r,ic] = solveeq(K,f,C,dof);

% Nodal displacements
Un = reshape(u,dof,size(X,1))';

% Calculate element forces and strains
[s,e] = Nbar(T,X,H,u);
```

The program activates the following processes. First the global load vector `f` is formed. Then the global stiffness matrix of the structure `K` is formed by the function `kbar` by collecting the stiffness contributions from all bar-elements. The function `solveeq` then solves the constrained equations, accounting for the support conditions. Finally the displacements are reshaped into vector components for each node in the matrix `Un`, and the internal forces `s` and strains `e` are obtained by post-processing.

The solution of the constrained equations in `solveeq` is obtained by a reformulation of the block equations (2.39)–(2.41) in terms of index sets. The structure in Fig. 2.46 has 7 nodes and thereby 14 degrees of freedom. Thus the index set containing all degrees of freedom is `ii = [1,2,···,13,14]`. At node 1 both displacement components are constrained, and at node 4 the second displacement component is constrained. Thus, the index set of the constraints is `ic = [1,2,8]`. In MATLAB the index set of the unconstrained degrees of freedom `iu` can then be obtained by the function call

```
iu = setdiff(ii,ic) = [3,4,5,6,7,9,10,11,12,13,14].
```

In MATLAB the sub-matrices are obtained by using the full matrices with the appropriate index set as index. Thus, `K(iu,iu)` extracts the sub-matrix $\mathbf{K}_{uu}$ etc. The equation (2.40) for the unconstrained displacement components then takes the form

```
K(iu,iu)*u(iu) = f(iu) - K(iu,ic)*u(ic)
```

and the reaction is found from the expression

```
r = K(ic,ii)*u(ii) - f(ic).
```

In the MATLAB syntax this gives the reactions `r` as a 3-component vector corresponding to the global degrees of freedom `ic = [1,2,8]`.

2.6 Exercises

Exercise 2.1. The figure shows a cantilever N-truss supporting a vertical load P acting at node F.

a) Determine the reaction forces.

b) Determine all bar forces by the method of joints.

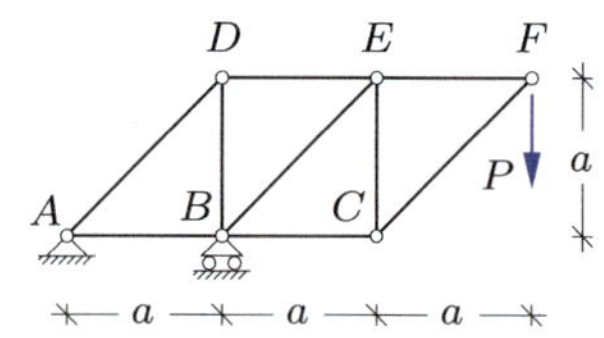

Exercise 2.2. The figure shows a symmetric and simply supported truss girder with vertical loads P acting at nodes B, C and D, respectively.

a) Determine the reaction forces.

b) Determine all bar forces by the method of joints.

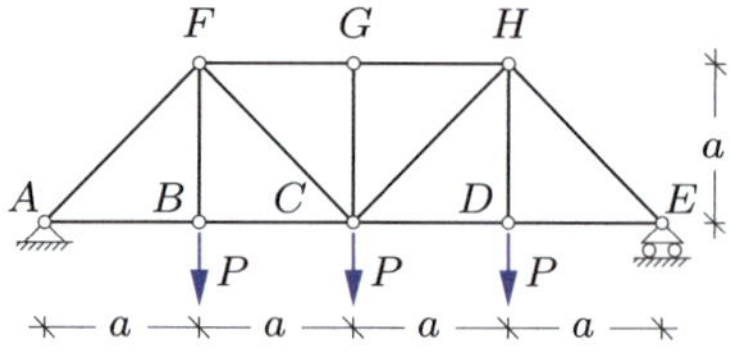

Exercise 2.3. The figure shows a truss structure with fixed simple supports at nodes A, B and C and a vertical load P at nodes D and E, respectively.

a) Illustrate the direction of the resulting reaction forces at the three supports.

b) Determine the reaction force in B by a single equilibrium equation and subsequently the bar force in BD.

c) Determine the all bar forces and the remaining reactions

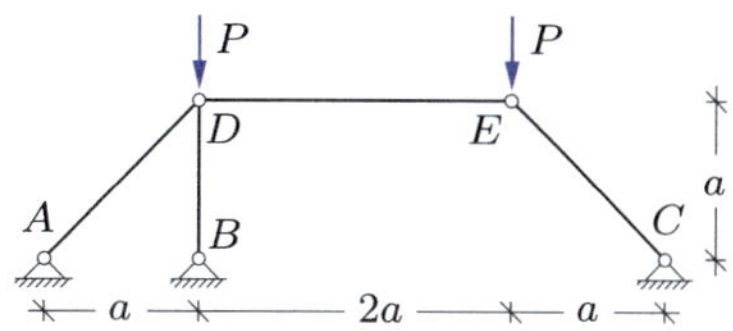

Exercise 2.4. The behavior of structures often depends on the support conditions. The two truss structures in the figure are identical, except that the supports with and without rollers have been interchanged.

a) Determine the reactions and bar forces for both structures.

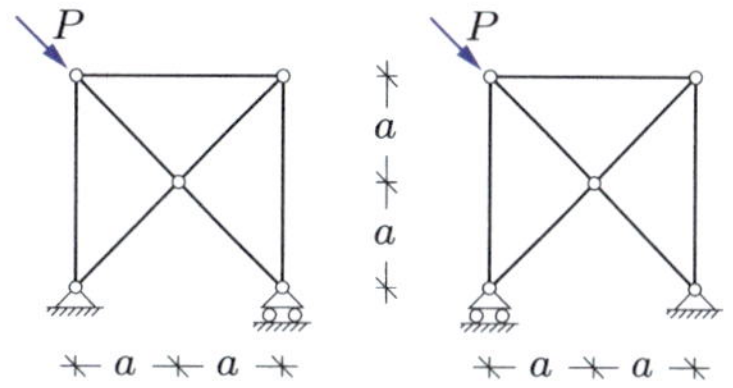

Exercise 2.5. The figure shows a simply supported V-truss, loaded by two vertical forces P_1 and P_2.

a) Determine the reactions and the bar forces for $P_1 = P_2 = P$.

b) Explain the influence of the relative height h/a of the truss girder on the bar forces in head and foot, and in the diagonals.

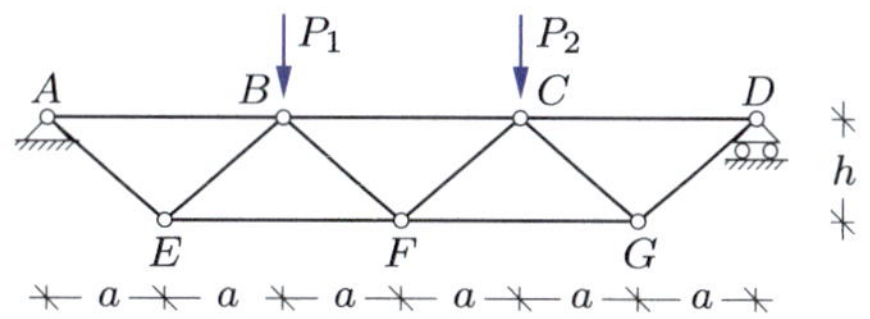

Exercise 2.6. Consider the simply supported V-truss from Exercise 2.5 with the load $P_1 = P$ and $P_2 = 0$.

a) Determine the reactions and the bar forces.

Exercise 2.7. The truss structure shown in the figure has a fixed simple support at A and a simple support permitting vertical motion at B. The truss is loaded by the vertical forces P and $2P$ and by the horizontal force P.

a) Determine the reaction forces.

b) Determine the bar forces.

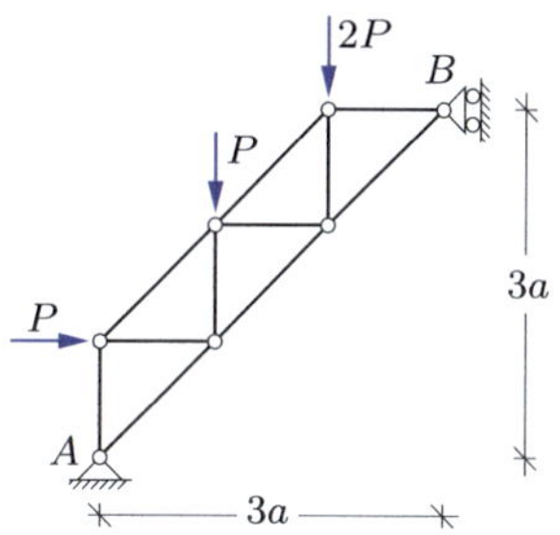

Exercise 2.8. The figure shows a truss structure in which all bars have the length a. The structure has a fixed simple support at A and a simple support permitting horizontal motion at B. The truss is loaded by vertical forces of magnitude P and $2P$.

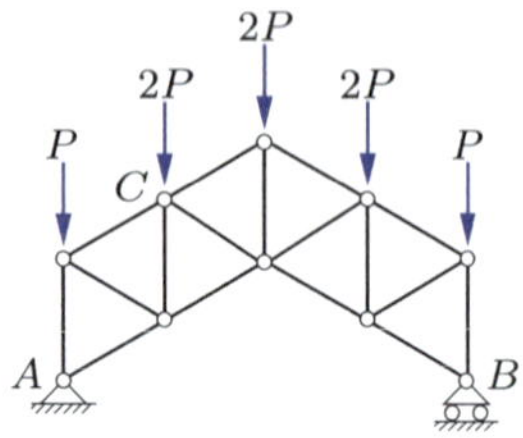

a) Determine the reaction forces.

b) Determine the forces in all the bars connected to node C.

Exercise 2.9. The figure shows a truss structure similar to that in Exercise 2.8. In the present truss the vertical middle bar has been removed, and the support in B is fixed in both directions.

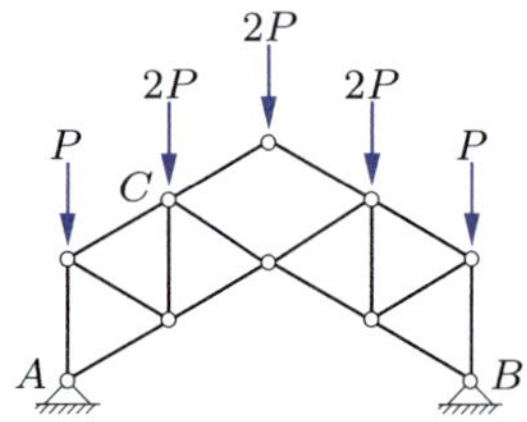

a) Determine the reaction forces.

b) Determine the forces in all the bars connected to node C.

Exercise 2.10. Determine the remaining bar forces in the roof truss from Example 2.3, and indicate all bar forces in a figure.

Exercise 2.11. The figure shows a simply supported roof truss of length a and height h with vertical loads P at nodes E, F and G, respectively. The symmetric structure of the interior bars is determined by the angles α and β.

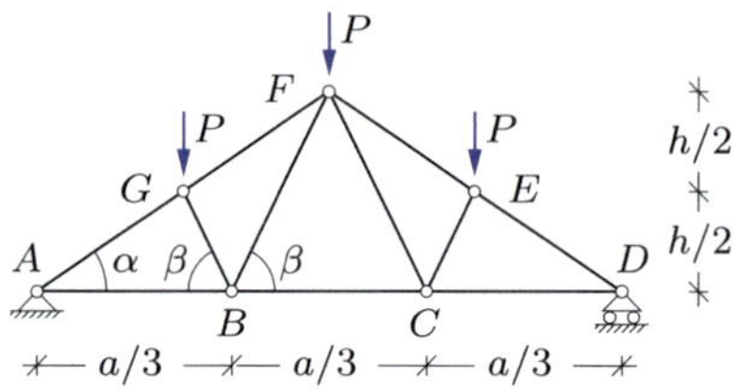

a) Determine the reaction forces.

b) Determine all bar forces by the method of joints.

Exercise 2.12. The figure shows a typical V-truss A–B from a building crane. In this exercise the truss is considered as plane. The truss is supported by a fixed hinge at A and an inclined tension bar DC, connected to the head of the truss at C. It carries the vertical load P at B.

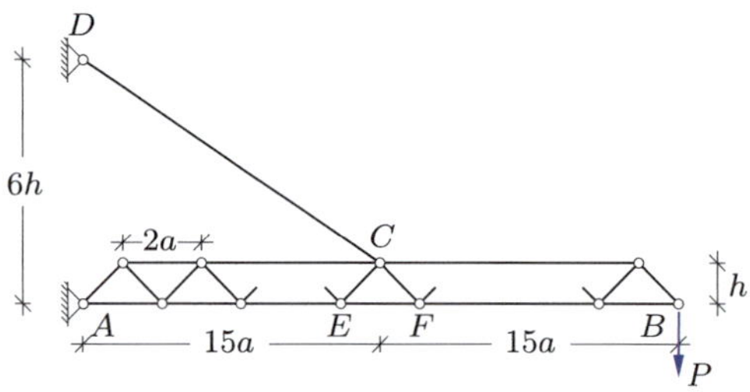

a) Determine all reaction components at A and D.

b) Determine the force in the inclined bar DC supporting the truss.

c) Find the bar forces N_{EF}, N_{CE} and N_{CF}.

Exercise 2.13. The figure shows a symmetric three-hinge frame formed by two identical truss frames AC and BC that are connected by a hinge at node C. The truss frames are supported by fixed simple supports at A and B, respectively. The load consists of a single vertical force P acting at E.

a) Determine all reaction components at A and B.

b) Determine the forces in the bars CD, CE, DE, DF and EF by the method of sections.

c) Check the result by e.g. isolating node D and verify that the node is in equilibrium.

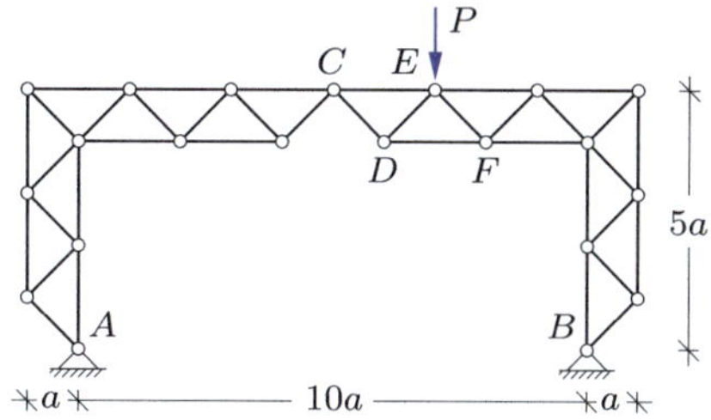

Exercise 2.14. A modified version of the truss frame from Exercise 2.13 is shown in the figure. A horizontal bar DD' is introduced to close the hinge at C, and rollers are added to the simple support in B, permitting horizontal motion and retaining the static determinacy of the structure.

a) Repeat questions a) to c) of Exercise 2.13 for the modified truss frame and determine also the force in the new bar DD'.

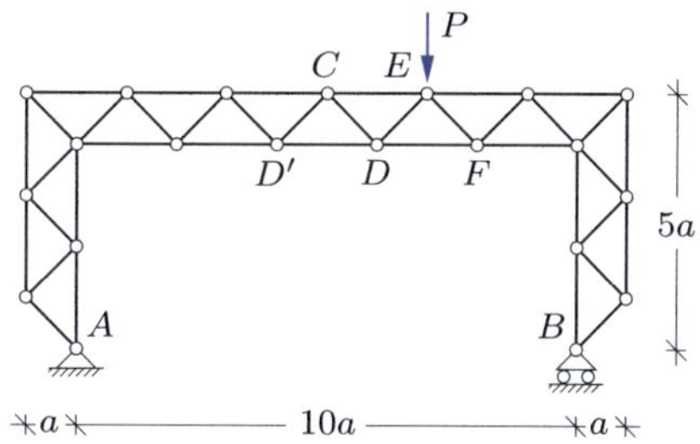

Exercise 2.15. The figure shows an N-type cantilever truss, loaded by concentrated vertical force of magnitude P at the tip joint D.

a) Find and show the reaction components in a figure.

b) Determine all the bar forces.

c) Find both vertical and horizontal displacement of the node D.

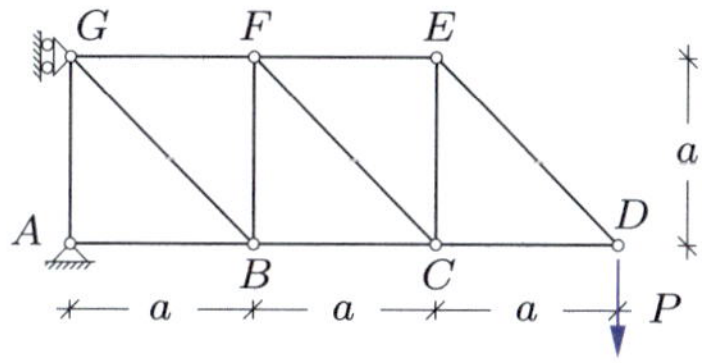

Exercise 2.16. The figure shows a simple cantilever truss, loaded by a concentrated force P at the joint E.

a) Find and show the reaction components in a figure.

b) Determine all the bar forces.

c) Find the vertical and horizontal displacement of the nodes E and D.

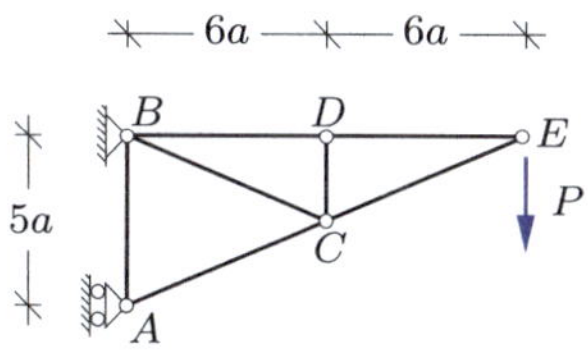

Exercise 2.17. The figure shows an N-truss girder, loaded by concentrated vertical forces of magnitude $3P$ at the joints F and J, and of magnitude $6P$ at the joints G, H and I. All bars are elastic with stiffness EA.

a) Find and show the reaction components on a drawing.

b) Determine all the bar forces.

c) Find the vertical displacement of the nodes F, G and H.

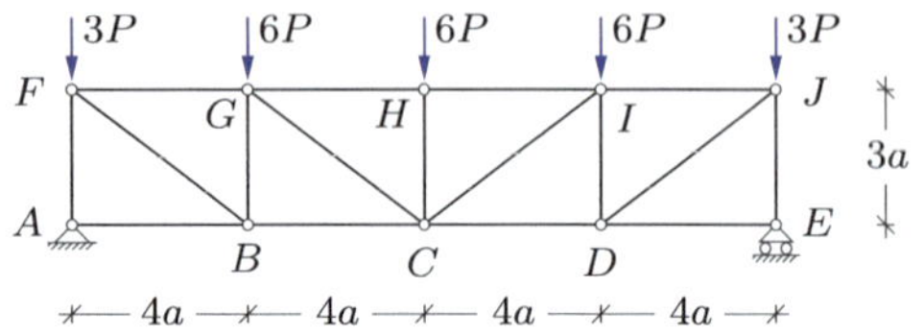

Exercise 2.18. The figure shows a V-truss girder with a vertical load in B. All bars are elastic with stiffness EA.

a) Determine the reaction and bar forces.

b) Determine the vertical displacement of nodes B and C.

c) Determine the horizontal displacement of node D.

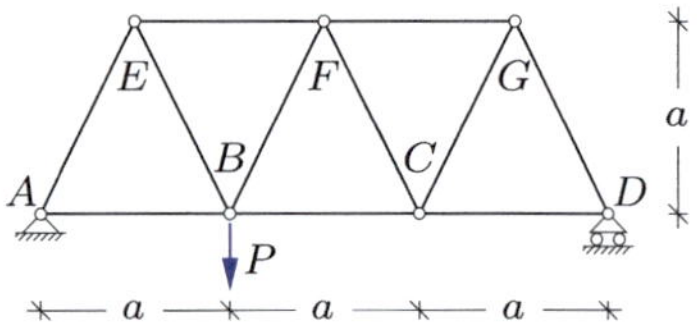

Exercise 2.19. Consider the W roof-truss shown in Fig. 2.46 and used in the description of the MiniTruss program in Section 2.5.3. Let the dimensions be $a = 12.0$ and $h = 4.0$, choose the properties so that $EA = 1$ and apply $P = 1$ at node 6.

a) Use the data file for the MiniTruss program and determine the reactions and bar forces. Compare with the results from Example 2.9.

Exercise 2.20. Consider the W roof-truss used in the description of the MiniTruss program in Section 2.5.3. Choose the dimensions and properties as in the Exercise 2.19, but let the load be symmetric as in Exercise 2.11.

a) Introduce the modified load case in a new data file `W_Truss02.m`.

b) Determine the reactions and bar forces by MiniTruss and compare with the results obtained in Exercise 2.11.

Exercise 2.21. Consider the simple triangular cantilever truss in Exercise 2.16, and choose $a = 1$, $EA = 1$ and $P = 1$.

a) Create a data file `Canti_Truss.m` to be used in MiniTruss.

b) Determine the vertical displacement of nodes D and E and compare with Exercise 2.16.

c) Determine the horizontal displacement of nodes D and E and compare the magnitude with the vertical displacements found in b).

d) Determine the reactions and bar forces and compare with the results obtained in Exercise 2.16.

Exercise 2.22. The figure shows a V-truss girder, loaded by two concentrated vertical forces $P_1 = P_2$.

a) Create a data file `V_Truss.m` for the V-truss to be used in MiniTruss. Introduce a and h as parameters in the data file. Choose the properties so that $EA = 1$ and let $P = 1$.

b) Use the MiniTruss program to determine the forces in the bars for $h = \sqrt{3}a$, i) when all all bars have the same length, and ii) for $h = a$, respectively.

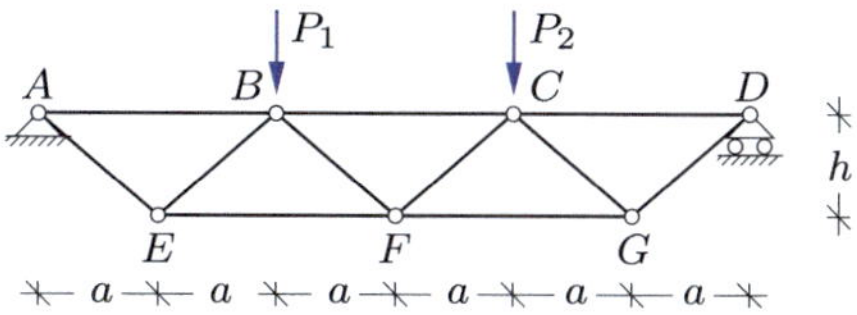

Exercise 2.23. Consider the simply supported N-truss girder in Example 2.6, where $a = 1$, $EA = 1$ and $P = 1$.

a) Create a data file `N_Truss.m` to be used in MiniTruss.

b) Determine the vertical displacement of node E.

c) Determine the bars with the largest tension and compression force, respectively.

d) Consider the load case where the only load is a vertical force of magnitude $7P$ acting in node E. Create a new data file and repeat questions b) and c).

3 Statics of Beams and Frames

In the case of trusses, discussed in the previous chapter, the statics of the individual bars forming the truss was fairly simple, because a bar can only carry a single force component – the axial force N – and furthermore this force is constant throughout the bar, because of the simplifying assumption that no loads are applied along the bar. The advantage of the truss structure is the material efficiency arising from loading the individual bar members only in tension-compression. However, a disadvantage of truss structures is the need for a considerable depth of the truss to reduce the magnitude of the forces.

An alternative way of carrying loads is by using frame structures, made up by connecting beams. It is an important characteristic of beams, that they can carry transverse loads by offering resistance to bending. The basic problem is illustrated in Fig. 3.1, where a simply supported beam carries a uniformly distributed transverse load. At the ends of the beam close to the supports the main action on the beam is produced by the concentrated reactions. This places the regions close to the supports in a state of shear. The state of shear is illustrated in the figure by a slice of the beam acted on by a transverse force of magnitude Q, called the shear force. Equilibrium of a thin slice indicates that an upward force of magnitude Q on one side of the slice corresponds to a downward force of magnitude approximately equal to Q on the other side of the slice. This is indicated in the figure together with a small parallelepiped

S. Krenk, J. Høgsberg, *Statics and Mechanics of Structures*,
DOI 10.1007/978-94-007-6113-1_3,

illustrating (in an exaggerated way) the deformation of the slice. Close to the end the moments will be small in the present case, due to the hinged supports. The shear force is the transverse component of an internal force associated with a specific section in the beam.

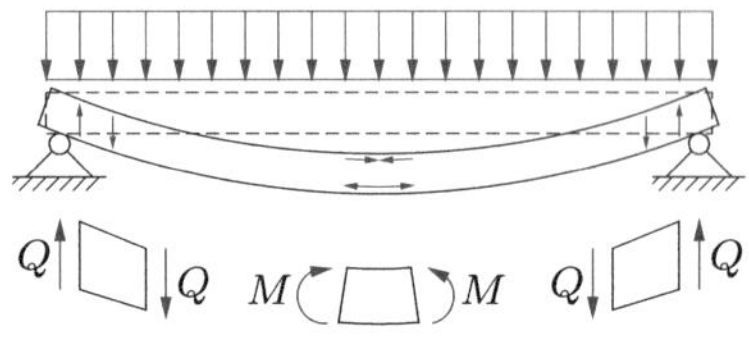

Fig. 3.1: Bending of simply supported beam.

At the center of the beam there is no shear force – as can be seen from the symmetry of beam and loading pattern. If dividing the beam into two parts by an imaginary vertical cut through its center, it follows from the equilibrium conditions studied in Chapter 1 that each part of the beam must be supported by a bending moment M at the central section. The figure shows these moments acting on the two sides of a thin slice at the center of the beam. The figure also indicates (in exaggerated form) the deformation of the central slice of the beam. The pattern of deformation is different from that at the ends of the beam. At the center the thickness of the slice becomes smaller at the top and larger at the bottom – indicating compression at the top and tension at the bottom. The moment considered here is a component of a general internal moment associated with a particular section of a beam. The distributions of internal forces and moments are closely related, and often the combined set of internal forces and internal moments are simply called the internal forces.

The internal forces in beams and frames constitute an important aspect of these structures, often called 'statics of beams and frames'. For a given structure and load distribution the section force distribution enables calculation of the deformed shape of the structure as well as the severity of the loading of the individual sections and thereby assessment of the strength of the structure. The deformation properties of elastic beams are described in Chapter 4 for statically determinate structures, and extended to three general analysis methods for statically indeterminate structures in Chapters 6–7. The strength properties constitute a part of 'Mechanics of Structures', where the behavior of the material in the individual sections is treated. This is dealt with in Chapters 8–11.

The present chapter gives a precise definition of section forces and presents analysis methods for the distribution of section forces in beams and frames. The main emphasis is on plane beams and frames, because most analyzes of the fully three-dimensional case is carried out by numerical methods, as presented in Chapter 7. First the concept of section forces (including section moments) is defined in precise form in Section 3.1. The next two sections describe the analysis of section forces in beams with concentrated and distributed load, respectively, and identify general relations between the distribution of loads and the distribution of section forces. Combination of different load cases is discussed in Section 3.4, and finally the results are extended to frames in Section 3.5.

3.1 Internal forces and moments

The idea of a section and the associated section forces plays a central role in the analysis of structures. The concept is first introduced for a plane beam with loads acting in the plane of the beam, and then extended to a fully three-dimensional setting.

Internal forces of planar beams

Figure 3.2 illustrates a plane beam with loads acting in the plane of the beam. The support is just included to give a more specific association with a beam. The beam is now separated into two parts by a hypothetical section as illustrated in the top figure. In the real beam the two parts of the beam exchange forces and moments – the left part of the beam acting on the right part and conversely. By the rule of identical action and reaction the forces and moments on the two sides of the section are equal in magnitude, but opposite in sign. In a three-dimensional context a general force has three components and similarly a general moment has three components. However, in the special case of a plane system of forces and moments, there are only three components in total: two force components and a single moment component. These components are shown for both parts of the beam in the lower part of Fig. 3.2.

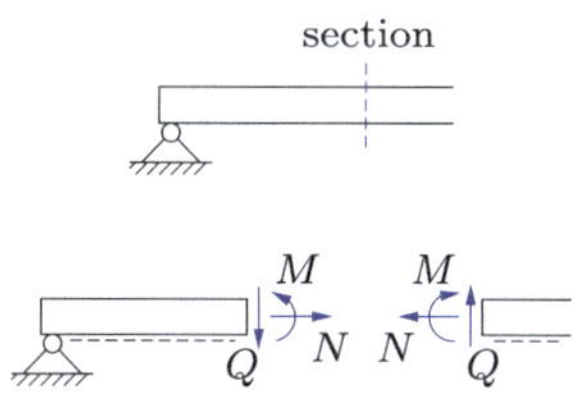

Fig. 3.2: Sign convention for internal forces in a plane beam.

The normal force N, the shear force Q, and the moment M are collectively called the section forces – or the internal forces – at the corresponding section. It is important to note that the concept of section forces includes the pair of mutually opposite components, acting on each side of the section. Thus, it is not relevant to discuss whether the section force Q points up or down, nor similar orientations of the section force N and section moment M. Instead a sign convention is needed that accounts for the orientation of the components on both sides of the section.

A common sign convention for planar beams – used throughout this book – is illustrated in Fig. 3.2. It is characterized by

- The normal force N is positive, when giving tension in the beam.
- The shear force Q is positive, when acing downward on the left part.
- The moment M is positive, when giving compression in the upper part.

These definitions make use of 'left part' and 'upper part'. For horizontal beams these concepts are evident, but for beams of general orientation the lower side of the beam is identified – typically by use of a dashed line as shown in the figure. Techniques for calculating the internal forces are discussed in the following sections.

General definition of internal forces

In a three dimensional beam formulation the beam is conveniently located in a three-dimensional xyz-coordinate system, typically defined with the x-axis along the beam, and the y- and z-axes defining a cross-section plane, as shown in Fig. 3.3. The six internal forces in 3D beam theory are the axial force N, the shear forces Q_y and Q_z, the bending moments M_y and M_z and the torsional moment M_x, respectively.

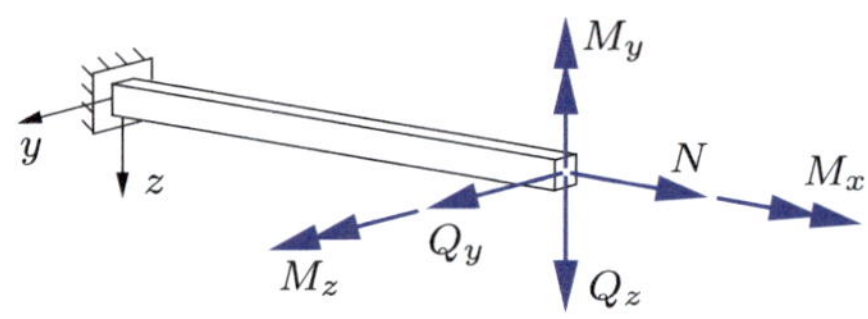

Fig. 3.3: Vector representation of section forces and section moments in a 3D beam.

The direction of the internal forces shown in Fig. 3.3 corresponds to force components in the coordinate directions, while the bending moments are chosen to recover the sign convention of planar beams in Section 3.1 when considering this special case. Figure 3.4 shows the internal forces and moments associated with the two individual coordinate planes, and it is seen that in both cases the sign convention agrees with the convention introduced in Fig. 3.2 for the planar case. Thus, this representation leads to a consistent sign convention in the individual xy and xz planes. However, the two planar cases shown in Fig. 3.4 both contain the normal force N, while the torsion moment M_x does not appear. This reflects the fact that not all phenomena in a 3D beam can be represented in terms of special planar cases, and in general three-dimensional analysis of beams and frames, typically carried out numerically, it is customary to let the section force components *and* the section moment components follow the corresponding coordinate axis directions.

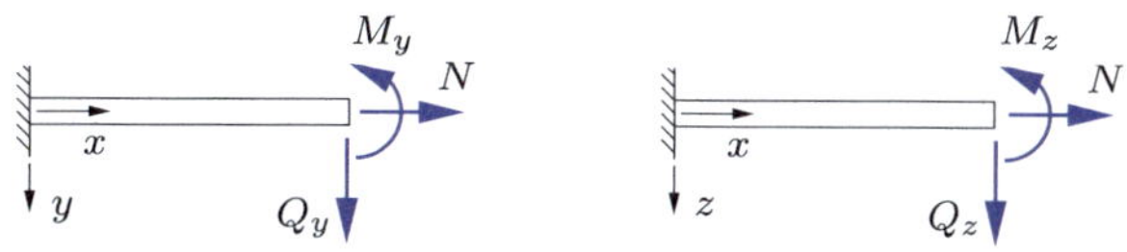

Fig. 3.4: Section forces in y- and z-plane.

3.2 Beams with concentrated loads

A statically determinate beam is a beam in which the internal forces – including moments – can be determined by use of equilibrium considerations alone. For a planar beam without hinges or other connections permitting some kind of relative motion there are three independent equilibrium conditions as discussed in Chapter 1. If the beam is supported via three static reactions, these can be determined by the equilibrium conditions. If the beam is then separated into two parts by a section as discussed above, each part of the beam provides three equilibrium equations for determination of the section forces N, Q and M. The conceptually simplest analysis procedure consists in introducing a section, the position of which is determined by a coordinate like x, and then determining the corresponding section forces $N(x)$, $Q(x)$ and $M(x)$ for this particular value of x. In many cases this procedure determines the section forces as function of the position x, at least for locations around the original position. This direct procedure is illustrated in the following two examples. Graphs of the internal forces give an impression of the statics of the beam, and identifies the parts that contribute most to the deformation of the beam, and the locations of concern with respect to the strength of the beam – subjects treated in later chapters.

Example 3.1. Cantilever with tip load. Figure 3.5 shows a horizontal cantilever beam of length ℓ, with a fixed support at the left end, and loaded by a vertical force P at the right end. The lower side of the beam is indicated by a dashed line, thereby identifying the sign convention of the section forces, shown in the lower part of the figure at a section identified by the distance x from the fixed end to the left.

In the present case it is convenient to determine the section forces $N(x)$, $Q(x)$ and $M(x)$ from equilibrium conditions for the right part of the beam. In principle, the section forces could equally well be determined from equilibrium of the left part of the beam, but this would involve prior determination and use of the three reaction components.

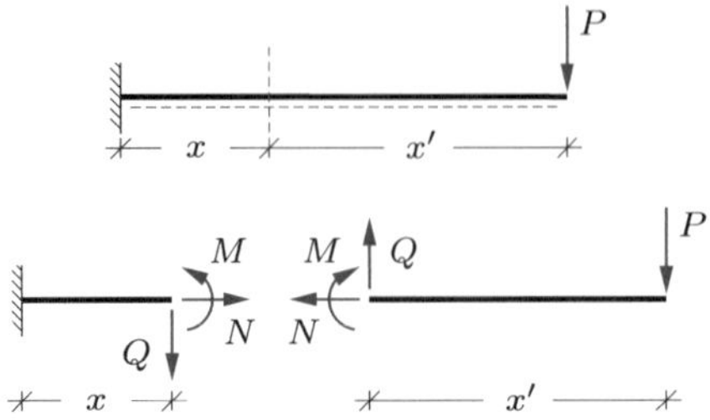

Fig. 3.5: Cantilever with vertical tip load P.

The first step consists in making a drawing showing all forces and moments acting on the part of the structure selected for the equilibrium calculation. This is shown in Fig. 3.6, where the load P is clearly indicated as well as *all* the section forces N, Q and M, shown in their positive direction. If any of the section forces turn out to have a different direction, this will show up in the form of a negative sign on the magnitude. Thus, there is no need to change the original figure, irrespective of the outcome of the calculation. Indeed, changes of basic definitions such as the sign convention during the computation may contribute uncertainty and lack of clarity to the computed results and should therefore be avoided.

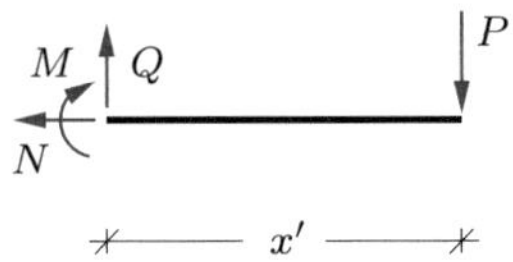

Fig. 3.6: Equilibrium for right part of cantilever.

In the present case it is convenient to introduce a 'conjugate' coordinate $x' = \ell - x$, that measures the distance from the free end at the right. The use of two coordinates x and x' indicating distances from the left and the right, respectively, will be found to be useful several times in the following.

The section forces on the selected part appears precisely as the reactions, that were determined in Chapter 1, and the same methods are used to determine them. First the normal force N is determined by horizontal projection,

$$\leftarrow \qquad N = 0\,.$$

The shear force is determined similarly by vertical projection,

$$\downarrow \qquad -Q + P = 0 \qquad \Rightarrow \qquad Q = P\,.$$

Finally, the section moment M is found by moment equilibrium about the section,

$$\curvearrowright \qquad M + x'P = 0 \qquad \Rightarrow \qquad M = -x'P\,.$$

There is no restriction on the location of the section within the beam, and thus the results express the section forces as function of the coordinate x, or its conjugate coordinate $x' = \ell - x$.

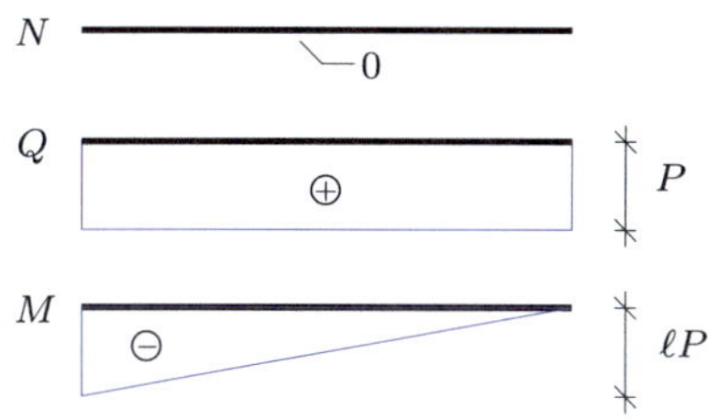

Fig. 3.7: Section forces in cantilever with tip load.

A calculation of section forces typically ends with a graphical representation of the results as shown in Fig. 3.7. Note, that reactions R'_A, R_A and M_A at the fixed support are determined by the internal forces at $x = 0$ as $R'_A = -N(0) = 0$, $R_A = Q(0) = P$, and $M_A = -M(0) = \ell P$. The reactions have already been determined in Example 1.4, with the sign convention shown in Fig. 1.26. □

Example 3.2. Simply supported beam. A typical and important example is the internal forces in a simply supported beam, loaded by a concentrated transverse force P as shown in Fig. 3.8. The length of the beam is ℓ, and the load is located at the distance a from the left end, corresponding to the distance $a' = \ell - a$ from the right end.

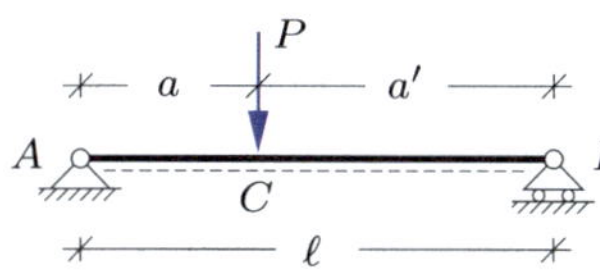

Fig. 3.8: Simply supported beam with vertical force P at point C.

The reactions are determined by horizontal projection, moment about B, and moment about A, respectively, as already demonstrated in Example 1.3. The results are

$$R'_A = 0, \qquad R_A = \frac{a'}{\ell} P, \qquad R_B = \frac{a}{\ell} P.$$

The section forces may exhibit discontinuity at the concentrated load, and are therefore determined in two steps: first by a section at distance $x \leq a$ to the left of the load, and then by a section at distance $x' \leq a'$ to the right of the load.

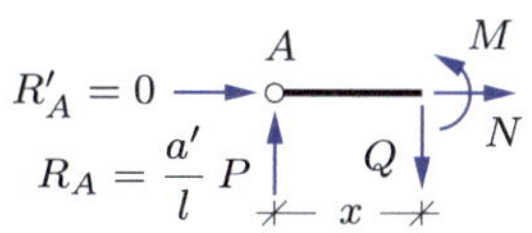

Fig. 3.9: Equilibrium for left part of simply supported beam.

Figure 3.9 shows the left part of the beam from the support A to a section at distance x from this support. The two reaction components R'_A and R_A are indicated in the figure together with the three section forces N, Q and M at x. The normal force N is determined by horizontal projection,

$$\rightarrow \qquad N = 0 \,.$$

The shear force is determined by vertical projection,

$$\downarrow \qquad Q - R_A = 0 \qquad \Rightarrow \qquad Q = R_A = \frac{a'}{l} P \,.$$

The section moment M is found by moment about the section,

$$\curvearrowright \qquad -M + x R_A = 0 \qquad \Rightarrow \qquad M = x R_A = \frac{a' x}{l} P \,.$$

These expressions are valid, when the section is to the left of the load, i.e. for $0 \le x < a$.

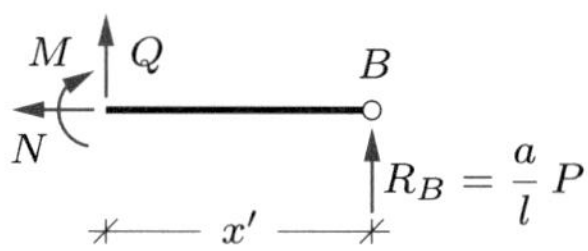

Fig. 3.10: Equilibrium for right part of simply supported beam.

The analysis of the section forces to the right of the load proceeds in a completely analogous way when introducing the parameter $a' = \ell - a$ denoting the distance of the load from the right end, and the coordinate $x' = \ell - x$ denoting the distance from the section to the right end. The part of the beam to the right of this section is shown in Fig. 3.10 together with the vertical reaction R_B and the section forces N, Q and M. The normal force N is determined by horizontal projection,

$$\leftarrow \qquad N = 0 \,.$$

The shear force is determined by vertical projection,

$$\uparrow \qquad Q + R_B = 0 \qquad \Rightarrow \qquad Q = -R_B = -\frac{a}{l} P \,.$$

The section moment M is found by moment about the section,

$$\curvearrowright \qquad M - x' R_B = 0 \qquad \Rightarrow \qquad M = x' R_B = \frac{a x'}{l} P \,.$$

These expressions are valid for a section to the right of the load defined by $0 \le x' < a'$, corresponding to $a < x \le \ell$.

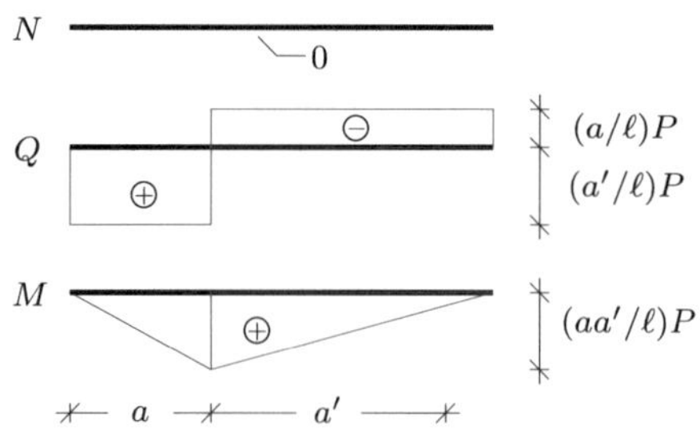

Fig. 3.11: Section forces in simply supported beam with concentrated force.

The distribution of the section forces is illustrated in Fig. 3.11, showing the graphs both to the left and to the right of the concentrated load. It is noted that at the location of the force P the shear force distribution $Q(x)$ exhibits a jump discontinuity of magnitude $-P$, while the moment distribution exhibits a discontinuity of the slope of magnitude $-P$. □

3.2.1 Variation of internal forces for concentrated loads

In Example 3.2 it was seen that the shear force distribution $Q(x)$ exhibits a jump discontinuity at the location of a concentrated transverse force, and similarly the moment distribution $M(x)$ exhibits a discontinuity in the slope. This is a special case of a general behavior exhibited by shear force and moment distributions for beams loaded by one or more concentrated forces. These properties are here discussed with reference to Fig. 3.12 showing a part of a beam around a concentrated transverse force P. The part of the beam is defined by a section at the distance x' to the left of the load, and a section at the distance x to the right. The figure also shows the section forces Q_- and M_- at the left end, and Q_+ and M_+ at the right end.

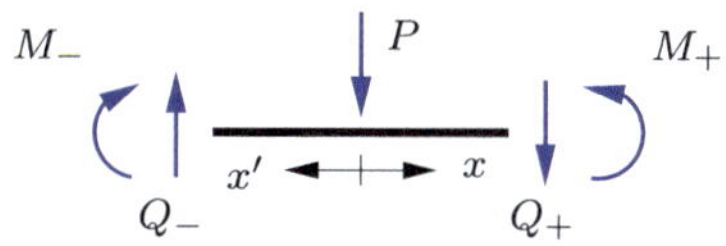

Fig. 3.12: Load P on beam with internal force.

It follows directly from projection of the forces on a transverse direction that

$$Q_+ - Q_- = -P. \tag{3.1}$$

This demonstrates that the shear force distribution exhibits a jump discontinuity of magnitude $-P$ at the point of action of a downward load of magnitude P. It is seen that this relation remains the same, if x' or x is changed. Thus, the shear force distribution is piecewise constant between transverse loads. It should be noted that reactions often produce a transverse load, and thus an internal support often leads to a discontinuity in the shear force distribution as illustrated in the following two examples. The jump in shear force $Q(x)$ between constant values is illustrated in Fig. 3.13a. The normal force exhibits a similar discontinuity at points of application of a concentrated axial force.

The moment distribution around the point of application of a transverse force P is determined by first considering moment equilibrium about the point of application of the force,

$$M_- - M_+ + x'Q_- + x\,Q_+ = 0. \tag{3.2}$$

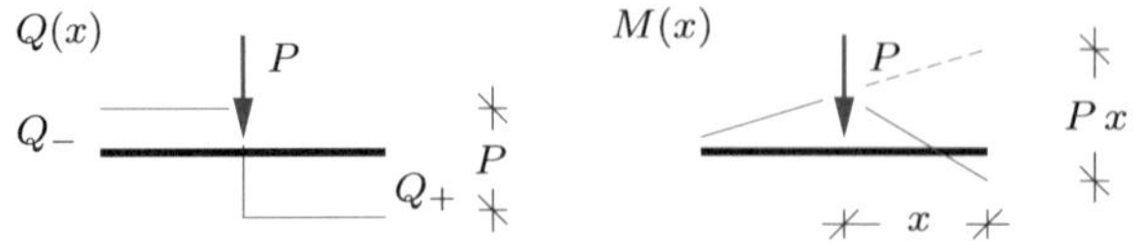

Fig. 3.13: Change in shear force $Q(x)$ and moment $M(x)$ at load P.

The moment distribution is continuous, when no concentrated moment is applied, and the moment at the point of application of the load is denoted M_0. The moment to the left of the load is then determined by setting $x = 0$ in this equation, whereby

$$M_- = M_0 - x' Q_- \qquad \Rightarrow \qquad \frac{dM_-}{dx} = Q_- \,. \tag{3.3}$$

It is observed that M_0 and Q_- are constants, and thus the moment distribution to the left of the load is linear with slope Q_-. A similar argument is used to determine the variation of the moment M_+ to the right of the load by using $x' = 0$. Moment equilibrium then gives

$$M_+ = M_0 + x\,Q_+ \qquad \Rightarrow \qquad \frac{dM_+}{dx} = Q_+ \,. \tag{3.4}$$

By use of the discontinuity equation (3.1) it then follows that the moment distribution exhibits a slope discontinuity:

$$\frac{dM_+}{dx} - \frac{dM_-}{dx} = -P \tag{3.5}$$

at the point of application of the load P. This slope discontinuity is illustrated in Fig. 3.13b, where the moment curve is shown as a solid line.

The results can be summarized as follows. For a beam that is loaded only by concentrated forces P_i the normal force distribution $N(x)$ and the shear force distribution $Q(x)$ are constant between the applied forces, and exhibit discontinuities that match the force with opposite sign at the point of the forces. The moment distribution $M(x)$ is continuous and linear between the points of application of transverse forces, and exhibits discontinuities of slope matching the transverse forces with opposite sign. These properties are important, as they permit construction of the complete section force curves from the values of the section forces at points where the structure is acted on by reactions or concentrated forces. The result that the derivative of the moment equals the shear force, $dM/dx = Q$, can be used either in the construction of the curves, or as a subsequent check of the correctness of these. This technique is illustrated in the following examples and the special cases shown in Table 3.1.

Example 3.3. Cantilevered beam with simple supports. Figure 3.14 shows a beam AC with a simple fixed support at the left end point A and a simple support with horizontal motion at an intermediate point B. A vertical force P acts on the free end C. The length of the simply supported part AB is a, and the length of the cantilever BC is b.

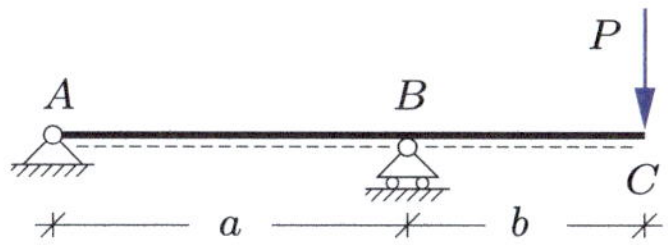

Fig. 3.14: Simply supported beam with cantilever and vertical tip force P.

The reactions are shown in Fig. 3.15. They are determined by horizontal equilibrium, moment about B and moment about A, leading to

$$R'_A = 0\,, \qquad R_A = -\frac{b}{a}P\,, \qquad R_B = \frac{a+b}{a}P\,.$$

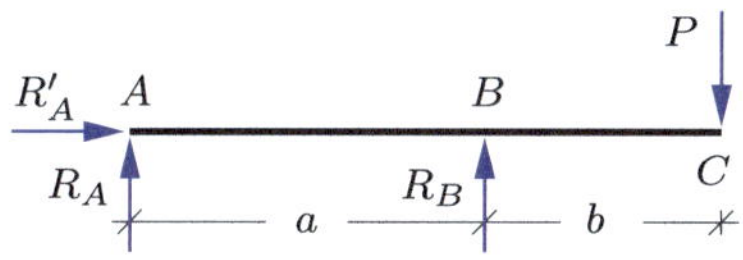

Fig. 3.15: Reaction forces.

The section forces are now determined by using their properties described above: piecewise constant normal force $N(x)$ and shear force $Q(x)$, and continuous piecewise linear moment $M(x)$.

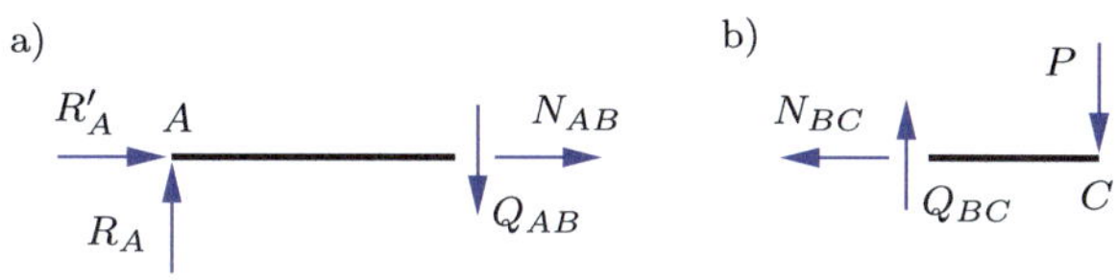

Fig. 3.16: Normal and shear force in AB and BC.

The piecewise constant property of $N(x)$ and $Q(x)$ implies that they can be determined from an arbitrary section in AB and BC, respectively. Figure 3.16a shows the left part of the beam when divided by a section in AB. Force projection in horizontal and vertical direction gives

$$N_{AB} = 0\,, \qquad Q_{AB} = R_A = -\frac{b}{a}P\,.$$

Similarly, Fig. 3.16b shows the right part, when the section is introduced in BC. Projection gives

$$N_{BC} = 0\,, \qquad Q_{BC} = P\,.$$

Hereby the determination of the distribution of the normal force and the shear force is completed.

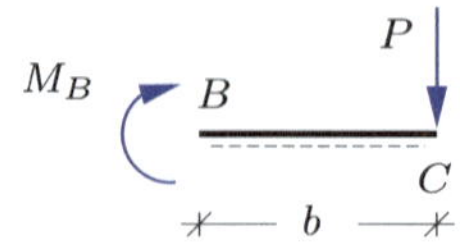

Fig. 3.17: Moment at B by equilibrium of BC.

The moment $M(x)$ is zero at the two ends of the beam, and the distribution is therefore fully determined by the moment M_B at B. Figure 3.17 shows the part BC of the beam, when a section is placed just to the right of the support at B. Moment about the section at B gives

$$M_B = -b\,P .$$

The distribution of the section forces is shown in Fig. 3.18. It is seen that the change in shear force at B corresponds to the reaction R_B, while the moment distribution exhibits a change in slope of magnitude R_B at the support. □

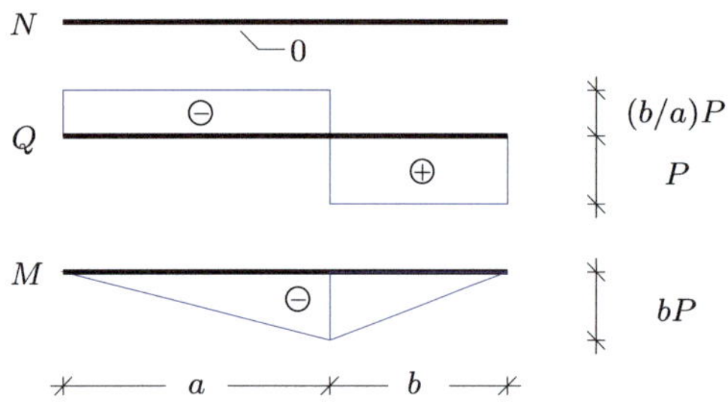

Fig. 3.18: Distribution of section forces.

Example 3.4. Hinged beam. Figure 3.19 shows a two span simply supported beam with a hinge in D. The load consists of a vertical force P acting at the center of DB, i.e. with distance $\frac{1}{2}b$ from the support at B.

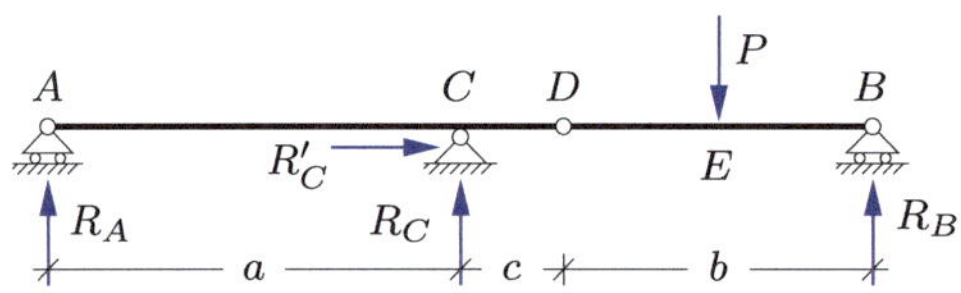

Fig. 3.19: Hinged beams with concentrated force.

The reactions were determined in Example 1.5 by considering DB as a simply supported beam, supported at B by the ordinary simple support, and at D which is fixed in space by the remaining structure ACD,

$$R_A = -\frac{c}{2a}\,P\,, \quad R_B = \tfrac{1}{2}P\,, \quad R_C = \frac{a+c}{2a}\,P\,, \quad R'_C = 0\,.$$

The internal forces are determined by suitable sections in the beam. It is seen immediately, that any section followed by horizontal projection of forces will give $N = 0$.

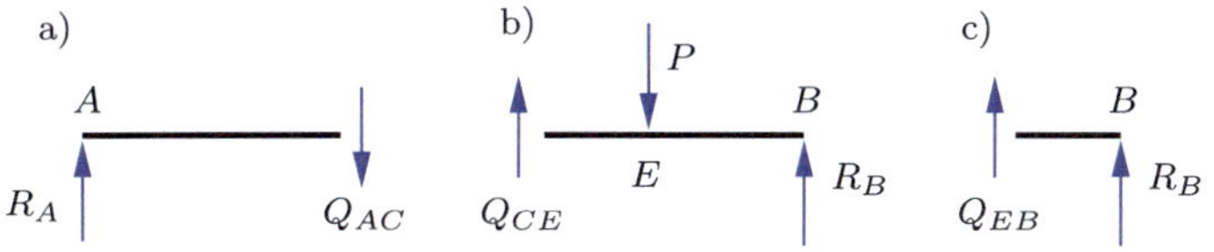

Fig. 3.20: Shear forces by section a) in AC, b) in CE, and c) in BE.

The shear force is constant between the location of transverse forces. Thus, the distribution of the shear force is determined by Q_{AC}, Q_{CE} and Q_{EB}. Figure 3.20a shows the left part after introducing a section in AC. The shear force Q_{AC} is determined by vertical force equilibrium,

$$\downarrow \qquad Q_{AC} - R_A = 0 \qquad \Rightarrow \qquad Q_{AC} = R_A = -\frac{c}{2a}P\,.$$

Figure 3.20b shows the right part of the structure after introducing a section between the support C and the point of load application E. Note, that it is immaterial, whether this section is to the left or to the right of the hinge at D, because the projection equation only depends on the transverse forces. This determines the shear force Q_{CE} by vertical force equilibrium,

$$\downarrow \qquad -Q_{CE} + P - R_B = 0 \qquad \Rightarrow \qquad Q_{CE} = P - R_B = \tfrac{1}{2}P\,.$$

Finally, the shear force Q_{BE} is determined by vertical force equilibrium of the right part of the structure, after introducing a section in BE as shown in Fig. 3.20c,

$$\downarrow \qquad -Q_{BE} - R_B = 0 \qquad \Rightarrow \qquad Q_{BE} = -R_B = -\tfrac{1}{2}P\,.$$

This completes the determination of the shear force distribution.

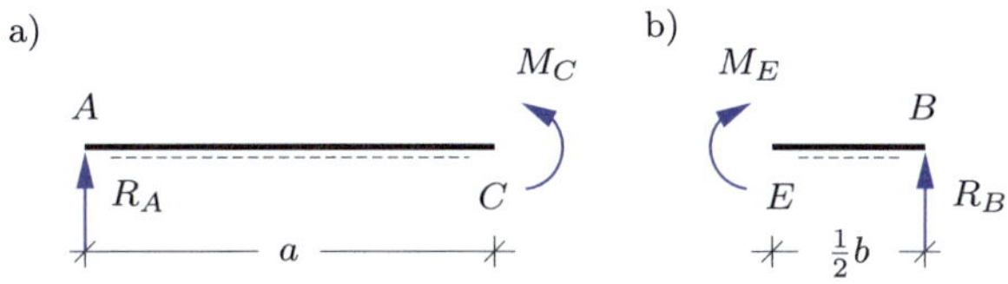

Fig. 3.21: Moment equilibrium, a) about C, and b) about E.

The moment M is piecewise linear between the points of transverse forces, i.e. the points A, C, E and B. The moment distribution is therefore determined from the moment at these four points. The moment is zero at the end points A and B due to the support conditions. Thus, the moment variation is fully determined from the two values M_C and M_E. After introducing a section at C moment about C for the left part of the beam shown in Fig. 3.21a gives

$$\curvearrowright \qquad -M_C + a\,R_A = 0 \qquad \Rightarrow \qquad M_C = a\,R_A = -\tfrac{1}{2}\,c\,P\,.$$

Similarly, a section at E followed by moment about E for the right part shown in Fig. 3.21b gives

$$\curvearrowright \qquad M_E - \tfrac{1}{2}\,b\,R_B = 0 \qquad \Rightarrow \qquad M_E = \tfrac{1}{2}\,b\,R_B = \tfrac{1}{4}\,b\,P\,.$$

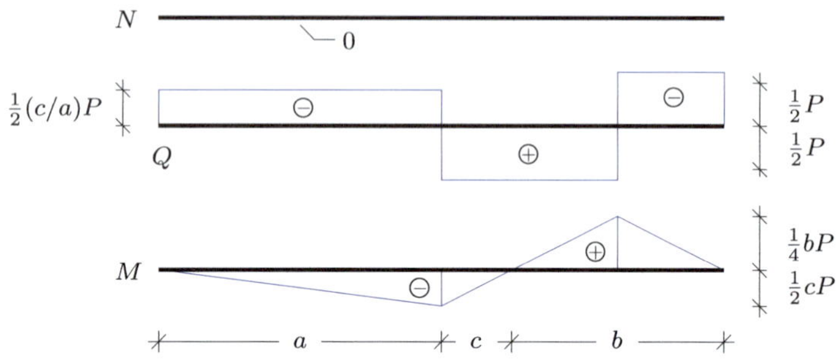

Fig. 3.22: Distribution of section forces.

The distribution curves for the shear force and the moment are shown in Fig. 3.22. The discontinuities should be checked against the load P and the reaction R_C, and the slopes of the moment curve against the corresponding value of the shear force Q. □

3.3 Beams with distributed load

Many beams are designed to carry distributed loads e.g. from their own weight, loads from wind and snow. The analysis procedures from the previous section for beams with concentrated loads must therefore be extended to cover distributed loads. Like in the previous section this is done in two steps: first by introduction of a generic section and direct computation of the internal forces on this section, and then by deriving and using a set of differential equations for the section forces in terms of the distributed load intensity. The first two examples correspond directly to those analyzed by the direct method for concentrated load in Examples 3.1 and 3.2. Then the differential equations for section forces are derived in Section 3.3.1. The important issue of the location and magnitude of the maximum internal moment is treated in Section 3.3.2 and illustrated by examples. A number of specific cases are summarized in Table 3.2.

Example 3.5. Cantilever with distributed load. Figure 3.23 shows a cantilever with a vertical distributed load with constant intensity p. The internal forces are determined by introducing a section at the distance $x' = \ell - x$ from the free end of the beam as indicated in the figure.

It is convenient to determine the section forces $N(x)$, $Q(x)$ and $M(x)$ by equilibrium of the right part of the beam, as this dos not require prior determination of the reactions. The right part with the load and all section forces indicated by their positive orientation is shown in the lower part of the figure. Note, that the load is shown in the form of an equivalent concentrated force of magnitude px', acting at the center of the distributed load. The replacement of distributed loads by an equivalent concentrated load was discussed extensively in Chapter 1 in connection with calculation of reactions.

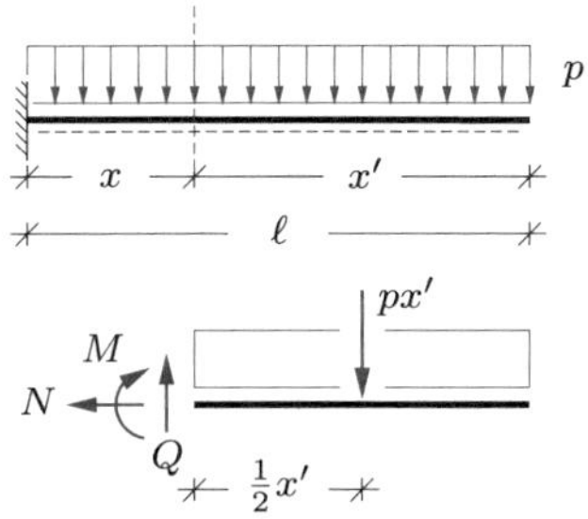

Fig. 3.23: Cantilever with distributed load p.

When representing the distributed load in terms of its equivalent concentrated force, located at the distance $\frac{1}{2}x'$ from the free end of the beam, the calculation proceeds as in Example 3.1. Horizontal equilibrium gives the normal force

$$\leftarrow \qquad N = 0\,.$$

The shear force Q is determined similarly by vertical projection,

$$\downarrow \qquad -Q + p\,x' = 0 \qquad \Rightarrow \qquad Q = p\,x' = p\,(\ell - x)\,.$$

Finally, the section moment M is found by moment equilibrium about the section,

$$\curvearrowright \qquad M + (p\,x')\,\tfrac{1}{2}x' = 0 \qquad \Rightarrow \qquad M = -\tfrac{1}{2}p\,(x')^2 = \tfrac{1}{2}\,p\,(l - x)^2\,.$$

Note, that the section forces are functions of the coordinate x. The shear force $Q(x)$ is of linear variation, while the moment $M(x)$ exhibits quadratic variation.

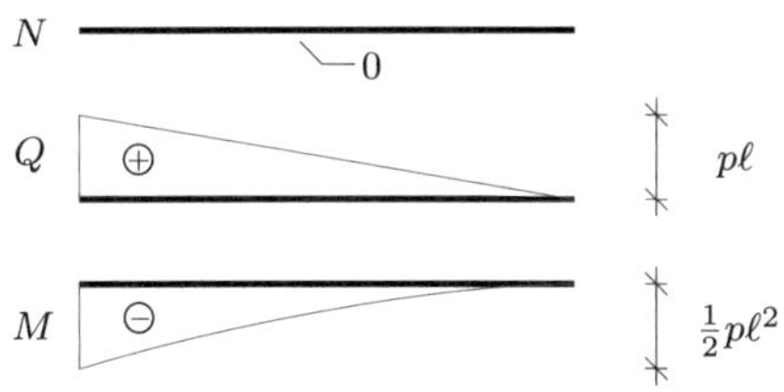

Fig. 3.24: Distribution of section forces in cantilever.

The distributions of the section forces are shown in Fig. 3.24. It is found that the numerically largest section forces occur at the support, where $x = 0$:

$$Q_{\max} = Q(0) = p\,\ell\,, \qquad M_{\max} = M(0) = -\tfrac{1}{2}p\,\ell^2\,.$$

The notation $M_{\max}$ is used for the numerically largest moment. The section forces at $x = 0$ determine the reactions as $R'_A = N(0)$, $R_A = Q(0)$, and $M_A = M(0)$, when using the sign convention shown in Fig. 3.23 also for the reactions. □

Example 3.6. Simply supported beam with distributed load. Figure 3.25 shows a simply supported beam of length ℓ with a vertical uniform distributed load with intensity p. The section forces N, Q and M are determined by introducing a section at the distance x from the left end of the beam, and considering equilibrium of the left part of the beam. The left part including the distributed load, the reactions and the section forces is shown in Fig. 3.25. The reactions R'_A, R_A and R_B are determined by horizontal force projection, moment about B, and moment about A, respectively, resulting in

$$R'_A = 0\,, \qquad R_B = R_A = \tfrac{1}{2}p\,\ell\,.$$

The reactions are seen to correspond to a total equivalent vertical force of magnitude $p\ell$, carried equally by the vertical supports.

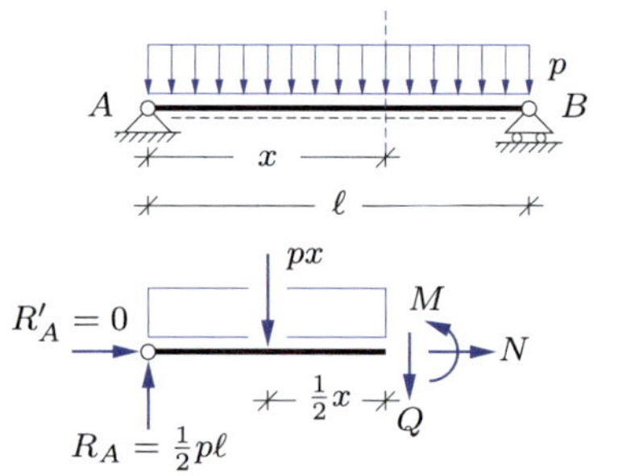

Fig. 3.25: Simply supported beam with distributed load p.

As shown in Fig. 3.25 the internal forces are determined by placing the section a distance x from the left support. Horizontal equilibrium gives the normal force as

$$\leftarrow \qquad\qquad N = 0\,.$$

The shear force Q is determined from vertical force equilibrium,

$$\downarrow \qquad Q - R_A + p\,x = 0 \qquad \Rightarrow \qquad Q = R_A - p\,x = p\,(\tfrac{1}{2}\ell - x)\,.$$

Finally, the moment M is determined from moment equilibrium about the section,

$$\curvearrowright \qquad -M - (p\,x)\,\tfrac{1}{2}x + R_A\,x = 0 \qquad \Rightarrow \qquad M = \tfrac{1}{2}\,p\,x\,(\ell - x)\,.$$

As in the cantilever in Example 3.5 the shear force $Q(x)$ depends linearly on x, while the moment $M(x)$ is quadratic.

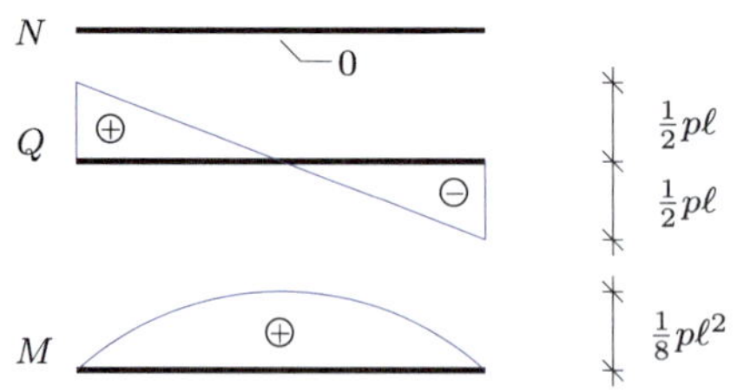

Fig. 3.26: Section forces in simply supported beam with uniform load.

The distribution of the section forces is shown in Fig. 3.26. It is noted that in this case the internal moment $M(x)$ is positive throughout the beam. It is found that the maximum shear force occurs at the supports, while the maximum moment appears at the center of the beam. The maximum values are

$$Q_{\max} = \tfrac{1}{2}p\,\ell\,, \qquad M_{\max} = \tfrac{1}{8}p\,\ell^2\,.$$

It is seen that these maximum values are considerably smaller than those found for the cantilever beam with uniformly distributed load. Therefore less strength is required of the beam in the present case, corresponding to the fact that the load is carried most efficiently, when supports are provided at both ends of the beam. □

3.3.1 Differential equations for internal forces

It was seen in the two last examples that for a constant load intensity p the shear force distribution $Q(x)$ is linear with slope equal to $-p$, while the moment distribution is quadratic with double derivative $-p$. As demonstrated here, these results are special cases of the differential equations for the section force distributions $N(x)$, $Q(x)$ and $M(x)$ in terms of the applied load intensities. The results for the normal force $N(x)$ are similar to those of the shear force $Q(x)$, and the following presentation will therefore concentrate on the differential equations for $Q(x)$ and $M(x)$.

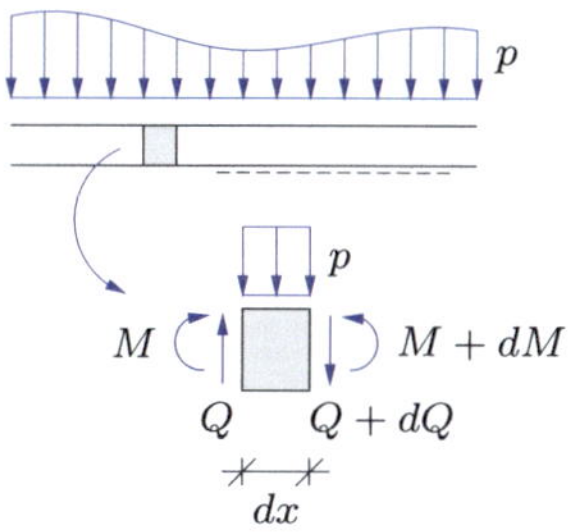

Fig. 3.27: Equilibrium of beam with distributed load.

Figure 3.27 shows a beam carrying a distributed load with intensity $p(x)$. A thin slice of infinitesimal thickness dx is cut from the beam, and the figure illustrates the section forces acting on the slice. For clarity of illustration the normal force has been left out, as it can be treated separately by elementary means. The left side of the slice corresponds to a section located at x, while the right side of the slice corresponds to section located at $x + dx$. Thus, the left side is acted on by the shear force Q and the moment M, while the right side is acted on by the slightly different values $Q + dQ$ and $M + dM$. In addition to these forces and moments the slice is acted on by a transverse load of magnitude $p\,dx$. Differential equations for the shear force distribution $Q(x)$ and the moment distribution $M(x)$ are now obtained by considering equilibrium of the slice of the beam.

Vertical equilibrium of the slice of thickness dx corresponds to the force projection equation

$$(Q + dQ) \; - \; Q \; + \; p \, dx \; = \; 0 \, . \tag{3.6}$$

The terms Q and $-Q$ cancel, leaving only the two differential terms dQ and $p\,dx$. Division by the differential length dx then gives

$$\frac{dQ}{dx} \; = \; - \, p(x) \, . \tag{3.7}$$

This equation implies that the slope of the shear force curve is equal to the intensity of the distributed load with opposite sign. This relation is illustrated in Fig. 3.28b. The change in shear force over a finite length of the beam can be found by integration of this relation as

$$Q_B \; - \; Q_A \; = \; - \int_A^B p(x) \; dx \, . \tag{3.8}$$

This relation is illustrated in Fig. 3.28a. Both figures illustrate that $Q(x)$ decreases for positive load intensity $p(x)$.

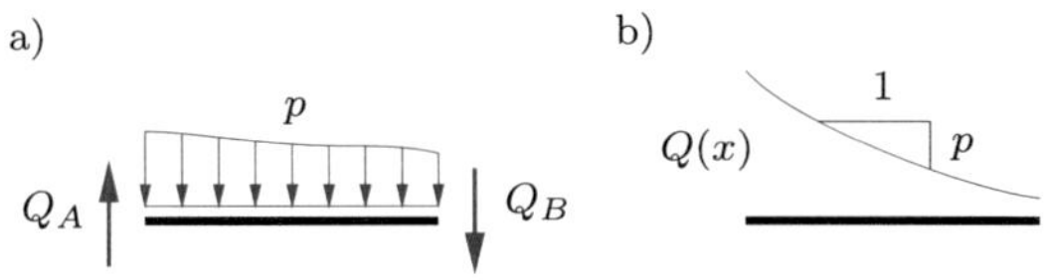

Fig. 3.28: Shear force distribution for a beam with distributed load.

Moment equilibrium about an internal point in the slice of thickness dx gives

$$(M + dM) \; - \; M \; - \; Q \, dx \; = \; 0 \, , \tag{3.9}$$

where it has been used that the moment of the load $p(x)$ is of order $(dx)^2$, and therefore vanishes from the equation in the limit of vanishing thickness of the slice. The moments M and $-M$ cancel, and division by dx yields

$$\frac{dM}{dx} \; = \; Q(x) \, . \tag{3.10}$$

The equation implies that the slope of the moment distribution is equal to the value of the shear force at that location. This relation is illustrated in Fig. 3.29b. The corresponding relation for a finite separation of the two sections is found by integration as

$$M_B \; - \; M_A \; = \; \int_A^B Q(x) \; dx \, . \tag{3.11}$$

This relation is illustrated in Fig. 3.29a, showing that the moment is increasing by the area under the shear force curve between A and B.

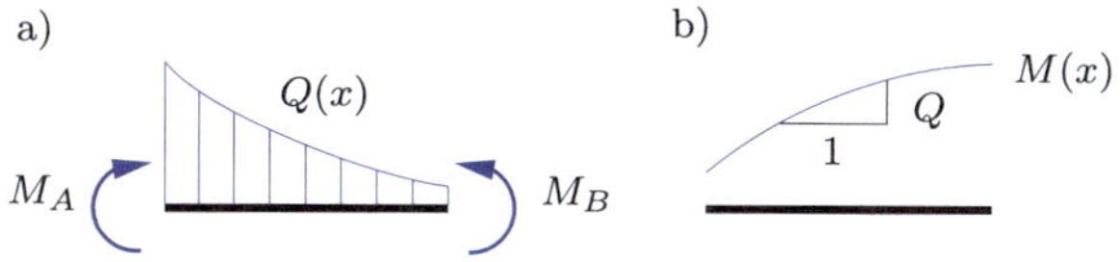

Fig. 3.29: Distribution of moment for beam with distributed load.

By combining the two differential relations (3.7) and (3.10) it is found that the second derivative of the moment is equal to the intensity of the distributed load with opposites sign,

$$\frac{d^2M}{dx^2} = -p(x)\,. \tag{3.12}$$

Figure 3.30 illustrates that a positive load intensity $p(x)$ leads to 'downward' curvature of the moment distribution $M(x)$.

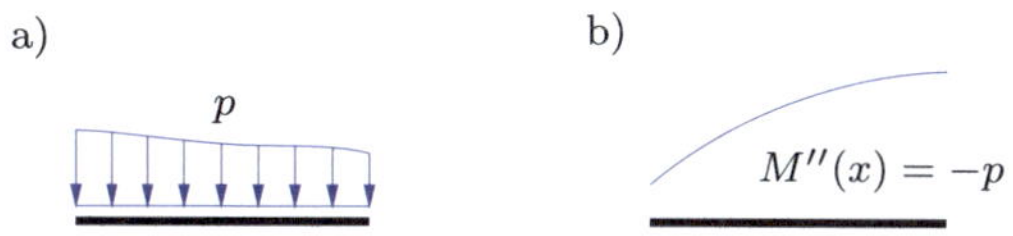

Fig. 3.30: Curvature of moment curve for beam with distributed load.

The differential relations for the shear force $Q(x)$ and the moment $M(x)$ derived above assist in determining the overall shape of the section force curves. Note, that the case of concentrated loads treated in the previous section corresponds to $p(x)$ identically equal to zero between the concentrated loads, and thus to constant shear force and linear moment variation. Similarly, the case of constant load intensity treated in the examples of this section correspond to p = const, and thereby to linear variation of the shear force and quadratic variation of the moment.

3.3.2 Maximum moment

The moment often constitutes the most important component in the loading of the material of the beam. It is therefore often important to determine the maximum value of the moment, $M_{\max}$, and the location $x_{\max}$ at which it occurs. In the case of concentrated loads the moment variation is linear between the loads, and thus the maximum moment must occur under concentrated loads or at supports. In the case of a distributed load it follows from the results of the previous section that the shear force $Q(x)$ is an integral of the load intensity $-p(x)$, and the moment $M(x)$ is an integral of the shear force

$Q(x)$. A condition for maximum – or minimum – moment within the span of the beam therefore follows from the zero-derivative condition in the form

$$\frac{dM}{dx} = Q = 0 \qquad \Rightarrow \qquad M_{\max} \ \text{for} \ Q(x_{\max}) = 0\,. \tag{3.13}$$

Clearly, the moment can also take its maximum or minimum value at the ends of the beam.

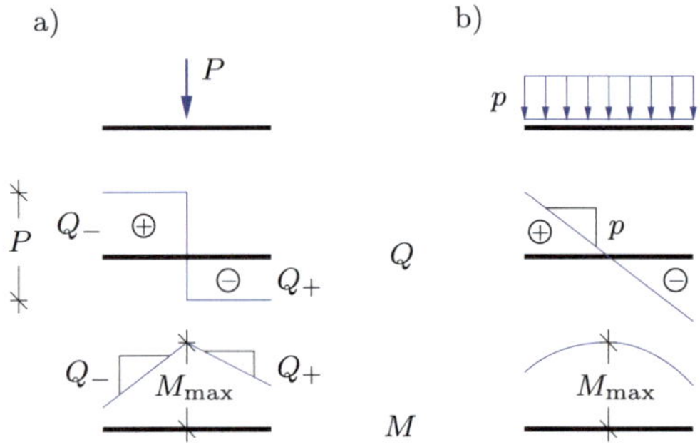

Fig. 3.31: $M_{\max}$ at $Q = 0$. a) Concentrated load, b) distributed load.

In a strict sense the maximum moment condition $Q(x_{\max}) = 0$ was derived for a continuous shear force distribution, as illustrated in Fig. 3.31b. However, the shear force is the derivative of the moment, $dM/dx = Q$, and thus a discontinuity in which the shear force changes sign implies a change of the slope of the moment distribution from increasing to decreasing, or conversely. This also corresponds to a maximum or minimum value of the moment, and thus the condition $Q(x_{\max}) = 0$ also covers the case of a combination of distributed and concentrated transverse loads. The case of a downward concentrated force is illustrated in Fig. 3.31a. It is seen that when the shear force $Q(x)$ changes from a positive to a negative value under the force, the moment exhibits a maximum. This illustrates that the present sign convention for moments shown in Fig. 3.2 makes a maximum moment under a force correspond to a downward force, while a minimum moment under the force corresponds to an upward force.

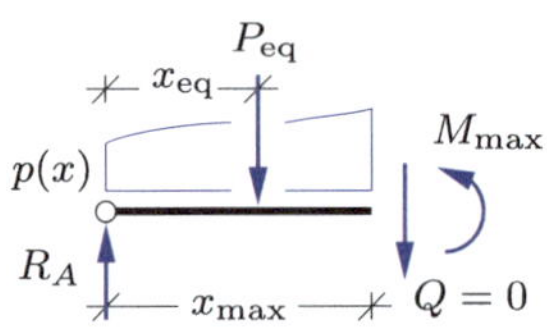

Fig. 3.32: Maximum moment in simply supported beam with distributed load.

The differential relations between the bending moment M, the shear force Q and the transverse load intensity p lead to a simple way of determining the maximum moment in a beam. The problem is illustrated in Fig. 3.32, showing the simply supported left end of a beam with distributed load of intensity $p(x)$. According to the condition (3.13) the maximum moment is located at $x_{\max}$, where the shear force vanishes. This in turn implies that the equivalent concentrated force $P_{\rm eq}$ is equal in magnitude but opposite in direction to the transverse reaction component R_A. Thus, the two forces constitute a force couple with moment $x_{\rm eq}R_A$, where $x_{\rm eq}$ denotes the distance of the equivalent load $P_{\rm eq}$ from the support at A. The maximum moment is therefore determined by

$$Q(x_{\max}) = 0 \quad \Rightarrow \quad P_{\rm eq} = R_A\,, \qquad M_{\max} = x_{\rm eq}R_A\,. \tag{3.14}$$

The distance $x_{\rm eq}$ of the equivalent load $P_{\rm eq}$ from the support depends on the distribution of the load, as illustrated in the following two examples.

Example 3.7. Partial loading of beam. Figure 3.33 shows a simply supported beam of length ℓ with a uniformly distributed load of intensity p extending the distance a from the left end A of the beam. The reactions are determined by horizontal force equilibrium, moment about B and moment about A, respectively, whereby

$$R'_A = 0\,, \qquad R_A = \Big(1 - \frac{a}{2l}\Big)ap\,, \qquad R_B = \frac{a}{2l}\,ap\,.$$

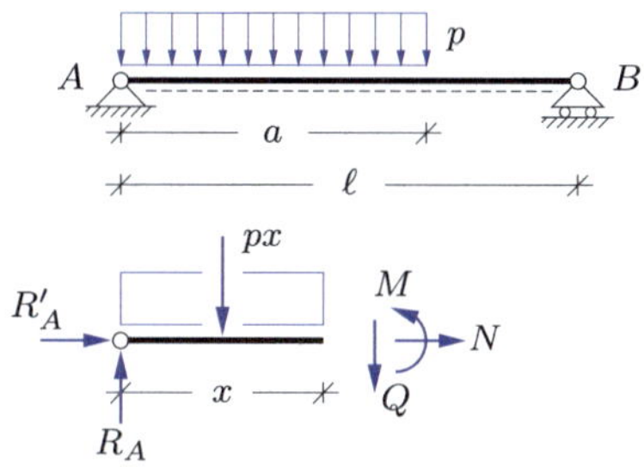

Fig. 3.33: Distributed load p on part of simply supported beam.

Horizontal equilibrium gives $N = 0$ at any section of the beam. The shear force $Q(x)$ for $x \le a$ is determined by considering vertical equilibrium of the part of the beam to the left of a section at x, as shown in the figure,

$$Q = R_A - px\,, \qquad x \le a\,.$$

The moment $M(x)$ attains its maximum, where the shear force vanishes, which occurs at

$$x_{\max} = \frac{R_A}{p} = \Big(1 - \frac{a}{2l}\Big)a\,.$$

It is seen that $x_{\max} < a$, corresponding to location of the maximum moment within the loaded part of the beam.

The condition that the shear force vanishes at the point of maximum moment leads to a simple expression for the maximum moment $M_{\max}$. When the shear force vanishes, the load on the part to the left of the section at $x_{\max}$ is equal to the reaction at A, and thus $R_A = x_{\max} p$. These two equal but opposite forces constitute a force couple. The forces have the distance $x_{\rm eq} = \frac{1}{2} x_{\max}$, and thus the maximum moment is

$$M_{\max} = \tfrac{1}{2} x_{\max} R_A = \frac{R_A^2}{2p} \, .$$

This result follows directly from local considerations around the end of the beam with constant load intensity, and thus remains valid even if other loads were present to the right of the section at $x_{\max}$.

The distribution of the shear force and moment is shown in Fig. 3.34. The shear force is $Q(0) = R_A$ and decreases with slope $-p$, whereby $Q(a) = R_A - ap = R_B$.

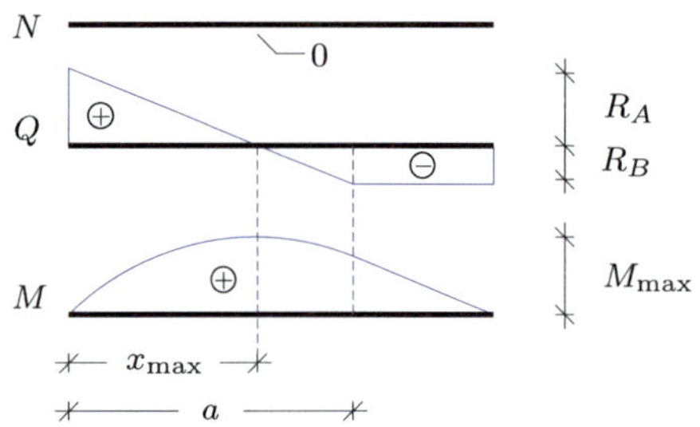

Fig. 3.34: Distribution of section forces in partly loaded beam.

The distributions of the section forces $N(x)$, $Q(x)$ and $M(x)$ are illustrated in Fig. 3.34 for $a = \frac{2}{3}\ell$. For this length of the loaded part of the beam the reactions are $R_A = \frac{2}{3} ap$ and $R_B = \frac{1}{3} ap$. The shear force is seen to decrease linearly from R_A at the left end with slope $-p$ to the constant value $-R_B$ for the unloaded part of the beam. This gives the location and magnitude of the maximum moment as

$$x_{\max} = \tfrac{2}{3}\, a = \tfrac{4}{9}\, \ell \, , \qquad M_{\max} = \tfrac{2}{9}\, a^2 p = \tfrac{8}{81}\, \ell^2 p \, .$$

The moment curve is parabolic in the loaded part of the beam and linear in the unloaded part. □

Example 3.8. Triangular load distribution. Figure 3.35a shows a simply supported beam with a linearly distributed load with maximum intensity p at the right end. As demonstrated in Section 1.3.3 and illustrated in Fig. 1.16 the equivalent load is $P = \frac{1}{2} p\ell$, acting on the beam at $\frac{2}{3}\ell$ from A. This is illustrated in Fig. 3.35b.

The reactions now follow from use of the equivalent load via horizontal force equilibrium, moment about B, and moment about A, respectively, whereby

$$R_A' = 0 \, , \qquad R_A = \tfrac{1}{6}\, \ell p \, , \qquad R_B = \tfrac{1}{3}\, \ell p \, .$$

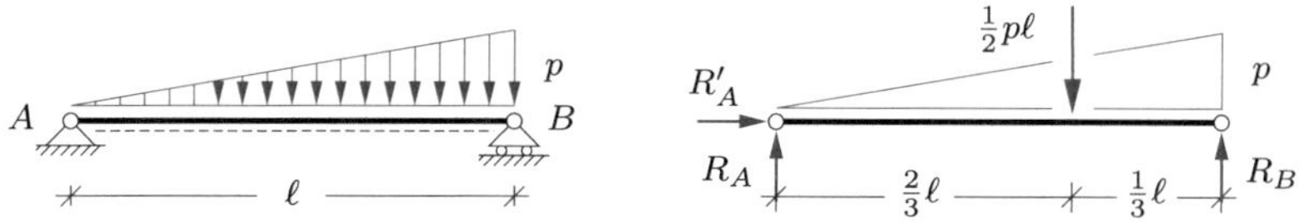

Fig. 3.35: Linearly distributed load on simply supported beam.

Because the intensity of the distributed load is varying linearly the shear force is varying quadratically, as illustrated in Fig. 3.36b.

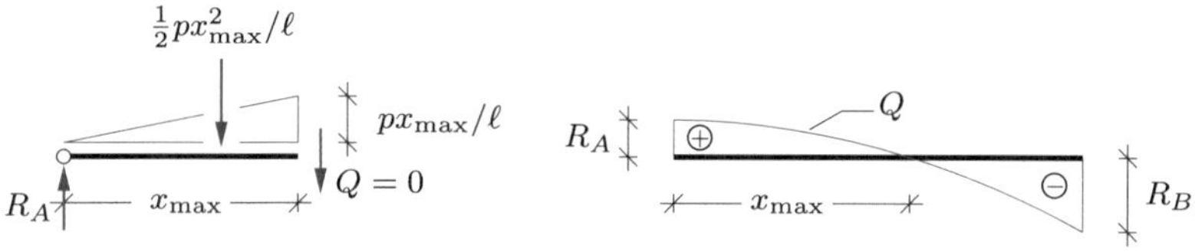

Fig. 3.36: Vanishing shear force at x_{max}.

The maximum moment M_{max} occurs where the shear force vanishes, i.e. at $Q(x_{max}) = 0$. In Fig. 3.36a the section is placed at x_{max}, which is to be determined. The intensity of the distributed load at the section is $p\,x_{max}/\ell$. Vertical equilibrium determines the shear force, which must vanish at x_{max},

$$Q(x_{max}) \;=\; R_A - \tfrac{1}{2}x_{max}(p\,x_{max}/\ell) \;=\; 0\,.$$

Substitution of the reaction R_A yields

$$x_{max}^2 \;=\; \tfrac{1}{3}\ell^2 \quad\Rightarrow\quad x_{max} \;=\; \ell/\sqrt{3} \;=\; 0.577\,\ell\,.$$

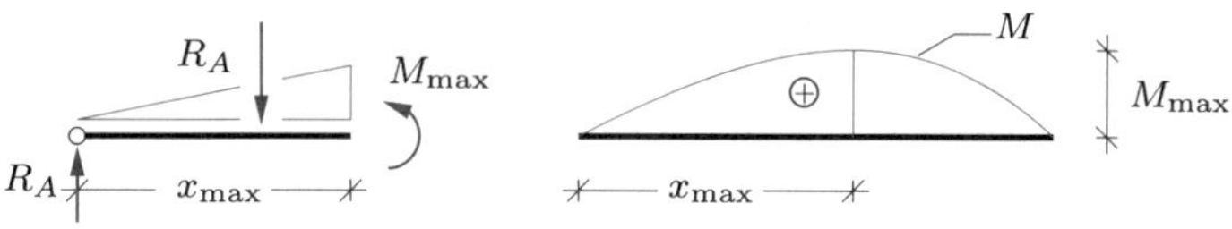

Fig. 3.37: Maximum moment by equilibrium and distribution of moment.

As illustrated in Fig. 3.37a M_{max} is found by moment equilibrium about the section at x_{max}. For the present triangular load intensity $x_{eq} = \frac{2}{3}x_{max}$, and thus gives

$$M_{max} \;=\; \tfrac{2}{3}x_{max}R_A \;=\; \frac{p\,\ell^2}{9\sqrt{3}}\,,$$

where the latter expression is found by substitution of the expressions for R_A and x_{max}, obtained previously. □

Table 3.1: Reactions and internal forces in beams with concentrated load.

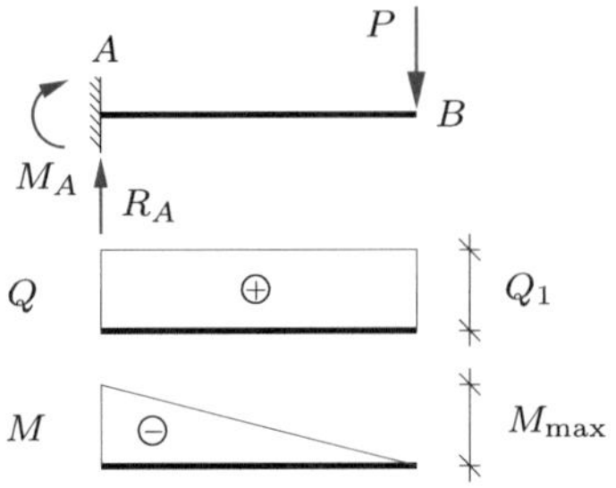

$$R_A = P \quad , \quad M_A = -\ell P$$

$$Q_1 = R_A = P$$

$$M_{\max} = -\ell P$$

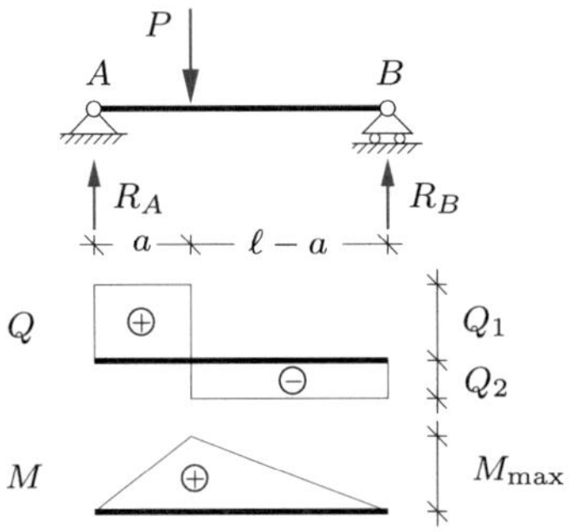

$$R_A = \frac{\ell - a}{\ell} P \quad , \quad R_B = \frac{a}{\ell} P$$

$$Q_1 = R_A \quad , \quad Q_2 = -R_B$$

$$M_{\max} = a\, R_A = \frac{\ell - a}{\ell}\, aP$$

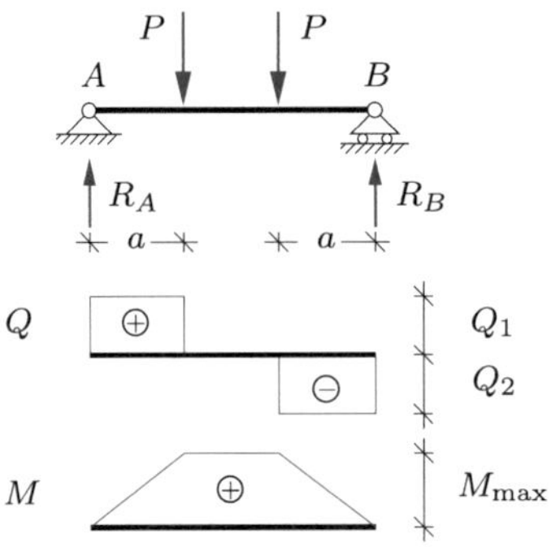

$$R_A = P \quad , \quad R_B = P$$

$$Q_1 = P \quad , \quad Q_2 = -P$$

$$M_{\max} = a\, R_A = a\, P$$

Note: $a \leq \frac{1}{2}\ell$

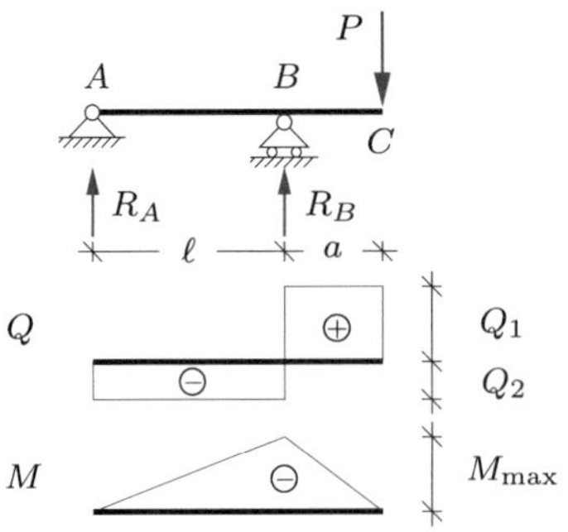

$$R_A = -\frac{a}{\ell} P \quad , \quad R_B = \frac{\ell + a}{\ell} P$$

$$Q_1 = -\frac{a}{\ell} P \quad , \quad Q_2 = P$$

$$M_{\max} = -aP$$

Table 3.2: Reactions and internal forces in beams with distributed load.

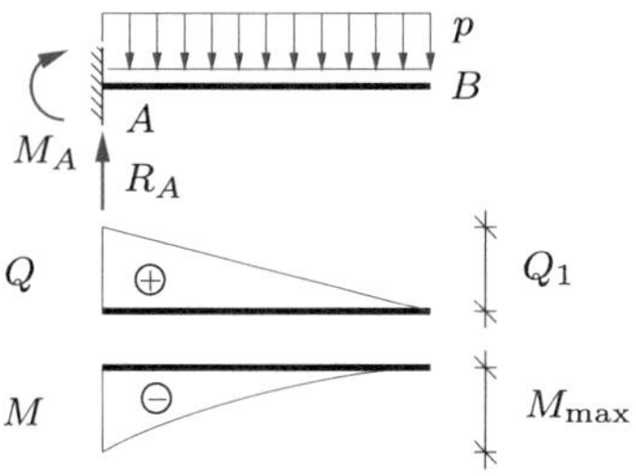

$$R_A = p\ell \quad , \quad M_A = -\tfrac{1}{2}p\ell^2$$

$$Q_1 = R_A = p\ell$$

$$M_{\max} = -\tfrac{1}{2}p\ell^2$$

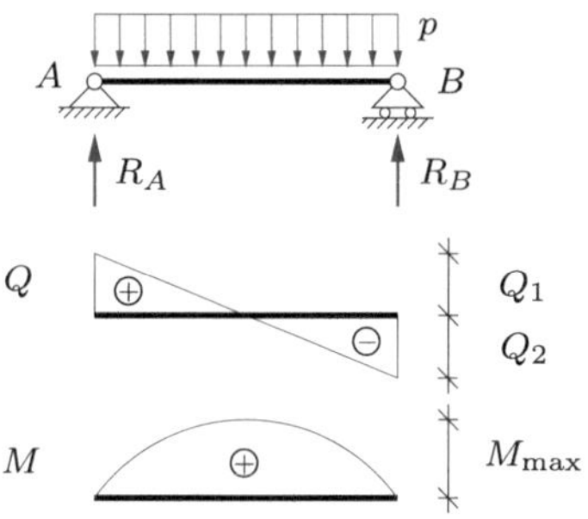

$$R_A = \tfrac{1}{2}p\ell \quad , \quad R_B = \tfrac{1}{2}p\ell$$

$$Q_1 = R_A \quad , \quad Q_2 = -R_B$$

$$M_{\max} = \tfrac{1}{8}\,p\,\ell^2$$

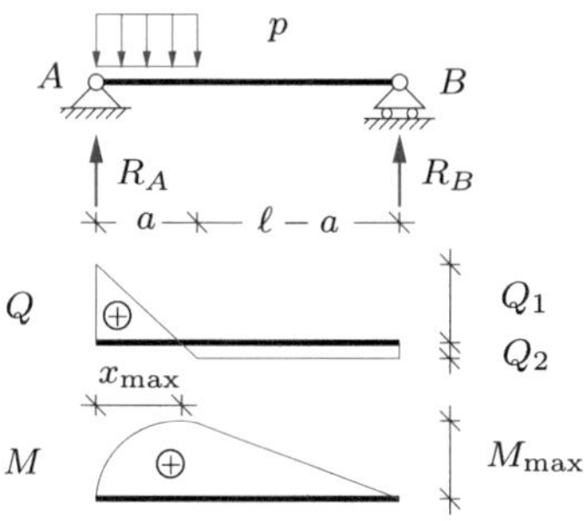

$$R_A = \left(1 - \frac{a}{2\ell}\right) ap$$

$$R_B = \frac{a}{2\ell}\, ap$$

$$Q_1 = R_A \quad , \quad Q_2 = -R_B$$

$$M_{\max} = \frac{R_A^2}{2p} \quad , \quad x_{\max} = \frac{R_A}{p}$$

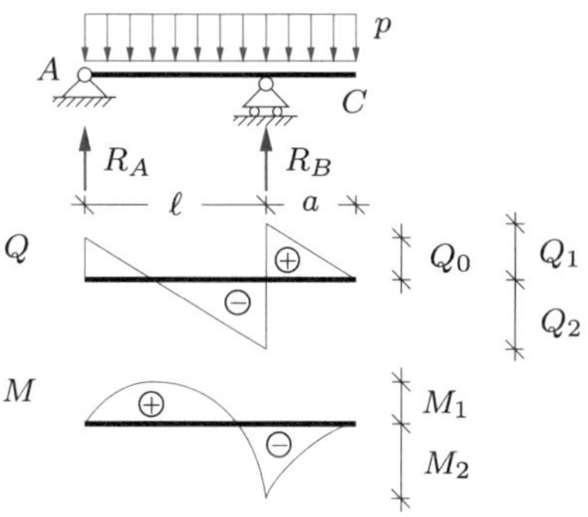

$$R_A = \frac{\ell^2 - a^2}{2\ell}\, p$$

$$R_B = \frac{(\ell + a)^2}{2\ell}\, p$$

$$Q_0 = R_A \quad , \quad Q_1 = ap$$

$$Q_2 = -\frac{\ell^2 + a^2}{2\ell}\, p$$

$$M_1 = \tfrac{1}{2}R_A^2/p \quad , \quad M_2 = -\tfrac{1}{2}pa^2$$

3.4 Combined loads

Structures are often analyzed for several loads acting simultaneously. For linear structures the combined effect of the individual loads can be obtained by direct addition – the so-called superposition. In connection with hand calculations it is often convenient to use a modified form of the superposition principle, in which all loads are included simultaneously, but the internal forces are computed in two steps as the sum of a 'global' and a 'local' part. These two approaches are illustrated in the following.

3.4.1 Superposition of load cases

The direct superposition principle is illustrated in Fig. 3.38, showing a beam with a combined load consisting of a uniform distributed load with intensity p and a concentrated load P. When the structure is statically determinate or linear elastic the total effect of the loads can be found by superposition, i.e. as the sum of the effect of the individual loads. First, this implies that the reactions can be found as the sum of the reactions of the individual load cases, e.g. the vertical reaction at the support A,

$$R_A = R_A^{(1)} + \cdots + R_A^{(n)} . \tag{3.15}$$

Similar relations apply for the other reaction components.

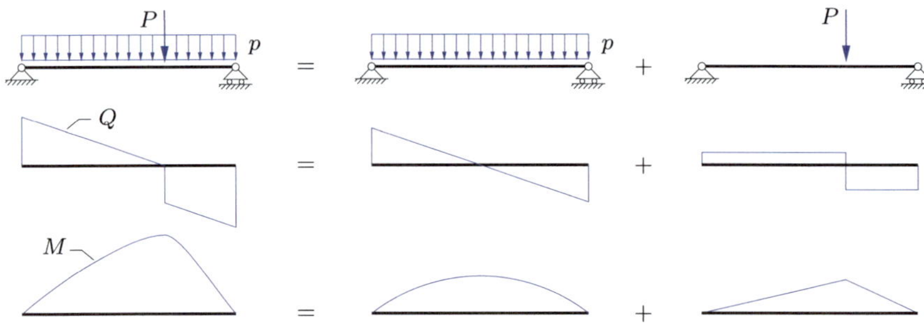

Fig. 3.38: Combining distributed and local load by superposition.

Internal forces are associated with any point of the structural members forming the structure. Thus, the internal forces are of the form $N(x)$, $Q(x)$ and $M(x)$, where x is a coordinate identifying the point. A particularly simple example is an individual beam as shown in Fig. 3.38, where x is simply the distance from the left end of the beam. As all operations are linear, the internal forces at any given cross-section are found as the sum of the contribution from each of the individual load cases, e.g. the moment

$$M(x) = M^{(1)}(x) + \cdots + M^{(n)}(x), \tag{3.16}$$

and similarly for the other internal forces. In the example in the figure the effect on the shear force distribution $Q(x)$ is to combine a linear variation from the distributed load with a piecewise constant part between the concentrated loads, while the total internal moment curve $M(x)$ follows from combination of the parabolic variation from the distributed load with the piecewise linear variation from the concentrated loads.

If there are n individual load cases, the direct superposition method consist in making n complete analyzes of the structure. The results, such as reactions and internal forces, can then be found by simple addition. Structures are often analyzed for several combinations of the same load cases, e.g. with different scaling. In those cases it is convenient to carry out the full analysis of the structure for the individual load cases. However, if a typical loading includes several individual load components that act together in a fixed combination it may be convenient to analyze this particular combination. Even in that case the principle of superposition may be used to advantage by permitting a 'global analysis' providing the internal forces at characteristic cross-sections, as if the load consisted only of concentrated parts, and then fill in the local effect of the distributed load afterwards, as explained below.

3.4.2 Superimposing the distributed load

In the direct superposition method the structure is analyzed for each load case separately, and the results are then added to form the final result of simultaneous application of all the load cases. In relation to analysis by hand another procedure is often used. The load may consist of different components, but it is the joint action that is the goal of the analysis. It is then convenient to organize the analysis as follows. First the reactions are determined including all loads, and e.g. representing the effect of distributed loads by their equivalent concentrated loads. The internal forces are then determined at selected cross-sections. These include cross-sections at supports, at joints and at locations of concentrated loads. Finally, the distribution of the internal forces between the selected cross-sections is calculated by using the local effect of the distributed load, known e.g. from the solution of the equilibrium differential equations.

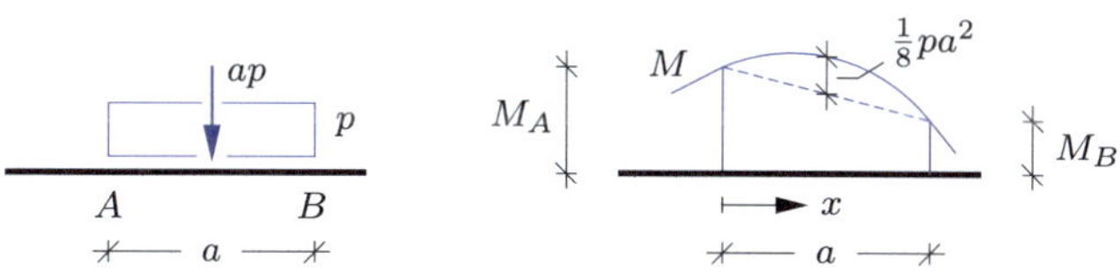

Fig. 3.39: Superimposing contribution to moment from distributed load.

The principle of superimposing local contributions to the moment distribution is illustrated in Fig. 3.39, where a distributed load with intensity p acts on the part AB of length a. The reaction forces of the structure are determined based on the equivalent load ap, and the internal moment is determined at key points on the structure, including M_A and M_B at the end points of the distributed load, as shown in Fig. 3.39b. An intermediate moment distribution is obtained by connecting the key point values by straight lines. For the parts of the structure without local distributed loading the piecewise linear distribution is correct, while e.g. for the part AB with distributed load the local parabolic effect must be superimposed and the linear curve is therefore only indicated by a dashed line. As demonstrated in Fig. 3.26 in Example 3.6 the local contribution to the moment distribution from an uniformly distributed load with intensity p is a symmetric parabola with maximum value $\frac{1}{8}p\ell^2$, where ℓ is the length of the interval. This means that in the present case in Fig. 3.39b the dashed line must be superimposed by a symmetric parabola with zero value at the ends of the interval and the local peak value $\frac{1}{8}pa^2$ at the center of the interval. The superimposed parabolic distribution is in Fig. 3.39b indicated by the solid curve between A and B. Note that because the symmetric parabola is superimposed on the (dashed) linear curve the actual local maximum is in general not located at the center of the interval. The resulting local maximum can be determined by inspection of the associated shear force distribution as demonstrated previously in Section 3.3.2. Alternatively, a parametric expression can be determined by superposition of the linear and parabolic expressions,

$$M_{AB}(x) = M_A\left(1-\frac{x}{a}\right) + M_B\frac{x}{a} + \tfrac{1}{2}pa^2\,\frac{x}{a}\left(1-\frac{x}{a}\right) \quad , \quad 0 \le x \le a\,.$$

The local maximum of the moment distribution is determined by $dM/dx = 0$ at $x = x_{\max}$, whereby

$$x_{\max} = \tfrac{1}{2}a + \frac{M_B - M_A}{pa}\,.$$

The last term represents the offset of the location of the local moment maximum due to the inclination of the linear (dashed) curve.

The principle of local member contributions to the internal forces from local loads on the corresponding member is a standard procedure in connection with finite element calculations as discussed in Chapter 7. The basic procedure is illustrated for beams in the following two examples and is extended to frames later.

Example 3.9. Simply supported beam with combined loads. Figure 3.40a shows a beam ACB loaded by a combination of a concentrated force P at C and a distributed load with intensity p on AC. The task is to find the distribution of the internal forces, and to determine the location and magnitude of the largest moment $M(x_{\max})$ in the beam.

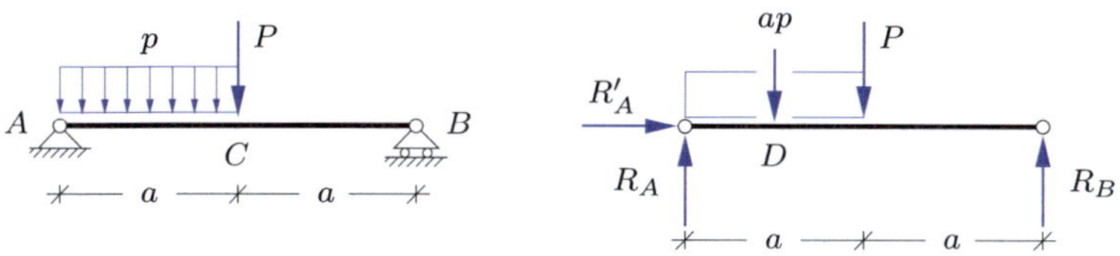

Fig. 3.40: Combined concentrated and distributed load.

For the purpose of determining the reactions, the distributed load is represented by an equivalent concentrated force of magnitude ap acting at D as shown in Fig. 3.40b. It follows immediately from horizontal projection that $R'_A = 0$. The reaction R_A follows from moment about B,

$$\overset{\frown}{B} \qquad 2aR_A - \tfrac{3}{2}a(ap) - aP = 0 \qquad \Rightarrow \qquad R_A = \tfrac{1}{2}P + \tfrac{3}{4}ap\,.$$

Similarly the reaction R_B follows from moment about A,

$$\overset{\frown}{A} \qquad 2aR_B - \tfrac{1}{2}a(ap) - aP = 0 \qquad \Rightarrow \qquad R_B = \tfrac{1}{2}P + \tfrac{1}{4}ap\,.$$

The shear force can now be plotted directly from its value at the supports at A and B and its discontinuity under the concentrated load at C,

$$Q_A = R_A\,, \qquad Q_B = -R_B\,, \qquad Q_C^+ - Q_C^- = -P\,.$$

The resulting shear force distribution is illustrated in Fig. 3.41 with $P = ap$, showing a linear decrease over AC and a discontinuous drop of magnitude P at C.

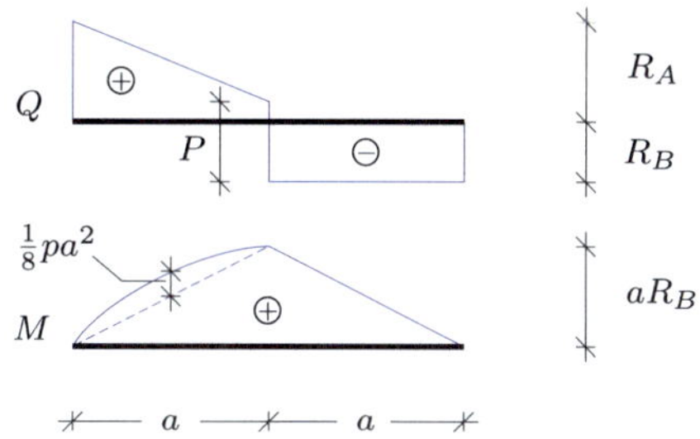

Fig. 3.41: Shear force and internal moment distribution.

The moment vanishes over the supports A and B, $M_A = M_B = 0$, and the moment M_C at mid-span follows from the reaction at B as

$$M_C = aR_B = \tfrac{1}{2}aP + \tfrac{1}{4}a^2p\,.$$

The moment distribution on the part AC with the distributed load is found from the linear interpolation between the moments M_A and M_C at the ends, plus a parabolic distribution with maximum value $\frac{1}{8}a^2p$ at the center from the distributed load. This is shown in Fig. 3.41 with the moment at D determined by

$$M_D = \tfrac{1}{2}M_C + \tfrac{1}{8}a^2p = \tfrac{1}{4}aP + \tfrac{1}{4}a^2p\,.$$

It is seen in the figure that for $P = ap$ the maximum moment is at the center of the beam with $M_C = \frac{1}{2}aP + \frac{1}{4}a^2p = \frac{3}{4}aP$.

If the intensity p of the distributed load is increased sufficiently, the maximum moment occurs within the part AC of the beam. The condition for a maximum moment is $Q(x_{\max}) = 0$. It therefore follows from the shear force curve in Fig. 3.41 that the condition $Q(x_{\max}) = 0$ to the left of the center C requires $R_B > P$, corresponding to $ap > 2P$. The location and value of the maximum moment then follow from the results of Example 3.7 as

$$x_{\max} = \frac{R_A}{p} = \frac{3ap + 2P}{4p}, \qquad M_{\max} = \tfrac{1}{2}x_{\max}^2 p,$$

where the expression for $M_{\max}$ is a convenient reformulation of the result from Example 3.7 for constant load intensity. □

Example 3.10. Cantilevered beam with concentrated and distributed loads. Figure 3.42a shows a cantilevered beam ABC loaded by a distributed load of intensity p on AB and a concentrated force P at C. Like in the previous example the objective is to analyze and illustrate the basic features of the distribution of the internal forces.

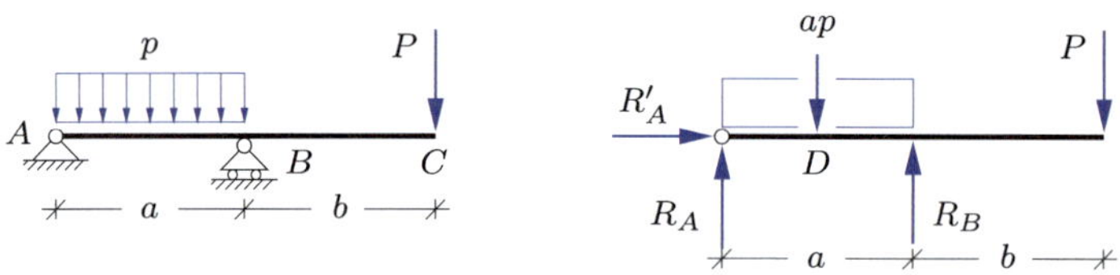

Fig. 3.42: Cantilever with combined concentrated and distributed load.

For the purpose of determining the reactions the distributed load is represented by an equivalent concentrated force of magnitude ap acting at D as shown in Fig. 3.42b. It follows immediately from horizontal projection that $R'_A = 0$, and the vertical reactions R_A and R_B are computed from moment about B and A, respectively,

$$R_A = \tfrac{1}{2}ap - \frac{b}{a}P, \qquad R_B = \tfrac{1}{2}ap + \frac{a+b}{a}P.$$

The shear force is plotted directly from its values at A and C and the discontinuity under the reaction at B,

$$Q_A = R_A, \qquad Q_{BC} = P, \qquad Q_B^+ - Q_B^- = R_B.$$

The resulting shear force distribution is illustrated in Fig. 3.43 with $P = \frac{3}{8}ap$, showing a linear decrease over AB and a discontinuous increase of magnitude R_B at B.

The moment vanishes at the beam ends A and C, $M_A = M_C = 0$, and at the support B the moment follows from the concentrated force P as $M_B = -bP$. In the beam AB the moment distribution also includes a parabolic contribution with maximum value $\frac{1}{8}a^2p$ at the center D. Thus, the moment M_D at the center of AB is

$$M_D = \tfrac{1}{2}M_B + \tfrac{1}{8}a^2p = -\tfrac{1}{2}bP + \tfrac{1}{8}a^2p.$$

The moment distribution along the beam AB is negative at B, and may contain a positive part from A, depending on the relative magnitude of loads and dimensions. The clearest

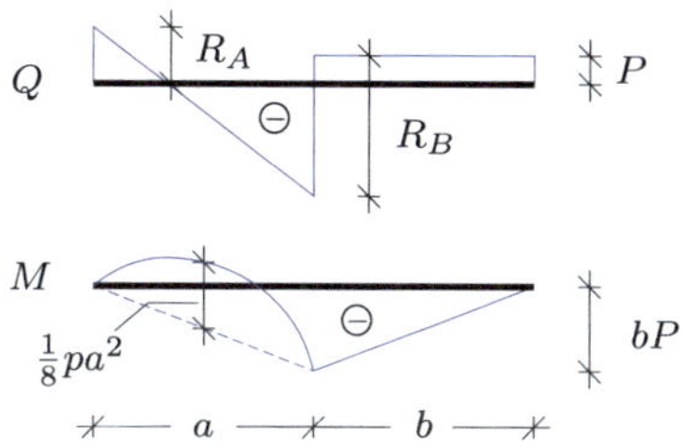

Fig. 3.43: Shear force and internal moment distribution.

way to illustrate the interplay of the parameters is to assume a local moment maximum at $x_{\max}$ and to calculate the location from the condition $Q(x_{\max}) = 0$, whereby

$$x_{\max} = \frac{R_A}{p} = \frac{a^2p - 2bP}{2ap}, \qquad M_{\max} = \tfrac{1}{2}x_{\max}^2 p\,.$$

It is seen that $0 < x_{\max} < a$ if $R_A > 0$, i.e. if there is an upward reaction in A. If the reaction can become negative, the beam should be constrained against uplift. □

3.5 Internal forces in frames

Frames consist of beams joined typically by rigid connections. However, they may also contain joints permitting motion, e.g. rotation, between the parts. Frames are typically used in buildings e.g. for housing, offices, storage etc. In multi-story buildings the frame will contain columns, connected to a primary system of horizontal beams. The floors may either be plates connected directly to the main beams, or supported by a system of secondary beams. While frames are generally three-dimensional, many buildings are laid out in the form of intersecting planes, and thus the conceptual analysis may be carried out using the theory of plane frames. Frames also find extensive use in engineering structures like bridges, cranes, towers, offshore structures etc. The difference between a frame and a truss is that the frame makes use of the bending resistance of the individual members, while trusses have a spatial layout, that enables the members to act only in tension or compression. Often the general layout of a frame borrows features from a similar truss structure, e.g. in the form of diagonal braces to resist horizontal loads.

It is a central feature of a frame that the individual members act as beams, and thus can support bending moments and shear forces in addition to the normal force, encountered in trusses. Frames can therefore also carry loads applied to the individual members. Conceptually this means that a loaded member in a frame must be able to support loads transmitted from other parts of the frame, the global action of the frame, as well as the effect of loads acting locally on the member. This split into local and global effects is similar to the superposition of the local effects of distributed loads on beams treated in Section 3.4.2, and is often used in hand calculation of internal forces in frames as explained and illustrated in the following.

The present section is limited to plane frames. Many frame structures are conceptually formed by intersecting planes, and thus an introductory analysis of the general behavior can often be based on one or more plane frames. A full analysis of three-dimensional frame structures will typically be carried out by use of a numerical method like the Finite Element Method, and a brief description of this is given in Chapter 7 together with a small computer program MINIFRAME for static analysis of frames. Here, some of the basic properties of the internal forces are explained and illustrated by simple examples.

Example 3.11. Angle frame. Figure 3.44 shows a simple angle frame ABC with a fixed support in A. The frame consists of the vertical beam AB and the beam BC, which is inclined by the angle α relative to horizontal. A vertical force P is acting at the tip C.

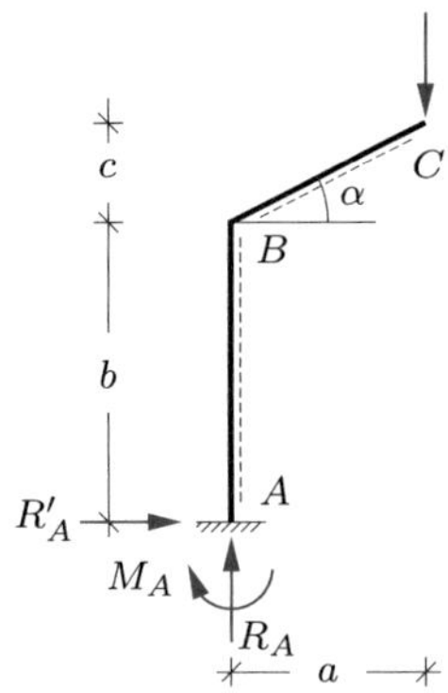

Fig. 3.44: Angle frame with fixed support at A and vertical tip load P.

The reactions are determined by horizontal and vertical force equilibrium, and moment about A. This yields

$$R'_A = 0\,, \qquad R_A = P\,, \qquad M_A = -aP\,.$$

The load is a concentrated force, and the normal force N and the shear force Q are therefore piecewise constant, while the moment M is varying linearly over each member of the frame. Thus, N and Q are determined by representative values for AB and BC, respectively, while the moment M is determined by its values at A, B and C. In Fig. 3.45 a section is introduced in B, and the internal force at the joint are determined by equilibrium for each of the two beams.

Figure 3.45a shows the vertical part AB, where the internal forces at B are determined by equilibrium as

$$N = -R_A = -P\,, \qquad Q = -R'_A = 0\,, \qquad M = M_A = -aP\,.$$

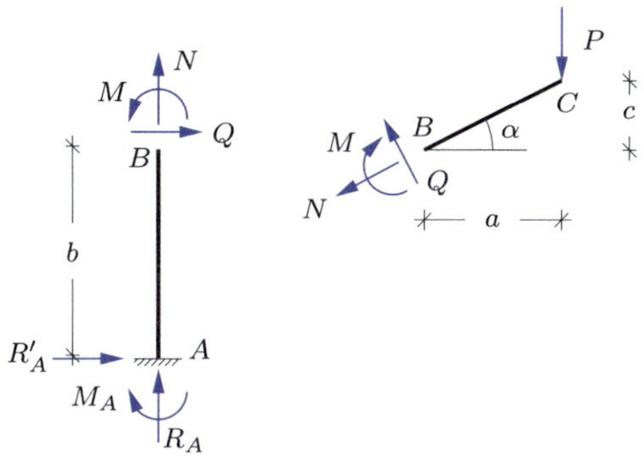

Fig. 3.45: Internal forces by section in B.

The normal and shear force in the inclined part BC, shown in Fig. 3.45b, are conveniently determined by projection on to the directions of the internal forces,

$$N = -P \sin\alpha\,, \qquad Q = P\cos\alpha\,.$$

The moment is continuous at A, and $M_A = -aP$ has already been determined. The distribution of the internal forces is shown in Fig. 3.46. □

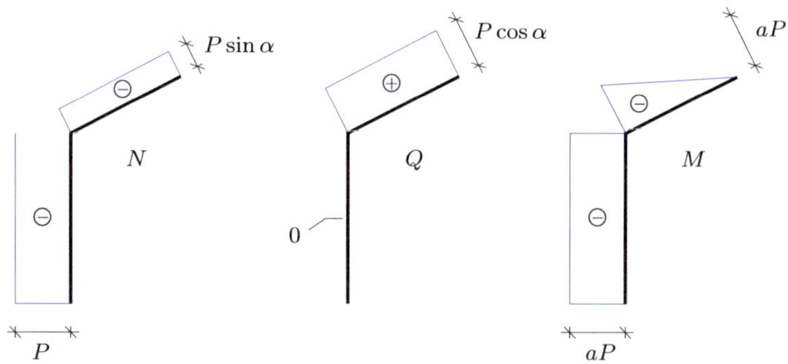

Fig. 3.46: Distribution of internal force in angle frame.

Example 3.12. Mast for electric train cables. An alternative to the angle frame of the previous example is shown in Fig. 3.47a. This type of mast is often used to carry cables for electric trains. The vertical mast ABD supports bars BC and DC, connected at the node C supporting the cable, here represented by a vertical force P.

The bar forces N_{BC} and N_{DC} are determined by introducing a vertical section to the left of C as shown in Fig. 3.47b,

$$N_{DC} = P/\sin\alpha\,, \qquad N_{BC} = N_{DC}\cos\alpha = P\,a/c\,.$$

Note, that a small angle α corresponding to a small value of c/a leads to large forces in the bars.

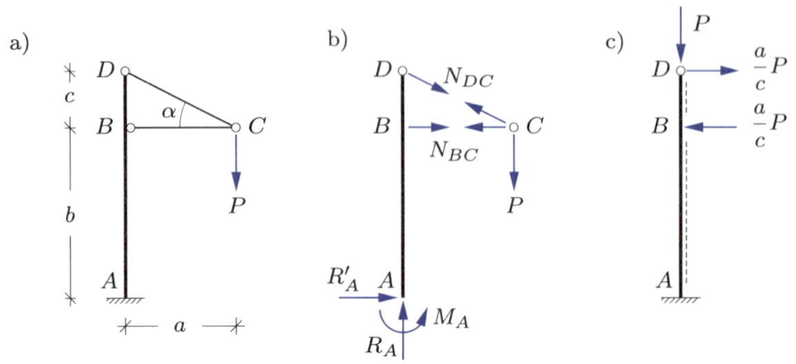

Fig. 3.47: Cable mast, fixed in A with vertical load in C.

The section forces N, Q and M in the mast are determined as for a beam loaded by concentrated forces in B and D, as shown in Fig. 3.48. The reactions can be found independently by equilibrium of the full structure as $R'_A = 0$, $R_A = P$ and $M_A = aP$. They are seen to correspond to the internal forces at the fixed support A. □

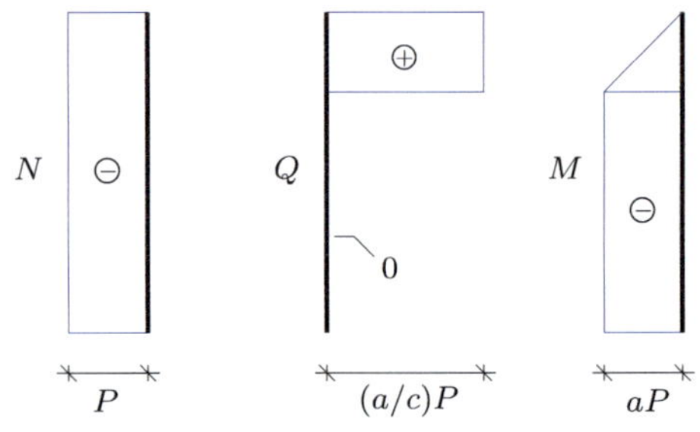

Fig. 3.48: Internal forces in cable mast.

3.5.1 Influence of load distribution

The calculation of internal forces by hand typically follows a standard procedure, in which the reactions are determined first, then the section forces are determined at supports, at joints and at concentrated loads, and finally the local influence of distributed loads on the section forces within the members is accounted for. An advantage of this procedure is that each of the three steps deals with a limited and precise problem and leads to results that can be checked for consistency and against static understanding of the structure. The procedure is a systematic extension of the method used for beams with combined concentrated and distributed components in Section 3.4.2.

The procedure is illustrated with reference to the simple angle frame shown in Fig. 3.49. The following three examples analyze the internal force distribution of this frame for three different loads: a concentrated vertical force, a distributed vertical force density, and finally a horizontal concentrated force. The method of analysis is quite similar in the three examples, and the most important point is to obtain familiarity with the main features of the results.

Example 3.13. Vertical concentrated load. Figure 3.49 shows a simply supported frame $ACDB$ of height a and width $2a$. The simple support at A is fixed, leading to two reaction components R'_A and R_A, while the support at B has horizontal rollers, leaving only a vertical reaction component R_B.

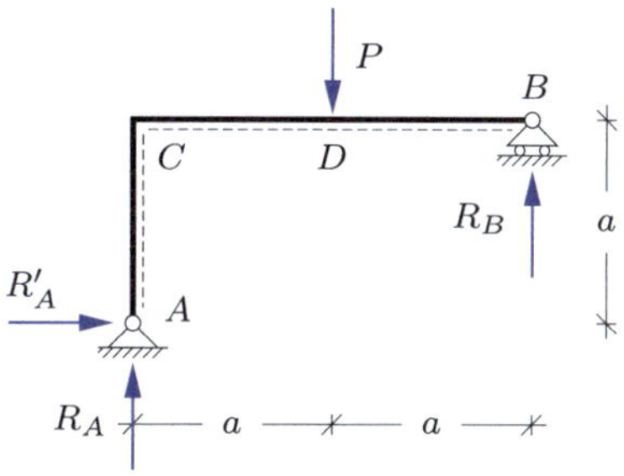

Fig. 3.49: Angle frame with vertical load P at center span.

In the present example the load consists of a concentrated vertical force P acting at the center D of the horizontal beam CDB. The reactions R'_A, R_A and R_B are determined by horizontal equilibrium, moment about B and moment about A, respectively. This leads to

$$R'_A = 0\,, \qquad R_A = R_B = \tfrac{1}{2}P\,.$$

It is observed that this is the same set of reactions that would have been obtained for a horizontal beam CDB with a fixed simple support at C and a horizontal roller bearing at B, see e.g. Example 3.2. The reason is, that the absence of a horizontal reaction component at A transfers the vertical reaction force R_A directly to C. This is an observation, helping to 'understand' the structure, and should supplement and not replace rigorous analysis.

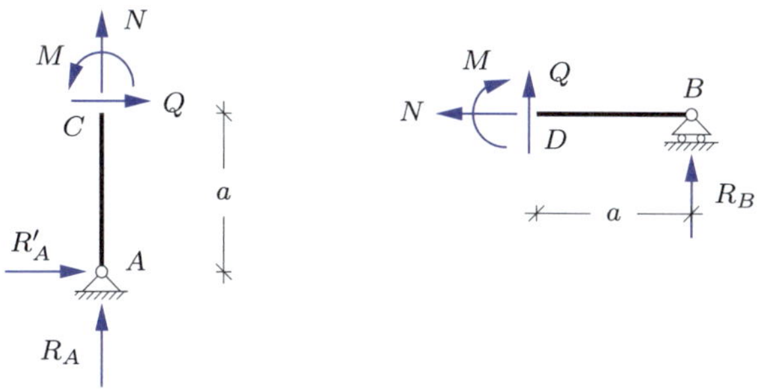

Fig. 3.50: Internal forces by sections at C and D, respectively.

The internal forces are constant or of linear variation between supports, joints and loads, and can therefore be determined by considering sections at C and D as illustrated in

Fig. 3.50. The internal forces N and Q in the vertical part and the moment M at the corner C are determined from equilibrium of the vertical member shown in Fig. 3.50a,

$$N = -R_A = -\tfrac{1}{2}P, \qquad Q = -R'_A = 0, \qquad M = -aR'_A = 0.$$

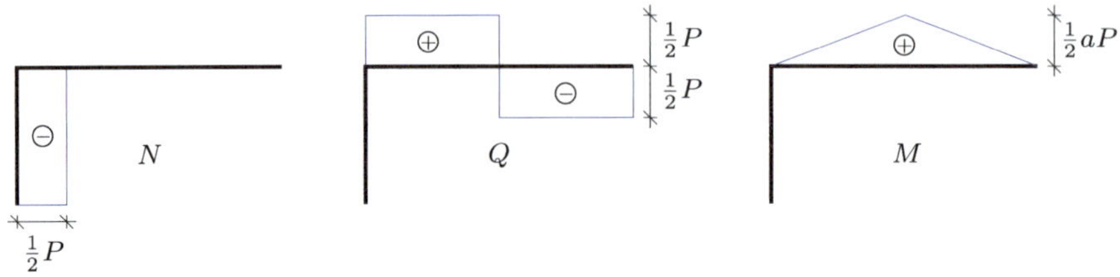

Fig. 3.51: Internal force distribution in angle frame.

In Fig. 3.50b the section is placed just to the right of D. Equilibrium gives

$$N = 0, \qquad Q = -R_B = -\tfrac{1}{2}P, \qquad M = aR_B = \tfrac{1}{2}aP.$$

The local force P introduces a discontinuity of magnitude P in the shear force at D. Thus, the distributions of the internal forces are fully determined, and they can be illustrated as shown in Fig. 3.51. Note, that the horizontal beam CDB acts as a horizontal, simply supported beam as discussed in connection with the reactions. □

Example 3.14. Vertical distributed load. Figure 3.52 shows the same frame as above, but with distributed load p on the horizontal beam CB. The reactions were found in Example 1.7, and can also be found by using the equivalent load $P = 2ap$ in the results of the previous example.

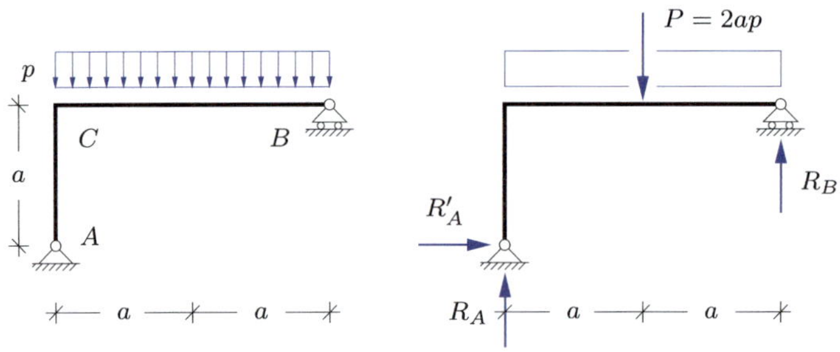

Fig. 3.52: Angle frame with distributed load.

The reactions determine the internal forces N, Q and M at the supports A and B, and at the joint C. The full distribution of the internal forces are then obtained by the following argument. There are no applied normal loads, and the normal force distribution is therefore constant within each of the beams AC and CB as shown in Fig. 3.53a. The shear force Q is constant in AC, as there is no transverse load on this part, and of linear variation over CB due to the constant transverse load intensity. This gives the shear force distribution shown in Fig. 3.53b. Finally, the moment distribution follows from the principle of adding

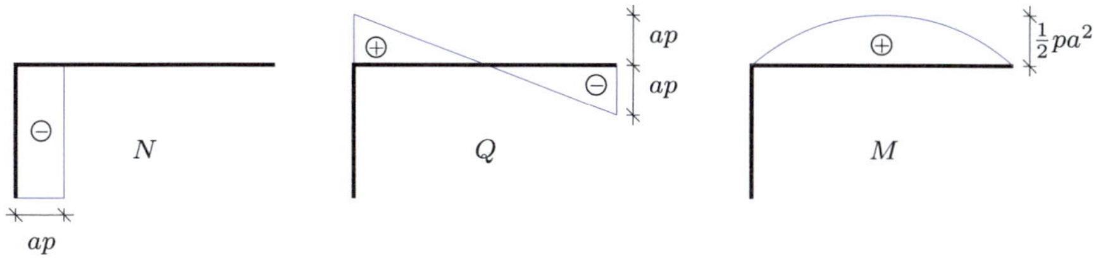

Fig. 3.53: Distribution of internal forces in angle frame with distributed load.

the effect of the distributed load as discussed in Examples 3.9 and 3.10. In the present case $M = 0$ at A, B and C, and thus the beam CB has a parabolic moment variation with center value $\frac{1}{8}p(2a)^2$ from the transverse load p distributed over the length $2a$ as shown in Fig. 3.53c. It is seen that the effect of distributing the load uniformly over the beam CB is a redistribution of the shear force Q and the moment M within the beam CB. □

Example 3.15. Horizontal concentrated load. Finally, the angle frame is analyzed for a horizontal load consisting of a concentrated force acting at the joint C as shown in Fig. 3.54a.

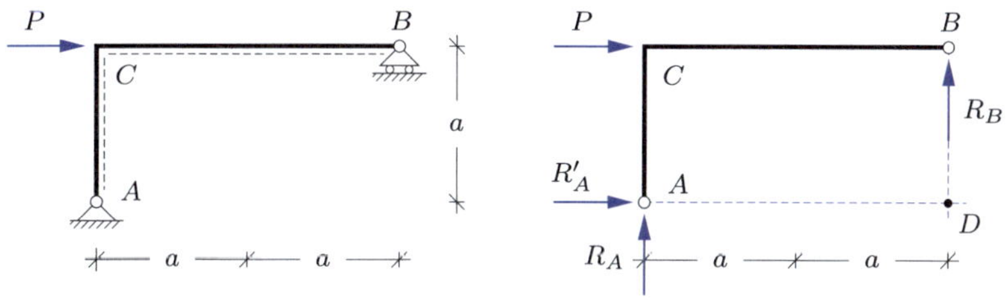

Fig. 3.54: Angle frame with horizontal load P at joint C.

The reactions are shown in Fig. 3.54b and determined by horizontal equilibrium, and moment about D and A, respectively,

$$R'_A = -P, \qquad R_A = -R_B = -\tfrac{1}{2}P.$$

Note, that the load P and the horizontal reaction R'_A constitute a force couple, which is counteracted by an opposite force couple formed by the vertical reactions R_A and R_B.

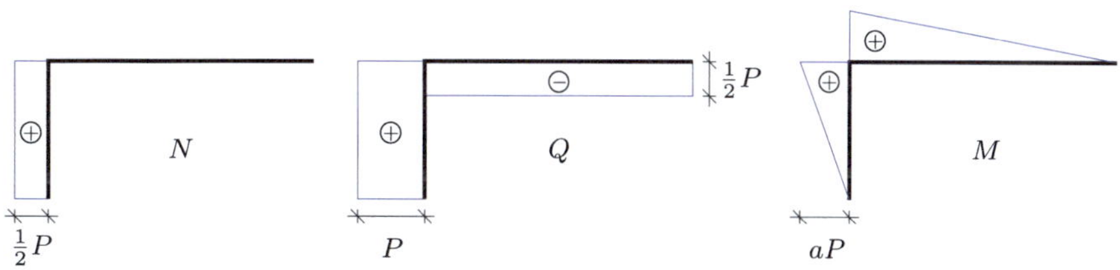

Fig. 3.55: Distribution of internal forces in angle frame with horizontal load.

As the load consists of a concentrated force at the joint C, the normal force N and the shear force Q are constant within each of the beams AC and CB, while the moment is continuous with linear variation over AC and CB. The resulting internal forces are shown in Fig. 3.55. □

3.5.2 Influence of support conditions

The distribution of the internal forces in a frame may be strongly influenced by the support conditions. By changing the supports the internal forces may be redistributed in the frame. This is illustrated in the following two examples, dealing with the same T-frame, but with two different sets of supports.

Example 3.16. T-frame with horizontal roller bearing. Figure 3.56 shows the T-frame with a horizontal roller bearing in B together with the positive definition of the reaction components R_A, R'_A and R_B.

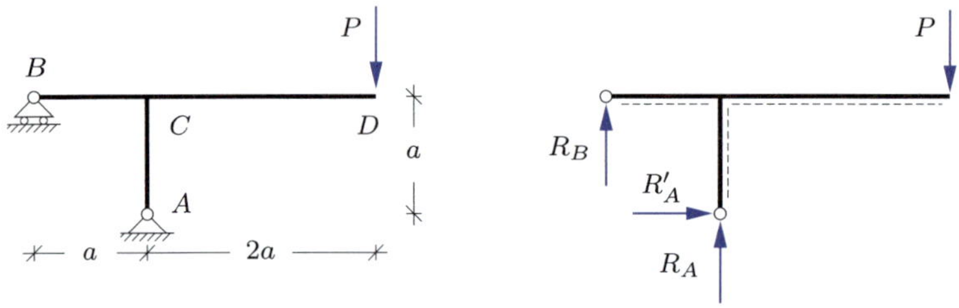

Fig. 3.56: T-frame with horizontal roller and vertical tip load P.

Reactions are determined by horizontal projection, and moment about B and A, respectively,

$$R'_A = 0\,, \qquad R_A = 3P\,, \qquad R_B = -2P\,.$$

The internal forces are determined by introducing a section in each of the members, joined at C, as shown in Fig. 3.57. For each of these members, now separated by a section, the section forces at C follow from equilibrium. The results are shown in Fig. 3.58.

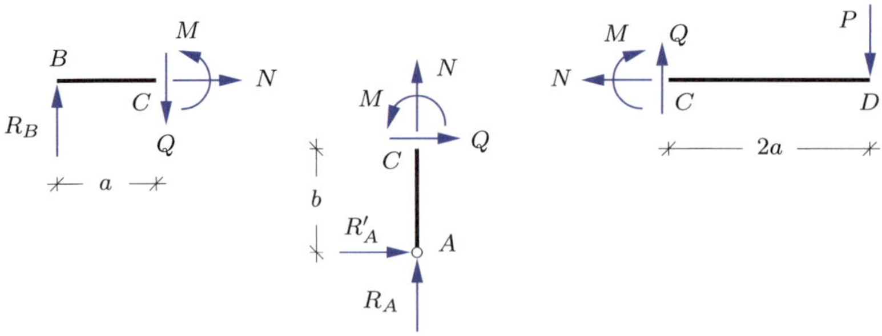

Fig. 3.57: Internal forces by sections in C.

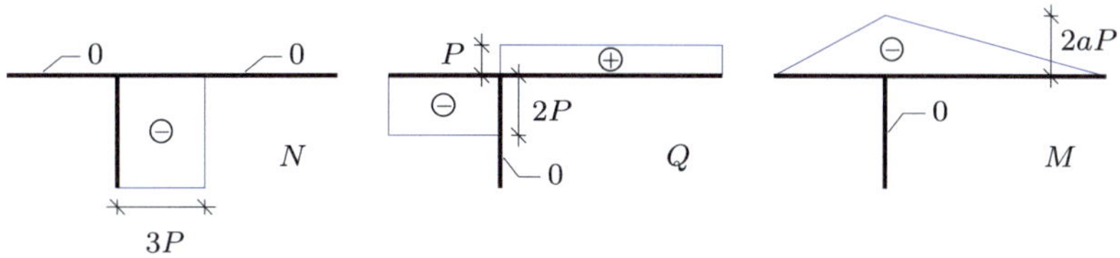

Fig. 3.58: Internal forces in T-frame with horizontal roller bearing.

The internal forces exhibit discontinuities at the T-joint. When introducing a section there is a set of internal force components acting on the beam, but also a set of components of equal magnitude but opposite direction acting on the joint. The non-vanishing internal force components acting on the joint C are shown in Fig. 3.59. It is seen that these components satisfy the three equilibrium conditions for the joint C. □

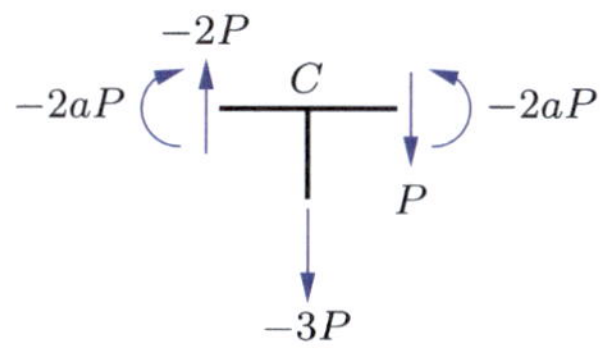

Fig. 3.59: Equilibrium of the T-joint at C.

Example 3.17. T-frame with simple vertical roller bearing. The distribution of internal forces change, when the support condition at B is changed to a vertical roller bearing as shown in Fig. 3.60. Reactions are determined by vertical projection, and moment about C and A, respectively,

$$R_A = P \quad , \quad R'_A = 2P \quad , \quad R'_B = -2P \, .$$

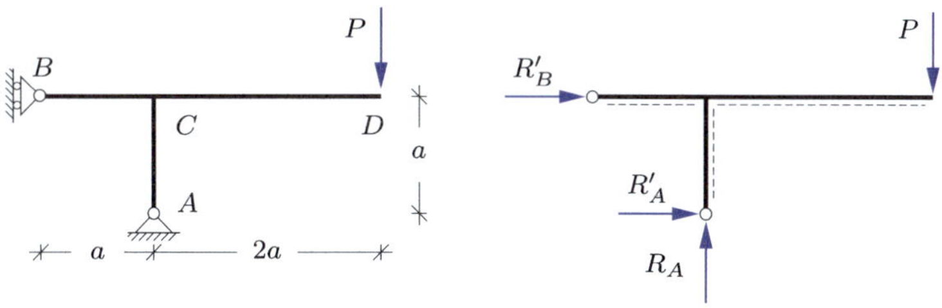

Fig. 3.60: T-frame with vertical roller and vertical tip load P.

The internal forces are determined by introducing a section in each of the beams joined at C and imposing three equilibrium conditions on each of the beams. The result is shown in Fig. 3.61. Note, that now there is bending and shear in AC and not in BC as in the previous example. □

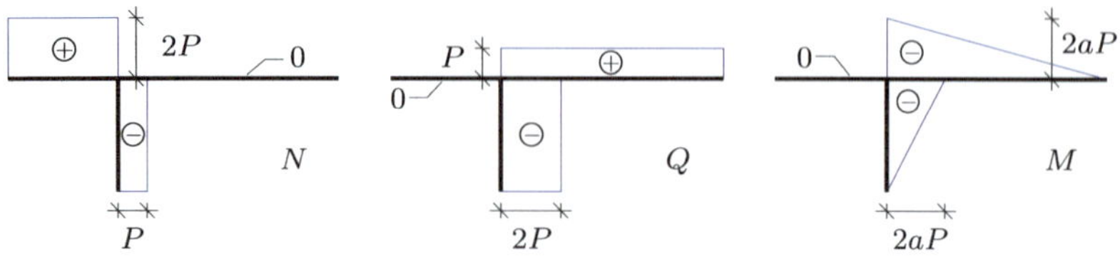

Fig. 3.61: Internal forces in T-frame with vertical roller bearing.

The two examples clearly demonstrate that a change of the support conditions may lead to a redistribution of the internal forces and the magnitude and direction of the reactions. Naturally, it is of considerable interest to be able to form a qualified opinion about the statics of a structure from basic considerations regarding its layout and support conditions. In the present case a good impression of the statics of the T-frame can be obtained by constructing the reactions geometrically, as illustrated for the three-hinge frame in Section 1.5.3. The geometric construction of the reactions for the two sets of support conditions is illustrated in Fig. 3.62.

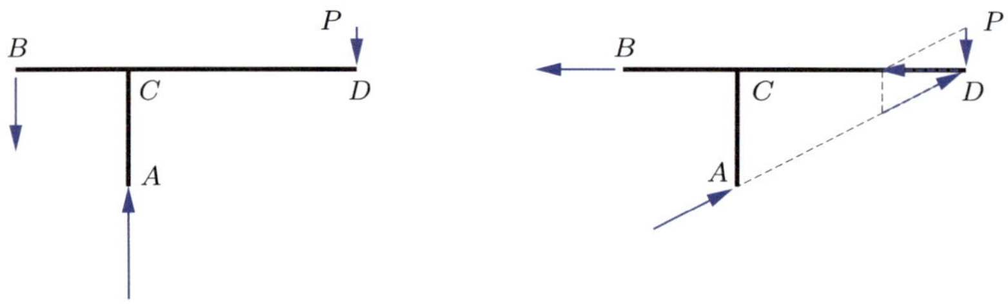

Fig. 3.62: Reactions on T-frame for different support conditions.

Both frames have a fixed simple support at A and a simple support with rollers at B. In the left figure the rollers at B permit horizontal motion, and the reaction R_B is therefore vertical. As the load is vertical, the reaction at the fixed support A must also be vertical. Thus, in this case the reaction forces can be found by resolving the load P into parallel forces R_A and R_B as shown in Fig. 3.62a. As there is no horizontal reaction in A, the column AC will have zero moment. In the right figure the line of action of the horizontal reaction R'_B intersects the load P, and the reactions can therefore be found by resolving the load P into components through the supports A and B as shown in the figure. This leads to larger reactions, and it is now the beam BC that does not have a transverse force, and therefore zero moment. Considerations like these may be helpful in identifying the basic characteristics of the way a frame carries the load, and may constitute a valuable supplement to the mere analysis of an assumed configuration.

3.5.3 Three-hinge frame

The three-hinge frame was introduced and discussed in connection with reactions in Section 1.5.3. The following two examples demonstrate the analysis of the internal forces for a concentrated and a uniformly distributed load, respectively.

Example 3.18. Three-hinge frame with concentrated load. The three-hinge frame has two fixed simple supports and an internal hinge, whereby the frame becomes statically determinate. Figure 3.63 illustrates a three-hinge frame with a concentrated vertical force P acting on the left roof beam DC at the *horizontal* distance x from the corner D, corresponding to a *horizontal* distance $x' = a - x$ from the central hinge at C.

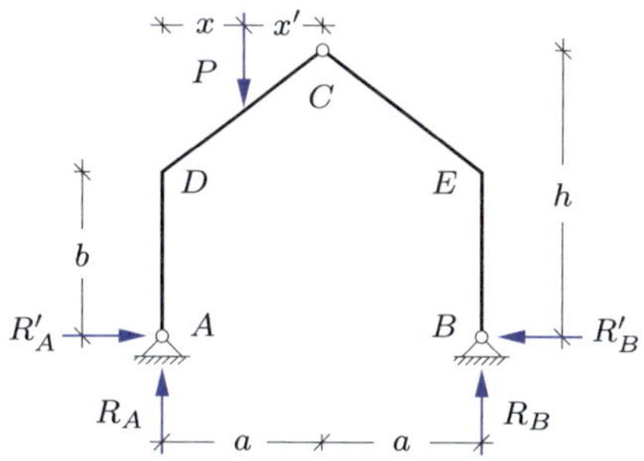

Fig. 3.63: Three-hinge frame with vertical load P.

The reactions have already been determined in Example 1.8. The vertical reactions follow from moment of all forces acting on the structure about the support points B and A, respectively,

$$R_A = \frac{2a - x}{2a} P, \qquad R_B = \frac{x}{2a} P.$$

The horizontal reaction components are equal and opposite, and R'_B is determined from moment about C for the right unloaded part of the frame,

$$R'_A = R'_B = \frac{a}{h} R_B = \frac{x}{2h} P.$$

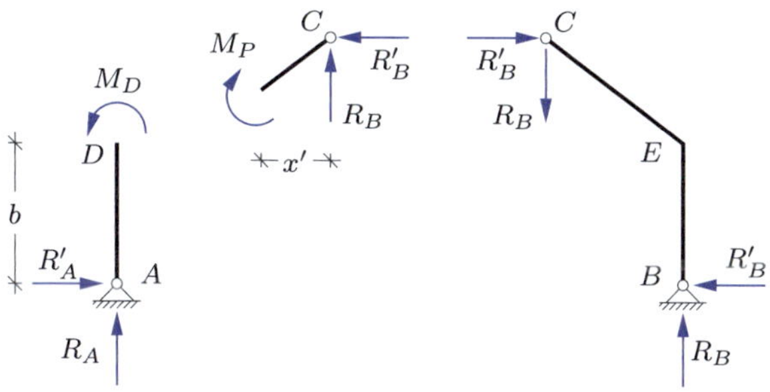

Fig. 3.64: Internal moment by sections in D and at force P.

It is noted that the vertical reaction components are independent of the height h of the central hinge, while the horizontal reactions are inversely proportional to h.

The internal moment varies linearly between the supports, the corners and the point of application of the load. The moment vanishes at the hinges, and thus the distribution is determined by calculating the section moment in the corners D and E, and under the force. The relevant parts of the frame are shown in Fig. 3.64. The moments at the corners D and E are equal, because they are given by the moment of the corresponding horizontal reaction components, $R'_A = R'_B$,

$$M_D = M_E = -b\,R'_A = -\frac{x}{2}\frac{b}{h}\,P\,.$$

Note, that the corner moments M_D and M_E depend on the relative height b/h of the corners.

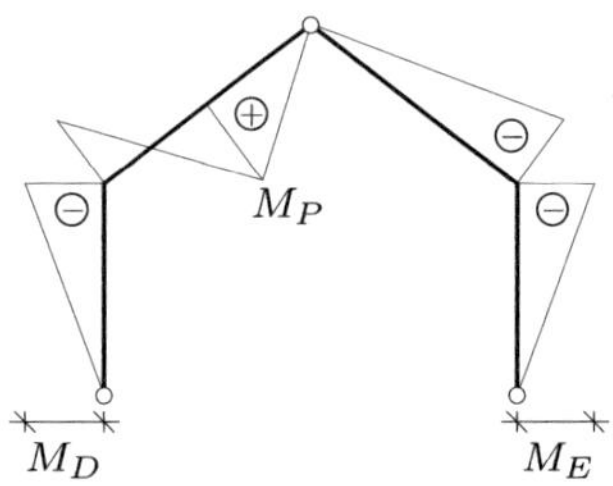

Fig. 3.65: Distribution of moment in three-hinge frame.

Finally, the moment M_P at the section of application of the force is found by considering the upper part of the left half of the frame as shown in the center part of Fig. 3.64. Moment about the point of application of the force gives

$$M_P = x'\,R_B + x'\,\frac{h-b}{a}\,R'_B \quad\Rightarrow\quad M_P = \Big(1 - \frac{b}{2h}\Big)\,\frac{x\,x'}{a}\,P\,.$$

Note, that in the absence of the parenthesis the result would correspond completely to a concentrated transverse force acting on a beam of length a, a problem treated in Example 3.2. The moment M_P attains its maximum value for $x = x' = \frac{1}{2}a$, i.e. when the force acts at the center of DC, and vanishes when the force acts at D or C. The moment distribution is shown in Fig. 3.65. □

Example 3.19. Three-hinge frame with uniformly distributed load. Three-hinge frames must often be analyzed for distributed roof load. The case of a uniformly distributed load is illustrated in Fig. 3.66. The load covers the full width $2a$ of the frame and is normalized as p per unit horizontal length. Thus, each half is acted on by a vertical force of magnitude ap as indicated in the figure. The vertical reactions follow from moment about A and B, respectively – or directly from symmetry – as

$$R_A = R_B = ap\,.$$

The horizontal reactions then follow from moment about the central hinge C for the appropriate half of the frame,

$$R'_A = R'_B = \frac{a^2}{2h}\,p\,.$$

Note, how the reactions are expressed in simple form in terms of the load intensity p per unit horizontal length and the vertical and horizontal frame dimensions a and h.

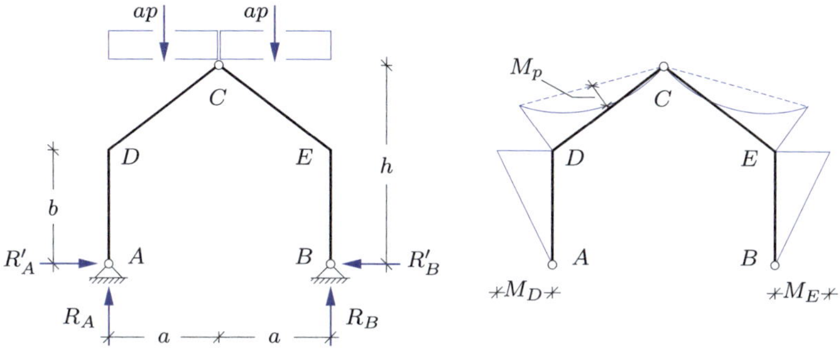

Fig. 3.66: Three-hinge frame with uniformly distributed load.

The moment at the corners D and E follow from the moment of the corresponding horizontal reaction,

$$M_D = M_E = -bR'_A = -\frac{ba^2}{2h}\,p\,.$$

The moment in the roof beams DC and EC consist of a part with linear variation between the zero moment at C and the moment just determined at the corner. In addition there is a parabolic contribution from the distributed load. This moment is determined by its maximum value M_p. The maximum moment is determined by observing that the length of the roof beams is $a/\cos\alpha$, where α is the angle of the roof with horizontal. The transverse load component is $p\cos\alpha$, and the intensity of the transverse load is therefore $(p\cos\alpha)\cos\alpha$. As a result the maximum moment from the uniformly distributed load is

$$M_p = \tfrac{1}{8}\Big(\frac{a}{\cos\alpha}\Big)^2 (p\cos\alpha)\cos\alpha = \tfrac{1}{8}\,a^2 p\,.$$

The moment M_c at the center of the roof beams DC and EC is then found as the sum of M_p and the center value of the linear contribution,

$$M_c = \tfrac{1}{2}M_D + M_p = \Big(1 - \frac{2b}{h}\Big)\,\tfrac{1}{8}\,a^2 p\,.$$

If the parenthesis is set to unity, this result corresponds to the moment at the center of a horizontal beam of length a, treated in Example 3.6. □

3.5.4 Principle of the arch

The discussion of beams and frames in the previous sections, and in particular the three-hinge frame, has illustrated that the distribution of the internal moment may be strongly influenced by the shape of the frame. This has been known since antiquity, in fact long before the development of precise theories for structures, and has inspired the use of arched structures, e.g. for bridges, and vaults and cupolas to cover large rooms. The basic idea is that a structural element carries the load most efficiently, if it acts in compression without appreciable moments. This is discussed in detail in Chapter 10. Here, the simple principle of the arch as used e.g. in bridges is briefly outlined in elementary form.

The idea of the arch is presented as a development from a simple beam. First a simple straight beam carrying a concentrated load is considered as shown in Fig. 3.67. The maximum moment at the center is $M_{\max} = \frac{1}{2}aP$ and the shear forces $Q = \pm\frac{1}{2}P$.

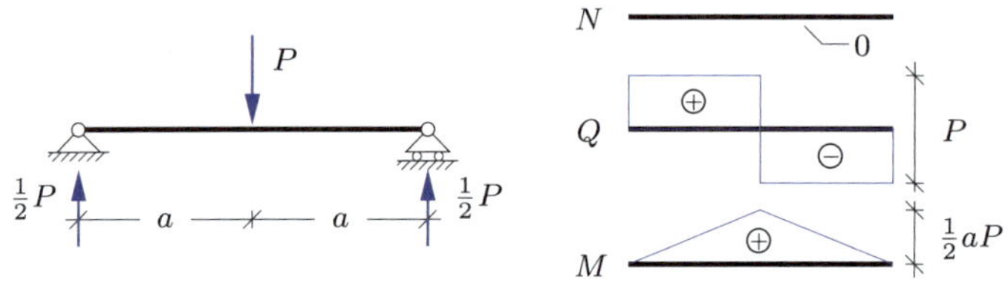

Fig. 3.67: Simply supported angle beam with central load.

To optimize the structure it is desirable to reduce the bending moment. A first attempt may be to incline the two parts of the beam to the left and to the right of the load by an angle α as illustrated in Fig. 3.68. It is seen from the figure that when the support conditions of the simple beam are retained, the reactions remain vertical, and the internal moment at the center is therefore as before. On the other hand the shear force is reduced, and a normal compression force is generated by the inclination of the to beams.

The key to reduction of the moment is the use of inclined members in connection with introduction of fixed simple supports as shown in Fig. 3.69. The structure is rendered statically determinate by a hinge at the center, and the two members then act as bars, each supporting the normal force $\frac{1}{2}P/\sin\alpha$. Note, that the shape of the moment-free structure follows the moment curve of the original beam structure. The mechanism is that the moment is now generated by a constant horizontal force acting via a moment arm, given by the local height of the inclined struts.

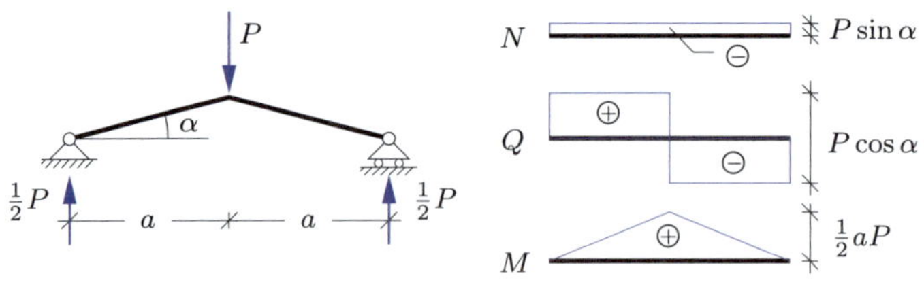

Fig. 3.68: Simply supported angle beam with central load.

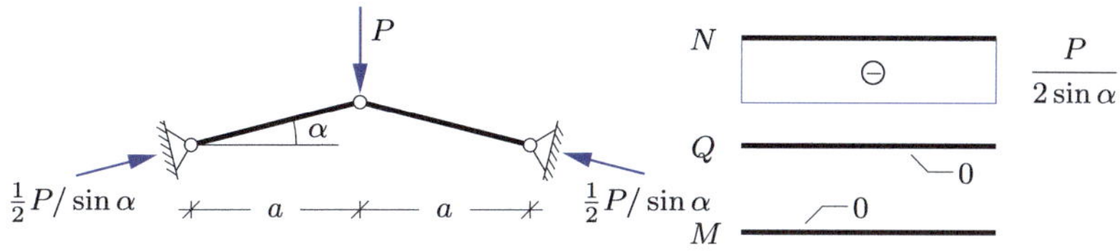

Fig. 3.69: Simply supported bars with center load.

In the case of bridges the weight of the structure often constitutes an important part of the total load, and the optimal shape is therefore suggested by studying the case of a distributed load, e.g. in the form of a uniform load distribution shown in Fig. 3.70. In this case the moment distribution is parabolic with maximum value $M_{\max} = \frac{1}{2}a^2p$.

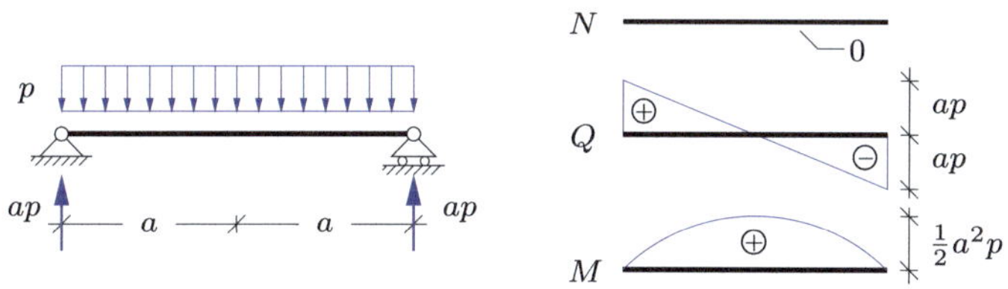

Fig. 3.70: Simply supported beam with uniformly distributed load.

The parabolic moment curve suggests a parabolic arc in connection with change of the supports to provide the necessary horizontal reaction. A parabolic arch with span $2a$ and height h supporting a uniformly distributed load p is illustrated in Fig. 3.71. A hinge is introduced at the center to render the arch statically determinate. The vertical reactions are determined from moment for the full structure about B and A, respectively,

$$R_A = R_B = ap\,. \tag{3.17}$$

The lower part of the figure shows the left half of the arch. It follows from horizontal projection that the horizontal components are equal and opposite, and the magnitude is determined by moment about B,

$$R'_B = R'_C = \frac{a^2}{2h}\,p\,. \tag{3.18}$$

A parabola has the property that when the tangent at an arbitrary point such as B is extended backwards, it intersects a horizontal line through the apex C at half the horizontal distance to B. This is the point where the equivalent concentrated load ap acts, and thus the combined reaction in B follows the direction of the tangent. This argument applies to any point on a uniformly loaded parabola, and thus the load is carried by the parabolic arch solely by

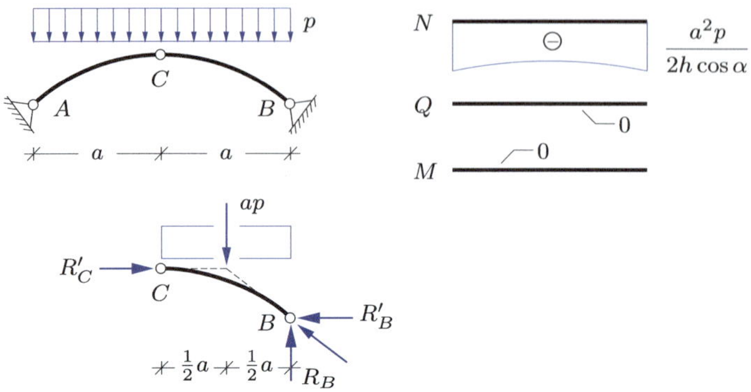

Fig. 3.71: Simply supported arch with uniformly distributed load.

a normal force N without any bending. The normal force has its minimum value at C and its maximum value at the supports A and B,

$$N_{\text{min}} = R'_C = \frac{a^2}{2h}p\,, \qquad N_{\text{max}} = \sqrt{R_B^2 + R_B'^2} = \sqrt{1 + \left(\frac{a}{2h}\right)^2}\,ap\,. \quad (3.19)$$

The variation of the normal force is indicated in the figure. Note, that the normal force in the arch increases with decreasing relative height h/a. Thus, shallow arches require strong horizontal supports. In practice the arch must support loads that deviate somewhat from the uniform distribution used here as illustration. This is typically accommodated by a shape where the height of the cross-section increases towards the supports.

3.6 Exercises

Exercise 3.1. The figure shows a cantilever beam with a vertical force P located at distance a from the support.

a) Determine the reaction components.
b) Determine the distribution of internal forces N, Q and M.

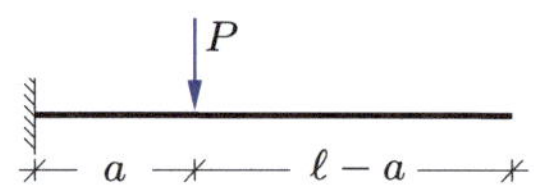

Exercise 3.2. The figure shows a simply supported beam with vertical forces $2P$ and P located at a and $2a$, respectively, from the left support.

a) Determine the reaction components.
b) Determine the distribution of internal forces N, Q and M.

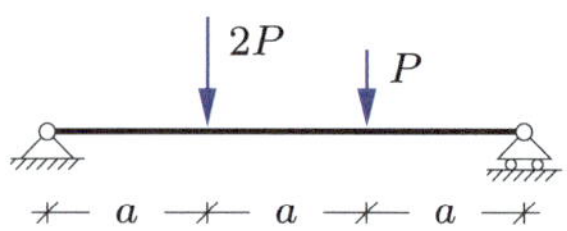

Exercise 3.3. The figure shows an inclined simply supported beam with vertical force P located at horizontal distance c from the left support.

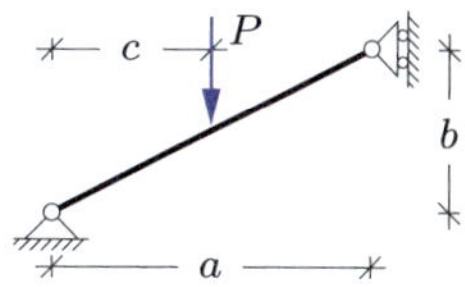

a) Determine the reaction components.

b) Determine the distribution of internal forces N, Q and M.

Exercise 3.4. The figure shows a three span simply supported beam. The two outer spans both have length a and are loaded by tip forces P. The inner span has length $4a$ and a single force $2P$ is acting on middle.

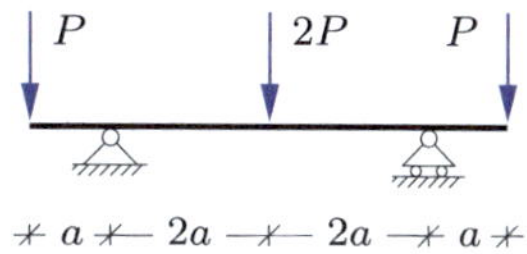

a) Determine the reaction components.

b) Determine the distribution of internal forces N, Q and M.

Exercise 3.5. The figure shows a two span simply supported beam, where the length of the inner beam AB is $4a$, while the length of the cantilever beam BC is $2a$. Vertical load $2P$ and P are acting on the middle of the beams AB and BC, respectively.

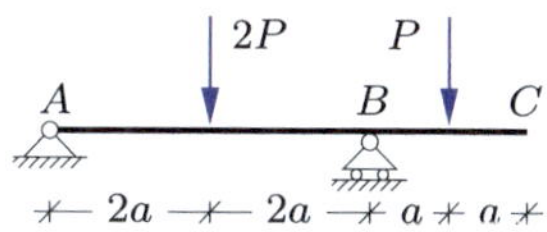

a) Determine the reaction components.

b) Determine the distribution of internal forces N, Q and M.

Exercise 3.6. The figure shows a simply supported beam with a cantilever. A moment M is acting on the tip of the beam.

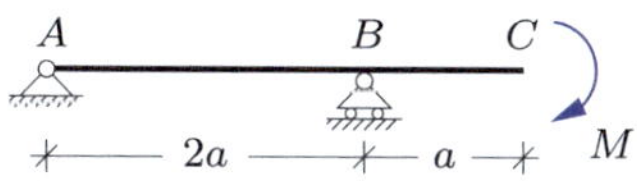

a) Determine the reaction components.

b) Determine the distribution of internal forces N, Q and M.

Exercise 3.7. The figure shows a two span simply supported beam with an 45° inclined support at the right end. A vertical force P is acting on the middle of the left span.

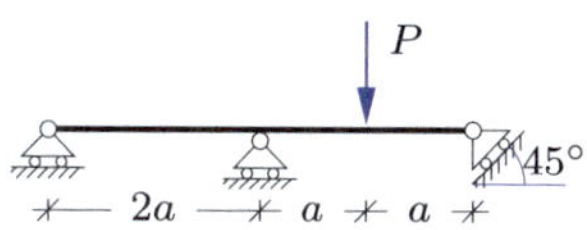

a) Determine the reaction components.

b) Determine the distribution of internal forces N, Q and M.

Exercise 3.8. The figure shows a two span simply supported beam similar to that in Exercise 3.3. A vertical distributed load with intensity p acts on both spans of the beam.

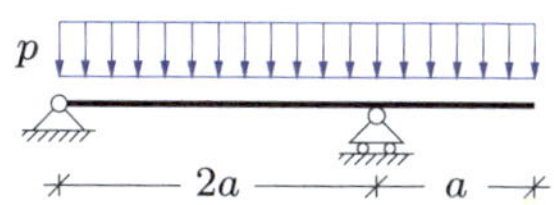

a) Determine the reaction components.

b) Determine the distribution of internal forces N, Q and M.

Exercise 3.9. The figure shows a symmetric three span beam similar to Exercise 3.5. In this case the load equally distributed with intensity p.

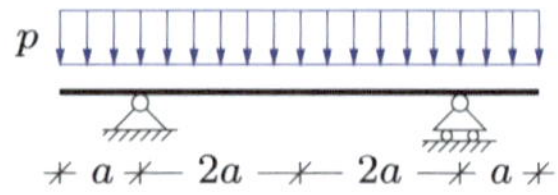

a) Determine the reaction components.

b) Determine the distribution of internal forces N, Q and M.

Exercise 3.10. The figure shows a simply supported beam with a vertical distributed load with intensity p acting over part of the beam. Structure and load is similar to that in Exercise 1.9, where the reactions have been determined.

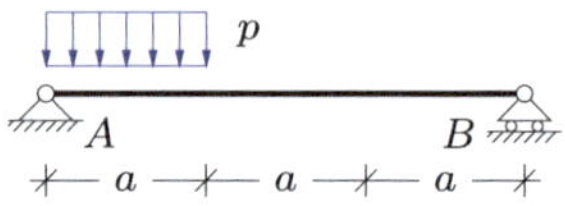

a) Determine the reaction components.

b) Determine the distribution of internal forces N, Q and M.

Exercise 3.11. The figure shows a hinged beam similar to that in Example 3.4. Two load cases are considered: i) distributed load p over the full length of the beam and ii) distributed load p only on AB.

a) Determine the reaction components for each of the two load cases.

b) Determine the distribution of internal forces N, Q and M for each of the two load cases.

c) Consider the sum of the two load cases in the figure as a load combination. This means that the intensity on AB is $2p$, while the intensity on BCD is p. Determine the distribution of the shear force Q and the maximum moment $M_{\max}$ for this load combination.

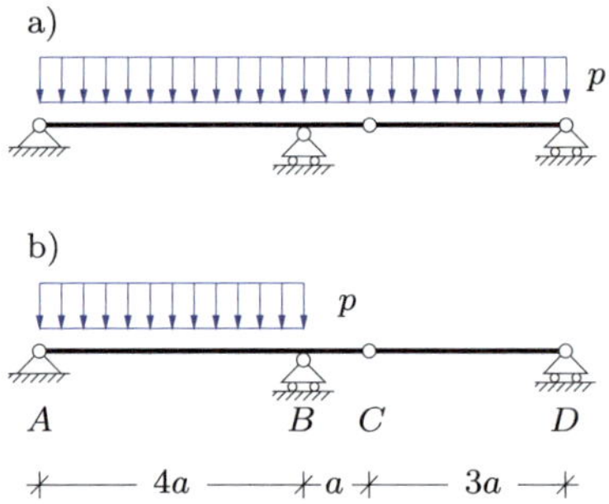

Exercise 3.12. The figure shows a beam of length $2a$, which is fixed in A and simply supported in C. The beam is furthermore hinged at the center in B. A vertical distributed load with intensity p is acting over the full length of the beam.

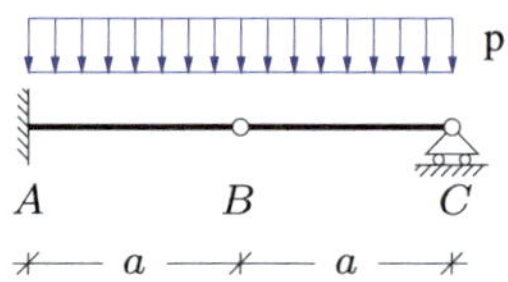

a) Determine the reaction components.

b) Determine the distribution of internal forces N, Q and M.

Exercise 3.13. The figure shows the cantilever frame previously considered in connection with Fig. 3.44. In this exercise the load is distributed on the inclined beam BC with intensity p. In the first load case the distributed load acts vertically, for example representing snow on the roof. In the second load case the distributed load acts normal to BC. This type of loading could arise from wind pressure.

a) Determine the reaction components for both load cases.

b) For the first load case find the transverse intensity per length of the beam.

c) Determine the distribution of the moment M for both load cases.

d) Find the maximum moment $M_{\max}$.

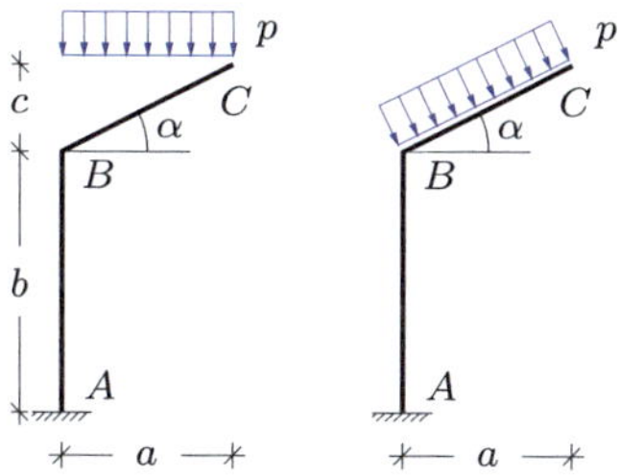

Exercise 3.14. The figure shows an angle frame with a fixed support at A. The height of the frame is $3a$, while the width is $2a$. A hinge is placed in the joint C and a bar element is connected to the two beam elements at B and D, respectively. A distributed transverse load with intensity p acts on the horizontal beam CE.

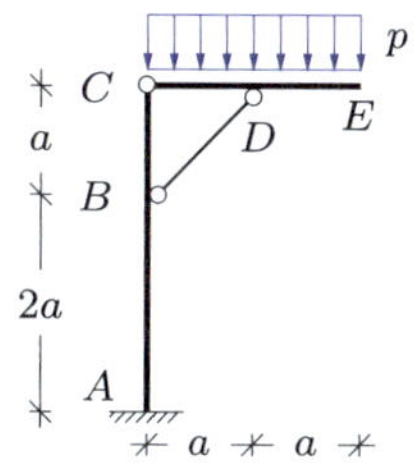

a) Determine the reaction components in A.

b) Determine the bar force N_{BD} by placing a section through C and BD.

c) Determine the distribution of the internal forces N, Q and M.

d) Find the maximum moment $M_{\max}$.

Exercise 3.15. The figure shows an angle frame with a simple support in A and a simple support in B permitting motion at an angle of 45°. The width of the frame is $2a$ and the height is a. A distributed load with intensity p acts vertically on the horizontal beam CB.

a) Determine the reactions. Note that the reaction in B is conveniently represented by its horizontal and vertical components $R'_B = R_B$.

b) Determine the internal forces N, Q and M at the sections A, B and C.

c) Determine the distribution of the internal forces N, Q and M.

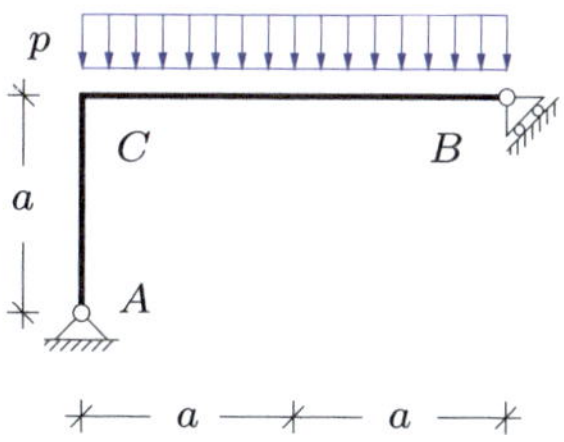

Exercise 3.16. The figure shows a T-frame with a simple sixed support in A and a simple roller bearing at B, permitting motion at 45°. A vertical force P is acting at the tip in D.

a) Determine the reaction components. Represent the inclined reaction in B by its horizontal and vertical components.

b) Determine the distribution of internal forces N, Q and M.

c) Isolate the joint C and verify force and moment equilibrium.

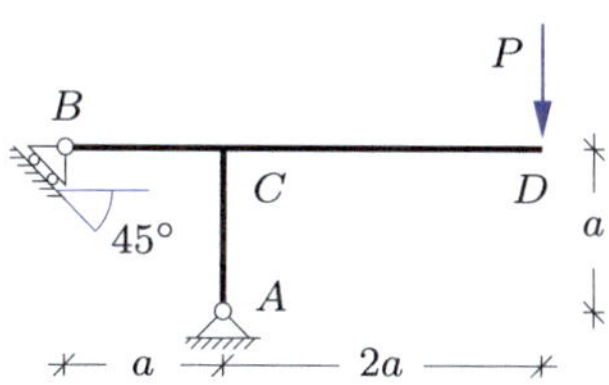

Exercise 3.17. The figure shows a T-frame with width $a + 2a$ and height b. As in Examples 3.16 and 3.17, where $b = a$, two different support conditions in B are considered: horizontal and vertical roller bearing, respectively. Find the reactions and determine the distri-

bution of the internal forces for both T-frames. Isolate the joint C and verify force and moment equilibrium.

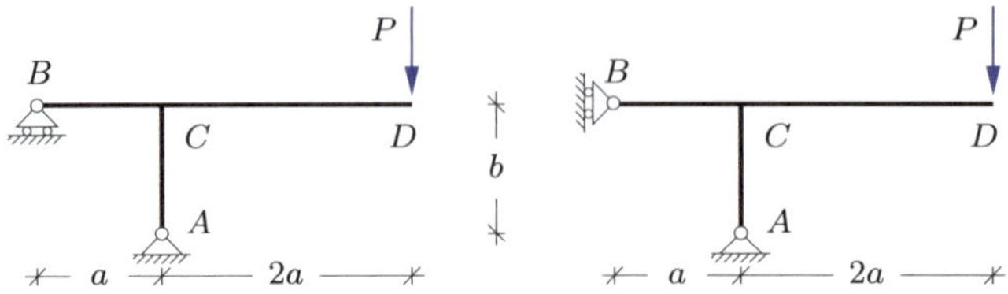

Exercise 3.18. The figure shows a three-hinge frame with an internal hinge placed in E at the center of the horizontal beam BC. The frame is loaded by a uniformly distributed load with intensity p on BC.

a) Find the reaction components.

b) Determine the internal forces N, Q and M at B.

c) Determine the distribution of the internal forces N, Q and M, and find the maximum moment $M_{\max}$.

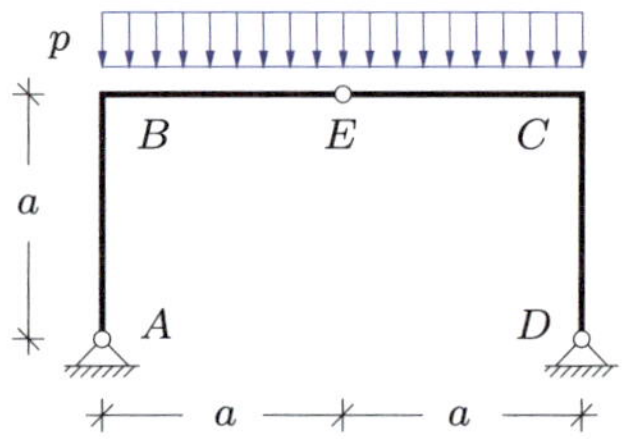

Exercise 3.19. The figure shows a three-hinge frame with an internal hinge placed in E at the center of the horizontal beam BC. In this case the frame is loaded by two horizontal forces P acting at the joints B and C.

a) Find the reaction components.

b) Determine the internal forces N, Q and M at the joint B.

c) Determine the distribution of the internal forces N, Q and M.

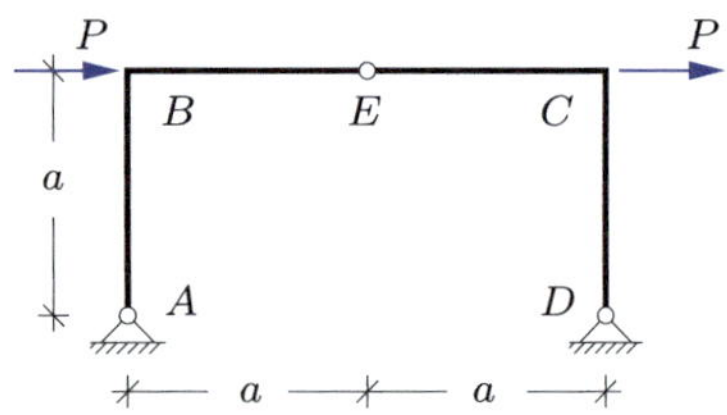

Exercise 3.20. The figure shows a three-hinge frame similar to that in Exercises 4.18 and 4.19. Consider in this case a load combination, consiting of the distributed load p from Exercise 3.18 and the two horizontal forces P from Exercise 4.19. Let $P = ap$. If Exercises 3.18 and 3.19 have been solved prior to this exercise, the concept of superimposing load cases, as described in Section 3.4.2, can be used.

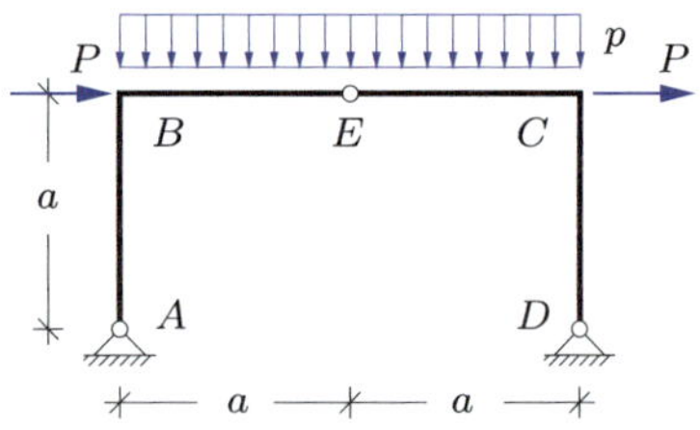

a) Determine the reaction components.

b) Determine the distribution of the internal forces N, Q and M in the frame.

Exercise 3.21. The figure shows a three-hinge frame with an internal hinge located at the top of the inclined roof. The frame is loaded by a uniformly distributed load with intensity p on the left part DC of the inclined roof. The load is acting in the vertical direction and the intensity p is force per horizontal length.

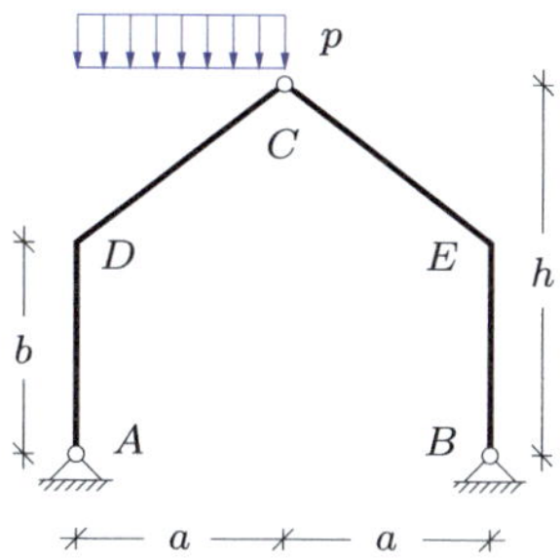

a) Determine the reaction components.
b) Determine the moment M at the joints D and E.
c) Determine the distribution of the internal moment M in the frame.

Exercise 3.22. The figure shows a three-hinge frame with the internal hinge located at the top of the inclined roof. The frame is loaded by a uniformly distributed load with intensity p on the left part DC of the inclined roof. The distributed is acting perpendicular on the beam DC and the intensity p is therefore per inclined length.

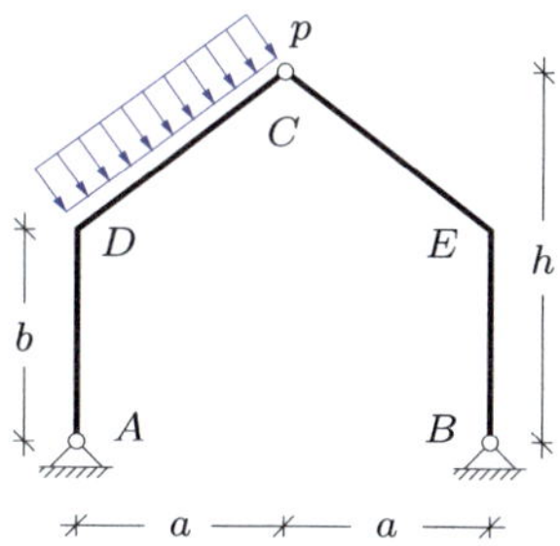

a) Determine the reaction components.
b) Determine the moment M at the joints D and E.
c) Determine the distribution of the internal moment M in the frame.

Deformation of Beams and Frames

4

The internal forces in beams and frames lead to deformations, and methods for analysis of these deformations are the subject of the present chapter. Typically the satisfactory performance of a structure will impose some limitation on the displacements. However, the deformation of beams also plays another important role, namely in the distribution of the internal forces in statically indeterminate structures. A statically indeterminate structure permits several ways of distributing the internal forces to carry a given load, and the actual distribution depends on the relative stiffness of the individual parts of the structure. These two roles of deformations and displacements are illustrated in Fig. 4.1 showing a horizontal beam supporting a distributed load. As discussed in the previous chapter the load creates a moment in the beam, and this moment in turn leads to curvature of the beam, resulting in transverse displacement. Satisfactory performance of the beam may impose a limit on the transverse displacement w. It is seen from the figure that the transverse displacements lead to rotation of the ends of the beam. If the beam were part of a larger structure, the connection to this structure via the ends of the beam would have to account for the rotation of the beam end. The transverse displacements along the beam and the rotation of the beam ends are parts of the same problem studied in this chapter. As indicated in the figure the deformation depends on the load p, the stiffness represented by the elastic modulus E, the beam length ℓ, the cross-section represented by the section height h, and possibly other parameters as well.

S. Krenk, J. Høgsberg, *Statics and Mechanics of Structures*,
DOI 10.1007/978-94-007-6113-1_4,

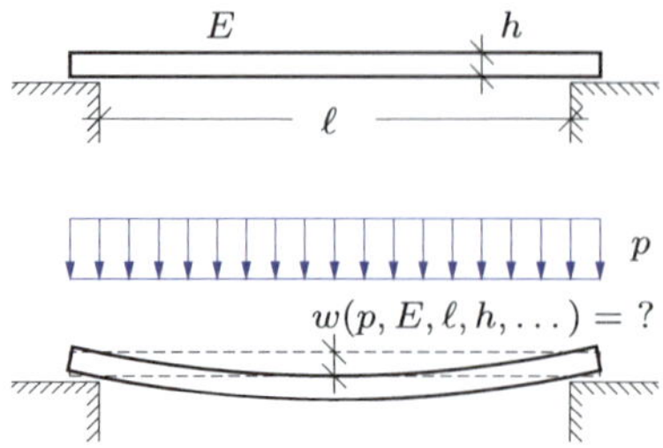

Fig. 4.1: Transverse deformation of simply supported beam.

The main mechanism in the deformation of beams is curvature generated by a bending moment M. The basic mechanism was identified and described quite clearly by Robert Hooke (1678). A bending moment will compress one side and extend the other as seen in the deformed beam in Fig. 4.1. For elastic beams the extension and compression are governed by the elastic relation between stress and strain from Section 2.4. A modern presentation of Hooke's theory of beam bending is presented in Section 4.1. In most cases of structural analysis the displacements are quite small, and a linearized form of the theory based on the undeformed geometry of the structure can then be used. The linear form of the theory is developed in the form of differential equations in Section 4.2.

In addition to bending, beams may also deform due to extension and shear, generated by the normal force N and the shear force Q, respectively. The extension due to a normal force has already been treated in Section 2.4 in connection with truss structures. The effect of deformation due to a shear force Q is treated in Section 4.3, where it is demonstrated, that this effect will often be negligible for slender beams.

In principle, the displacements of elastic beam and frame structures can be determined by solving the corresponding differential equations. However, for frames this would involve special transition conditions at joints, and even for beams the integration of the differential equations may be quite laborious for non-trivial load distributions and non-homogeneous beams. Therefore the principle of virtual work plays an indispensable role in the analysis of displacements of structures. In Section 4.4 the principle of virtual work – encountered in a limited form in connection with bars and trusses in Section 2.4.3 – is extended to beams and frames.

4.1 Bending of elastic beams

The theory for bending of elastic beams is most easily developed by first considering the case of homogeneous bending of a beam, i.e. the bending of a beam by applying moments of equal magnitude but opposite orientation at the two ends of a homogeneous beam. This simple set-up serves to identify the mechanism of bending in a precise way. The theory is then linearized corresponding to 'small displacements' and extended to nonhomogeneous bending.

4.1.1 Homogeneous bending

Figure 4.2a illustrates a straight simply supported homogeneous beam with a cross-section that is symmetric with respect to the plane of the figure. The beam is loaded by moments of equal magnitude but opposite orientation at the beam ends. The two moments are in equilibrium, and there is therefore no reactions. The upper side of the beam is compressed and the lower extended, and the beam therefore deforms as shown in Fig. 4.2b. Somewhere between the upper and lower part of the beam there must be a plane where the beam retains its original length. This is the so-called neutral plane. The axis of the beam is taken as the intersection of the neutral plane and the plane of symmetry, i.e. the plane of the figure, and s denotes the arc-length along this axis. When introducing a section at an arbitrary point s, equilibrium implies that the internal moment is equal to the imposed moment M, and thus

$$M(s) \equiv M \tag{4.1}$$

for any value of s. All cross-sections have the same internal moment and thereby the same state of deformation. Therefore the beam axis is bent into a circle. The center C and the radius R of the circle are indicated in the figure.

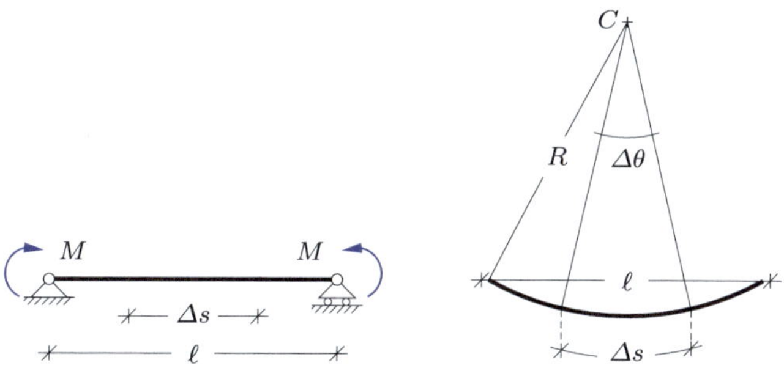

Fig. 4.2: Homogeneous bending of beam.

In connection with bending of beams it is convenient to describe the deformation by the curvature $\kappa = 1/R$, and in order to obtain a quantitative theory the relation between the bending moment M and the curvature κ must be established. Figure 4.3a shows a slice of the beam with initial length Δs. The neutral axis retains its original length, and in the deformed state

$$\kappa = \frac{1}{R} = \frac{\Delta\theta}{\Delta s}, \tag{4.2}$$

where $\Delta\theta$ is the angle between the two cross-sections. Now, consider a 'fiber' located at the distance z below the neutral axis. The initial length of this fiber is Δs. After deformation it is bent into a circle with radius $R + z$ and corresponds to the center-angle $\Delta\theta$. Thus the length of this fiber after deformation is

$$\Delta s_* = (R + z)\,\Delta\theta = (R + z)\,\kappa\,\Delta s\,, \tag{4.3}$$

where the angle $\Delta\theta$ was substituted from (4.2). The elongation corresponds to the normal strain

$$\varepsilon = \frac{\Delta s_* - \Delta s}{\Delta s} = \kappa\, z\,. \tag{4.4}$$

Thus, the normal strain is proportional to the curvature κ and to the distance z from the neutral axis.

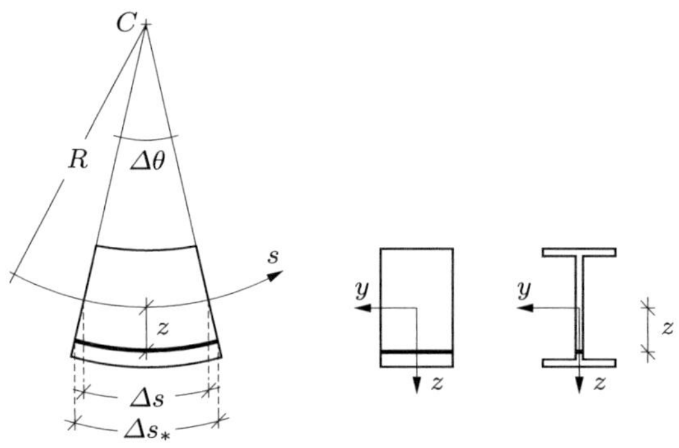

Fig. 4.3: Extension of fiber in beam bending.

If the material is linear elastic with modulus of elasticity E the normal stress follows from the strain relation (4.4)

$$\sigma = E\,\varepsilon = E\,\kappa\, z\,. \tag{4.5}$$

The stress σ and the modulus of elasticity are expressions of force per unit area, typically expressed in terms of Pascal, $[\mathrm{Pa}] = [\mathrm{N/m^2}]$. This expression for the normal stress leads to a condition that determines the location of the neutral axis and an expression for the bending moment in terms of the curvature κ.

In the present problem the loading is assumed to be pure bending. Thus, there is no normal force in the beam, and it follows from the stress relation (4.5) that this corresponds to

$$N = \int_A \sigma \, dA = \kappa \int_A E \, z \, dA = 0 \,. \tag{4.6}$$

If this condition is not satisfied the origin of the z-axis is not on the neutral axis. If the section is symmetric about the plane of bending but does not have up-down symmetry, the location of the neutral axis is easily determined by introducing a preliminary transverse coordinate z'. The coordinate z, centered at the neutral axis is then determined from the condition (4.6), when substituting $z' = z + z_0$ instead of z. A general analysis of non-symmetric and inhomogeneous cross-sections is presented in Chapter 10.

The bending moment generated by the normal stress σ from (4.5) is determined by adding the contributions from all infinitesimal areas dA. The force on the infinitesimal area is σdA, and the corresponding moment then is $z(\sigma dA)$, when z is the distance to the neutral axis. This gives the bending moment as

$$M = \int_A z \, \sigma \, dA = \kappa \int_A z^2 \, E \, dA \,. \tag{4.7}$$

It is seen from this relation that the contribution from the area element dA is weighted by the elastic stiffness modulus E. It is not a severe complication to include variable elasticity modulus in the cross section integral, but in order to illustrate the basic theory with minimal complication here, the modulus of elasticity is here assumed constant over the cross-section, and the extension to non-homogeneous material properties is postponed to the discussion of general cross-section analysis in Chapter 10. Thus, for a constant modulus of elasticity the elastic bending relation takes the form

$$M = E \, I_z \, \kappa \,. \tag{4.8}$$

In this relation EI_z is the bending stiffness of the beam. It consists of the product of the modulus of elasticity E and the geometric quantity

$$I_z = \int_A z^2 \, dA \,, \tag{4.9}$$

constituting the moment of inertia about the neutral axis. The moment of inertia of the cross-section depends on the shape as well as the size of the cross-section as illustrated in the following example.

Example 4.1. Moment of inertia for rectangle and I-profile. Figure 4.4a shows the distribution of the coordinate z over the cross-section of a rectangular beam of height h and width b. The neutral axis is along the horizontal axis of symmetry, and the bending moment of inertia of the rectangular cross-section then is

$$I_z = \int_A z^2 \, dA = b \int_{-h/2}^{h/2} z^2 \, dz = \tfrac{1}{12} h^3 b = \tfrac{1}{12} h^2 A \,.$$

The rectangular beam has the largest bending stiffness, when oriented such that $h \geq b$, as normally seen in structures. The result for the rectangular cross section can be used to determine the moment of inertia for various types of cross sections composed of rectangular parts, as demonstrated in Exercise 4.1 for a thin-walled box section.

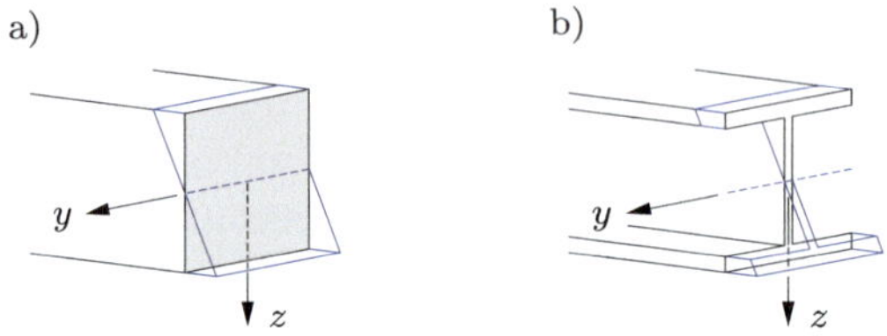

Fig. 4.4: Plane bending deformation: a) Rectangular beam, b) I-section beam.

Figure 4.4b shows an I-beam, consisting of two flanges connected by a web. The web is usually thin and contributes only little to the area and bending moment of inertia. It is therefore neglected in the derivations below. The neutral axis is the axis of symmetry, and if the area is assumed to be mainly located in the two flanges of thickness t, the bending moment of inertia is

$$I_z = \int_A z^2\, dA \simeq 2b \int_{(h-t)/2}^{(h+t)/2} z^2\, dz = \frac{b}{12}\Big((h+t)^3 - (h-t)^3\Big) \simeq \tfrac{1}{2}h^2 tb = \tfrac{1}{4}h^2 A\,,$$

where higher order terms in the thickness, i.e. terms containing the powers t^2 or t^3, are omitted because $t \ll h, b$ for thin-walled cross sections. It is seen that in the I-beam a given area provides three times greater bending stiffness than in a rectangular section of the same height. The stress distribution is also more favorable in the I-beam, which is common in structures. □

4.1.2 Linear kinematic relations

For a general plane curve the curvature is defined as the rate of change of the angle of the tangent with respect to a fixed direction, as illustrated in Fig. 4.5. Thus, the curvature is defined by the limit

$$\kappa(s) = \frac{1}{R(s)} = \lim_{\Delta s \to 0} \frac{\Delta\theta}{\Delta s} = \frac{d\theta}{ds}\,. \tag{4.10}$$

In linear beam theory the displacements are considered 'small' with respect to the dimensions of the structure. This implies that the change of angle is small, and a linearized representation of the curvature can then be used.

Figure 4.6 illustrates the deformed state of an initially straight beam. The x-axis defines the original beam axis, and the transverse displacement $w(x)$ is positive in the direction of the z-axis. The angle θ is defined via the relation

$$\sin\theta = -\frac{dw}{ds}\,. \tag{4.11}$$

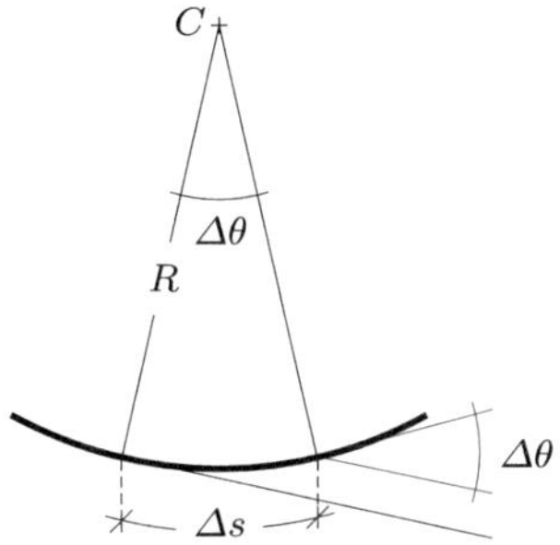

Fig. 4.5: Curvature as change in rotation.

Differentiation of this equation with respect to s gives the following expression for the curvature

$$\kappa = \frac{d\theta}{ds} = -\frac{1}{\cos\theta}\frac{d^2w}{ds^2}. \tag{4.12}$$

When the slope of the deformed beam is limited $dx/ds = \cos\theta \simeq 1$, and the original x-coordinate can be used instead of the arc-length s. This leads to the linearized expressions

$$\theta \simeq -\frac{dw(x)}{dx}, \qquad \kappa \simeq \frac{d\theta(x)}{dx} \simeq -\frac{d^2w(x)}{dx^2}. \tag{4.13}$$

These expressions are used with equality sign in technical 'small displacement' beam theory.

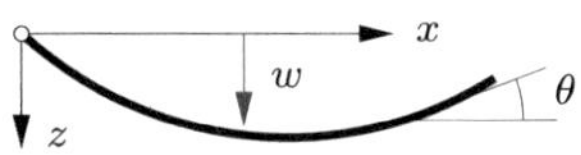

Fig. 4.6: Sign convention of beam deformation.

The kinematic relations (4.13) give approximate definitions of the angle θ to the tangent of the beam axis and the corresponding curvature κ. However, the axial strain from beam bending is defined in terms of the relative rotation of two neighboring cross-sections as illustrated for homogeneous bending in Fig. 4.3. In the case of homogeneous bending the cross-sections remain orthogonal to the beam axis in the deformed state. Thus, the relative rotation of

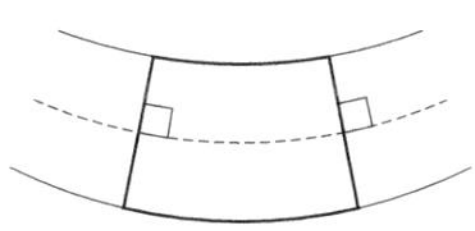

Fig. 4.7: Cross-section perpendicular to deformed beam axis.

neighboring cross-sections is determined by the rotation of the tangent to the beam axis between two sections, as shown in Fig. 4.7. In non-homogeneous bending the change of the moment along the beam generates a shear force, which in turn leads to an inclination of the cross-sections. However, the effect of the deviation of the cross-sections from the orthogonal position is often small, and therefore normal cross-sections are often assumed in the development of technical beam theories. This approach is taken in the following section on the so-called Bernoulli beam theory, while the effect of shear flexibility is addressed in Section 4.3.

Example 4.2. Linearization error for homogeneous bending. According to the exact beam theory a beam with constant internal moment M bends into a circular shape with radius

$$R = \frac{1}{\kappa} = \frac{EI}{M},$$

as illustrated in Fig. 4.2. The angle θ between the cross-sections at the supports follows from trigonometry as

$$\sin(\tfrac{1}{2}\theta) = \frac{\ell}{2R} = \tfrac{1}{2}\kappa\ell\,.$$

The angle θ may be expanded in a Taylor series as

$$\tfrac{1}{2}\theta = \arcsin(\tfrac{1}{2}\kappa\ell) = (\tfrac{1}{2}\kappa\ell) + \tfrac{1}{6}(\tfrac{1}{2}\kappa\ell)^3 + \cdots,$$

showing that for small curvature $\theta \simeq \kappa\ell$.

The displacement w in the middle follows from the circular deformed shape as

$$w = R\big[1 - \cos(\tfrac{1}{2}\theta)\big] = \tfrac{1}{2}R(\tfrac{1}{2}\theta)^2\big[1 - \tfrac{1}{12}(\tfrac{1}{2}\theta)^2 + \cdots\big],$$

where the Taylor expansion of the cosine function has been introduced. Substitution of the expansion for $\frac{1}{2}\theta$ in terms of $(\frac{1}{2}\kappa\ell)$ gives

$$w = \tfrac{1}{8}\kappa\ell^2\big[1 + \tfrac{1}{4}(\tfrac{1}{2}\kappa\ell)^2 + \cdots\big].$$

The first term is the result from a linearized analysis,

$$w_{\rm lin} = \tfrac{1}{8}\kappa\ell^2 = \frac{\ell^2 M}{8EI}\,.$$

This is the result that would be obtained from the approximate theory based on the linearized rotation and curvature relations (4.13).

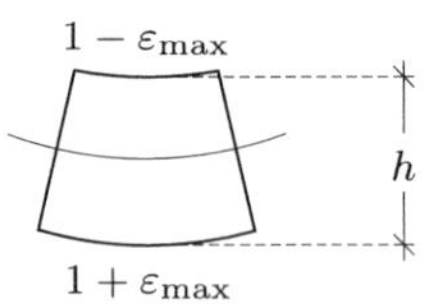

Fig. 4.8: Bending of symmetric unit beam element.

The relative error is represented by the second term in the square brackets, $\frac{1}{4}(\frac{1}{2}\kappa\ell)^2$. This term can be related directly to the local deformation of the beam. The top of the beam is in compression and the bottom in tension as illustrated in Fig. 4.8. The curvature can be expressed in terms of the beam height and the maximum strain by use of the relation (4.4). The maximum strain $\varepsilon_{\rm max}$ occurs at the distance $\frac{1}{2}h$ from the beam axis, and thus $\kappa = 2\varepsilon_{\rm max}/h$. The leading relative error term can then be expressed as

$$\frac{w - w_{\rm lin}}{w_{\rm lin}} \simeq \left(\varepsilon_{\rm max}\frac{\ell}{2h}\right)^2$$

Thus, for a limited maximal strain in the material – e.g. $\varepsilon_{\rm max} \simeq 0.002$ – the error in the linearized theory will only be important for very slender beams. □

4.2 Bernoulli beam theory

A full beam theory must deal with the statics and kinematics of the beam. The statics deals with internal forces and their equilibrium with the imposed loads. Kinematics is the description of displacements and the related measures of deformation such as the curvature. The deformation generated by the internal forces depends on the mechanical behavior of the material of the beam, e.g. linear elasticity. Finally, the beam must be adequately supported to permit transfer of the loads to the supports. This section presents these four aspects in the context of the so-called Bernoulli beam theory, which is a technical theory of elastic beams in which shear deformations are neglected. The extension to shear-flexible beams is discussed in Section 4.3. In the linear theory of beams the normal force and the associated axial deformation is uncoupled from the transverse displacements generated by bending. The normal force and axial deformation has been dealt with in connection with bars, and the presentation here is therefore limited to the beam bending problem in terms of transverse displacements and bending moments and shear forces, as illustrated in Fig. 4.9.

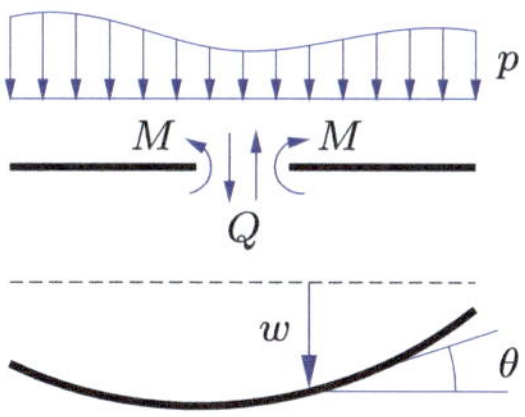

Fig. 4.9: Static and kinematic parameters in Bernoulli beam theory.

Statics

The statics of the Bernoulli beam theory consists of the equilibrium equations for the shear force $Q(x)$ and the internal bending moment $M(x)$. These equilibrium equations were treated in detail in Section 3.3.1 and are just summarized here. The shear force must secure vertical equilibrium, whereby

$$\frac{dQ(x)}{dx} = -p(x). \tag{4.14}$$

Similarly the moment must secure equilibrium for rotation of a slice of the beam, leading to

$$\frac{dM(x)}{dx} = Q(x). \tag{4.15}$$

Bernoulli beam theory is based on the simplifying assumption that the shear force $Q(x)$ does not contribute directly to the deformation of the beam. It may therefore sometimes be of interest to eliminate the shear force from the equilibrium equations to obtain a direct relation between the internal moment and the external load,

$$\frac{d^2M(x)}{dx^2} + p(x) = 0. \tag{4.16}$$

For statically determinate beams the boundary conditions permit solution of these differential equations for the statics without consideration of the kinematic relations. However, in general a full description of the beam also requires determination of the kinematic variables, i.e. displacement $w(x)$, rotation $\theta(x)$ and curvature $\kappa(x)$.

Kinematics

In the Bernoulli beam theory the kinematics, i.e. the displacement and deformation, of the beam is described completely by the transverse displacement $w(x)$. Within the approximation of the Bernoulli beam theory the rotation of the cross section is equal to the rotation of the tangent of the beam axis, given by the linearized expression

$$\theta(x) = -\frac{dw(x)}{dx}. \tag{4.17}$$

The curvature is the rate of change of the rotation angle $\theta(x)$, and within the linearized theory this is

$$\kappa(x) = \frac{d\theta(x)}{dx} = -\frac{d^2w(x)}{dx^2}. \tag{4.18}$$

The angle θ may appear in the support conditions, and the curvature $\kappa(x)$ is related to the internal moment $M(x)$.

Elasticity

In the Bernoulli beam theory the deformation from the shear force is neglected, leaving only the curvature generated by the bending moment. For linear material behavior, such as linear elasticity, this implies proportionality between the curvature $\kappa(x)$ and the bending moment $M(x)$. It was demonstrated by the case of homogeneous bending in Section 4.1 that this relation has the form

$$M(x) = EI_z\,\kappa(x) = -\,EI_z\,\frac{d^2w(x)}{dx^2}\,. \tag{4.19}$$

Relations between the static quantities and the deformation measures are often called constitutive relations. In this relation the bending stiffness EI_z consists of a material parameter, representing the elastic modulus E of the beam material, and a geometric parameter I_z, which for a homogeneous material distribution is the moment of inertia about the neutral axis. The more general case of non-homogeneous material distribution over the beam cross-section is treated in Chapter 10.

Support conditions

The beam must be supported by at least the number of constraints that prevent free motion. However, many structures have more supports than strictly needed to fix the structure in the plane or in space. Extra supports and extra connections between structural elements typically increase the stiffness of the structure, while rendering the structure statically indeterminate.

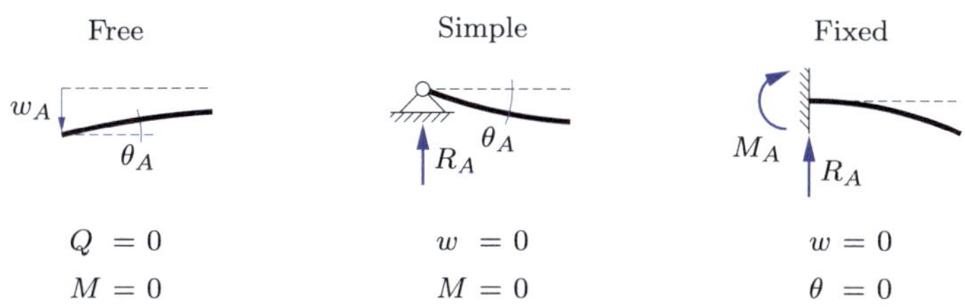

Fig. 4.10: Support conditions for Bernoulli beams.

Three typical support conditions are illustrated in Fig. 4.10 with reference to the left end A of a beam. Additional supports have been discussed previously in connection with Fig. 1.20. The first support condition in Fig. 4.10 is a free end, at which the shear force and the moment vanish, $Q = 0$ and $M = 0$. In the case of a statically determinate beam $Q(x)$ and $M(x)$ are the unknown functions, and this support condition is expressed directly in terms of these functions at the point A. For a statically indeterminate beam the problem is formulated in terms of the transverse displacement $w(x)$, and the boundary conditions must therefore be expressed in terms of the transverse displacement. The expression for the moment follows directly from the constitutive relation (4.19) as

$$M_A = -\,EI_z\,\frac{d^2w}{dx^2}\bigg|_A\,. \tag{4.20}$$

The expression for the shear force follows from differentiation of the moment equation according to (4.15),

$$Q_A = -\frac{d}{dx}\Big(EI_z\,\frac{d^2w}{dx^2}\Big)_A . \tag{4.21}$$

For a concentrated load at A the moment and/or the shear force are simply defined by the concentrated external load.

Figure 4.10 also illustrates the simple support, in which the transverse displacement vanishes, $w_A = 0$, together with the moment, $M_A = 0$. For problems formulated in terms of the transverse displacement $w(x)$, the moment is expressed by (4.20). Finally, the figure also illustrates the fixed end, representing constraint of the transverse displacement, $w_A = 0$, and constraint of the rotation, $\theta_A = 0$.

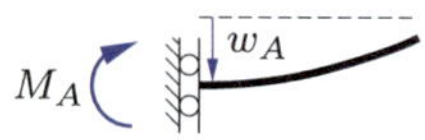

Fig. 4.11: Fixed support with vertical rollers.

The supports in Fig. 4.10 can be combined, as demonstrated by the fixed support with vertical rollers in Fig. 4.11. The vertical rollers represent the unconstrained transverse displacement of the free end in Fig. 4.10a and the constrained rotation of the fixed support in Fig. 4.10c.

4.2.1 Statically determinate beams

In a statically determinate beam the support conditions permit that the distribution of the internal forces $Q(x)$ and $M(x)$ can be determined without reference to the kinematics of the beam. Thus, a statically determinate beam is typically analyzed by first determining $Q(x)$ and $M(x)$ – either from the differential relations (4.14)–(4.16) or by the direct equilibrium based methods developed in Chapter 3. Once the statics part of the problem has been solved, the displacement $w(x)$ can be determined by integration of the moment relation (4.19). The principle may be described as solution of the second order static equation

$$\frac{d^2M(x)}{dx^2} + p(x) = 0\,, \tag{4.22}$$

followed by solution of the second order kinematic equation

$$\frac{d^2w(x)}{dx^2} + \frac{M(x)}{EI_z} = 0\,, \tag{4.23}$$

although the static solution is often obtained by a static procedure based on sections. In particular concentrated or discontinuous loads may introduce complications for a purely mathematical integration. The analysis of statically determinate beams is illustrated by the following examples.

Example 4.3. Simply supported beam with end moment. Figure 4.12 shows a simply supported beam of length ℓ with a moment M_B acting at the right support in B. There is no load acting along the beam, and the moment distribution is therefore linear as shown in Fig. 4.12. The expression for the moment is

$$M(x) = M_B \frac{x}{\ell}.$$

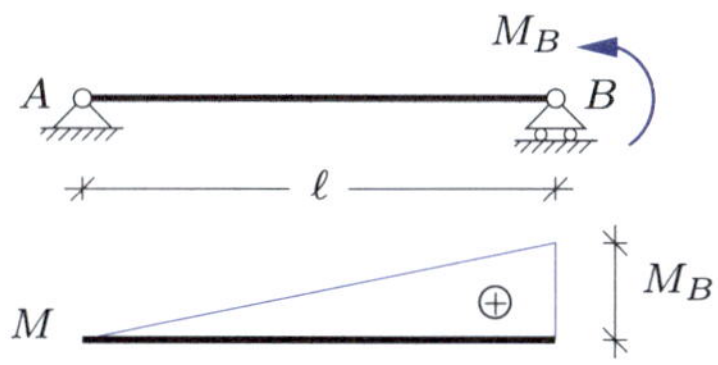

Fig. 4.12: End moment on simply supported beam.

The curvature is given by the constitutive relation (4.23) as

$$\frac{d^2w(x)}{dx^2} = -\frac{M(x)}{EI_z} = -\frac{M_B}{EI_z}\frac{x}{\ell}.$$

The displacement $w(x)$ then follows by double integration as

$$w = -\frac{1}{6}\frac{M_B\ell^2}{EI_z}\left(\frac{x}{\ell}\right)^3 + C_0 + C_1 x,$$

where C_0 and C_1 are the two arbitrary constants of integration. It is seen that a linear moment distribution implies a cubic solution for the displacement. The two constants are determined by the two kinematic boundary conditions for the simple supports in A and B,

$$w(0) = w(\ell) = 0.$$

The first condition $w(0) = 0$ directly yields that $C_0 = 0$. Hereafter, the second condition $w(\ell) = 0$ gives

$$-\frac{1}{6}\frac{M_B\ell^2}{EI_z} + C_1\ell = 0 \qquad \Rightarrow \qquad C_1 = \frac{1}{6}\frac{M_B\ell}{EI_z}.$$

Now that the two arbitrary constants have been determined the final expression for the displacement can be written as

$$w(x) = \frac{1}{6}\frac{M_B\ell^2}{EI_z}\frac{x}{\ell}\left[1 - \left(\frac{x}{\ell}\right)^2\right].$$

The first factor gives the magnitude of the displacement, while the second factor represents the distribution. The rotation is obtained from the derivative of the displacement

$$\theta(x) = -\frac{dw(x)}{dx} = -\frac{1}{6}\frac{M_B\ell}{EI_z}\left[1 - 3\left(\frac{x}{\ell}\right)^2\right].$$

This gives the rotation at the beam ends as

$$\theta_A = \theta(0) = -\frac{1}{6}\frac{M_B \ell}{EI_z} \quad , \quad \theta_B = \theta(\ell) = \frac{1}{3}\frac{M_B \ell}{EI_z}.$$

It is noted that the magnitude of the rotation θ_B at the end with the applied moment is the double of the rotation at the other end.

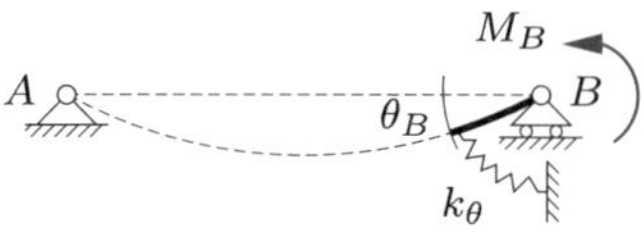

Fig. 4.13: Equivalent rotation spring with stiffness k_θ.

The expression for the rotation θ_B can be written as $M_B = k_\theta \theta_B$, where k_θ is an equivalent rotation spring stiffness parameter given by

$$k_\theta = \frac{3EI_z}{\ell},$$

as illustrated in Fig. 4.13. The rotational stiffness k_θ has the dimension of a moment with unit [N m], and is seen to be inversely proportional to the beam length ℓ.

The maximum displacement $w_{\max}$ occurs for $\theta(x) = 0$. It then follows from the expressions for $\theta(x)$ and $w(x)$ that

$$x_{\max} = \frac{\ell}{\sqrt{3}}, \qquad w_{\max} = w(x_{\max}) = \frac{1}{9\sqrt{3}}\frac{M_B \ell^2}{EI_z}.$$

A maximum permissible displacement $w_{\max}$ is hereby translated into a maximum permissible moment. □

Example 4.4. Simply supported beam with distributed load. Figure 4.14 shows a simply supported beam of length ℓ with uniformly distributed load p. It has two static boundary conditions $M_A = M_B = 0$ and the moment distribution can therefore be determined first, either by introducing a section and using equilibrium or by integration of the moment differential equation (4.22),

$$\frac{d^2 M(x)}{dx^2} = -p.$$

The load intensity p is constant, and the moment distribution $M(x)$ is therefore parabolic with value zero at A and B due to the static boundary conditions. It is easily seen that the moment distribution that satisfies these conditions has the form

$$M(x) = \tfrac{1}{2}\, p\, x(\ell - x),$$

as already determined directly from statics in Example 3.6.

The transverse displacement $w(x)$ is found from the relation (4.23) between curvature and moment,

$$\frac{d^2 w(x)}{dx^2} = -\frac{M(x)}{EI_z} = -\frac{1}{2}\frac{p\ell^2}{EI_z}\frac{x}{\ell}\Big(1 - \frac{x}{\ell}\Big).$$

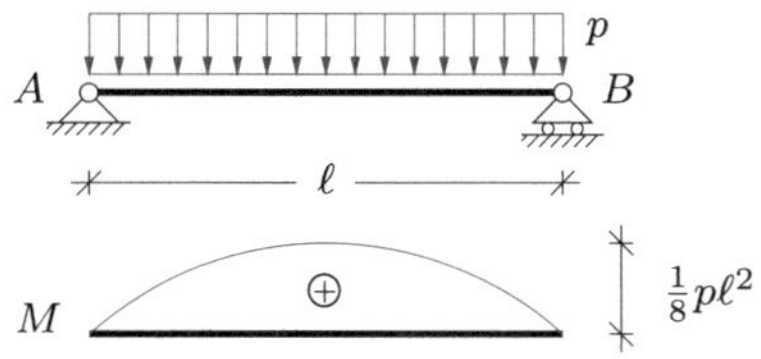

Fig. 4.14: Parabolic moment distribution for simply supported beam.

The displacement $w(x)$ is obtained by double integration as

$$w(x) = -\frac{1}{2}\frac{p\ell^4}{EI_z}\left[\frac{1}{6}\Big(\frac{x}{\ell}\Big)^3 - \frac{1}{12}\Big(\frac{x}{\ell}\Big)^4\right] + C_0 + C_1 x\,,$$

with two arbitrary integration constants C_0 and C_1. The kinematic boundary conditions for the simple supports are

$$w(0) = w(\ell) = 0\,,$$

where the first condition directly implies that $C_0 = 0$. The second condition $w(\ell) = 0$ yields

$$-\frac{1}{24}\frac{p\ell^4}{EI_z} + C_1\ell = 0 \qquad \Rightarrow \qquad C_1 = \frac{1}{24}\frac{p\ell^3}{EI_z}\,,$$

whereby the final expression for the transverse displacement becomes

$$w(x) = \frac{1}{24}\frac{p\ell^4}{EI_z}\frac{x}{\ell}\left[1 - 2\Big(\frac{x}{\ell}\Big)^2 + \Big(\frac{x}{\ell}\Big)^3\right].$$

The displacement $w(x)$ is symmetric with respect to the center of the beam, and the maximum displacement therefore occurs at $x = \frac{1}{2}\ell$,

$$w_{\max} = \frac{5}{384}\frac{p\ell^4}{EI_z}\,.$$

The rotation follows from the derivative of the displacement $w(x)$ as

$$\theta(x) = -\frac{dw(x)}{dx} = -\frac{1}{24}\frac{p\ell^3}{EI_z}\left[1 - 6\Big(\frac{x}{\ell}\Big)^2 + 4\Big(\frac{x}{\ell}\Big)^3\right],$$

and the rotations at the supports are then found to be

$$\theta_B = -\theta_A = \frac{1}{24}\frac{p\ell^3}{EI_z}\,.$$

Note, that symmetry implies same magnitude and opposite sign. □

So far the examples have not involved concentrated loads or load discontinuities within the beam span. In the case of a concentrated load on the beam, the moment distribution will be described by different analytical expressions on the two sides of the beam. The solution must therefore be obtained by integration in both intervals and combined by suitable continuity conditions. The following example illustrates the procedure for the case of a concentrated transverse load. It is demonstrated that the discontinuities generated by the concentrated load severely complicates the derivation of the solution.

In practice, problems with concentrated loads are therefore most often solved by using a method based on the principle of virtual work, presented in Section 4.4.

Example 4.5. Simply supported beam with local force. The present example considers a simply supported beam with a concentrated transverse force P acting at distance a from the left support as shown in Fig. 4.15. It is convenient to introduce the notation $a' = \ell - a$ for the distance of the load from the right support, and similarly letting $x' = \ell - x$ denote the coordinate from the right support.

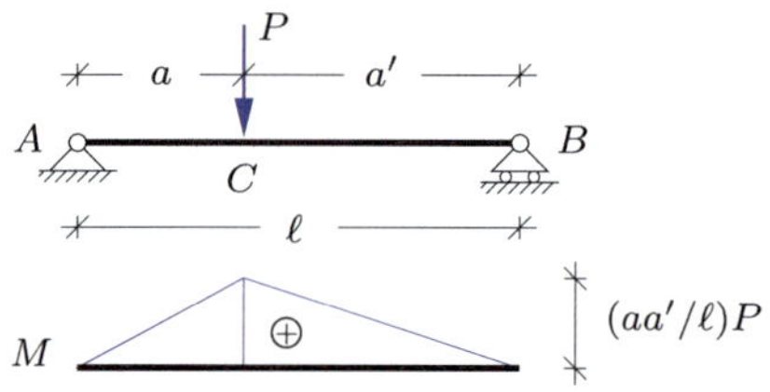

Fig. 4.15: Simply supported beam with concentrated load.

The beam is statically determinate and the moment distribution consists of two linear parts connecting the supports with the maximum moment $M(a) = (aa'/\ell)P$ at $x = a$. The analytical expressions for the moment distribution are,

$$M(x) = \begin{cases} x\,\dfrac{a'}{\ell}\,P\,, & x \le a \\[2ex] x'\,\dfrac{a}{\ell}\,P\,, & x \ge a\,. \end{cases}$$

The curvature relation (4.23) can be expressed in either the variable x or x',

$$\frac{d^2w(x)}{dx^2} = -\frac{M(x)}{EI_z}\,, \qquad \frac{d^2w(x')}{dx'^2} = -\frac{M(x')}{EI_z}\,.$$

Substitution of the moment for the left and the right part of the beam followed by double integration gives

$$w(x) = \begin{cases} -\dfrac{x^3a'}{6\ell}\,\dfrac{P}{EI_z} + C_0 + C_1x\,, & x \le a \\[2ex] -\dfrac{x'^3a}{6\ell}\,\dfrac{P}{EI_z} + D_0 + D_1x'\,, & x \ge a\,. \end{cases}$$

It is seen that the solution contains two arbitrary integration constants for each of the two integration intervals, making a total of four. The boundary conditions $w(x=0) = 0$ and $w(x=\ell) = w(x'=0) = 0$ determine the constants $C_0 = D_0 = 0$.

The remaining two constants C_1 and D_1 are to be determined from the conditions that $w(x)$ and $dw(x)/dx$ are continuous at $x = a$, corresponding to

$$w(a_-) = w(a_+)\,, \qquad \left.\frac{dw}{dx}\right|_{a-} = -\left.\frac{dw}{dx'}\right|_{a+}\,.$$

The minus in the equation for the derivatives is due to the change of variable from x to x'. The displacement continuity yields the equation

$$-\frac{a^3a'}{6\ell}\frac{P}{EI_z} + C_1\,a \;=\; -\frac{a'^3a}{6\ell}\frac{P}{EI_z} + D_1\,a'\,,$$

while continuity of the derivatives leads to

$$-3\frac{a^2a'}{6\ell}\frac{P}{EI_z} + C_1 \;=\; 3\frac{a'^2a}{6\ell}\frac{P}{EI_z} - D_1\,.$$

The constant D_1 can be eliminated between the two equations by multiplying the second equation by a' and then taking the sum,

$$-(a^2+3aa')\frac{aa'}{6\ell}\frac{P}{EI_z} + (a+a')C_1 \;=\; 2a'^2\frac{a'a}{6\ell}\frac{P}{EI_z}\,.$$

When introducing the identity $a+a'=\ell$ it is found that

$$C_1 \;=\; (a^2+3aa'+2a'^2)\,\frac{a'a}{6\ell^2}\frac{P}{EI_z}\,.$$

This expression can be reduced further by introducing $a=\ell-a'$ in the parenthesis, whereby

$$C_1 \;=\; (\ell+a')\,\frac{a'a}{6\ell}\frac{P}{EI_z} \;=\; (\ell^2-a'^2)\,\frac{a'}{6\ell}\frac{P}{EI_z} \;=\; \left[1-\left(\frac{a'}{\ell}\right)^2\right]\frac{P\,\ell\,a'}{6EI_z}\,.$$

The formulation is symmetric, and the expression for D_1 can therefore be found by interchanging a' and a in the expression for C_1,

$$D_1 \;=\; \left[1-\left(\frac{a}{\ell}\right)^2\right]\frac{P\,\ell\,a}{6EI_z}\,.$$

The final solution is obtained by substitution of the constants C_1 and D_1 into the expressions for the left and right intervals, respectively,

$$w(x) \;=\; \begin{cases} \dfrac{P\ell a'x}{6EI_z}\left[1-\left(\dfrac{a'}{\ell}\right)^2-\left(\dfrac{x}{\ell}\right)^2\right], & x\le a \\[2ex] \dfrac{P\ell ax'}{6EI_z}\left[1-\left(\dfrac{a}{\ell}\right)^2-\left(\dfrac{x'}{\ell}\right)^2\right], & x\ge a\,. \end{cases}$$

The solution is seen to be in a fairly systematic form with full symmetry between marked and unmarked variables. However, it is difficult to use this property to simplify the derivation further than shown here. □

4.2.2 Statically indeterminate beams

In the case of statically indeterminate beams there are more than two kinematic support conditions, and the problem must therefore be solved by use of the kinematic variables. In practice this means formulating the problem in terms of the transverse displacement $w(x)$ right from the start. The transverse displacement defines the moment via (4.23), and the moment must satisfy the equilibrium equation (4.22). Combination of these two conditions gives the following fourth order differential equation for the transverse displacement $w(x)$,

$$\frac{d^2}{dx^2}\Big(EI_z\,\frac{d^2w(x)}{dx^2}\Big) - p(x) = 0\,. \tag{4.24}$$

Integration of this fourth order differential equation introduces four arbitrary integration constants, that are determined from the four boundary conditions representing the supports. Typical boundary conditions are illustrated in Fig. 4.10. They consist of combinations of relations involving the transverse displacement w and its derivative dw/dx, and the static conditions expressed in terms of the moment and shear force as

$$M(x) = -\,EI_z\,\frac{d^2w(x)}{dx^2}\,, \qquad Q(x) = -\,\frac{d}{dx}\Big(EI_z\,\frac{d^2w(x)}{dx^2}\Big)\,. \tag{4.25}$$

The integration is illustrated in the following simple example of a beam with a fixed and a simple support carrying a distributed load of constant intensity. In the case of concentrated loads or non-constant bending stiffness EI_z complications similar to those illustrated in Example 4.5 will occur, and methods presented in Section 4.4, based on the principle of virtual work, may be preferable.

Example 4.6. Cantilever beam with extra end support. Figure 4.16 shows a homogeneous beam AB of length ℓ with a fixed end at A and a simple support at B. There are three reaction components M_A, R_A and R_B associated with the beam bending problem, and the beam is therefore statically indeterminate. The displacement function $w(x)$ is obtained from the differential equation (4.24) with constant load intensity,

$$\frac{d^4w(x)}{dx^4} = \frac{p}{EI_z}\,.$$

The boundary conditions are

$$w(0) = 0\,, \qquad dw(0)/dx = 0\,, \qquad w(\ell) = 0\,, \qquad d^2w(\ell)/dx^2 = 0\,,$$

where the last follows from the moment condition via (4.25a).

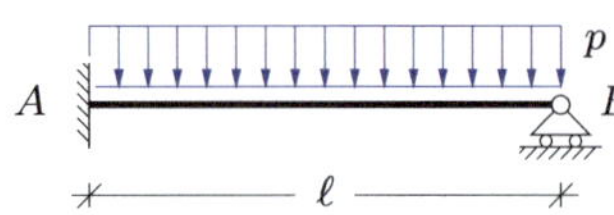

Fig. 4.16: Cantilever beam with uniform load and extra support.

When integrating the differential equation four times the result is the following fourth degree polynomial,

$$w(x) = C_0 + C_1x + C_2x^2 + C_3x^3 + \frac{1}{24}\,\frac{p}{EI_z}\,x^4\,.$$

The first four terms containing the integration constants C_0, C_1, C_2 and C_3 represent the solution to the homogeneous equation, while the last term is a particular solution that accounts for the distributed load.

It is convenient first to apply the two boundary conditions at $x = 0$, which immediately give $C_0 = C_1 = 0$. Hereby the expression for the displacement is reduced to

$$w(x) = C_2 x^2 + C_3 x^3 + \frac{1}{24} \frac{p}{EI_z} x^4 .$$

The third boundary condition $w(\ell) = 0$ gives the equation

$$C_2 + C_3 \ell + \frac{1}{24} \frac{p}{EI_z} \ell^2 = 0 .$$

The vanishing moment condition $d^2 w(\ell)/dx^2 = 0$, gives the equation

$$C_2 + 3 C_3 \ell + \frac{1}{4} \frac{p}{EI_z} \ell^2 = 0 .$$

The constant C_2 is eliminated by forming the difference between the two equations, whereby C_3 is determined. Substitution of C_3 into any of the original equations then determines C_2. This leads to

$$C_2 = \frac{3}{48} \frac{p\ell^2}{EI_z} , \qquad C_3 = -\frac{5}{48} \frac{p\ell}{EI_z} .$$

The resulting solution for the displacement can then be expressed in normalized form as

$$w(x) = \frac{1}{48} \frac{p\ell^4}{EI_z} \left(\frac{x}{\ell}\right)^2 \left[3 - 5 \frac{x}{\ell} + 2 \left(\frac{x}{\ell}\right)^2 \right] .$$

The moment is given via (4.25) as

$$M(x) = -EI_z \frac{d^2 w}{dx^2} = -\frac{p\,\ell^2}{8} \left[1 - 5 \frac{x}{\ell} + 4 \left(\frac{x}{\ell}\right)^2 \right] ,$$

and the moment at the fixed support at $x = 0$ is

$$M_A = M(0) = -\tfrac{1}{8}\, p\, \ell^2 .$$

The moment $M(x)$ is illustrated in Fig. 4.17, normalized by the magnitude of the negative moment $M = M_A$ at the fixed support.

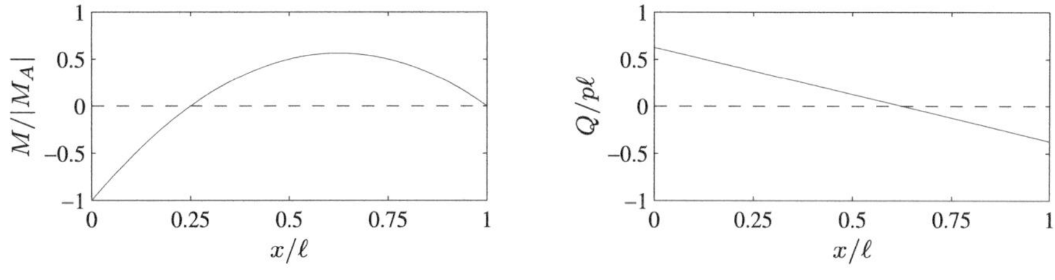

Fig. 4.17: Moment and shear force distribution.

The shear force is obtained as the derivative of the moment,

$$Q(x) = \frac{dM(x)}{dx} = -\frac{p\,\ell}{8}\Big[-5+8\Big(\frac{x}{\ell}\Big)\Big].$$

A local maximum moment occurs where $Q(x) = 0$. It is found that

$$x_{\max} = \tfrac{5}{8}\ell, \qquad M_{\max} = M(x_{\max}) = \tfrac{9}{128}\,p\,\ell^2 .$$

Note, that the magnitude of the moment at the fixed end is larger, since $|M_{\max}/M_A| = \frac{9}{16} < 1$. The shear force also determines the vertical reactions

$$R_A = Q(0) = \tfrac{5}{8}p\,\ell, \qquad R_B = -Q(\ell) = \tfrac{3}{8}p\,\ell .$$

It is observed that the sum of the vertical reactions is equal to the imposed load, $R_A+R_B = p\,\ell$. It is also observed that the extra stiffness provided by the fixed support increases the vertical reaction R_A relative to the value $\frac{1}{2}p\,\ell$ in a similar simply supported beam. □

4.3 Shear flexible beams

The basic assumption of the Bernoulli beam theory is that the only deformation mechanism is curvature. This mechanism was identified with reference to constant bending moment, as illustrated in Fig. 4.3. In this particular case the rotation of the cross-section is identical to the rotation of the tangent of the beam axis, and thus sections initially orthogonal to the beam axis will remain orthogonal to the beam axis in the deformed state. However, for non-homogeneous bending this is an approximation, in which the effect of deformation generated by the shear force is neglected. Shear deformation can be illustrated by placing a paperback book on a desk and pushing the top cover towards the front of the book. All pages retain their original size but slide a little bit in the direction of the push. Thereby the originally vertical lines along the sides of the book form an angle with vertical – the shear angle. Figure 4.18 illustrates the bending and the shear deformation modes for a beam.

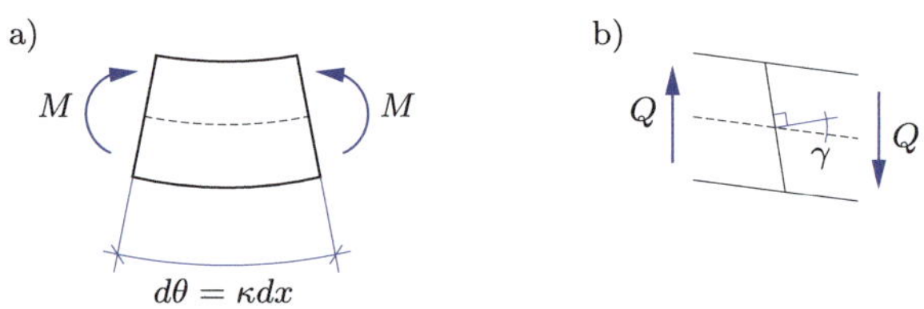

Fig. 4.18: a) Bending and b) shear deformation mechanisms.

Shear stresses and strains

Before addressing shear deformation of beams a few basic concepts in relation to shear stress and shear strain are briefly introduced. The basic shear mechanism is illustrated in Fig. 4.19 for a cube of unit side length. The top and bottom faces are loaded by forces of magnitude τ in their own plane towards the right and the left, respectively. The forces are in-plane and normalized per unit area, and therefore called shear stresses. Equilibrium requires that the left and right faces are loaded with a downward and an upward shear stress of the same magnitude to prevent rotation of the cube. This is the basic shear load. The shear load will rotate the sides of the cube as shown in the figure as γ_1 and γ_2. The change of the angle between the faces of the cube is the shear strain

$$\gamma = \gamma_1 + \gamma_2 \,. \tag{4.26}$$

The shear strain is a change of angle *between* the faces, and is independent of any overall rotation of the cube.

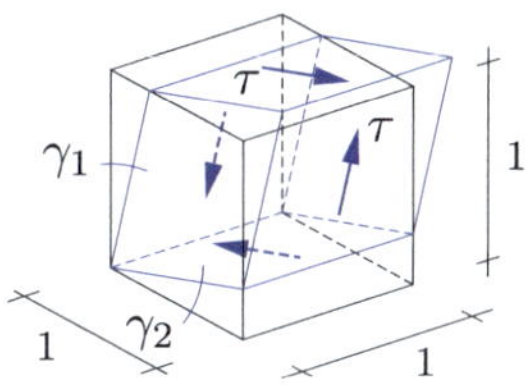

Fig. 4.19: Unit cube with shear stress τ and shear strain γ.

In a linear elastic material the shear stress τ is proportional to the shear strain γ,

$$\tau = G\gamma \,. \tag{4.27}$$

The coefficient G is a material parameter, called the shear modulus. It is similar to the modulus of elasticity and also has the dimension of force per area. For an isotropic material – that is a material in which the properties are independent of the direction in the material – there is a connection between the modulus of elasticity E and the shear modulus G. The relation implies that

$$\tfrac{1}{3}E < G < \tfrac{1}{2}E \,. \tag{4.28}$$

A typical value is $G \simeq 0.4\,E$. In composite materials the effective shear modulus may be much lower, see e.g. Jones (1999). A more detailed discussion of general states of stress and strain is given in Chapters 8–9.

Equations of shear flexible beams

In a theory for beams with shear deformation it is important to make a distinction between the rotation of the cross-section $\theta(x)$ and the rotation of the tangent of the beam axis $-dw(x)/dx$. In bending, fibers parallel to the beam axis change length due to a change of the cross-section rotation $\theta(x)$ along the beam. The change of angle between neighboring cross-sections is $d\theta = \kappa\, dx$. Thus, the parameter κ associated with beam bending is the change in cross-section rotation per unit length along the beam. The shear mechanism accounts for the fact that a shear force Q in the beam will introduce shear strains as illustrated in Fig. 4.19, and the 'average' shear strain γ appears as an angle between the beam axis tangent and the cross-section normal as shown in Fig. 4.18b. Thus, the kinematic relations for a beam theory including shear deformations consist of a definition of κ in terms of the angle θ and a definition of the angle θ in terms of the transverse displacement w and the shear strain γ. The shear mechanism in Fig. 4.18b shows that the rotation of a cross-section is $-dw/dx$ due to rotation of the beam axis plus an additional rotation γ due to shear straining,

$$\theta(x) = -\frac{dw(x)}{dx} + \gamma(x)\,. \tag{4.29}$$

It is seen that for $\gamma \equiv 0$ the cross-section rotation angle θ is defined solely by the derivative of the beam axis dw/dx as in Bernoulli beam theory.

The theory of beams with shear flexibility – often called Timoshenko beam theory – has the same basic ingredients as the Bernouilli theory for beams without shear flexibility. There are two internal forces, the shear force $Q(x)$ and the internal moment $M(x)$. They must satisfy transverse and rotation equilibrium, whereby

$$\frac{dQ(x)}{dx} = -p(x)\,, \qquad \frac{dM(x)}{dx} = Q(x)\,. \tag{4.30}$$

The kinematics of the beam introduces two measures of deformation, the curvature $\kappa(x)$ and the shear strain $\gamma(x)$. They are defined in terms of the transverse displacement $w(x)$ and the cross-section rotation $\theta(x)$ as

$$\kappa(x) = \frac{d\theta(x)}{dx}\,, \qquad \gamma(x) = \frac{dw(x)}{dx} + \theta(x)\,. \tag{4.31}$$

The internal forces M and Q and the deformation measures κ and γ are connected by the constitutive relations that describe the mechanical behavior of the material of the beam.

In elastic beams the generalized strains κ and γ are proportional to the moment M and the shear force Q, respectively, as expressed by the relations

$$M(x) = EI_z\,\kappa(x)\,, \qquad Q(x) = GA_z\,\gamma(x)\,, \tag{4.32}$$

where EI_z and GA_z are the bending and shear stiffness, respectively. The bending stiffness is expressed in terms of the elastic modulus E and the moment of inertia I_z of the cross-section about its neutral axis. The shear stiffness is expressed in terms of the elastic shear modulus G and an equivalent 'shear area' of the cross-section A_z. Shear stresses are non-uniformly distributed over the cross section, and this implies that the shear area A_z is smaller than the full cross-section area A. For a rectangular cross-section $A_z = \frac{5}{6}A$, and for an I-section the shear area A_z is approximately equal to the area of the web. The analysis of the shear stress distribution and the associated cross-section stiffness is treated in Chapter 11.

The boundary conditions of a beam are expressed in terms of the kinematic parameters w and θ, or the internal forces Q and M. When shear flexibility is included, it is often most convenient to integrate the basic equilibrium and kinematic relations directly. This procedure is illustrated in the following two examples.

Example 4.7. Shear flexible cantilever. The effect of shear flexibility is illustrated by the cantilever beam loaded with a concentrated force P at the end as shown in Fig. 4.20. The cantilever is statically determinate, and the internal moment and shear force,

$$M(x) = (x-\ell)\,P\,, \qquad Q(x) = P\,,$$

are shown below the beam in the figure.

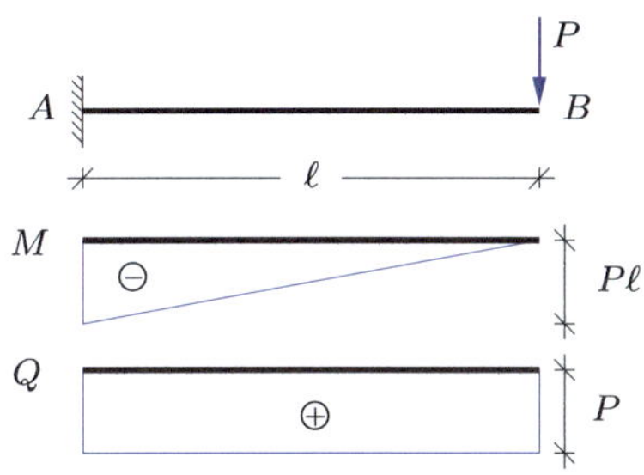

Fig. 4.20: Cantilever with shear flexibility.

The bending relation for the cross-sections rotation $\theta(x)$,

$$\frac{d\theta}{dx} = \frac{M(x)}{EI_z} = \frac{P}{EI_z}\,(x-\ell)$$

is integrated to give

$$\theta = \frac{P}{EI_z}\left(\tfrac{1}{2}x^2 - x\ell\right) + C_1\,.$$

The arbitrary constant C_1 is the value of θ at the support, and thus $C_1 = 0$.

The transverse displacement $w(x)$ is now determined by integrating the shear strain relation

$$\frac{dw}{dx} = -\theta + \gamma = \frac{P}{EI_z}\left(x\ell - \tfrac{1}{2}x^2\right) + \frac{P}{GA_z},$$

where the shear strain γ has been expressed in terms of the shear force P. Integration gives

$$w = \frac{P}{EI_z}\left(\tfrac{1}{2}x^2\ell - \tfrac{1}{6}x^3\right) + \frac{P}{GA_z}\,x + C_2 .$$

The arbitrary constant C_2 is the displacement w at the support, whereby $C_2 = 0$. Thus, the displacement of a shear flexible cantilever is

$$w = \frac{P\ell^3}{6EI_z}\left(\frac{x}{\ell}\right)^2\left[3 - \frac{x}{\ell}\right] + \frac{P\ell}{GA_z}\frac{x}{\ell} .$$

It is seen that the rotation of the cross-sections $\theta(x)$ is independent of the shear flexibility, while the displacement $w(x)$ consists of two additive contributions, a contribution from bending flexibility identical to that for Bernoulli beams, and a contribution from shear deformation. This additive form of the displacement remains valid for other statically determinate load cases. For most long and slender beams the displacement contribution from shear flexibility is fairly small. This is discussed further in Section 4.4.2. □

In the deformation and finite element methods distributed loads are included via their equivalent concentrated loads on the nodes. These concentrated loads correspond to the reactions at the ends of a rigidly supported beam, and it is therefore of particular interest to investigate the influence of shear flexibility on the reactions of a rigidly supported beam.

Example 4.8. Shear flexible beam with fixed ends. Figure 4.21 shows a rigidly supported homogeneous beam with uniformly distributed load p. In this case it is convenient to use a coordinate x with origin at the center of the beam. The solution is obtained by sequential integration of the basic relations.

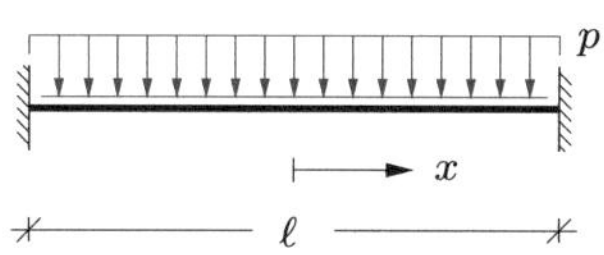

Fig. 4.21: Fixed beam with uniformly distributed load.

The shear force $Q(x)$ is obtained by integration of the transverse equilibrium equation,

$$\frac{dQ}{dx} = -p \qquad \Rightarrow \qquad Q(x) = -p\,x ,$$

where the symmetry condition $Q(0) = 0$ has been used to eliminate an arbitrary constant. Integration of the moment relation then gives

$$\frac{dM}{dx} = Q = -p\,x \qquad \Rightarrow \qquad M(x) = -\tfrac{1}{2}p\,x^2 + M_0 ,$$

where the arbitrary constant M_0 is the moment at the center of the beam. The bending relation then gives

$$\frac{d\theta}{dx} = \frac{M}{EI_z} = \frac{1}{EI_z}\big(-\tfrac{1}{2}p\,x^2 + M_0 \big) \quad \Rightarrow \quad \theta(x) = \frac{1}{EI_z}\big(-\tfrac{1}{6}p\,x^3 + M_0 x \big)\,,$$

where the symmetry condition $\theta(0) = 0$ has been used to eliminate an arbitrary constant. The moment M_0 is determined by the boundary condition $\theta(\frac{1}{2}\ell) = 0$, whereby $M_0 = \frac{1}{24}p\ell^2$. The center moment M_0, and thereby the full distribution of internal forces, is seen to be independent of shear flexibility. The internal moment distribution is illustrated in Fig. 4.22. It is observed that the moment at the supports is $M_{A,B} = -\frac{1}{12}p\ell^2$, i.e. negative and of double magnitude compared to that at the center.

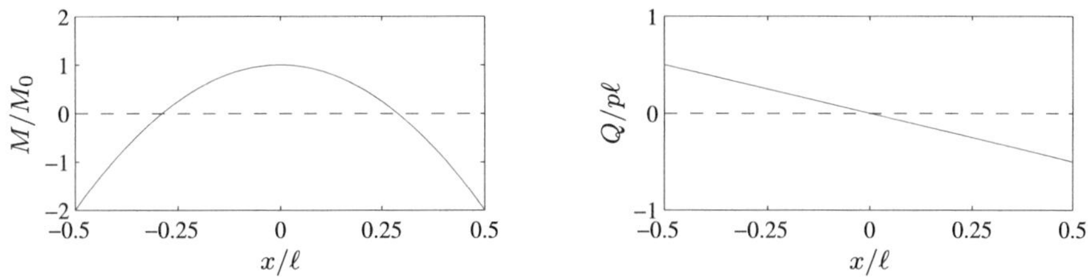

Fig. 4.22: Moment and shear force distribution.

Finally, the displacement field follows from integration of the shear relation, with $\gamma = Q/GA_z$. This gives

$$\frac{dw}{dx} = -\theta + \gamma = -\frac{p}{24EI_z}\big(x\ell^2 - 4x^3\big) - \frac{p\,x}{GA_z}\,,$$

from which

$$w(x) = -\frac{p}{24EI_z}\big(\tfrac{1}{2}x^2\ell^2 - x^4\big) - \frac{p\,x^2}{2GA_z} + w_0\,.$$

The arbitrary constant w_0, representing the displacement at the center, is determined from the boundary condition $w(\frac{1}{2}\ell) = 0$,

$$w_0 = \frac{p\,\ell^4}{384EI_z} + \frac{p\,\ell^2}{8GA_z}\,.$$

Substitution of this gives the displacement field

$$w(x) = \frac{p\,\ell^4}{384EI_z}\Big[1 - 2\Big(\frac{2x}{\ell}\Big)^2 + \Big(\frac{2x}{\ell}\Big)^4\Big] + \frac{p\,\ell^2}{8GA_z}\Big[1 - \Big(\frac{2x}{\ell}\Big)^2\Big].$$

Also in this case the displacement field is the sum of a bending and a shear deformation contribution. At the center the relative magnitude of the shear contribution is $w_G/w_E = 48EI_z/(GA_z\ell^2)$, and thus the shear contribution to the deformation is determined by the non-dimensional parameter $EI_z/(GA_z\ell^2)$. □

It is seen from these two examples that a full analysis of a beam with shear flexibility becomes quite extensive, when applying a direct integration approach. It is demonstrated in the following section that the shear flexibility effect can be included rather easily when using virtual work principles as basis of the analysis. That approach is also used when developing a simple beam element with shear flexibility in Chapter 7.

4.4 Virtual work and displacements of beams

The equilibrium equations for beams and frames constitute conditions by which each part of the beam is in local equilibrium. This equilibrium implies that if the beam or frame is subjected to a hypothetical small displacement the associated work must vanish. The principle of virtual work was derived for truss structures in Section 2.4.3, and it was demonstrated that it could be used to determine displacements of elastic truss structures from an expression of internal work. Here the same ideas are extended to beams and frames. The derivations are described in detail for plane beams and frames, but are easily generalized to three-dimensional beams and frames. First, the equation of virtual work is derived for a beam. The virtual work equation is an identity by which the external virtual work performed by the loads on the beam is identical to the internal virtual work formed by the section forces via the virtual deformation of the beam. For elastic beams this results in an expression for local displacements in terms of internal work. This result constitutes an important alternative to calculation of beam displacements by direct integration of the differential equations. After presenting the results for single beams a simple generalization to frames is presented.

4.4.1 Principle of virtual work

The principle of virtual work will be derived for plane beams and frames, and in order to obtain full generality the formulation includes the normal force N, the shear force Q and the bending moment M. The beams can then later be joined to form frames.

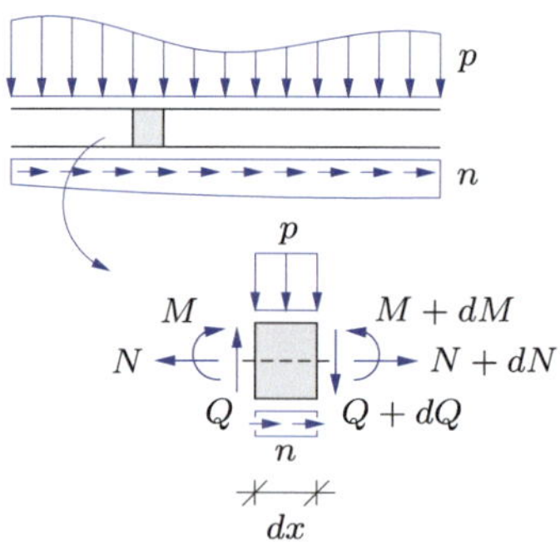

Fig. 4.23: Equilibrium of beam with section forces N, Q and M.

Figure 4.23 shows a plane beam with normal force $N(x)$, shear force $Q(x)$ and moment $M(x)$. The beam is loaded by an axial load $n(x)$, a transverse load $p(x)$ and a distributed moment load $m(x)$. Typically, a distributed moment $m(x)$ will occur, if the axial load $n(x)$ is not acting along the beam axis but at some transverse distance. The equilibrium equations are obtained by considering a small slice of the beam of thickness dx as in Section 3.3.1. The resulting three equilibrium equations are

$$\frac{dN}{dx} + n = 0\,, \qquad \frac{dQ}{dx} + p = 0\,, \qquad \frac{dM}{dx} - Q + m = 0\,. \tag{4.33}$$

Section force distributions $N(x)$, $Q(x)$ and $M(x)$ that satisfy the equilibrium conditions and appropriate static boundary conditions are said to satisfy the static conditions. The loads are here represented by continuous load densities. Concentrated loads can be considered as local load densities of very high intensity over a very short length of the beam.

The beam kinematics is expressed in terms of an axial displacement $u(x)$, a transverse displacement $w(x)$, and the cross-section rotation $\theta(x)$. Together the three functions are called the generalized displacements of the beam, or simply the displacements. The displacements of a beam typically lead to deformation, described in terms of the axial strain $\varepsilon(x)$, the shear strain $\gamma(x)$, and the curvature $\kappa(x)$. These kinematic quantities describing the deformation are called the generalized strains. As discussed previously they are defined as

$$\varepsilon = \frac{du}{dx}\,, \qquad \gamma = \frac{dw}{dx} + \theta\,, \qquad \kappa = \frac{d\theta}{dx}\,. \tag{4.34}$$

These relations between the (generalized) displacements and the (generalized) strains are called the kinematic conditions of the beam.

The idea of virtual work is to consider a beam in equilibrium, i.e. a beam that satisfies the equilibrium equations (4.33). The beam is then subjected to a hypothetical displacement described by the virtual displacements $\delta u(x)$, $\delta w(x)$ and $\delta\theta(x)$. A small unbalance in the axial equilibrium condition would lead to virtual work through the virtual axial displacement $\delta u(x)$, and similarly for virtual transverse motion and rotation. The total virtual work over the whole beam is expressed by integration over the length of the beam. This virtual work is expressed as

$$\int_0^\ell \Big\{ \delta u\Big(\frac{dN}{dx} + n\Big) + \delta w\Big(\frac{dQ}{dx} + p\Big) + \delta\theta\Big(\frac{dM}{dx} - Q + m\Big)\Big\} dx = 0\,. \tag{4.35}$$

For a beam in equilibrium each of the expressions within the parentheses vanishes identically, and thus the integral must be equal to zero for any choice of the virtual displacements.

Each of the parentheses contains a derivative of a generalized section force and a term representing a load intensity. The products containing derivatives of the generalized section forces can be reformulated by use of integration by parts. This procedure generates terms at the interval end-points, and derivatives of the virtual displacements. When arranged with the external load terms first, the result takes the form

$$\int_0^\ell (\delta u\, n + \delta w\, p + \delta\theta\, m)dx + \Big[\delta u\, N + \delta w\, Q + \delta\theta\, M\Big]_0^\ell$$
$$- \int_0^\ell \Big(\frac{d(\delta u)}{dx} N + \Big(\frac{d(\delta w)}{dx} + \delta\theta\Big) Q + \frac{d(\delta\theta)}{dx} M\Big) dx = 0 . \tag{4.36}$$

In this formula the first integral represents the virtual work of the applied load along the beam, and the terms in the square brackets represent the virtual work of the forces and moments acting at the ends of the beam. This is illustrated in Fig. 4.24.

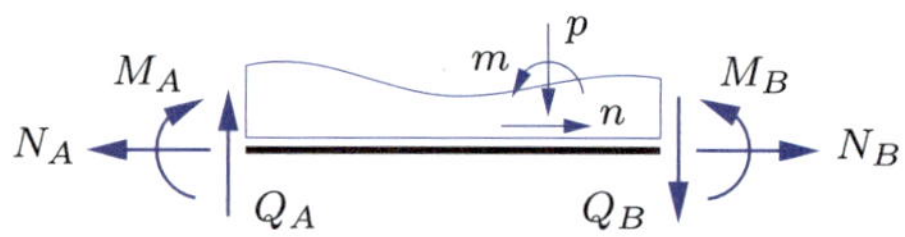

Fig. 4.24: Components of external virtual work.

The kinematic quantities in the last integral are seen to correspond to the virtual strains defined via the kinematic relations (4.34). When the notation for the corresponding virtual strains is introduced the virtual work equation takes the form

$$\Big[\delta u\, N + \delta w\, Q + \delta\theta\, M\Big]_0^\ell + \int_0^\ell (\delta u\, n + \delta w\, p + \delta\theta\, m)dx$$
$$= \int_0^\ell (\delta\varepsilon\, N + \delta\gamma\, Q + \delta\kappa\, M)dx . \tag{4.37}$$

In this equation the terms on the left side represent the external virtual work, i.e. the virtual work of the loads acting on the beam,

$$\delta V_{\rm ex} = \Big[\delta u\, N + \delta w\, Q + \delta\theta\, M\Big]_0^\ell$$
$$+ \int_0^\ell (\delta u\, n + \delta w\, p + \delta\theta\, m)dx . \tag{4.38}$$

The integral on the right side of the virtual work equation represents the internal virtual work, i.e. the work of the internal forces through the virtual strains.

$$\delta V_{\rm in} = \int_0^\ell (\delta\varepsilon\, N + \delta\gamma\, Q + \delta\kappa\, M)dx . \tag{4.39}$$

The internal work of the beam is associated with three deformation mechanisms – extension, shear and bending – illustrated in Fig. 4.25.

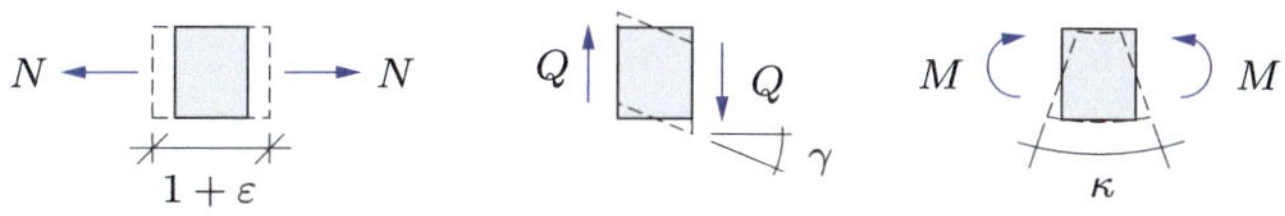

Fig. 4.25: Axial, shear and bending deformation of a beam.

With these definitions the virtual work equation (4.37) is an expression of the identity of external and internal virtual work,

$$\delta V_{\rm ex} = \delta V_{\rm in} \,. \tag{4.40}$$

This virtual work equation serves several useful purposes. It is based on the equilibrium equations (4.33) for the beam. By the reformulation via integration by parts it uniquely defines the virtual strains that correspond to the internal forces. Thus, the kinematic relations (4.34) actually *follow* from the assumption of the existence of a virtual work equation, and need not be defined independently. By making suitable assumptions about material behavior and deformation characteristics the virtual work equation serves to determine deformation and stiffness properties of beams.

In particular, the virtual work equation for Bernoulli beams follows as a special case of vanishing shear flexibility. In the case of Bernoulli beams the shear strain vanishes identically, $\gamma \equiv 0$, and the shear force term disappears from the internal work. This special formulation is often used in practice, because the effect of shear flexibility deformation is insignificant in many structures. However, as demonstrated in the following, it is often fairly simple to include the shear flexibility effect in virtual work based calculations, if needed.

4.4.2 Displacements in elastic beams

The virtual work equation is an identity of the internal work and the external work for a combination of an actual static state with given loads and a virtual displacement field. For an elastic beam it is of interest to reverse the roles such that the virtual displacement field represents the actual displacement field of interest. On the other hand, the static field is selected to serve a specific diagnostic role, e.g. focusing on the force/displacement component at a specific point. A special case of this has been considered in Section 2.4.3 for truss structures, in which bending and shear do not occur.

The general procedure is illustrated by a specific example in Fig. 4.26, showing a simply supported elastic beam with a load distribution $p(x)$. The cor-

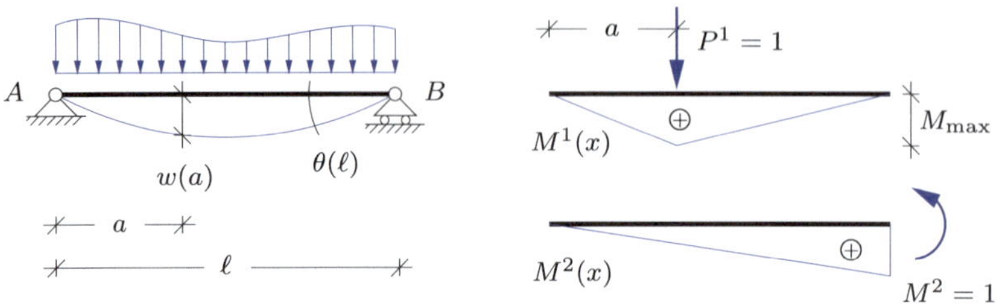

Fig. 4.26: Actual load P and unit test loads P^1 and M^2.

responding displacements are $u^0(x)$, $w^0(x)$ and $\theta^0(x)$. The superscript 0 indicates that these distributions correspond to the actual situation. These displacements define a deformation field described by an axial strain $\varepsilon^0(x)$, a shear strain $\gamma^0(x)$ and a curvature $\kappa^0(x)$, given via the kinematic relations (4.34). The beam is now assumed to be linear elastic. This implies linear constitutive relations between the internal forces and the strains for the three deformation mechanisms illustrated in Fig. 4.25,

$$\varepsilon^0 = \frac{N^0}{EA}, \qquad \gamma^0 = \frac{Q^0}{GA_z}, \qquad \kappa^0 = \frac{M^0}{EI_z}. \tag{4.41}$$

The extension stiffness EA combines the elastic modulus E with the area of the beam, the shear stiffness GA_z combines the shear modulus G with a modified 'shear area' of the beam cross-section, and finally the bending stiffness EI_z combines the elastic modulus with the moment of inertia about the neutral axis of the cross-section. The elastic relations (4.41) imply that the actual beam kinematics can be characterized by the internal forces $N^0(x)$, $Q^0(x)$ and $M^0(x)$ corresponding to the actual loads.

The static components of the virtual work equation are taken to represent a suitable test case of particular interest. Two cases are illustrated in Fig. 4.26b: a concentrated unit transverse force $P^1 = 1$ at the distance a from the left end, and a concentrated unit counterclockwise moment $M^2 = 1$ at the right support, respectively.

The external work for test case 1 consists of the single term $w^0 P^1 = w^0$, corresponding to the actual displacement of the beam in the direction of the test force. According to the virtual work equation (4.40) this is equal to the internal virtual work, expressed as

$$w^0 = w^0\, P^1 = \int_0^\ell \Big(\varepsilon^0(x)\, N^1(x) + \gamma^0(x)\, Q^1(x) + \kappa^0(x)\, M^1(x)\Big)\, dx. \tag{4.42}$$

When the actual strains are expressed via the elastic relations (4.41), the virtual work equation takes the form

$$w^0 = \int_0^\ell \Big(\frac{N^0(x)\,N^1(x)}{EA} + \frac{Q^0(x)\,Q^1(x)}{GA_z} + \frac{M^0(x)\,M^1(x)}{EI_z} \Big) dx. \qquad (4.43)$$

This relation permits calculation of a displacement component w^0 by introducing a unit test load $P^1 = 1$ at the point of the desired displacement and then evaluating the appropriate integrals involving the distribution of internal forces from the actual load, and from the test load.

The rotation of a cross-section in the beam can be determined by introducing a unit test moment. Figure 4.26b illustrates the application of a unit test moment $M^2 = 1$ at the right support of the beam. The corresponding external work is $\theta^0 M^2 = \theta^0$, and the equation of virtual work takes the form

$$\theta^0 = \theta^0\, M^2 = \int_0^\ell \Big(\varepsilon^0(x)\, N^2(x) + \gamma^0(x)\, Q^2(x) + \kappa^0(x)\, M^2(x) \Big) dx. \quad (4.44)$$

When the actual strains are represented in terms of the internal forces by use of the elastic relations (4.41), the following expression for the cross-section rotation is obtained,

$$\theta^0 = \int_0^\ell \Big(\frac{N^0(x)\,N^2(x)}{EA} + \frac{Q^0(x)\,Q^2(x)}{GA_z} + \frac{M^0(x)\,M^2(x)}{EI_z} \Big) dx. \qquad (4.45)$$

It is observed that the only difference between the procedures for computing a displacement and a cross-section rotation is the choice of test load. A displacement is determined from the internal forces generated by a unit test force in the direction of the desired displacement, while a cross-section rotation is determined from the internal forces generated by application of a unit moment at the cross-section.

Calculation of a displacement component by the virtual work equation implies three steps:

i) calculation of the internal forces $N^0(x)$, $Q^0(x)$ and $M^0(x)$ from the actual loads,

ii) calculation of the internal forces $N^j(x)$, $Q^j(x)$ and $M^j(x)$ from the unit test load in case j,

iii) evaluation of the integrals defining the internal work.

The two first tasks have been dealt with in considerable detail in the previous chapters. In particular it was found in Chapter 3 that distributed loads lead to linear variation of $N(x)$ and $Q(x)$ while the internal moment $M(x)$ has parabolic variation. For these types of distributions of the internal forces the integrals can be obtained in terms of the principal measures of the corresponding internal force distributions. A collection of basic results are summarized in Table 4.1. The use of the table is illustrated in the following examples.

Table 4.1: Product integrals for simple shapes.

$\int_0^\ell f(x)\,g(x)\,dx$	$f(x)$	$g(x)$
$\frac{\ell}{3}\,A\,B$		
$\frac{\ell}{6}\,A\,B$		
$\frac{\ell}{6}\Big(2AC+2BD+AD+BC\Big)$		
$\frac{\ell}{3}\,A\Big(B+C\Big)$	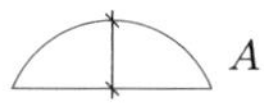	
$\frac{\ell}{12}\,A\Big(3B+5C\Big)$		
$\frac{\ell}{12}\,A\Big(B+3C\Big)$	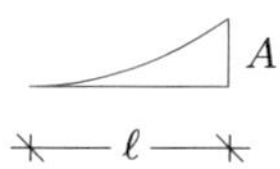	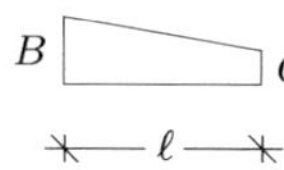

Example 4.9. Displacement of cantilever with tip load. Figure 4.27a shows a cantilever with a transverse force P at the tip. The goal is to calculate the transverse displacement $w(x)$ as a function of x and to evaluate the influence of shear flexibility on the solution. The similar problem was considered by integration of the differential equations in Example 4.7.

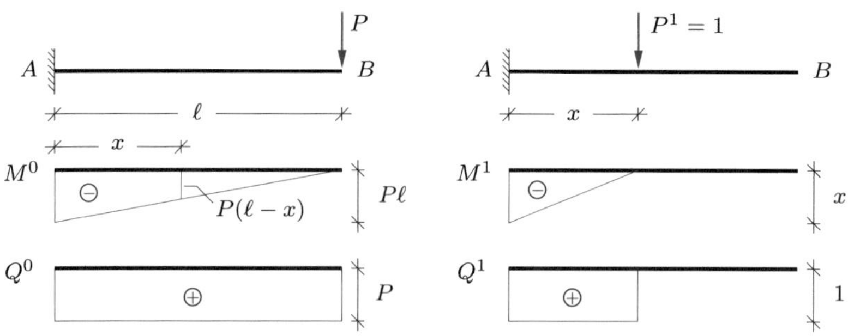

Fig. 4.27: Displacement of cantilever: Actual and test load distributions.

The normal force vanishes identically along the beam, $N^0 \equiv 0$. The distribution of the shear force $Q^0(s)$ and the moment $M^0(s)$ is shown in the lower part of Fig. 4.27a. A vertical test load in the form of a transverse unit force $P^1 = 1$ at the distance x from the support is shown in Fig. 4.27b together with the corresponding shear force and moment distributions $Q^1(s)$ and $M^1(s)$. The displacement $w(x)$ under the test load is then determined by the virtual work relation (4.43) in the form

$$w(x) = \int_0^x \Big(\frac{M^0(s)}{EI_z}\, M^1(s) + \frac{Q^0(s)}{GA_z}\, Q^1(s) \Big) ds\,.$$

Note, that the integration interval only extends over the interval from the support to the applied test load, as $M_1(s) = 0$ and $Q_1(s) = 0$ for $s > x$.

The integral is computed by the integral formulas in Table 4.1. The moment integral is computed from the linear distributions in the third row, where it is indicated in the figure that the actual moment at the location of the test force is $M^0(x) = P(\ell - x)$. The shear forces are constant. Thus, the integral is

$$w(x) = \frac{1}{EI_z}\frac{x}{6}\Big[2\,P\ell\, x + P(\ell - x)\, x \Big] + \frac{1}{GA_z}\, x\, P\,.$$

After a slight rearrangement this takes the normalized form

$$w(x) = \frac{1}{6}\frac{P\ell^3}{EI_z}\Big(\frac{x}{\ell}\Big)^2\Big[3 - \frac{x}{\ell} \Big] + \frac{P\ell}{GA_z}\frac{x}{\ell}\,.$$

This result agrees with the result obtained by solving the differential equation in Example 4.1. The rotation $\theta(x)$ can be determined similarly by applying a local moment $M^1 = 1$ at x. This problem is considered in the exercises. □

The importance of including shear flexibility in the beam theory can be estimated by comparing bending and shear contributions, w^E and w^G, to the displacement at the tip of the beam. As seen from the solution

$$w^E = \frac{P\ell^3}{3EI_z}\,, \qquad w^G = \frac{P\ell}{GA_z}\,.$$

Thus, the relative importance of the shear contribution in the present case is determined by the ratio

$$\frac{w^G}{w^E} = \frac{3EI_z}{GA_z\ell^2} \sim \frac{3}{10}\frac{E}{G}\frac{h^2}{\ell^2}\,, \tag{4.46}$$

where the last expression is representative of a rectangular cross-section with height h. For isotropic beams $G \simeq 0.4E$, whereby the two first factors combine to a number around 0.75, and thus the relative magnitude of the shear flexibility contribution is approximately $(h/\ell)^2$. For slender beams this is small and the shear flexibility effect will be negligible. However, for short beams and for composite beams where the shear stiffness may be much smaller than the axial stiffness, $G \ll E$, the shear flexibility effect may be significant.

Example 4.10. Cantilever with discontinuous bending stiffness. The virtual work is efficient when evaluating the influence of local changes in the stiffness of the structural elements. Figure. 4.28 shows a cantilever of length ℓ with a transverse force P acting at the tip. The moment increases linearly towards the support, and a smaller cross-section on the exterior part of the beam may therefore be sufficient from a strength perspective. The present example investigates the influence of reducing the bending stiffness from EI_z to $\frac{1}{2}EI_z$ in the exterior half of the beam. The beam is considered as slender, and the effect of shear deformation is therefore neglected.

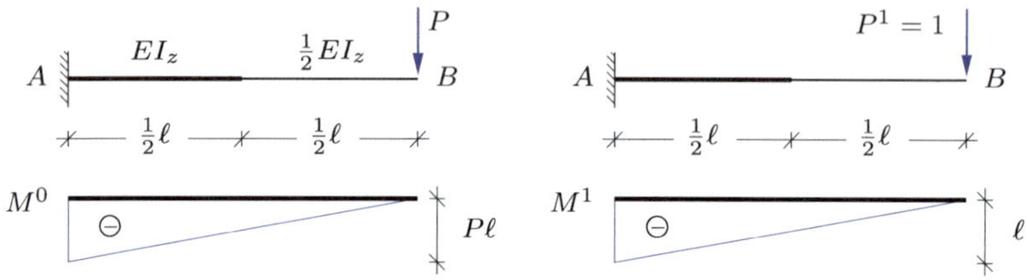

Fig. 4.28: Tip displacement of cantilever: Actual and test load distributions.

The tip displacement w_B is found by the virtual work equation (4.43), and the integral is divided into two parts due to the change in bending stiffness,

$$w_B = \frac{1}{EI_z}\int_0^{\ell/2} M^0\, M^1\, dx + \frac{2}{EI_z}\int_{\ell/2}^{\ell} M^0\, M^1\, dx.$$

The formula for integrating the product of two triangular densities is considerably simpler than the general formula for densities with linear variation. It is therefore convenient to absorb half of the last integral into the first by extending the interval of integration to the full beam length,

$$w_B = \frac{1}{EI_z}\int_0^{\ell} M^0\, M^1\, dx + \frac{1}{EI_z}\int_{\ell/2}^{\ell} M^0\, M^1\, dx.$$

Hereby both integrals relate to triangles with zero value at the right end. They are both computed by the formula in the first row of Table 4.1,

$$w_B = \frac{\ell}{3EI_z}(-P\ell)(-\ell) + \frac{\frac{1}{2}\ell}{3EI_z}(-\tfrac{1}{2}P\ell)(-\tfrac{1}{2}\ell) = (1+\tfrac{1}{8})\,\frac{P\ell^3}{3EI_z} = \frac{3}{8}\,\frac{P\ell^3}{EI_z}.$$

The additional displacement is given by the last term, giving a relative increase of $\frac{1}{8}$. The reason for the rather modest increase is that the section was deliberately reduced in the part of the beam that carries only a small moment. □

Example 4.11. Midpoint displacement of simply supported beam. The simply supported beam with uniform load shown in Fig. 4.29 is a common structural element. This example investigates the displacement at the center and the contribution from shear flexibility. The beam is statically determinate and the moment distribution, determined in Example 3.6, is parabolic with a maximum of $M_{\max} = \frac{1}{8}p\ell^2$, as shown in Fig. 4.29a. This is the actual moment distribution, denoted by $M^0(x)$.

The transverse displacement at the middle of the beam is determined by the virtual work equation (4.43). The test load consists of a transverse unit force $P_1 = 1$, acting at the center of the beam as shown in Fig. 4.29b. The center displacement is then expressed as

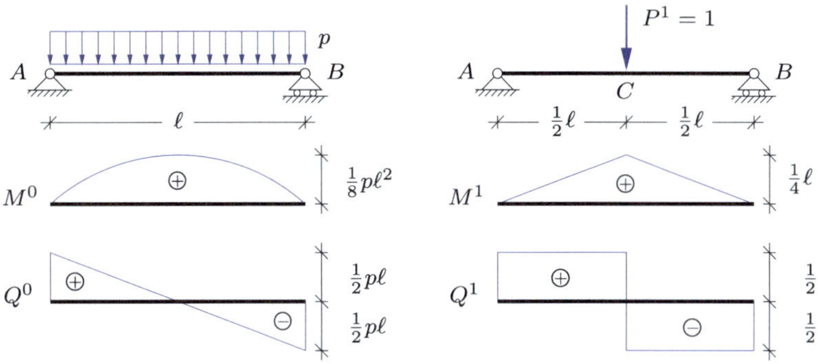

Fig. 4.29: Displacement of midpoint: Actual and test load distributions.

$$w_C = \int_0^\ell \Big(\frac{M^0(x)}{EI_z}\, M^1(x) + \frac{Q^0(x)}{GA_z}\, Q^1(x) \Big) dx$$
$$= 2 \int_0^{\ell/2} \Big(\frac{M^0(x)}{EI_z}\, M^1(x) + \frac{Q^0(x)}{GA_z}\, Q^1(x) \Big) dx.$$

The moment and shear force distributions $M^0(x)$, $M^1(x)$ and $Q^0(x)$, $Q^1(x)$ are shown in Fig. 4.29, and the integral can be computed by using the integral formulas in Table 4.1. The moment integral is evaluated by the formula in the fifth row of Table 4.1 and the shear force integral is evaluated as a simple mean value,

$$w_C = 2\,\frac{\ell}{2}\Big(\frac{1}{EI_z}\,\frac{5}{12}\,\frac{p\ell^2}{8}\,\frac{\ell}{4} + \frac{1}{GA_z}\,\frac{1}{2}\,\frac{p\ell}{2}\,\frac{1}{2} \Big) = \frac{5}{384}\,\frac{p\ell^4}{EI_z} + \frac{1}{8}\,\frac{p\ell^2}{GA_z}\,.$$

The first term is the result obtained in Example 4.4 by integrating the differential equation for a beam without shear flexibility. Also in this case the relative magnitude of the contribution from shear flexibility is given by the parameter $EI_z/(GA_z\ell^2)$ identified in Example 4.9. The conclusion in this case is the same, namely that shear flexibility is mainly important for rather short beams and beams with low shear modulus. □

Example 4.12. Support rotation of simply supported beam. In this example the task is to determine the rotation at the support of the uniformly loaded beam shown in Fig. 4.30a. The test load is a unit moment $M^1 = 1$, applied at the support as shown in Fig. 4.30b.

The rotation follows from the virtual work relation (4.45). It is observed from the figure that the actual shear force distribution $Q^0(x)$ is anti-symmetric, while the shear force distribution $Q^1(x)$ from the test load is symmetric. Thus, the corresponding integral vanishes and there is no contribution from shear flexibility to the rotation at the support. The expression for the rotation therefore takes the simplified form

$$\theta_B = \int_0^\ell \frac{M^0(x)}{EI_z}\, M^1(x)\, dx\,.$$

The bending stiffness EI_z is constant, and the result then follows from the product of a parabolic and a triangular distribution, given in the fourth row in Table 4.1,

$$\theta_B = \frac{1}{EI_z} \int_0^\ell M^0(x)\, M^1(x)\, dx = \frac{1}{EI_z}\,\frac{2\ell}{3}\,\frac{p\ell^2}{8}\,\frac{1}{2} = \frac{1}{24}\,\frac{p\ell^3}{EI_z}\,.$$

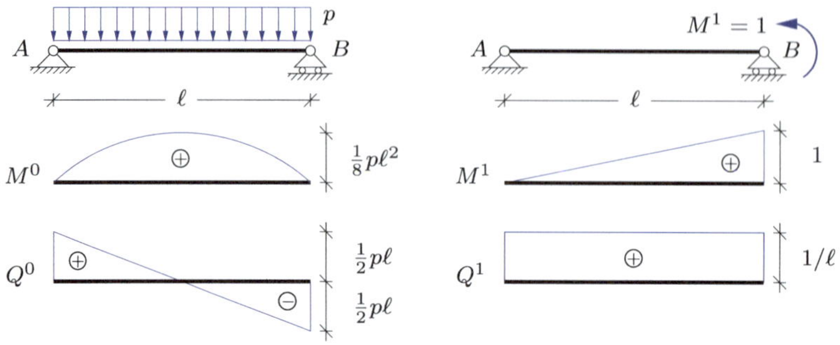

Fig. 4.30: Rotation at support: Actual and test load distributions.

Note, that because the parabolic distribution is symmetric, only the symmetric part of the triangular distribution contributes to the integral. The symmetric part of the triangle is a constant intensity of $1/2$, and thus the integral represents half the area of the parabolic distribution. Symmetry arguments like this can often be used to reduce the computation of the virtual work integral. □

Example 4.13. Deformation of beam with two load components. Structures are often analyzed for a number of load combinations, and it may be advantageous to treat the effect of simple load components individually, and then combine the results. The problem is illustrated for a simply supported beam extending beyond one of the supports as shown in Fig. 4.31. The load is a combination of a concentrated force P acting at C and a distributed load with intensity p on AB. The present example determines the tip displacement by superposition of the tip displacements for the individual load cases. The effect of shear deformation is neglected.

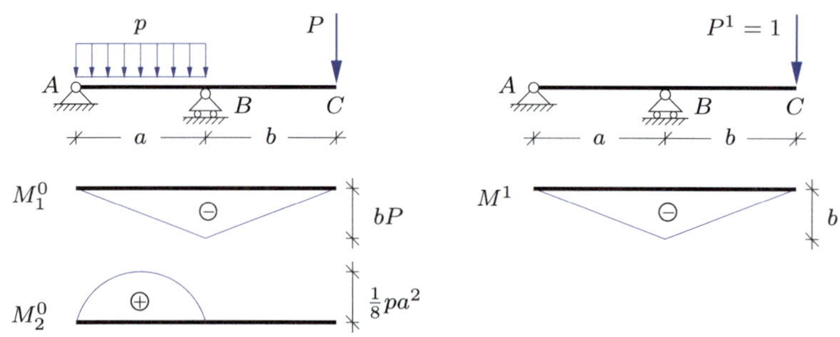

Fig. 4.31: Beam with combined concentrated and distributed load.

The actual moment distribution M^0 consists of the separate moment distributions M_1^0 and M_2^0 for tip load P and for the distributed load p, respectively, as indicated in Fig. 4.31a. The tip displacement for each of the two load cases is determined by the virtual work equation (4.43) by using the moment distribution M^1 from a unit test force $P^1 = 1$ acting at C as illustrated in Fig. 4.31b. The moment curves for the individual load components are seen to be simple shapes, easily integrated by use of the results in Table 4.1. The tip displacement from the concentrated load is obtained from the moment distribution $M_1^0(x)$ as

$$w_{C_1} = \int_{ABC} \frac{M_1^0\, M^1}{EI_z}\, ds = \frac{(a+b)}{3EI_z}\,(-bP)\,(-b) = \frac{1}{3}\,\frac{Pb^2(a+b)}{EI_z}\,,$$

while the tip displacement from the distributed load is obtained from the moment distribution $M_2^0(x)$ as

$$w_{C_2} = \int_{AB} \frac{M_2^0\, M^1}{EI_z}\, ds = \frac{a}{3EI_z}\,\frac{1}{8}\,pa^2\,(-b) = -\,\frac{1}{24}\,\frac{pa^3 b}{EI_z}\,.$$

The total tip displacement for the load combination is found as the sum of the individual contributions,

$$w_C = w_{C1} + w_{C2} = \frac{b}{3EI_z}\Big(Pb(a+b) - \tfrac{1}{8}pa^3\Big)\,.$$

Zero tip displacement $w_C = 0$ is obtained for

$$Pb(a+b) - \tfrac{1}{8}pa^3 = 0 \qquad \Rightarrow \qquad P = \frac{1}{8}\,\frac{pa^3}{b(a+b)}\,.$$

For spans of same length zero tip displacement corresponds to a distributed load of magnitude $pa = 16P$. □

4.4.3 Virtual work and displacements in frames

The virtual work equation (4.40) for beams can be extended to frame structures, formed by joining beams. The procedure is here illustrated for plane frames, but applies also to three-dimensional frame structures. The frame consists of a number of beams, that each have a set of internal forces that satisfy equilibrium with the external load. The frame is now subjected to a virtual displacement field, by which joints are moved and beams are deformed. In this process the principle of virtual work applies to each of the individual beams, and therefore also to the sum of the external and internal work contributions from all the beams,

$$\sum_{\text{beams}} \delta V_{\text{ex}} = \sum_{\text{beams}} \delta V_{\text{in}}\,. \tag{4.47}$$

The internal virtual work δV_{in} in each of the beams follows directly from its definition (4.39) in terms of the internal forces and the generalized virtual strains $\delta\varepsilon$, $\delta\gamma$ and $\delta\kappa$ for a single beam.

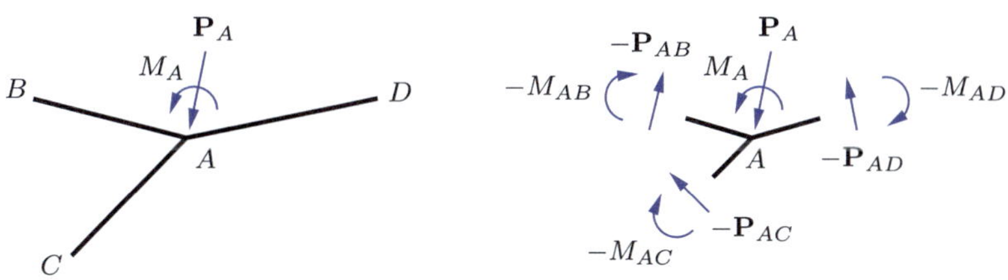

Fig. 4.32: Balance between internal and external forces at joint.

The external virtual work $\delta V_{\rm ex}$ relating to the beams consists of the virtual work of the loads on the beams plus the external virtual work of the section forces at the ends of the beams, illustrated in Fig. 4.24. The virtual work of the loads inside the beams are retained as contributions relating to the individual beam, but the end loads formed by the section forces at the beam ends are reformulated. The principle is illustrated in Fig. 4.32 showing three beams AB, AC and AD joined at A. Equilibrium of the joint implies that the external load on the joint is balanced by the section forces from the beams connected to the joint. The section forces act in the opposite direction of those acting on the beams, and thus the equilibrium conditions for the joint can be written as

$$\mathbf{P}_A - \sum \mathbf{P}_{A*} = \mathbf{0}\,, \qquad M_A - \sum M_{A*} = 0\,, \tag{4.48}$$

where the subscript $A*$ denotes any of the connected beams AB, AC, etc. By these equilibrium conditions the sum of all contributions from the beams connected to a joint is equal to the external load applied to the joint. Thus, the contribution from all the forces and moments from beams connected to a joint j can be replaced by $\delta\mathbf{u}_j^T\mathbf{P}_j$ and $\delta\theta_j M_j$. After replacing the contributions from the beam end sections with the contribution from the corresponding joints, the virtual work equation for frames takes the form

$$\begin{aligned}\sum_{\text{joints}} \Big[\,\delta\mathbf{u}_j^T\mathbf{P}_j + \delta\theta_j\, M_j\Big] &+ \sum_{\text{beams}} \int_{\ell_i} \big(\delta\mathbf{u}^T\mathbf{p} + \delta\theta\, m\big)ds \\ &= \sum_{\text{beams}} \int_{\ell_i} \big(\,\delta\varepsilon\, N + \delta\gamma\, Q + \delta\kappa\, M\,\big)ds.\end{aligned} \tag{4.49}$$

The virtual work equation for frames is a straightforward extension of the similar result (4.37) for beams, and its use for calculation of displacements in elastic frames is illustrated in the following examples.

Example 4.14. Displacements in simple frame. Figure 4.33 shows a simple angle frame with a uniformly distributed vertical load with intensity p acting on the horizontal beam CDB. The effect of extension/compression effects due to the normal force on the vertical displacement at the center D is illustrated in the context of slender beams, where it was demonstrated in Example 4.11 that the shear flexibility effect can be neglected.

The vertical displacement is determined by the virtual work equation (4.43) involving an integral over the entire frame,

$$w_D = \int_{ACB} \Big(\frac{M^0 M^1}{EI_z} + \frac{N^0 N^1}{EA}\Big)ds.$$

The moment distributions are only present over CB and symmetric, while the normal force only contributes over AC. Thus, the virtual work relation can be written as

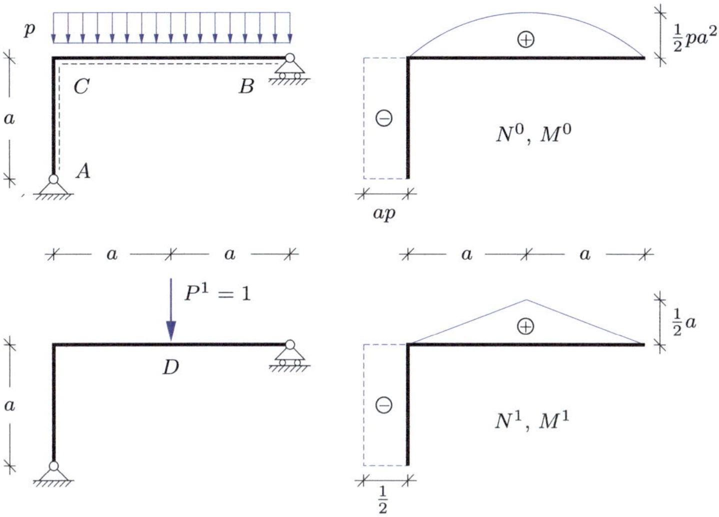

Fig. 4.33: Angle frame: Actual and test load distributions, M (–), N (- -).

$$w_D = \frac{2}{EI_z}\int_{CD} M^0 M^1\, ds + \frac{1}{EA}\int_{AC} N^0 N^1\, ds\,.$$

The first integral involves a half parabola and a triangle, given by the fifth row of Table 4.1, while the normal force distributions are piecewise constant as indicated in Fig. 4.33b by the dashed lines. The integrals then give

$$w_D = \frac{2}{EI_z}\,\frac{5a}{12}\,\frac{pa^2}{2}\,\frac{a}{2} + \frac{a}{EA}\,pa\,\frac{1}{2} = \frac{5}{24}\,\frac{pa^4}{EI_z} + \frac{1}{2}\,\frac{pa^2}{EA}\,.$$

The relative importance of the axial deformation is determined by the parameter combination

$$\frac{EI_z}{EA\ell^2} \sim \frac{1}{12}\,\frac{h^2}{\ell^2}\,.$$

where $\ell = 2a$, and the numerical factor corresponds to a rectangular cross-section. The relative magnitude of the axial deformation is proportional to $(h/\ell)^2$ as for the effect of shear flexibility. Therefore the axial deformation is often neglected in hand-calculation analysis of elastic frames. □

Example 4.15. Horizontal loading of simple frame. In the present example the loading of the simple frame in the previous example is changed, so that a distributed load with intensity p is acting in the horizontal direction on the vertical beam AC. The horizontal displacement w_C of the junction is determined by the virtual work equation. The reactions can for instance be determined by horizontal projection, moment about B and moment about A:

$$R'_A = ap\,, \qquad R_A = -\tfrac{1}{4}ap\,, \qquad R_B = \tfrac{1}{4}ap\,.$$

The moment is zero at the supports in A and B and varies linearly in BC and as a parabola in AC. Thus, the full moment distribution is determined by the moment M_C at the joint C and the local maximum of the moment in AC. By section in BC at C it is found that

$$\curvearrowright \qquad M_C - 2aR_B = 0 \qquad \Rightarrow \qquad M_C = \tfrac{1}{2}pa^2\,.$$

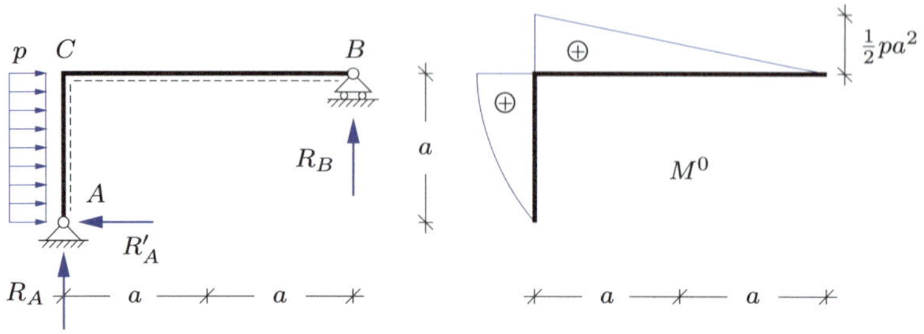

Fig. 4.34: Frame with horizontal distributed load on vertical beam.

The local maximum of the moment in AC is found via the shear force Q. Due to the roller support in B there is no normal force in CB, and thus the shear force in AC vanishes at the joint C. This implies that the moment distribution in AC has the maximum value M_C and zero slope in C. Hereby, the moment distribution M^0 is determined and it is shown in Fig. 4.34b.

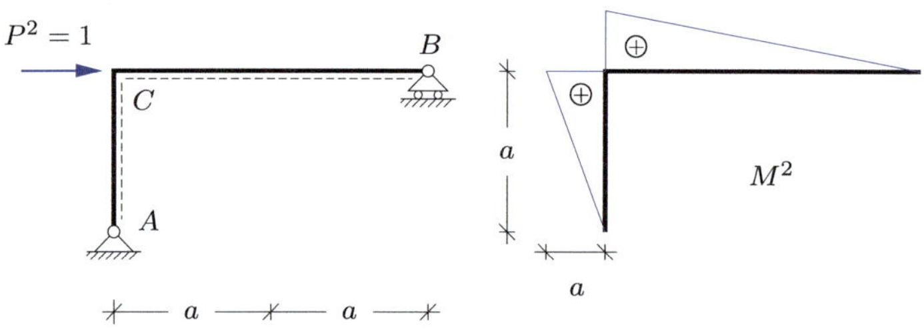

Fig. 4.35: Virtual moment distribution corresponding to the horizontal displacement of C.

The horizontal displacement of the joint C is determined from the extended form of the virtual work relation (4.43) integrating the product of the actual curvature and the moment from the test load in Fig. 4.35 over the whole frame,

$$w_C = \int_{ACB} \kappa^0 M^2 \, ds = \int_{AC} \frac{M^0 M^2}{EI_z} \, ds + \int_{CB} \frac{M^0 M^2}{EI_z} \, ds \, .$$

The integral for AC is solved by the integral formula in the fifth row, while the integral for CB is determined by the triangles in the first row of Table 4.1,

$$w_C = \frac{a}{12EI_z} \frac{pa^2}{2} 5\,a + \frac{2a}{3EI_z} \frac{pa^2}{2} a = \frac{13}{24} \frac{pa^4}{EI_z} \, .$$

This displacement is considerably larger than the vertical displacement found in the previous example. This is primarily due to the rolling support in B. □

Example 4.16. Displacement of a cable supported beam. The virtual work equation is also a convenient tool for determining displacements in structures with different member types. Figure 4.36 shows a cable supported beam as found e.g. in canopy roofs and simple bridges. The beam ADC has a fixed simple support at A and is supported by the cable BD at its center. The structure is statically determinate, and the internal hinge at D allows for four support reactions as shown in the figure. The length of the beam ADC is $8a$, and the height AB is $3a$. These simple proportions give the cable length BD as $5a$. The goal is to calculate the vertical displacement of the beam, represented by the vertical displacement of the tip w_C. The calculation is based on the bending deformation of the beam ADC with stiffness EI_z and the extension of the cable BD with stiffness EA. The cable cross-section is typically much smaller than the area of the beam cross-section, and the extension of the cable can therefore provide a considerable contribution to the displacement w_C.

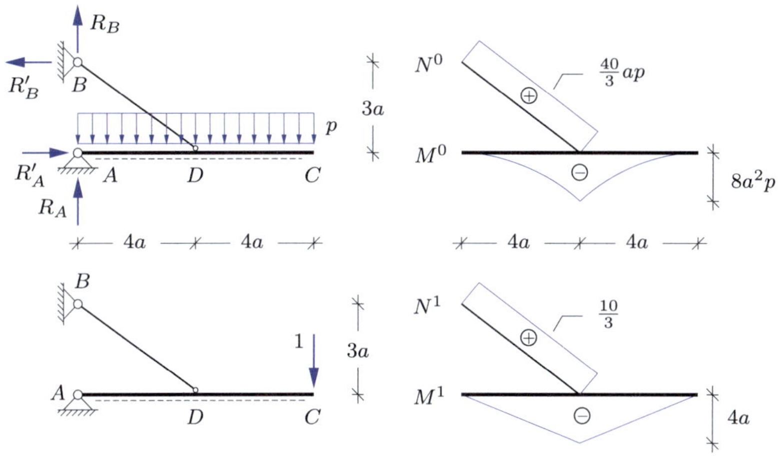

Fig. 4.36: Beam with simple and cable supports.

The horizontal reaction components are of equal magnitude and determined by moment about A or B,

$$R'_A = R'_B = \frac{32}{3}\, ap\,.$$

The vertical reaction component R_B must lead to a total reaction at B in the direction of the cable, whereby the reactions in B are the horizontal and vertical projections of the bar force N_{BD}, which implies the following equilibrium relations

$$R_B = \tfrac{3}{4} R'_B = 8ap\,.$$

The reaction component R_A is determined from moment equilibrium of the beam about the center D, whereby $R_A = 0$. The tension in the cable N^0_{BD} follows e.g. from vertical projection of node B,

$$N^0_{BD} = \tfrac{5}{3} R_B = \tfrac{40}{3} ap\,.$$

The moment curve for the beam ADC is zero at the ends where it has horizontal tangent as the shear force vanishes. The maximum value occurs at the center D, where

$$M_D = 2a\,(4ap) = 8a^2 p\,.$$

The moment distribution in the beam and the axial force in the cable are shown in the top right part of Fig. 4.36.

The virtual system corresponding to the vertical displacement w_C of the beam tip is shown in the lower part of Fig. 4.36, indicating a vertical unit force $P^1 = 1$ acting in C. The moment in the beam ADC is zero at the ends, and varies linearly to a moment $M^1_D = 4a$

at the center. The cable force is found e.g. by calculating the horizontal reaction in B by moment for the full structure about A, and then obtaining the cable force via horizontal projection of the forces acting on the node B. This determines the cable force $N^1_{BD} = \frac{10}{3}$.

The vertical displacement w_C is now determined from the virtual work equation including the terms representing bending of the beam ADC and extension of the cable BD. This can be formulated as

$$w_C = \int_{ADC} \frac{M^0(s)M^1(s)}{EI_z}\, ds + \int_{BD} \frac{N^0(s)N^1(s)}{EA}\, ds\,.$$

The moment integral is evaluated by splitting it into equal integrals over AD and DC and then using the last row in Table 4.1. The normal forces are constant and this integral therefore follows immediately from the length $5a$ of BD,

$$w_C = 2\frac{4a}{EI_z}\frac{1}{4}(8a^2p)(4a) + \frac{5a}{EA}\frac{40ap}{3}\frac{10}{3} = 64\frac{a^4p}{EI_z} + \frac{200}{9}\frac{a^2p}{EA}\,.$$

The first term is the contribution from beam bending while the second is the effect of cable extension. It is seen that a small cable cross-section area A relative to the beam cross-section may lead to a considerable contribution from extension of the cable.

It is worth noting that the effect of flexibility of e.g. the support B can easily be included in the form of a local spring that would be loaded by the corresponding reaction at B. □

4.5 Exercises

Exercise 4.1. The figure shows a rectangular box section with height h, width b and wall thickness t. As demonstrated in Example 4.1 the moment of inertia of a massive rectangular cross section is $I_z = \frac{1}{12}h^3b$. This result can be used to determine the moment of inertia of the present box section. The cross section is assumed to be thin-walled, whereby $t \ll h$ and b. Note that dimensions h and b are with respect to the centerlines of the individual flanges, which e.g. means that the total height of the box section is $h + t$.

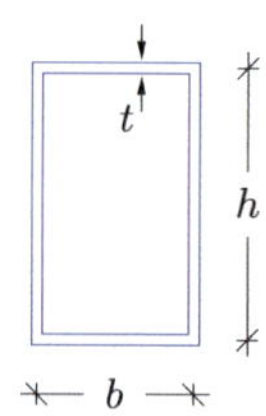

a) Determine the moment of inertia I_z^{out} for a massive rectangular cross section with dimensions corresponding to the outer dimensions of the box section.

b) Determine the moment of inertia I_z^{in} for a massive rectangular cross section with dimensions corresponding to the inner dimensions of the box section.

c) Use the results in a) and b) to determine the moment of inertia I_z of the box section.

Exercise 4.2. The figure shows a cantilever beam of length ℓ with constant bending stiffness EI_z. A concentrated moment M_B and a vertical force P_B act at the tip of the beam in B.

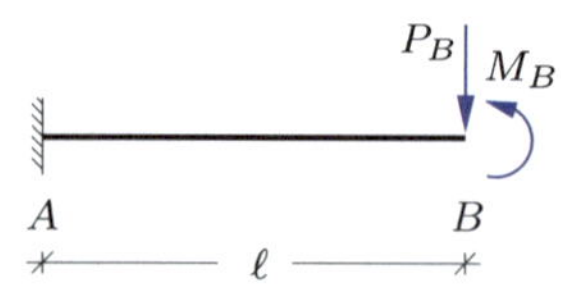

a) Find an expression for the moment $M(x)$.

b) Determine the expression for $w(x)$.

c) Find the displacement $w(\ell)$ and rotation $\theta(\ell)$ at the tip.

Exercise 4.3. The figure shows a cantilever beam of length ℓ with constant bending stiffness EI_z. The beam is loaded by a transverse uniformly distributed load with intensity p.

a) Find an expression for the moment $M(x)$.

b) Determine the expression for $w(x)$.

c) Find the displacement $w(\ell)$ and rotation $\theta(\ell)$ at the tip.

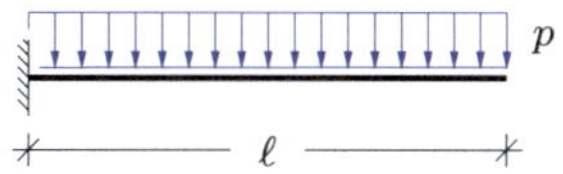

Exercise 4.4. Consider the cantilever beam in Fig. 4.27, but let the force P act at distance a from the left support instead of at the tip. Determine an expression for the displacement $w(x)$ in the two intervals $x < a$ and $x > a$, respectively.

Exercise 4.5. The figure shows a cantilever beam of length ℓ with constant bending stiffness EI_z. The beam is loaded by a transverse distributed load with linearly increasing intensity $p(x) = px/\ell$, whereby p is the tip intensity.

a) Solve the differential equation (4.16) to find the expression for the moment $M(x)$.

b) Determine the expression for the displacement $w(x)$ and the rotation $\theta(x)$.

c) Find the displacement $w(\ell)$ and rotation $\theta(\ell)$ at the tip.

d) Find the magnitude of the reactions.

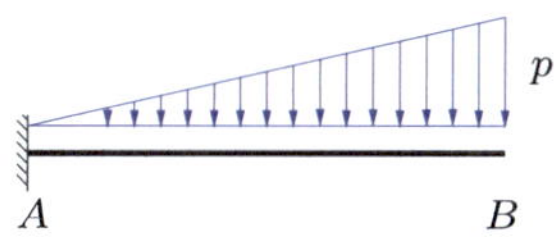

Exercise 4.6. The figure shows a beam of length ℓ with constant bending stiffness EI_z. The beam is fixed in A and simply supported in B, which makes it statically indeterminate. It is loaded by a transverse distributed load with linearly increasing intensity $p(x) = px/\ell$, where p is the intensity at the simple support.

a) Setup the fourth order differential equation governing the transverse displacement $w(x)$, and find the solution containing four arbitrary constants.

b) Use the four boundary conditions to determine the expression for the displacement $w(x)$ and the rotation $\theta(x)$.

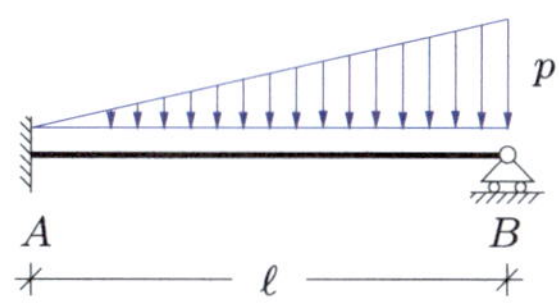

c) Determine the expression for the moment $M(x)$ and the shear force $Q(x)$ and find the magnitude of the reactions.

d) Determine the location $x_{\max}$ and the magnitude $M_{\max}$ of the maximum moment.

Exercise 4.7. The figure shows a simply supported beam of length ℓ and with constant bending stiffness EI_z. A concentrated vertical load P acts at distance a from the left support. This problem is solved in Example 4.5 via the differential equation, whereby the beam must be divided into two parts, which complicates the analysis. In this exercise the problem is solved by the principle of virtual work, and hopefully it is observed how comparatively easy the results are obtained.

a) Draw the actual moment distribution M^0.

b) Use the virtual work equation to find the transverse displacement w_C at the location of the force.

c) Use the virtual work equation to find the rotations at the supports.

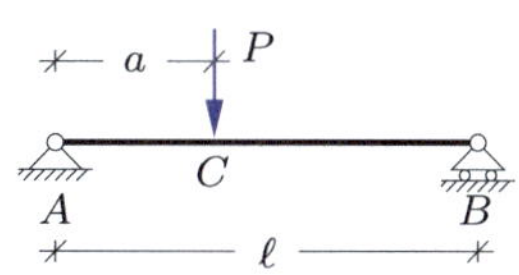

Exercise 4.8. The figure shows a beam of length $a + b$ with a simple fixed support in A and a simple support permitting horizontal motion in B. The beam is linear elastic with constant bending stiffness EI_z, and a vertical force P acts at the tip of the beam in C. Solve the following problems by the principle of virtual work.

a) Determine the displacement w_C at the tip of the beam.

b) Determine the rotation θ_C at the tip of the beam.

c) Determine the rotation θ_A at the simple support in A.

d) Determine the displacement w_D at the center D of AB.

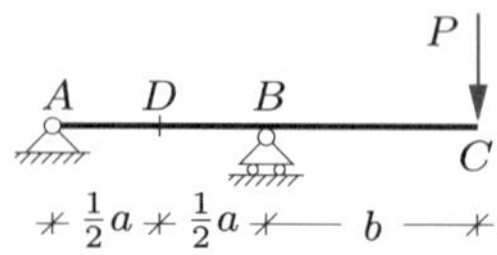

Exercise 4.9. The figure shows a two span simply supported beam similar to that in the previous exercise. The beam is linear elastic with bending stiffness EI_z, and a vertical uniformly distributed load with intensity p acts on both spans of the beam. Solve the following problems by the principle of virtual work.

a) Determine the displacement w_C at the tip of the beam.

b) Determine the rotation θ_A at the simple support in A.

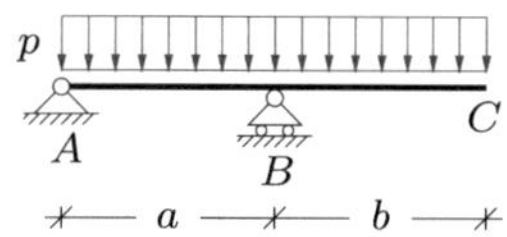

Exercise 4.10. The figure shows a three span simply supported beam with constant bending stiffness EI_z. The two exterior spans AC and BE both have length a, while the inner span ADB has length $4a$. Vertical tip forces P act at C and E, while a force $2P$ acts at the center of the beam in D. Solve the following problems by the principle of virtual work.

a) Determine the displacement w_C of the tip.

b) Find the rotation θ_A at the left support.

c) Determine the transverse displacement w_D at the center of the beam.

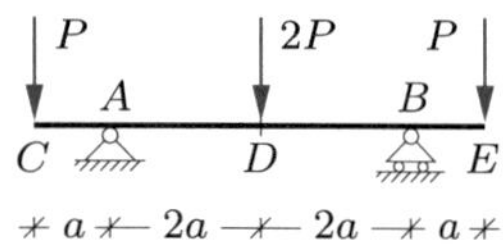

Exercise 4.11. The figure shows a beam of length $2a$ which is fixed in A and simply supported in B. The beam is furthermore hinged at the center C. The beam is linear elastic with constant bending stiffness EI_z, and it is loaded by a uniformly distributed load with intensity p. Solve the following problems by the principle of virtual work.

a) Determine the displacement w_C at the location of the hinge.

b) Determine the rotation θ_B at the simple support.

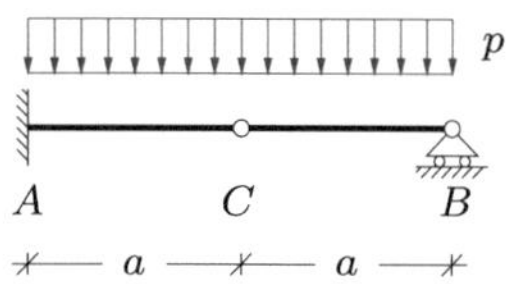

Exercise 4.12. Consider the simple frame in Example 4.15. Determine the rotation θ_C of the joint and sketch the deformation form of the frame.

Exercise 4.13. The figure shows a simple frame similar to that in Example 4.14, with width $2a$ and height a. The supports are fixed simple in A and simple with a vertical roller in B. All beams are linear elastic with constant bending stiffness EI_z. The frame is loaded by a uniformly distributed load with intensity p on the horizontal beam BC. Solve the following problems by the principle of virtual work.

a) Determine the vertical displacement w_B at the support in B.

b) Determine the rotation θ_C at the joint.

c) Determine the rotation θ_A at the fixed simple support in A.

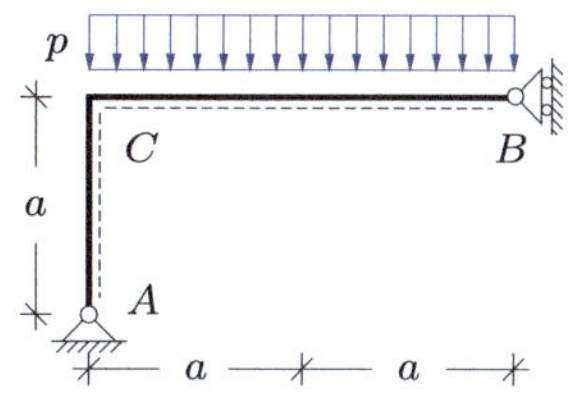

Exercise 4.14. The figure shows a T-frame with width $a + 2a$ and height b. A vertical force P is acting at the tip in D. The distribution of the internal forces has previously been determined in Exercise 3.17. All beams in the frame are linear elastic with constant bending stiffness EI_z. Solve the following problems by the principle of virtual work.

a) Determine the vertical displacement w_D at the location of the force.

b) Determine the rotation θ_C at the joint.

c) Determine the vertical displacement w_B at the left support in B.

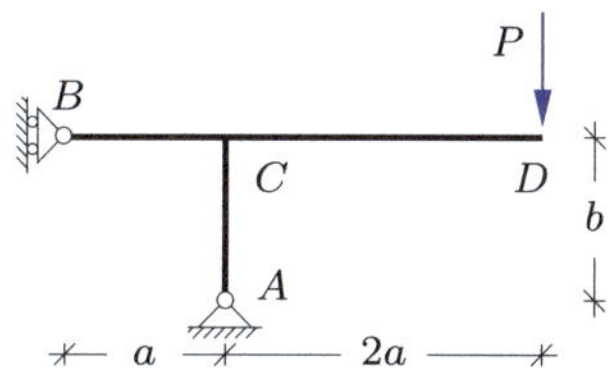

Exercise 4.15. The figure shows a T-frame similar to that in the previous exercise, with width $a+2a$ and height b. Note that the roller in B permits horizontal motion. A uniformly distributed vertical load with intensity p acts on the cantilever part CD. All beams in the frame are linear elastic with constant bending stiffness EI_z. Solve the following problems by the principle of virtual work.

a) Determine the vertical displacement w_D at the tip D.

b) Determine the rotation θ_C at the joint C.

c) Determine the horizontal displacement w_B at the left support in B.

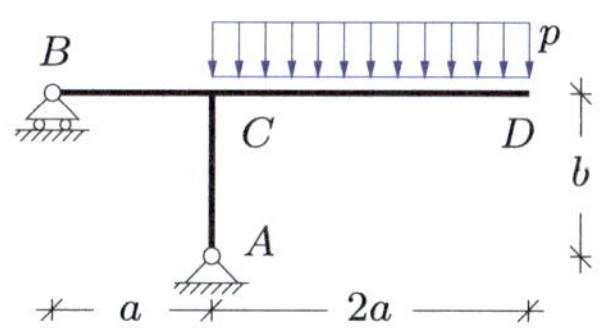

Exercise 4.16. The figure shows a three-hinge frame with an internal hinge placed in E at the center of the horizontal beam BC. The frame is loaded by a uniformly distributed load with intensity p on BC. The distribution of the internal forces has previously been determined in Exercise 3.18. All beams in the frame are linear elastic with constant bending stiffness EI_z. Solve the following problems by the principle of virtual work.

a) Determine the vertical displacement at the hinge in E.

b) Determine the rotation θ_B at the left joint.

c) Determine the rotation θ_A at the left support.

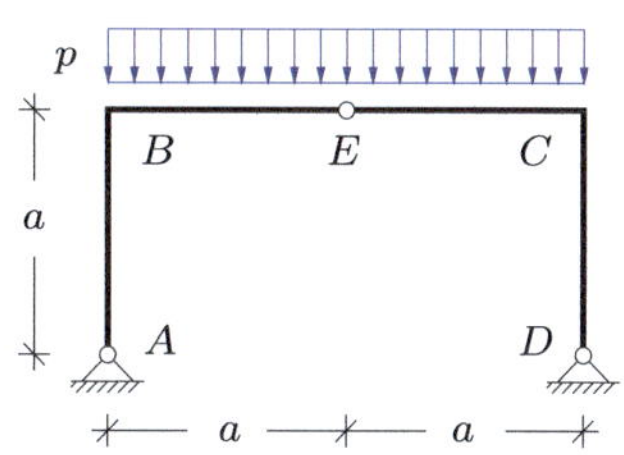

Exercise 4.17. The figure shows a three-hinge frame with an internal hinge placed in E at the center of the horizontal beam BC. The frame is similar to that in the previous exercise, but is now exposed to horizontal forces P acting at the joints B and C, respectively. The distribution of the internal forces has previously been determined in Exercise 3.19.

a) Determine the sum of the horizontal displacements in B and C, and use this result to obtain the individual horizontal displacements in B and C.

b) Determine the rotation θ_B at the left corner B.

c) Determine the rotation θ_A at the left support.

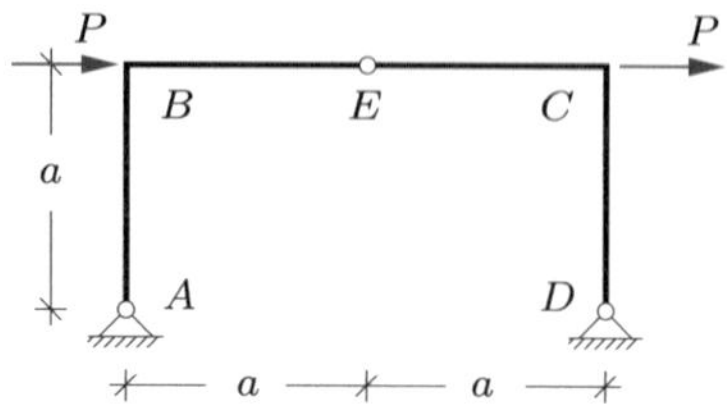

5 Column Stability

A central assumption of the theories for trusses, beams and frames developed in the previous chapters is that the equilibrium conditions have been formulated with reference to the original geometry, i.e. neglecting the fact that the structure via its deformation takes on a somewhat modified geometric configuration. The basic problem of a curve with elastic stiffness – the so-called 'elastica' – loaded by a concentrated force at the end was treated by Leonhard Euler (1707–1783), who gave a very extensive and general analysis of this special problem. In many cases the effect of the normal force on the stability of beam and frame structures can be analyzed by using a somewhat simpler theory including small, but finite, displacements of the original configuration.

In this chapter the theory is developed for a beam with a non-trivial normal force. For simplicity – and because this is often the case – the normal force is assumed to be given, or to be a parameter to be determined by the specific problem in question. The key point of the theory is, that when the beam is displaced $w(x)$ in the transverse direction, the normal force acting in the beam is also displaced. When the normal force has a sufficient magnitude, this

S. Krenk, J. Høgsberg, *Statics and Mechanics of Structures*,
DOI 10.1007/978-94-007-6113-1_5,

effect becomes important, and in the case of a compressive force the effective stiffness of the structure may be reduced, possibly leading to instability of the structure. In many cases of practical interest the normal force may be deduced from the load, and therefore considered as a known quantity. This simplified stability problem is the subject of the present chapter.

The present chapter is devoted to the basic properties of the 'linearized' stability problem of a single member. First, the simple bending theory of beams is extended to include the effect of a normal force in Section 5.1, and it is demonstrated that a compressive normal force leads to reduced stiffness. For a sufficiently large normal force there is no stiffness to resist bending and the column becomes unstable and buckles. The magnitude of the load, at which buckling occurs, is called the critical load, and Section 5.2 considers the stability problem and the associated critical load and buckling shape for ideal columns. In practice, columns are not ideally straight and the load is not only axial. Buckling of real columns is therefore a gradual process, in which the displacements increase and eventually become virtually unbounded. The magnitude of the displacements before reaching the critical load is determined by imperfections in the initial geometry and by bending loads, causing initial curvature. A column design procedure based on a combination of material strength and the influence of initial imperfections is developed in Section 5.3. In the specific column problems treated in this chapter the shear deformation is of minor importance, and the column theory is therefore developed as a generalization of Bernoulli beam theory without including shear flexibility. The effects of normal force and shear flexibility can be combined in a convenient approximate way as discussed in Chapter 7.

The focus in this chapter is on the development of the basic principles in the context of the single structural element – the column. Many structures contain beams carrying a substantial normal force. These members are called beam-columns, and they combine the properties of a beam with the particular features of a column. In beam-columns the effect of the normal force is to change the deformation stiffness. A convenient way of representing this effect, suitable for numerical analysis of beams and frames, is presented in Chapter 7.

5.1 Beam with normal force

The basis for the theory of elastic beam-columns is the equilibrium equations, formulated for the deformed state of the beam. Figure 5.1 shows the deformed state of a beam-column with distributed transverse load $p(x)$. The displacements, and in particular the rotations due to the displacement gradients, are assumed to be small, and thus no distinction will be made between the length increment ds along the beam axis in the deformed state and its projection dx on the line of the initial beam axis. The figure shows a slice of thickness $ds \simeq dx$, and the forces and moments acting on it. The internal force vector on a section of the beam has the components N and Q in the axial and transverse directions, respectively. Note, that in the present formulation of a linearized beam-column theory the normal force N is taken as the component along the direction of the original beam axis, and the shear force Q is in a direction normal to this. In the linearized theory the normal force N is treated as prescribed or as an unknown parameter, and the axial equilibrium and extension of the beam is therefore not treated separately.

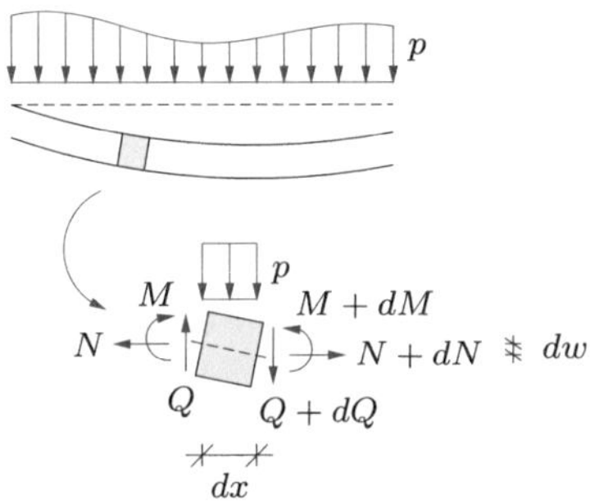

Fig. 5.1: Equilibrium of beam-column in deformed state.

Equilibrium in the transverse direction requires that the projection of all forces on this direction have the sum zero. With the present definition of shear force and normal force, only the shear force Q contributes to the transverse equilibrium, where the sum of internal and external contributions is

$$(Q + dQ) - Q + p\,dx = 0\,. \tag{5.1}$$

The terms $\pm Q$ cancel, and division by dx leads to the differential equation

$$\frac{dQ}{dx} = -p\,. \tag{5.2}$$

The sum of moments must also vanish. In the present case the shear force Q contributes as a force couple with distance dx, and the normal force N contributes as a force couple with distance dw. This gives the moment equilibrium equation

$$(M + dM) - M + N\,dw - Q\,dx = 0\,. \tag{5.3}$$

After cancelation of $\pm M$, division by dx leads to the differential equation

$$\frac{dM}{dx} + N\frac{dw}{dx} = Q\,. \tag{5.4}$$

The two first-order differential equations (5.2) and (5.4) must be satisfied irrespective of the material properties of the beam. The shear force can be eliminated, resulting in the second-order differential equation

$$\frac{d^2M}{dx^2} + \frac{d}{dx}\Big(N\,\frac{dw}{dx}\Big) + p = 0\,. \tag{5.5}$$

This differential equation contains the moment M, the normal force N and the displacement derivative dw/dx. The dependence on the displacement gradient excludes the possibility of determining the internal forces from statics alone, and it is therefore necessary to express the moment M in terms of the deformation of the beam.

In the linear beam bending theory developed in Chapter 4 the cross-section rotation θ and the curvature of the beam axis κ were introduced as

$$\theta = -\,\frac{dw}{dx}\,, \qquad \kappa = \frac{d\theta}{dx} = -\,\frac{d^2w}{dx^2}\,. \tag{5.6}$$

The relation between the moment M and the curvature κ is not changed by the presence of the normal force, and thus the moment is expressed in terms of the bending stiffness EI_z and the curvature by (4.18),

$$M = EI_z\,\kappa = -\,EI_z\,\frac{d^2w}{dx^2}\,. \tag{5.7}$$

The difference between the bending theory and the present theory including the displacement of the normal force is found in the relation for the shear force Q. The shear force Q is determined from the equilibrium equation (5.4). When the moment is expressed in terms of the constitutive relation (5.7), the shear force equation takes the form

$$Q = -\,\frac{d}{dx}\Big(EI_z\,\frac{d^2w}{dx^2}\Big) + N\,\frac{dw}{dx}\,. \tag{5.8}$$

Note the occurrence of the normal force N in the expression for the shear force.

The equilibrium equation follows from substitution of the shear force (5.8) into the transverse equilibrium equation (5.2),

$$\frac{d^2}{dx^2}\Big(EI_z\,\frac{d^2w}{dx^2}\Big) - \frac{d}{dx}\Big(N\,\frac{dw}{dx}\Big) - p = 0\,. \tag{5.9}$$

For a single beam or column this equation must be solved in connection with two boundary conditions at each end. The kinematic boundary conditions are expressed in terms of the displacement w and the rotation $\theta = -dw/dx$, while static boundary conditions are expressed in terms of the moment M and the shear force Q as given by (5.7) and (5.8).

5.1.1 Stiffness reduction from normal force

An important effect of a normal force in a beam is that it changes the effective stiffness of the beam. This effect is here illustrated by a simple example but is of general nature.

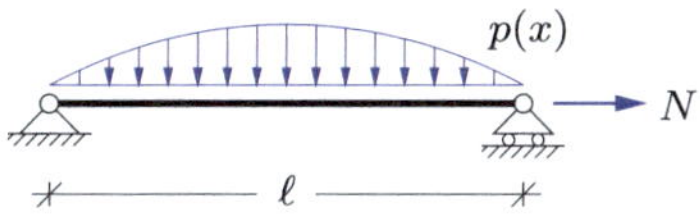

Fig. 5.2: Beam with distributed transverse load and normal force N.

Figure 5.2 shows a simply supported beam-column with normal force N, loaded by a distributed transverse load of intensity

$$p(x) = p_0 \sin\left(\pi \frac{x}{\ell}\right). \tag{5.10}$$

The simple supports at the ends imply that $w = 0$ and $d^2w/dx^2 = 0$ at $x = 0$ and $x = \ell$. These boundary conditions are satisfied by the function

$$w(x) = w_c \sin\left(\pi \frac{x}{\ell}\right), \tag{5.11}$$

and the displacement w_c at the center is determined by substitution into the differential equation (5.9),

$$EI_z \left(\frac{\pi}{\ell}\right)^4 w_c + N \left(\frac{\pi}{\ell}\right)^2 w_c - p_0 = 0. \tag{5.12}$$

This determines the center displacement as

$$w_c = \frac{p_0}{EI_z\left(\dfrac{\pi}{\ell}\right)^4 + N\left(\dfrac{\pi}{\ell}\right)^2} = \frac{1}{1 + \dfrac{N}{EI_z}\left(\dfrac{\ell}{\pi}\right)^2} \frac{p_0}{EI_z}\left(\frac{\ell}{\pi}\right)^4. \tag{5.13}$$

The last factors represent the displacement in the corresponding beam problem with $N = 0$,

$$w_c^0 = \frac{p_0}{EI_z}\left(\frac{\ell}{\pi}\right)^4. \tag{5.14}$$

The first factor is an amplification factor, containing the effect of the normal force. It is seen that the amplification becomes infinite at a compressive force $N = -P_E$ of magnitude

$$P_E = EI_z\left(\frac{\pi}{\ell}\right)^2. \tag{5.15}$$

This particular load is called the Euler load, a reference to the original work of Euler on columns. In column problems it is often convenient to consider compressive axial forces as positive. This is handled by introducing the notation $P = -N$, whereby P denotes an axial force with positive values corresponding to compression.

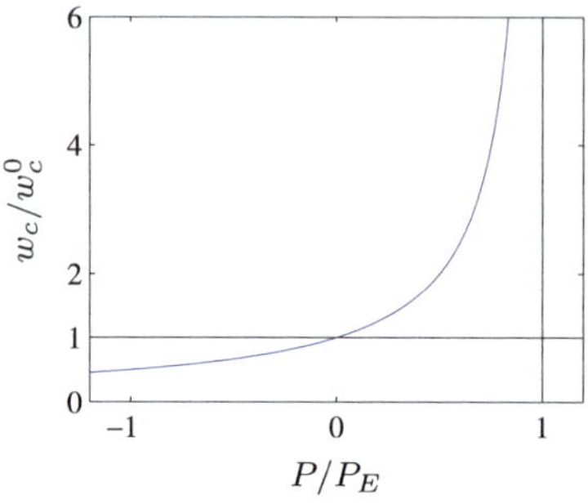

Fig. 5.3: Load-deflection curve for simply supported beam.

In terms of the two characteristic parameters w_c^0 and P_E the center displacement formula takes the form

$$w_c = \frac{w_c^0}{1 - P/P_E}. \tag{5.16}$$

This relation is shown graphically in Fig. 5.3. Although the solution is particularly simple in the present case due to the special choice of load distribution, the amplification behavior illustrated in Fig. 5.3 applies to most beam-column problems. It is seen that the application of axial compression ($P > 0$) leads to an increase of the deformations, while axial tension ($P < 0$) reduces the deformations. The effect of axial compression is much more dramatic than axial tension, and for $P = P_E$ the beam-column has lost its stiffness completely, leading to column instability, discussed in the following section.

5.2 Stability of the ideal column

In the previous section it was found that for a sufficiently large compressive axial force a simply supported beam could obtain arbitrarily large transverse deformations, even for a very small transverse load. This axial load, often called the Euler load, can be identified directly, without applying a transverse load. The idea is to consider an ideally straight column as shown in Fig. 5.4a. A compressive axial load $P = -N > 0$ is then applied, Fig. 5.4b. Hereby the column becomes slightly shorter, but in most cases of practical interest, this shortening is negligible. The main point is, that because the column is ideally straight and there is no transverse load, it will remain straight under a limited axial load. If the axial load is increased, a magnitude P_E is reached, at which two solutions exist: a straight configuration, and a buckled form as shown in Fig. 5.4c. This problem has the form of an eigenvalue problem, and the associated critical load P_E is found as an eigenvalue.

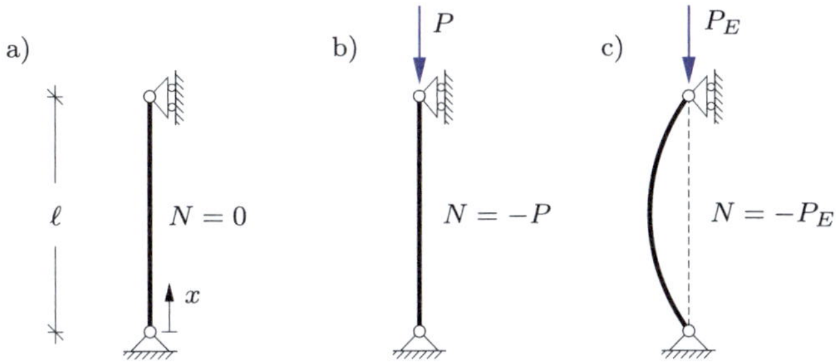

Fig. 5.4: Euler column.

Consider a column of length ℓ and constant bending stiffness EI_z. There is no transverse load, and the differential equation (5.9) then takes the homogeneous form

$$\frac{d^4w}{dx^4} + \frac{P}{EI_z}\frac{d^2w}{dx^2} = 0\,. \tag{5.17}$$

In this equation the axial force P only appears via the coefficient to the second term. This coefficient has the dimension [length^{-2}], and it is therefore advantageous to introduce the parameter k with dimension [length^{-1}], defined by

$$k^2 = \frac{P}{EI_z}\,. \tag{5.18}$$

It is noted that the direct interpretation of k as real-valued assumes a compression force, $P \geq 0$. If the effect of a tension force on the beam bending problem is to be investigated, a modified notation can be used, or the results based on the present k-parameter can be translated into real-valued form.

The differential equation now takes the normalized form

$$\frac{d^4w}{dx^4} + k^2\frac{d^2w}{dx^2} = 0\,. \tag{5.19}$$

The general solution to this homogeneous 4'th order differential equation is

$$w(x) = C_1 + C_2\,kx + C_3\cos(kx) + C_4\sin(kx)\,. \tag{5.20}$$

In this form the coordinate x only appears in the non-dimensional combination kx. The moment follows from (5.7) as

$$\frac{M(x)}{EI_z} = -\frac{d^2w}{dx^2} = C_3\,k^2\cos(kx) + C_4\,k^2\sin(kx)\,, \tag{5.21}$$

and the shear force from (5.8),

$$\frac{Q(x)}{EI_z} = -\frac{d^3w}{dx^3} - k^2\frac{dw}{dx} = -C_2\,k^3\,. \tag{5.22}$$

These relations are used to formulate static boundary conditions. First the general solution is obtained for the simply supported ideally straight column – the so-called Euler column. This solution is used as a reference, and the influence of alternative support conditions is illustrated by examples.

The Euler column

The boundary conditions of the Euler column shown in Fig. 5.4 are

$$w(0) = w(\ell) = 0\,, \qquad M(0) = M(\ell) = 0\,. \tag{5.23}$$

Both the differential equation (5.19) and the boundary conditions (5.23) are homogeneous. Thus, the solution will be $w(x) \equiv 0$, except for particular values k_n of the parameter k that permit a nontrivial solution. These values k_n are the eigenvalues, and to each eigenvalue corresponds an eigenfunction $w_n(x)$. The eigenfunctions describe the buckled shape of the column and are often called the buckling modes.

The boundary conditions at the end $x = 0$ give the equations

$$\begin{aligned} w(0) &= C_1 + C_3 = 0\,, \\ w''(0) &= \quad\; - k^2\,C_3 = 0\,, \end{aligned} \tag{5.24}$$

where the notation $w'' = d^2w/dx^2$ has been used for the second derivative. These equations determine the parameters $C_1 = C_3 = 0$. The boundary conditions at $x = \ell$ then give

$$\begin{aligned} w(\ell) &= k\ell\,C_2 + \sin(k\ell)\,C_4 = 0\,, \\ w''(\ell) &= \quad\; - k^2\sin(k\ell)\,C_4 = 0\,. \end{aligned} \tag{5.25}$$

These equations imply that

$$k\ell\,C_2 = 0\,, \qquad k^2\sin(k\ell)\,C_4 = 0\,. \tag{5.26}$$

A nontrivial solution requires $k\ell > 0$, and thus the first equation gives $C_2 = 0$. This leaves the final equation (5.26b). Naturally this equation can be satisfied by $C_4 = 0$, but this would reduce the solution to $w(x) \equiv 0$.

A nontrivial solution with $C_4 \neq 0$ is found by selecting the parameter k such that

$$\sin(k\ell) = 0\,. \tag{5.27}$$

This equation has the positive roots

$$k\ell = \pi, 2\pi, 3\pi, \cdots \quad \text{or} \quad k_n = n\,\frac{\pi}{\ell}\,,\; n = 1, 2, 3, \cdots \tag{5.28}$$

These roots correspond to the axial loads

$$P_n = EI_z\, k_n^2 = n^2\Big(\frac{\pi}{\ell}\Big)^2 EI_z\,, \quad n = 1, 2, 3, \cdots \tag{5.29}$$

The smallest of these loads is called the Euler load,

$$P_E = EI_z\,\Big(\frac{\pi}{\ell}\Big)^2. \tag{5.30}$$

At this load the column can buckle into a non-straight mode of deformation, given by the transverse displacement

$$w_E(x) = C\,\sin\Big(\pi\frac{x}{\ell}\Big)\,, \tag{5.31}$$

illustrated in Fig. 5.4c.

The transverse displacement of a beam-column with a transverse load will grow towards infinity, as the axial compression force P approaches the Euler load P_E as demonstrated for a special case in Section 5.1.1. Thus, it appears that a beam-column gradually looses its bending stiffness with increasing normal compression.

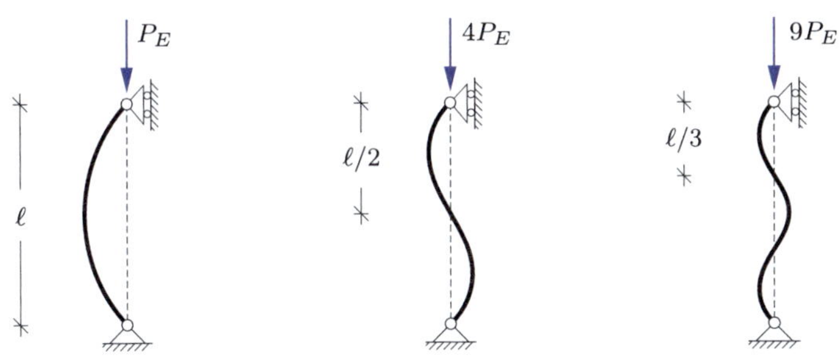

Fig. 5.5: First three buckling forms of the Euler Column.

In general the buckling modes of the Euler column and the corresponding buckling loads are given by

$$\left.\begin{aligned} P_n &= n^2\Big(\frac{\pi}{\ell}\Big)^2 EI_z \\ w_n(x) &= C_n\,\sin\Big(n\pi\frac{x}{\ell}\Big) \end{aligned}\right\}\quad n = 1, 2, 3, \cdots \tag{5.32}$$

The first three buckling modes are shown in Fig. 5.5. In practice it will be difficult to increase the load beyond the smallest buckling load P_E, if the column is only supported at the ends. However, the higher buckling modes correspond to the buckling modes of columns with equally spaced intermediate supports.

The column theory expressed by the linear differential equation (5.9) is only approximate. In its derivation it was assumed that the rotations are 'small', and that the length along the deformed beam can be represented by its projection, $ds \simeq dx$. These approximations reduce the problem to the form of a linear eigenvalue problem, but also limit the scope of the solution to the onset of instability, where the deformation is small. Thus, the theory is useful in establishing a reference value, such as P_E, for the onset of instability, while description of the development of the load and displacements after the onset of instability requires a non-linear theory, see e.g. Dym (1974) or Bažant and Cedolin (2010).

Example 5.1. Built-in column. Figure 5.6 shows a column of length ℓ with one fixed end, supporting an axial compression force P at the free end. The general solution is given by (5.20), where the arbitrary constants and the parameter k are to be determined by the boundary conditions as in the case of the simply supported column treated above.

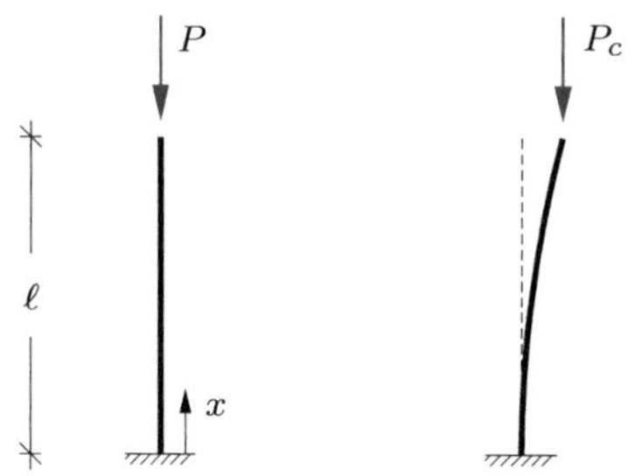

Fig. 5.6: Buckling of built-in column.

In the present problem the boundary conditions at the fixed end are

$$w(0) = C_1 + C_3 = 0\,,$$
$$w'(0) = k\,C_2 + k\,C_4 = 0\,.$$

These equations give $C_3 = -C_1$, $C_4 = -C_2$, and the solution (5.20) reduces to the form

$$w(x) = C_1[\,1 - \cos(kx)\,] + C_2[\,kx - \sin(kx)\,]\,.$$

The boundary conditions at the top of the column are $M(\ell) = 0$ and $Q(\ell) = 0$. By (5.22) the condition $Q(\ell) = 0$ gives $C_2 = 0$, and by (5.21) the condition $M(\ell) = 0$ then is

$$w''(\ell) = k^2\,\cos(k\ell)\,C_1 = 0\,.$$

From this equation the eigenvalues are determined as

$$k\ell = \tfrac{1}{2}\pi, \tfrac{3}{2}\pi, \tfrac{5}{2}\pi, \cdots \quad \text{or} \quad k_n = n\Big(\frac{\pi}{2\ell}\Big), \; n = 1, 3, 5, \cdots$$

corresponding to the axial loads

$$P_n = EI_z\, k_n^2 = n^2\Big(\frac{\pi}{2\ell}\Big)^2 EI_z\,, \quad n = 1, 3, 5, \cdots$$

and the buckling modes

$$w_n(x) = C_n[\,1 - \cos(k_n x)\,]\,, \quad k_n = n\Big(\frac{\pi}{2\ell}\Big), \; n = 1, 3, 5, \cdots$$

Note, that these modes correspond to the symmetric buckling modes of a simply supported Euler column of length 2ℓ, obtained by extending the actual column symmetrically below the fixed support. □

Example 5.2. Column combining a fixed and a simple support. In Fig. 5.7 a simple support has been added to the column of Example 5.1. This does not change the solution procedure, but the result can no longer be given explicitly. The general solution is given by (5.20), and after imposing the boundary conditions at the fixed end as in Example 5.1 the solution takes the form

$$w(x) = C_1[\,1 - \cos(kx)\,] + C_2[\,kx - \sin(kx)\,]\,.$$

The boundary conditions at the top of the column are $w(\ell) = 0$ and $M(\ell) = 0$, whereby

$$\begin{aligned} w(\ell) &= [\,1 - \cos(k\ell)\,]\,C_1 + [\,k\ell - \sin(k\ell)\,]\,C_2 = 0\,,\\ w''(\ell) &= k^2 \cos(k\ell)\,C_1 + k^2 \sin(k\ell)\,C_2 = 0\,. \end{aligned}$$

A nontrivial solution to this pair of equations can only be obtained, if the determinant of the equation system vanishes, i.e. if

$$[\,1 - \cos(k\ell)\,]\sin(k\ell) - [\,k\ell - \sin(k\ell)\,]\cos(k\ell) = 0\,.$$

This equation can be reformulated as

$$\tan(k\ell) = k\ell\,.$$

This is a transcendental equation. The left and right hand sides are shown in Fig. 5.8.

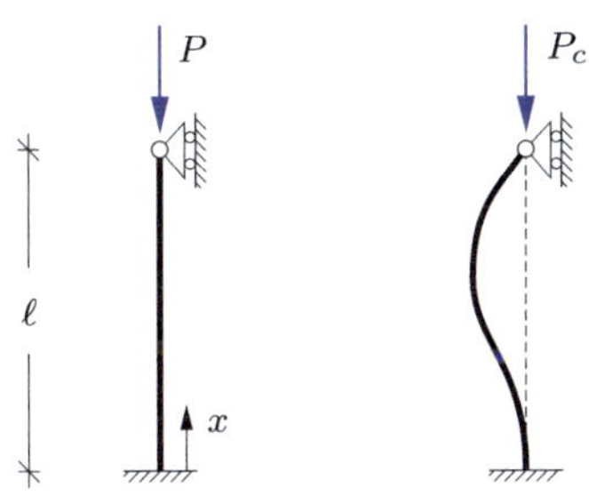

Fig. 5.7: Column with fixed and simple supports.

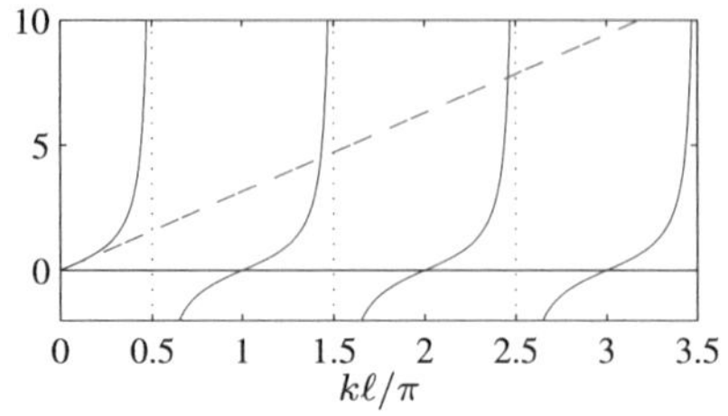

Fig. 5.8: Stability equation: $\tan(k\ell)$ (–) and $k\ell$ (- -).

The roots of the transcendental stability equation are given by the abscissae $k_1\ell, k_2\ell, \cdots$ of the points of intersection. These abscissae can be found by iteration, starting from the values at which $\tan(k\ell)$ has a vertical asymptote,

$$k_n^{(0)}\ell = (n + \tfrac{1}{2})\,\pi \qquad , \qquad n = 1, 2, 3, \cdots$$

Note, that the root $k_0\ell = 0$, corresponding to $k_0^0 = \frac{1}{2}\pi$ is without interest. The iteration procedure can be formulated as

$$k_n^{(i+1)}\ell = n\,\pi + \tan^{-1}(k_n^i\ell)\,.$$

The two first steps and the final value of the parameter $k_n\ell$ are given below.

n	1	2	3	4
$k_n^{(0)}\ell$	4.7124	7.8540	10.9956	14.1372
$k_n^{(1)}\ell$	4.5033	7.7273	10.9049	14.0665
$k_n\,\ell$	4.4934	7.7253	10.9041	14.0662
P_n/P_E	2.0457	6.0468	12.0471	20.0472

The table also gives the buckling loads, conveniently determined by

$$\frac{P_n}{P_E} = \frac{\ell^2}{\pi^2}\frac{P_n}{EI_z} = \left(\frac{k_n\ell}{\pi}\right)^2.$$

The parenthesis is the ratio of the non-dimensional parameter $k_n\ell$ of the actual column to its value π for the first buckling mode of the Euler column. This is a convenient form, as $k_n\ell$ is the unknown iteration parameter.

The table illustrates that, apart from the first two roots, the remaining roots are given to within 1 pct. by the formula $k_n\ell \simeq (n + \frac{1}{2})$, $n = 3, 4, \cdots$, used as the start value in the iteration. However, in a technical context it is often the first root that is of interest.

The buckling modes are determined by the values $k_n\ell$ that have just been determined. It follows from the boundary condition $w''(\ell) = 0$ that

$$C_1 = -\tan(k\ell)\,C_2 = -k\ell\,C_2\,,$$

where the last relation follows from the determinant equation. When this relation is used to eliminate C_1 the buckling modes can be written as

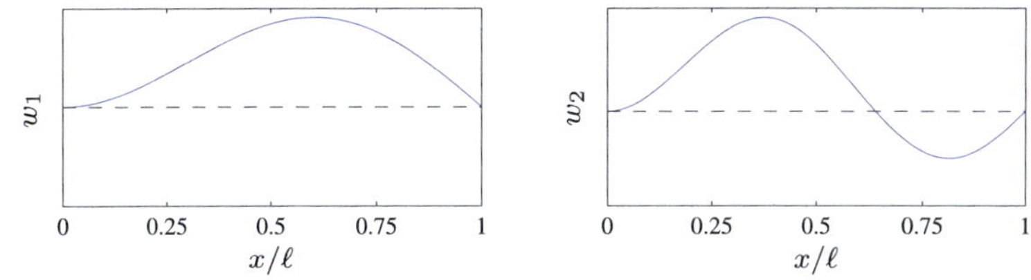

Fig. 5.9: First and second buckling modes of fixed-simple support column.

$$w_n(x) = C_n\left[k_n\ell\big(\cos(k_nx) - 1\big) + k_nx - \sin(k_nx)\right],$$

where C_2 is replaced by C_n, indicating that the solution contains a single unknown scaling factor for each mode n. The first buckling mode corresponding to $k_1\ell = 4.49$ is shown in Fig. 5.9a. The displacement is zero at the supports, and the slope is zero at the fixed support to the left in the figure. The maximum of the buckling mode occurs at $x_{\max} = 0.6\,\ell$. Figure 5.9b shows the second buckling mode, associated with $k_2\ell = 7.73$. This mode satisfies the same homogeneous boundary conditions and has an additional zero crossing. In practice this buckling mode will only occur, if the column is supported at the point of the zero crossing. However, an internal support in the neighborhood of this point will lead to a slight modification of buckling load and buckling form. □

Example 5.3. Instability of column supported beam. Figure 5.10a shows a beam BCD of length $3a$ with a uniformly distributed load of intensity p. The beam is simply supported at B and supported at C by the column CA. The column is simply supported at A and is connected to the beam through a hinge in C, whereby no moment is transferred between beam and column.

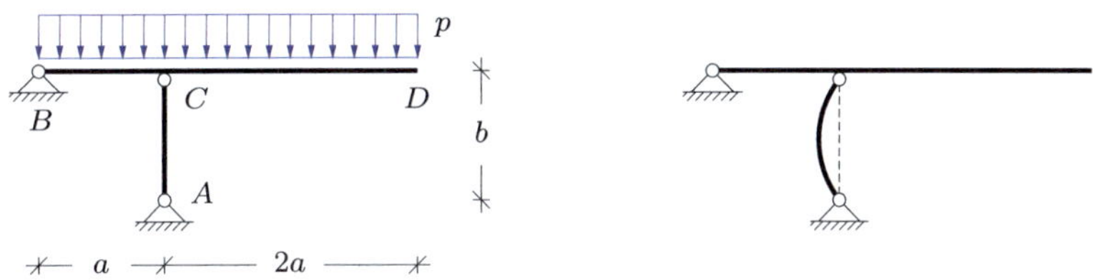

Fig. 5.10: Beam with supporting column.

Moment about C for the column gives zero horizontal reaction in A, and horizontal projection then gives zero horizontal reaction in B. The vertical reactions in A and B are found by moment for the entire structure about B and A, respectively,

$$R_A = \tfrac{9}{2}ap\,, \qquad R_B = -\tfrac{3}{2}ap\,.$$

This implies that the normal force in the column is

$$N_{AC} = -\tfrac{9}{2}ap\,.$$

The minus shows that the column is in compression. The column AC is an Euler column of length b, with zero moment at both ends. Thus the critical magnitude p_c of the intensity of the distributed load is reached, when the magnitude of the normal force is equal to the Euler force, i.e. $|N_{AC}| = P_E$. This gives

$$\tfrac{9}{2}ap = EI_z\left(\frac{\pi}{b}\right)^2 \qquad \Rightarrow \qquad p_c = \frac{2\pi^2 EI_z}{9ab^2}\,.$$

The buckling mode is a sine half-wave, as shown in Fig. 5.10b. □

Often a column will be rigidly connected to a support or structure in a way that imposes an elastic partial constraint against rotation. The elastic constraint gives a restraining moment to the column. This constraint will increase the critical column load and form an intermediate state between the free hinge and a full constraint. The general problem is dealt with in detail in Chapter 7, while the simpler problem of a spring supported column is dealt with in the following example.

Example 5.4. Simply supported column with rotation spring. Figure 5.11 shows a column of length ℓ with simple supports. At the support B the column is also supported by a rotation spring with stiffness parameter $k_\theta = M_B/\theta_B$. The critical load P_c depends on the spring stiffness, and by increasing the spring stiffness the critical load increases from the Euler value P_E to that of the column with one hinged and one fixed support, i.e. approximately $2P_E$.

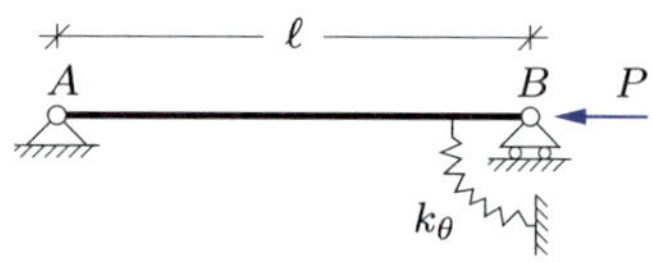

Fig. 5.11: Simply supported column with rotation spring.

The expression for the transverse displacement $w(x)$ is given in (5.20),

$$w(x) = C_1 + C_2\,kx + C_3\cos(kx) + C_4\sin(kx)\,.$$

The boundary conditions in $x = 0$ are $w(0) = 0$ and $M(0) = 0$, and imply that $C_1 = C_3 = 0$. The full solution can then be reduced to

$$w(x) = C_2\,kx + C_4\sin(kx)\,.$$

The third boundary condition is $w(\ell) = 0$, which leads to

$$w(\ell) = C_2\,k\ell + C_4\sin(k\ell) = 0\,.$$

The final boundary condition represents the balance between the internal moment at the right support and the moment introduced by the rotation spring, i.e. $M(\ell) = -k_\theta\theta(\ell) = k_\theta w'(\ell)$. When introducing the moment relation $M(\ell) = -EI_z w''(\ell)$ this boundary condition can be written as

$$EI_z w''(\ell) + k_\theta w'(\ell) = \Big(EI_z k^2\sin(k\ell) + k_\theta k\cos(k\ell)\Big)C_4 + k_\theta k C_2 = 0\,.$$

The boundary conditions at $x = \ell$ are conveniently written on matrix form

$$\begin{bmatrix} k\ell & \sin(k\ell) \\ \alpha & \alpha\cos(k\ell) - k\ell\sin(k\ell) \end{bmatrix}\begin{bmatrix} C_2 \\ C_4 \end{bmatrix} = \begin{bmatrix} 0 \\ 0 \end{bmatrix},$$

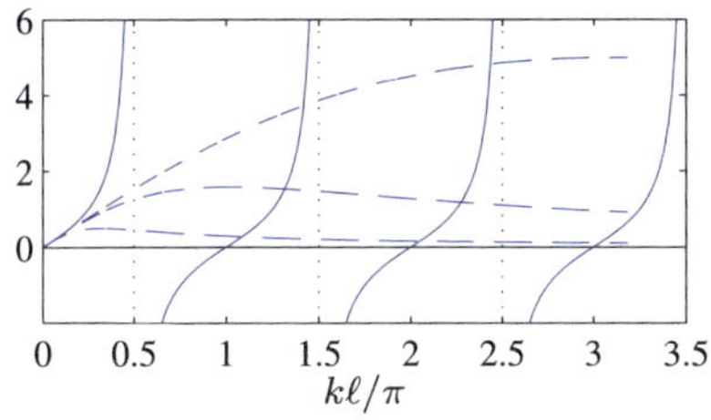

Fig. 5.12: Stability equation: $\tan(k\ell)$ (–) and $k\ell/(1+(k\ell)^2/\alpha)$ (- -).

where $\alpha = k_\theta \ell / EI_z$ is a non-dimensional rotation spring stiffness. The matrix equation is homogeneous and non-trivial solutions requires that the determinant of the matrix is zero. This gives the characteristic equation

$$k\ell\Big(\alpha\cos(k\ell) - k\ell\sin(k\ell)\Big) - \alpha\sin(k\ell) = 0,$$

which can be reduced to the following stability equation

$$\tan(k\ell) = \frac{k\ell}{1+(k\ell)^2/\alpha}.$$

For $\alpha \to 0$ the second term in the denominator tends to infinity and thus the right hand side vanishes, whereby $\tan(k\ell) = 0$ and $k\ell = n\pi$. This corresponds to the solution for the Euler column as expected. On the other hand $\alpha \to \infty$ implies that $\tan(k\ell) = k\ell$, which agrees with the stability equation found for the column in Example 5.2 with a fixed and a simple support. The intermediate behavior of the stability equation can be illustrated by plotting the two components of the equation. In Fig. 5.12 the solid curve is the tangent function while the dashed curves are the right hand side of the stability equation for $\alpha = 1$, 10 and 100, respectively. The solutions are represented by the intersections of the curves. □

5.2.1 Equivalent column length

The general solution to the homogeneous differential equation for a straight column was given in (5.20) as

$$w(x) = C_1 + C_2\,kx + C_3\cos(kx) + C_4\sin(kx). \tag{5.33}$$

This solution consists of a linear part, represented by the two first terms, and a trigonometric part, represented by the two last terms. The critical load, at which instability occurs, is determined by the curvature of the buckled shape. The linear part of the solution does not contribute to the curvature, which is determined by the parameter k^2. In fact, it follows from the definition (5.18) of the parameter k that the critical load is determined by

$$P_c = k_c^2\,EI_z, \tag{5.34}$$

where k_c denotes the value associated with the critical load P_c, corresponding to instability.

The critical load of the Euler column, i.e. the lowest instability load for a simply supported ideal column, was given by (5.30) as

$$P_E = \left(\frac{\pi}{\ell}\right)^2 EI_z\,. \tag{5.35}$$

This formula gives the critical load P_E in terms of the bending stiffness EI_z and the length ℓ between the supports of the column. The Euler column has simple supports at both ends, and the column length ℓ is therefore also the length of a trigonometric half-wave, spanning between the supports. This length is also characterized as the length between the inflection points of the buckling form, where an inflection point is defined as a point where the curvature changes sign. It is this property that leads to the role of ℓ in the Euler column formula (5.35).

For general support conditions the parameter k_c is characterized by the length between the inflection points – i.e. a trigonometric half-wave of the buckling form. This length is called the effective column length and is denoted ℓ_e. It follows from its definition as the half-wave length of the column solution that it is related to the parameter k_c as $k_c = \pi/\ell_e$, whereby the trigonometric part of the solution is of the form $\sin(k_c x) = \sin(\pi x/\ell_c)$. Hereby the general critical load formula (5.34) takes the form

$$P_c = \left(\frac{\pi}{\ell_e}\right)^2 EI_z\,. \tag{5.36}$$

Thus, the concept of an equivalent column length ℓ_e translates the Euler column formula into a general format.

The role of the effective column length can be further illustrated by combining the general formula (5.36) for the critical load with the Euler formula (5.35),

$$\frac{P_c}{P_E} = \left(\frac{\ell}{\ell_e}\right)^2. \tag{5.37}$$

This formula illustrates the influence of the boundary conditions in changing the critical load by changing the effective column length via the support conditions.

The importance of the concept of the equivalent column length ℓ_e is two-fold: it gives a compact form of the stability load by generalizing the Euler formula, and it provides a visual interpretation of the column length for columns with general support conditions. This latter property is illustrated in Fig. 5.13 showing the lowest buckling mode for the support conditions treated above. All columns are shown for the same actual length ℓ. The corresponding effective column length ℓ_e is indicated to the right in each of the sub-figures. It is remarkable that even for the column combining a fixed

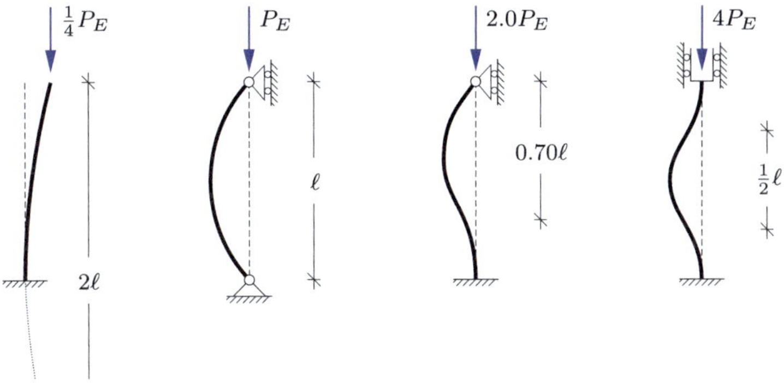

Fig. 5.13: Buckling load and effective column length.

and a hinged support the effective length $\ell_e \simeq 0.7$ can be estimated with fair accuracy from visual inspection of a sketch of the buckled shape.

5.2.2 Buckling direction and intermediate supports

Columns may have cross-sections with different properties with regard to bending and buckling in the transverse y- and z-direction. The problem is illustrated in Fig. 5.14 for a rectangular cross-section with dimensions $a \times b$ with a in the y-direction and b in the z-direction as shown.

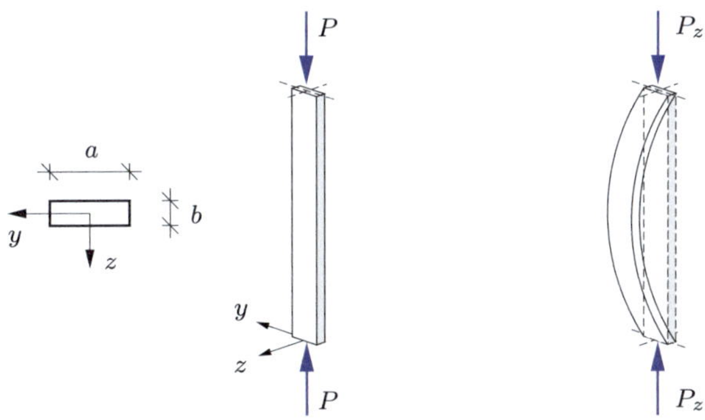

Fig. 5.14: Buckling about the weakest axis.

The area of the cross-section is $A = ab$, and the moments of inertia follow from Example 4.1 as

$$I_y = \int y^2 \, dA = \tfrac{1}{12}a^2 A \,, \quad I_z = \int z^2 \, dA = \tfrac{1}{12}b^2 A \,. \tag{5.38}$$

The column is simply supported at both ends with respect to displacements in both the y- and the z-direction. Thus, the two corresponding buckling loads are

$$P_E^y = \left(\frac{\pi}{\ell}\right)^2 EI_y = \frac{\pi^2}{12}\frac{a^2}{\ell^2}EA\,, \quad P_E^z = \left(\frac{\pi}{\ell}\right)^2 EI_z = \frac{\pi^2}{12}\frac{b^2}{\ell^2}EA\,. \tag{5.39}$$

Clearly, for $a > b$ the bending stiffness in the y-direction is larger, and accordingly this buckling load is larger, $P_E^y > P_E^z$. The argument depends on the relative magnitude of the moments of inertia I_y and I_z, and is not restricted to rectangular sections, but merely assumes symmetry axes.

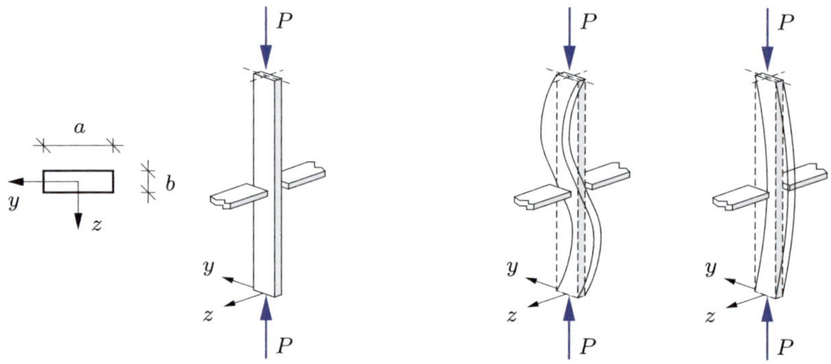

Fig. 5.15: Intermediate support in the z-direction.

Columns are often used as the vertical members of plane frames. This gives the possibility of supporting the column at intermediate locations against motion in the plane of the frame as illustrated in Fig. 5.15, showing an intermediate support at the mid-point providing a constraint against transverse motion in the z-direction. This support increases the buckling load P_E^z associated with buckling in the z-direction. A balanced design can be obtained as illustrated in the following example.

Example 5.5. Column with intermediate support. Consider a column of length ℓ with simple supports in both transverse directions at both ends. The column forms part of a plane frame and is supported against transverse displacement at its mid-point in the z-direction as shown in Fig. 5.15. This support reduces the effective column length for buckling in the z-direction to $\ell_e = \frac{1}{2}\ell$. The cross-section is a rectangle with dimensions a and b in the y- and the z-direction, respectively. The critical loads for buckling in the two directions follow from (5.39), when accounting for the reduced effective column length $\ell_e = \frac{1}{2}\ell$ for buckling in the z-direction,

$$P_E^y = \left(\frac{\pi}{\ell}\right)^2 EI_y = \frac{\pi^2}{12}\frac{a^2}{\ell^2}EA\,, \quad P_E^z = \left(\frac{\pi}{\ell_e}\right)^2 EI_z = \frac{\pi^2}{3}\frac{b^2}{\ell^2}EA\,.$$

The two buckling loads will be identical, $P_E^y = P_E^z$, provided the cross-section dimensions satisfy $a = 2b$. This corresponds closely to building practice for simple wooden frames, where the transverse cross-section dimension is often the double of the in-plane dimension. □

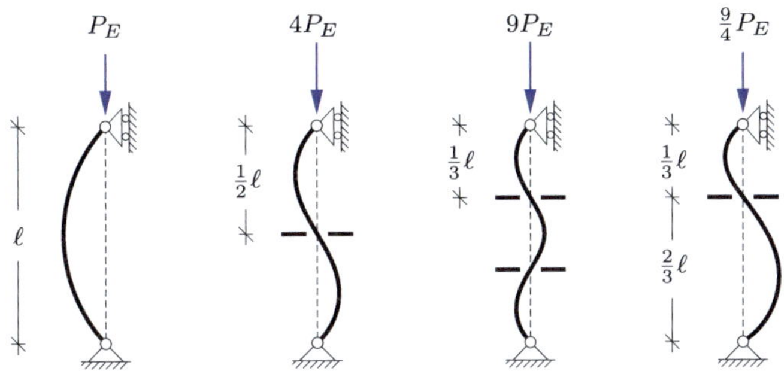

Fig. 5.16: Conservative effective column length estimates.

The effective column length is reduced by the introduction of intermediate supports. For the simply supported columns illustrated in Fig. 5.16 the effective column length is simply $\ell_e = \ell/n$, when the column is divided into n equal parts. If the parts are not of equal length the shorter parts constrain the larger parts. A conservative estimate of the effective column length is $\ell_e \simeq \ell_{\max}$, where $\ell_{\max}$ is the longest distance between neighboring supports or points of inflection.

5.3 Design of columns

The primary goal of column design is to provide sufficient resistance to withstand the load, here primarily in the form of a compressive axial load P. While the previous sections have concentrated on determination of the critical load P_c for an ideal column under various support conditions, design of columns must account for the effect of other factors such as the strength of the material and typical imperfections in column shape. Two extreme cases are illustrated in Fig. 5.17. The left figure shows buckling of a long slender simply supported column at the critical load P_E corresponding to elastic instability. The right figure shows the opposite scenario, in which the capacity of a short column is determined solely by the material strength represented by the yield load P_y.

Most real columns are in a parameter interval somewhere between the two extreme cases shown in the figure, and it is important to develop a column design procedure that provides a smooth transition between the two extremes of ideal elastic buckling and pure material strength. This is done in the following by combining three steps. The first, dealt with in Section 5.3.1, identifies a suitable characterization of column length, whereby the capacity of 'long' columns is dominated by instability, while the capacity of 'short' columns is dominated by material strength. The following Section 5.3.2 describes the effect of the fact that real columns are not ideally straight. The deviation from ideal straightness is called the geometric imperfection of the column, and

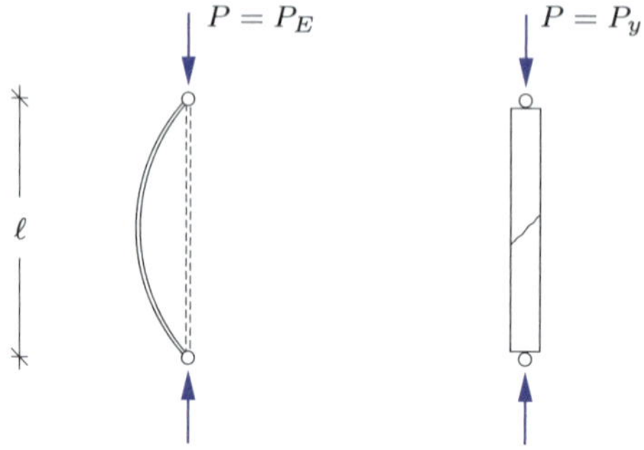

Fig. 5.17: Collapse mechanisms for long and short columns, respectively.

the imperfections lead to a change from buckling at a precise critical load, e.g. P_E, to a more gradual deformation of the column. During this process stresses develop in the column cross-sections as described in Section 5.3.3. Finally, these aspects are combined into a direct column design procedure in Section 5.3.4.

5.3.1 Column length and slenderness

The purpose of classifying a column as 'long' or 'short' is to indicate whether its capacity is primarily governed by stability or strength considerations. Thus, it is clear that the length characterization of a column can not just be a geometric measure, e.g. of actual length relative to a characteristic transverse dimension of the cross-section, but must include some reference to stability and strength parameters for the column.

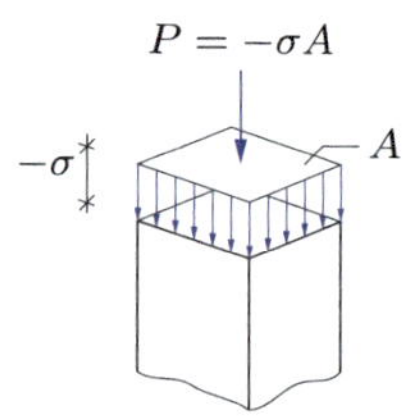

Fig. 5.18: Homogeneous distribution of normal stress σ.

Figure 5.18 illustrates a uniform stress distribution over the cross-section of a short column, acted on by a central compressive force P, as shown in Fig. 5.17b. The stresses are uniformly distributed over the cross-section area A, whereby the constant stress intensity σ is given as

$$\sigma = -\frac{P}{A}\,. \tag{5.40}$$

Note the sign convention with P positive in compression and σ positive in tension.

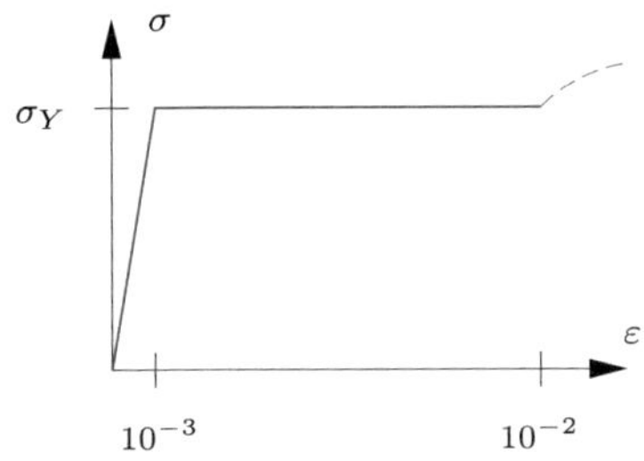

Fig. 5.19: Idealized uniaxial stress-strain curve.

The material can only withstand normal stresses up to a certain magnitude, here denoted σ_Y, where the subscript Y denotes the yield stress. The concept of material yield refers to a behavior observed in most metals, particularly steel. This behavior is illustrated for steel in Fig. 5.19 showing the relation between the normal stress σ and the corresponding longitudinal strain ε in a uniaxial tension or compression test. For limited strain, typically $\varepsilon \lesssim 0.002$, the stress and strain are proportional, corresponding to the linear elastic relation

$$\sigma = E\varepsilon . \tag{5.41}$$

When the strain exceeds the elastic limit, the stress remains at the yield stress, i.e. $|\sigma| \leq \sigma_Y$.

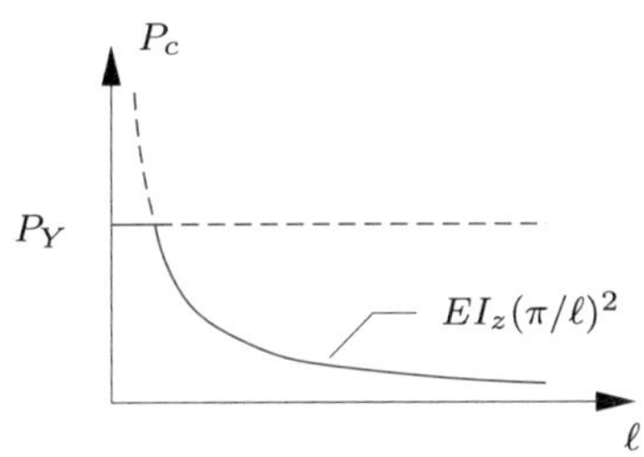

Fig. 5.20: Collapse curve for column.

The collapse of a short column is associated with yielding, and the maximum axial load is determined by the material yield stress as

$$P_Y = \sigma_Y A . \tag{5.42}$$

This strength is independent of the equivalent column length ℓ_e, when this length is sufficiently short. This is indicated by a horizontal line in Fig. 5.20, which shows the critical load P_c as function of the equivalent column length ℓ_e. On the other hand, the capacity of an ideal elastic column is given by the buckling formula in (5.36),

$$P_{el} = \left(\frac{\pi}{\ell_e}\right)^2 EI_z \,. \tag{5.43}$$

The subscript *el* refers to *elastic* column instability, as treated in Section 5.2. For an Euler column $\ell_e = \ell$, and P_{el} in (5.43) recovers the Euler load P_E. It follows from (5.43) that the stability load P_{el} decreases with ℓ_e^{-2}. This is also indicated in Fig. 5.20. The two idealized criteria in terms of strength P_Y and elastic instability P_{el}, respectively, define an intersection point, which is used to define the transition from a short to a long column.

Column slenderness

In the further discussion of column capacity it is convenient to shift focus from the axial force P to the normal stress σ. At this point there is no bending, and thus the stress σ is simply a normalized form of the axial force as given by (5.40). When introducing this normalization in the buckling formula (5.43), the following expression is obtained for the mean compressive stress σ_{el} at elastic column instability,

$$\sigma_{el} = \frac{P_{el}}{A} = \frac{EI_z}{A}\left(\frac{\pi}{\ell_e}\right)^2 . \tag{5.44}$$

In this expression the influence of the cross-section is represented by the ratio I_z/A. This ratio has the dimension [length^2] and defines a length r_z, called the radius of gyration, as

$$r_z^2 = \frac{I_z}{A} \,. \tag{5.45}$$

The radius of gyration is a characteristic distance from the neutral axis of the cross-section. If the area A was split into two parts and these were concentrated at $\pm r_z$, this equivalent cross-section would have the same bending stiffness as the original.

Example 5.6. Radius of gyration for rectangle and I-profile. The bending of a beam with rectangular cross-section of height h and width b is illustrated in Fig. 4.4a. The moment of inertia was calculated in Example 4.1 as $I_z = \frac{1}{12}h^2 A$. The corresponding radius of gyration then follows from (5.45) as

$$r_z = \sqrt{\frac{I_z}{A}} = \frac{h}{2\sqrt{3}} \simeq 0.289h \,.$$

Similarly, the moment of inertia of an I-section of height h shown in Fig. 4.4b with flange width b and flange thickness t was calculated in Example 4.1 as $I_z = \frac{1}{4}h^2 A$, when neglecting the contribution from the web. For this cross-section the radius of gyration is

$$r_z = \sqrt{\frac{I_z}{A}} = \tfrac{1}{2}h \,,$$

corresponding to the fact that the flanges already have concentrated the cross-section area at the distances $\pm\frac{1}{2}h$ from the neutral axis. □

The compression stress σ_{el} at elastic instability can now be expressed by introducing the substitution $I_z = r_z^2 A$ into (5.44),

$$\sigma_{el} = E\left(\frac{\pi\, r_z}{\ell_e}\right)^2 = E\left(\frac{\pi}{\lambda}\right)^2, \tag{5.46}$$

where the non-dimensional slenderness parameter

$$\lambda = \frac{\ell_e}{r_z} \tag{5.47}$$

has been introduced. The slenderness parameter λ is a purely geometric quantity, expressing column length relative to the cross-section radius of gyration r_z.

The formula gives the instability stress σ_{el} in terms of the elastic modulus E and the slenderness parameter λ. It is often more convenient to express σ_{el} relative to the yield stress σ_Y. This ratio is given by

$$\frac{\sigma_{el}}{\sigma_Y} = \frac{E}{\sigma_Y}\left(\frac{\pi}{\lambda}\right)^2 = \frac{1}{\lambda_r^2}, \tag{5.48}$$

where the last equality defines the *relative* slenderness as

$$\lambda_r = \frac{\lambda}{\pi}\sqrt{\frac{\sigma_Y}{E}} = \frac{\ell_e}{\pi r_z}\sqrt{\frac{\sigma_Y}{E}}. \tag{5.49}$$

It follows from the equations in (5.48) that the intersection point between the strength and the stability curves in Fig. 5.20 is characterized by $\sigma_{el} = \sigma_Y$, corresponding to $\lambda_r = 1$.

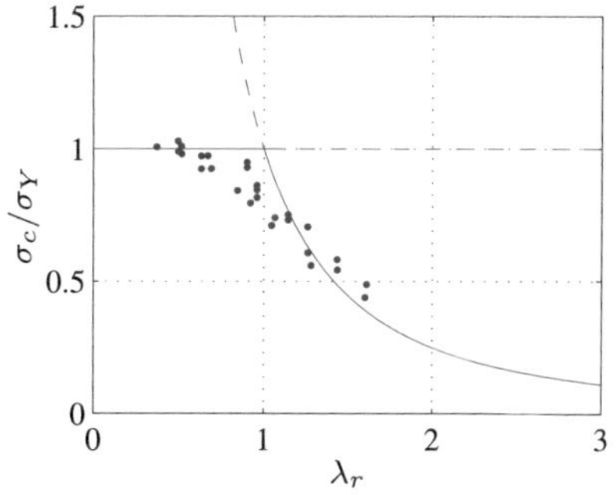

Fig. 5.21: Critical stress as function of relative slenderness.

Figure 5.21 shows the critical column stress in normalized form σ_c/σ_Y plotted against the relative slenderness λ_r. The figure also shows some experimental results for steel columns from Dowling et al. (1988). It is seen that there is a gradual transition from the material strength criterion $\sigma_c \simeq \sigma_Y$ for $\lambda \ll 1$ to the elastic stability criterion $\sigma_c \simeq \sigma_{el}$ for $\lambda \gg 1$. The graph illustrates that the relative slenderness λ_r is an important parameter for characterizing columns, and that a reduction of the idealized column capacity, as material yield *or* instability occurs, must be developed for the transition range around $\lambda_r = 1$. It follows from (5.49) that the transition occurs at the column length

$$\ell_* = \pi\, r_z \sqrt{E/\sigma_Y}\,. \tag{5.50}$$

This formula is convenient for checking, whether a column is 'short' or 'long'.

Example 5.7. Transition length for steel columns. For a steel column the elastic modulus is typically $E = 210\,\text{GPa}$ and for a yield stress $\sigma_Y = 0.3\,\text{GPa}$ this gives the transition length

$$\ell_* = 83.12\, r_z\,.$$

Consider a rectangular cross-section with height h and width b. It was found in Example 5.6 that buckling in the z-direction has a radius of gyration $r_z = h/(2\sqrt{3})$. For an I-section with height h and flange width b and thickness t the radius of gyration was found as $r_z = \frac{1}{2}h$, corresponding to all the material of the cross-section being concentrated at $\pm\frac{1}{2}h$. Substitution of r_z into the expression for the transition length gives

$$\frac{\ell_*}{h} = \begin{cases} 24.0 &, \text{rectangle}\,, \\ 41.6 &, \text{I-section}\,. \end{cases}$$

This indicates that the transition from a short to a long column occurs for significantly longer columns for the I-section than for the rectangular cross-section. □

5.3.2 Geometric imperfections

Real columns are not ideally straight, and as it turns out the effect of deviations from the ideal straightness provides an explanation of the reduction of column capacity for relative slenderness in the transition interval as illustrated by the data points in Fig. 5.21.

Figure 5.22a shows an imperfect simply supported column, where the (small) deviation from a straight line in the unloaded state is described by the function $w_0(x)$. The length of the column is ℓ and the constant bending stiffness is EI_z. When loaded by an axial compression force P, as shown in Fig. 5.22b, an *additional* transverse displacement $w(x)$ occurs. The moment is determined by this additional displacement as

$$M = -EI_z \frac{d^2w}{dx^2}\,. \tag{5.51}$$

The effect of the normal force on the equilibrium is through the *total* displacement $w(x) + w_0(x)$, and thus the equilibrium equation takes the form

$$\frac{d^2}{dx^2}\Big(EI_z\,\frac{d^2w}{dx^2}\Big) - \frac{d}{dx}\Big(N\,\frac{dw+dw_0}{dx}\Big) = 0\,. \tag{5.52}$$

For constant axial force $N = -P$ and bending stiffness EI_z this equation can be written in normalized form as

$$\frac{d^4w}{dx^4} + k^2\,\frac{d^2w}{dx^2} = -k^2\,\frac{d^2w_0}{dx^2}\,, \tag{5.53}$$

where the parameter $k^2 = P/EI_z$ has been introduced in (5.18). The boundary conditions for the simply supported column are $w(0) = w(\ell) = 0$ and $M(0) = M(\ell) = 0$.

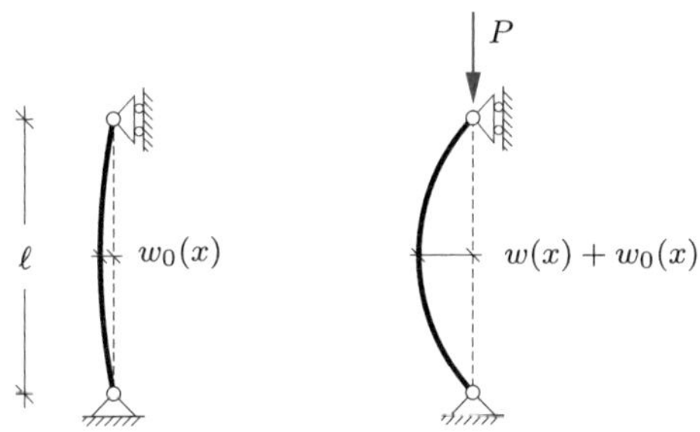

Fig. 5.22: Column with initial imperfection $w_0(x)$.

It is seen that the initial imperfection $w_0(x)$ introduces a non-homogeneous term in the column equation, and thus the imperfect column problem is not an eigenvalue problem. The complete solution to the non-homogeneous differential equation (5.53) is obtained by representing $w(x)$ as a series expansion in terms of the eigenfunctions of the corresponding homogeneous equation. For the present problem the corresponding eigenvalue problem was solved in Section 5.2 and the solution given in (5.32). The series representation of the displacement $w(x)$ then takes the form

$$w(x) = \sum_{n=1}^{\infty} w_n\,\sin(k_nx)\,, \qquad k_n = n\,\frac{\pi}{\ell}\,. \tag{5.54}$$

Substitution of this expansion into the non-homogeneous equilibrium equation (5.53) gives

$$\sum_{n=1}^{\infty}\big(k_n^4 - k^2k_n^2\big)\,w_n\sin(k_nx) = -\,k^2\,\frac{d^2w_0}{dx^2}\,. \tag{5.55}$$

The unknown displacement coefficients w_n are determined by representing the initial imperfection $w_0(x)$ as a series similar to the expansion (5.54) for $w(x)$,

$$w_0(x) = \sum_{n=1}^{\infty} w_n^0 \sin(k_n x)\,, \qquad k_n = n\,\frac{\pi}{\ell}\,. \tag{5.56}$$

The coefficients w_n^0 in this expansion are determined by use of the orthogonality relation for the sine function as

$$w_n^0 = \frac{2}{\ell}\int_0^{\ell} w_0(x)\sin(k_n x)\,dx\,, \quad n = 1, 2, \cdots \tag{5.57}$$

When the series expansion (5.56) for $w_0(x)$ is substituted into (5.55), the displacement coefficients are found to be

$$w_n = \frac{w_n^0}{k_n^2/k^2 - 1} = \frac{w_n^0}{P_n/P - 1}\,, \quad n = 1, 2, \cdots \tag{5.58}$$

This result shows that for increasing tension, $P \to -\infty$, the column becomes increasingly straight, $w_n \to -w_0^n$. Conversely, if an applied compression force $P > 0$ approaches any of the buckling loads P_n the corresponding component of the initial imperfection is amplified. In principle infinite amplification is obtained for $P = P_n$.

The present analysis has been applied to the special case of a simply supported column. However, the method of expanding both the unknown displacement function $w(x)$ and the initial imperfection function $w_0(x)$ in terms of the buckling modes of the corresponding homogeneous problem corresponding to an ideal straight column is also valid for other boundary conditions, and in each case the buckling modes also satisfy a suitable orthogonality relation. This leads to the general conclusion, that the application of a central axial compression load P on an imperfect column will lead to amplification of the contributions of the different buckling mode components in the initial imperfection function $w_0(x)$. In practice the imperfection component corresponding to the first buckling mode will most often dominate the deformation.

Example 5.8. Column with initial curvature. Consider a special case of the imperfect simply supported column shown in Fig. 5.22, in which the imperfection consists of a single sine half-wave of amplitude $w_1^0 = e$. It follows from (5.58) that an axial load P gives the transverse displacement

$$w(x) = \frac{e}{P_E/P - 1}\sin\left(\pi\frac{x}{\ell}\right),$$

where the Euler load is given by (5.30) as $P_E = EI_z(\pi/\ell)^2$. The total displacement for an axial compression force P then is

$$w(x) + w_0(x) = \left(\frac{e}{P_E/P - 1} + e\right)\sin\left(\pi\frac{x}{\ell}\right) = \frac{w_0(x)}{1 - P/P_E}\,.$$

This result shows that the initial displacement $w_0(x)$ is amplified by the same amplification factor for all points on the column. The amplification of the displacements is illustrated in Fig. 5.23 showing the increase of the center displacement for an initial center eccentricity e. It is noted that the amplification factor here is the same as that determined in Section 5.1.1 for the amplification of the displacement from a transverse load.

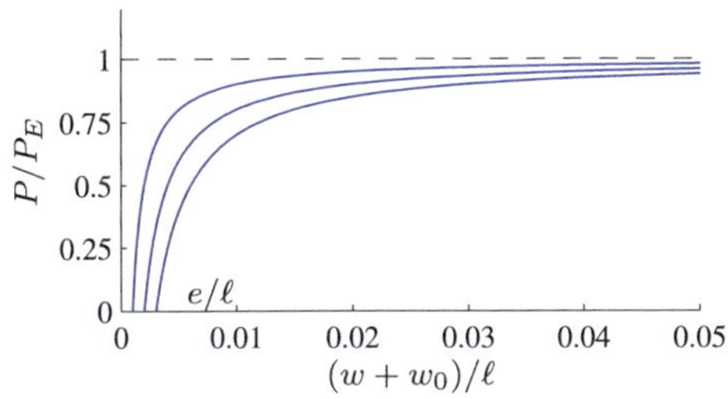

Fig. 5.23: Displacement amplification from $e/\ell = 0.001,\ 0.002,\ 0.003$.

The moment can be found either by the second derivative from (5.51) or from equilibrium of part of the column defined by a section at the position x,

$$M = (w + w_0)\,P = = \frac{w_0(x)\,P}{1 - P/P_E}\,.$$

Thus, the moment exhibits the same amplification as shown in Fig. 5.23. □

5.3.3 Stresses in column cross-sections

The strength of a column depends on the magnitude of the stresses that develop in it. The individual cross-section supports a compressive force P from the imposed axial load and a bending moment M from the transverse displacement $w_{tot} = w + w_0$ of the beam cross-section. The corresponding stress distribution is illustrated in Fig. 5.24. The normal force P produces a uniformly distributed state of compressive stress given by (5.40),

$$\sigma = -\frac{P}{A}\,. \tag{5.59}$$

The bending moment M produces a stress distribution with linear variation over the cross-section, found by combining the strain relation (4.5) with the elastic curvature relation (4.8),

$$\sigma = \frac{M}{I_z}\,z\,. \tag{5.60}$$

The total stress in the cross-section is obtained by addition of these two contributions as illustrated in Fig. 5.24,

$$\sigma = -\frac{P}{A} + \frac{M}{I_z}\,z\,. \tag{5.61}$$

It is seen that the combination of a compressive force and a moment leads to a stress distribution with large compressive stresses at one side of the cross-section and a moderate stress level in either compression or tension at the other side.

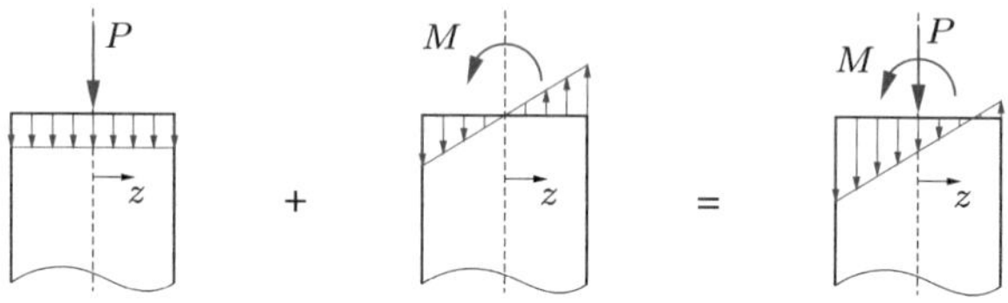

Fig. 5.24: Combination of normal stress from normal force and moment.

In the present context the moment is caused by eccentricity of the axial compressive force P, and it is of interest to investigate the influence of this eccentricity on the stress distribution. If the force P acts close to the neutral line it will generate compressive stresses over the full cross-section. However, for larger eccentricity of the force tension stresses will occur at the side opposite to the eccentric force. The area, within which a normal compressive force will not produce tension in the cross-section, is called the kernel of the cross-section.

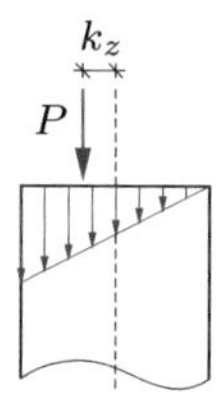

Fig. 5.25: Offset of force P by kernel radius k_z.

The limiting case, where the axial stress vanish at the side of the cross-section, is shown for a cross-section that is symmetric about the plane of the paper in Fig. 5.25. The figure illustrates the case in which the eccentric force acting at $z = -k_z$ gives zero axial stress at the opposite side of the cross-section, defined by $z = z_{\max}$. In this case the moment is $M = k_z P$, and the stress distribution formula (5.61) then gives the condition

$$0 = -\frac{P}{A} + \frac{k_z P}{I_z} z_{\max} \, . \tag{5.62}$$

This condition determines the so-called kernel radius,

$$k_z = \frac{I_z}{A \, z_{\max}} = \frac{r_z^2}{z_{\max}} \, , \tag{5.63}$$

in the symmetry plane of a cross-section.

As demonstrated in Example 4.1 the moment of inertia is $I_z = \frac{1}{4}h^2 A$ for an idealized I-section of height h, in which the influence of the web is neglected. This means that the kernel radius in the web-direction is

$$k_z = \frac{r_z^2}{z_{\max}} \simeq \frac{h^2}{4}\frac{2}{h} = \tfrac{1}{2}h\,. \tag{5.64}$$

Thus, the kernel of the I-section contains the full height h of the section.

Kernel area of a rectangle

The kernel of a cross-section is an area within the cross-section, here illustrated by the rectangle with height h and width b in Fig. 5.26. The planes of symmetry contain the y- and z-axis. When the height of the section is along the z-axis the kernel radius along the two axis are

$$k_y = \frac{r_y^2}{y_{\max}} = \frac{b^2}{12}\frac{2}{b} = \tfrac{1}{6}b\,, \qquad k_z = \frac{r_z^2}{z_{\max}} = \frac{h^2}{12}\frac{2}{h} = \tfrac{1}{6}h\,. \tag{5.65}$$

Thus, for a rectangle the kernel extends along the central third of the cross-section width along the axes as illustrated in Fig. 5.26.

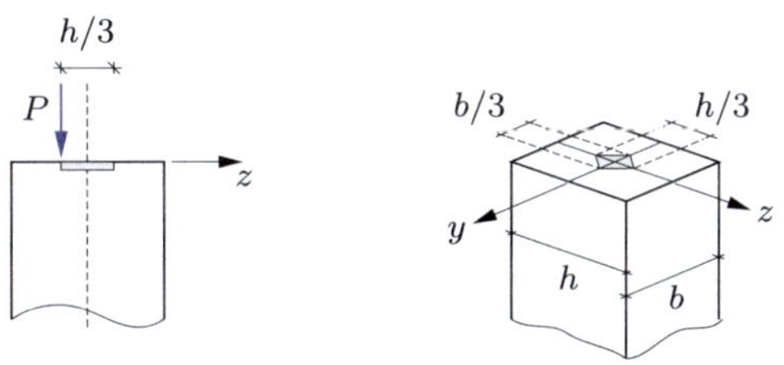

Fig. 5.26: Kernel area for rectangular cross-section.

The shape of the kernel is determined by combining the effect of moments about the y- and the z-axis of the cross-section. It follows from superposition that the normal stress formula (5.61) generalizes to

$$\sigma = -\frac{P}{A} + \frac{M_y}{I_y}\,y + \frac{M_z}{I_z}\,z\,. \tag{5.66}$$

The moments are $M_y = \pm k_y P$ and $M_z = \pm k_z P$, and the kernel condition then generalizes to

$$0 = -\frac{P}{A} \pm \frac{k_y P}{I_y}\,y_{\max} \pm \frac{k_z P}{I_z}\,z_{\max}\,. \tag{5.67}$$

After division by P/A, this condition takes the form

$$\pm\,\frac{y_{\max}}{r_y^2}\,k_y \,\pm\, \frac{z_{\max}}{r_z^2}\,k_z \,=\, 1\,. \tag{5.68}$$

This is a set of four equations for the lines interpolating the point (k_y, k_z) on the kernel boundary between the four points on the axes, that have already been determined. The result is the hatched area shown in Fig. 5.26.

5.3.4 Perry-Robertson's column design criterion

The capacity of columns depends on several factors. The importance of material strength σ_Y and the bending stiffness EI_z have been analyzed in Section 5.3.1, but geometric imperfections, non-ideal material behavior and residual stresses left from the fabrication process are also important. Several of these factors depend on the type of column and the fabrication process, e.g. the residual stresses left by hot rolling as discussed in considerable detail by Beedle and Tall (1960). The non-uniform axial stress distribution implies that yield of the section occurs in a more gradual fashion than shown in the idealized stress-strain relation in Fig. 5.19.

It turns out that it is possible to combine the influence of stiffness EI_z, strength σ_Y and an imperfection parameter e to generate a family of curves for capacity of columns that represent experimental results well over the whole range of column slenderness. The central result is the Perry-Robertson column capacity formula. The idea of the Perry-Robertson column capacity formula is that the column has an initial deflection represented by the eccentricity parameter e, and that the capacity of the column is exhausted when the maximum stress in the most severely loaded cross-section reaches the ultimate stress represented by σ_Y. A historical account of the Perry-Robertson formula has been given by Heyman (1998).

The assumption is that the maximum compressive stress reaches σ_Y. For a column with positive transverse displacement this stress will occur at the negative side of the section at $z = -z_{\max}$. Thus, it follows from (5.61) that the maximum compressive stress is given by

$$\sigma_Y \,=\, \frac{P}{A} + \frac{M}{I_z}\,z_{\max}\,. \tag{5.69}$$

In this formula the following three substitutions are made:

a) The load is represented by the critical stress $\sigma_c = P/A$.

b) The factor $z_{\max}$ is expressed by the kernel radius $k_z = I_z/Az_{\max}$.

c) The moment is introduced via its amplified value $M = eP/(1 - P/P_{el})$, as determined in Example 5.8 (with P_E replaced by P_{el}).

With these substitutions the condition (5.69) takes the form

$$\sigma_Y = \sigma_c + \frac{e}{k_z} \frac{\sigma_c}{1 - \sigma_c/\sigma_{el}} . \tag{5.70}$$

Multiplication of this equation with the denominator gives the product format

$$(\sigma_Y - \sigma_c)(\sigma_{el} - \sigma_c) = \frac{e}{k_z} \sigma_{el}\, \sigma_c . \tag{5.71}$$

In this format it is seen that, if there is no eccentricity, i.e. for $e = 0$, the equation is simply a product of the strength and the elastic stability criteria, while $e > 0$ leads to an interpolation between these two criteria. Figure 5.27 illustrates the influence of imperfections by showing the critical stress for $e/k_z = 0.00$, 0.02 and 0.06.

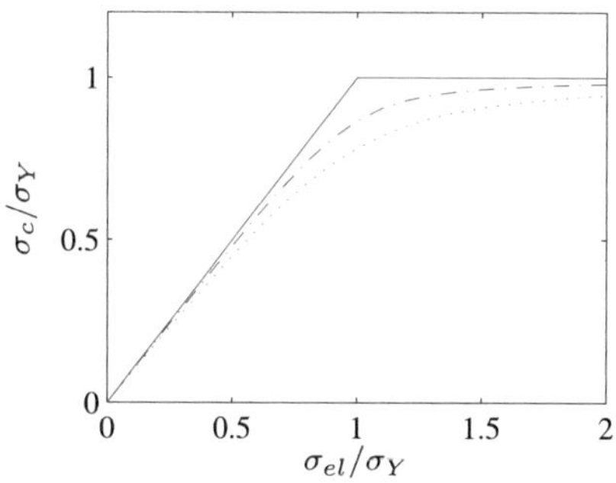

Fig. 5.27: Critical stress for $e/k_z = 0.00$ (—), 0.02 (- -) and 0.06 (···).

The explicit solution to the criterion (5.71) is found by expressing the mean stress σ_{el} at elastic instability in terms of the relative slenderness, defined in (5.48) by

$$\frac{\sigma_{el}}{\sigma_Y} = \frac{1}{\lambda_r^2} . \tag{5.72}$$

This is substituted into the Perry-Robertson equation (5.71), which then takes the following form of a quadratic equation,

$$\lambda_r^2 \left(\frac{\sigma_c}{\sigma_Y}\right)^2 - \left(\lambda_r^2 + 1 + \frac{e}{k_z}\right)\frac{\sigma_c}{\sigma_Y} + 1 = 0 . \tag{5.73}$$

The solution to this equation is conveniently expressed as

$$\frac{\sigma_c}{\sigma_Y} = \beta - \sqrt{\beta^2 - \lambda_r^{-2}} , \tag{5.74}$$

where the non-dimensional parameter β has been defined as

$$\beta = \frac{1}{2\lambda_r^2}\Big(\lambda_r^2 + 1 + \frac{e}{k_z}\Big). \tag{5.75}$$

These relations determine the critical stress σ_c in terms of the yield stress σ_Y, the relative slenderness λ_r, and the relative imperfection e/k_z. The relative slenderness follows by (5.72) from the elastic instability stress σ_{el} and the yield stress σ_Y as

$$\lambda_r = \sqrt{\sigma_Y/\sigma_{el}}\,, \tag{5.76}$$

while the magnitude of the imperfection parameter depends on a number of factors.

In practice, the imperfection depends on the length of the column, and structural codes often assume a direct relation between e/k_z and the relative slenderness λ_r. The simplest relation is linear proportionality,

$$\frac{e}{k_z} = \alpha\,\lambda_r\,. \tag{5.77}$$

The non-dimensional parameter α represents the total effect of imperfections, residual stresses etc. According to the theory α is determined by

$$\alpha = \frac{e}{k_z}\,\frac{1}{\lambda_r} = \pi\,\frac{e}{\ell_e}\,\frac{r_z}{k_z}\sqrt{\frac{E}{\sigma_Y}}\,. \tag{5.78}$$

If the material is linear up to the yield limit, the corresponding yield strain is $\varepsilon_Y = \sigma_Y/E$. For steel the yield strain is around $\varepsilon_Y \simeq 0.002$. A typical imperfection of a steel column is $e \simeq 0.001\ell$, see e.g. Timoshenko and Gere (2009). For steel columns a representative value of α is then of the order

$$\alpha \simeq \pi\,\frac{0.001}{\sqrt{0.002}}\,\frac{r_z}{k_z} = 0.070\,\frac{r_z}{k_z} = \begin{cases} 0.070 & \text{I-section} \\ 0.122 & \text{rectangle} \end{cases}, \tag{5.79}$$

with the kernel radii from (5.63) and (5.65), and the gyration radii from Example 5.6. It is seen that among the other factors the parameter α also

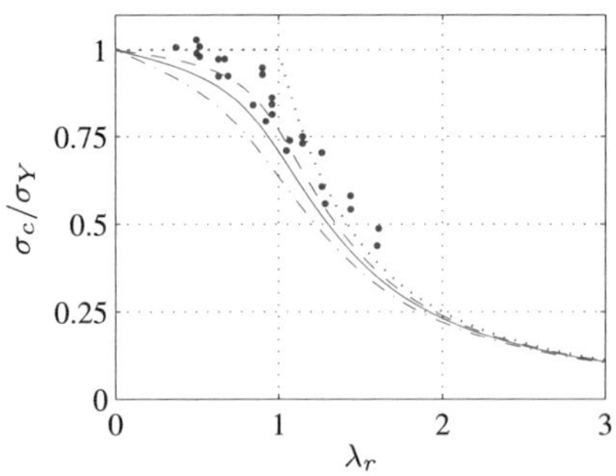

Fig. 5.28: Critical stress as function of relative slenderness.

depends on the cross-section shape. Figure 5.28 shows the critical stress for the imperfection e/k_z given by (5.77) with $\alpha = 0.07$, 0.122 and 0.21.

Simulated column capacity results, obtained by statistical combination of material properties and imperfections, demonstrate that a curve through a lower fractile – e.g. 5 pct. – is similar in shape to the curves generated by the Perry-Robertson formula, see e.g. Chen and Han (1985). This supports a design procedure in which the critical design stress is obtained by increasing the parameter α beyond its representative mean value.

Example 5.9. Critical stress for column with tubular cross-section. This example illustrates the design procedure for the simply supported column shown in Fig. 5.29. The cross-section is tubular with radius a and wall thickness t, whereby $A = 2\pi a t$ and $I_z = \pi t a^3$. The wall thickness is $t = a/10$, and the length is $\ell = 100a$. The imperfection e is represented via (5.77) with $\alpha = 0.2$. The material is steel with elastic modulus $E = 210\,\mathrm{GPa}$ and yield stress $\sigma_Y = 300\,\mathrm{MPa}$.

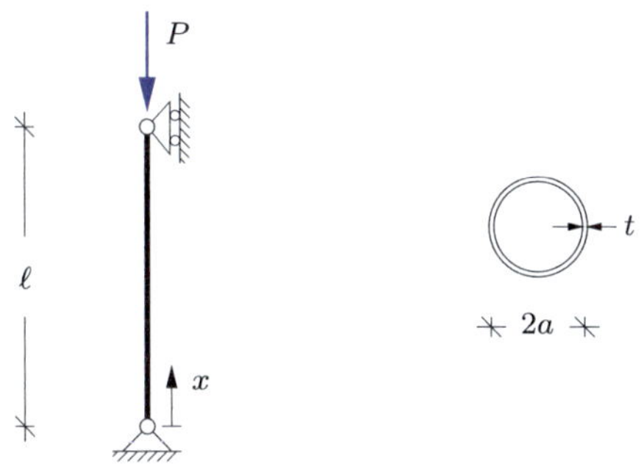

Fig. 5.29: Euler column with tubular cross-section.

First the instability stress is calculated from (5.44),

$$\sigma_{el} = P_E/A = \tfrac{1}{2}(\pi a/\ell)^2 E = 106.6\,\mathrm{MPa}.$$

The relative slenderness then follows from (5.76),

$$\lambda_r = \sqrt{\sigma_Y/\sigma_{el}} = \sqrt{300/106.6} = 1.70\,.$$

As $\lambda_r > 1$, this is a 'long' column. The parameter β is now calculated from (5.75) after substituting e/k_z from (5.77) with $\alpha = 0.2$,

$$\beta = \frac{1}{2\lambda_r^2}\left(\lambda_r^2 + 1 + \alpha\lambda_r\right) = 0.73\,.$$

The relative magnitude of the critical stress is then found from (5.74)

$$\frac{\sigma_c}{\sigma_Y} = \beta - \sqrt{\beta^2 - \lambda_r^{-2}} = 0.296 \quad \Rightarrow \quad \sigma_c = 88.8\,\mathrm{MPa}\,.$$

This is a substantial reduction relative to the yield stress because of the length of the column. The magnitude of the critical stress relative to the instability stress is $\sigma_c/\sigma_{el} = 0.86$, where the reduction shows the effect of the imperfection. □

5.4 Exercises

Exercise 5.1. The figure shows a simply supported beam of length ℓ with constant bending stiffness EI_z. It is loaded by a uniformly distributed transverse load with intensity p and a horizontal compression force P at the right support, which produces a constant normal force $N = -P$.

a) Set up the differential equation for the transverse displacement $w(x)$, and find the solution with four arbitrary constants.

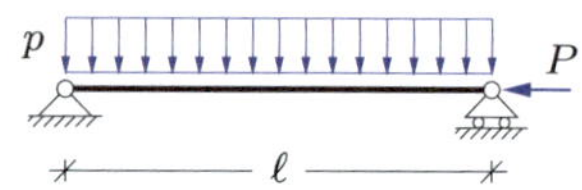

b) Use the four boundary conditions to determine the expression for the displacement $w(x)$.

c) Find an expression for the maximum displacement $w_{\text{max}} = w(\frac{1}{2}\ell)$.

d) Find the critical load P_c corresponding to $w_{\text{max}} \to \infty$.

Exercise 5.2. The figure shows a cantilever of length ℓ with constant bending stiffness EI_z. It is loaded by a vertical force P_v and a horizontal force P at the tip. Note that the latter produces a constant positive normal force $N = P$.

a) Set up the differential equation for the transverse displacement $w(x)$, and show that the solution can be written as

$$w(x) = C_1 + C_2 kx + C_3 \cosh(kx) + C_4 \sinh(kx)\,.$$

b) Use the four boundary conditions to determine the expression for the displacement $w(x)$.

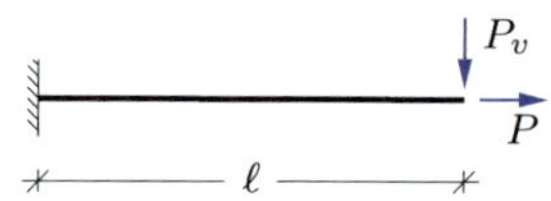

c) Find the transverse tip displacement w_{tip}, and the tip displacement w^0_{tip} for $P = 0$.

d) Find the magnitude of the axial load P required to reduce the transverse tip deflection to $w_{\text{tip}} = \frac{1}{2} w^0_{\text{tip}}$.

Exercise 5.3. The figure shows a cantilever of length ℓ with constant bending stiffness EI_z. It is loaded by a uniformly distributed transverse load with intensity p and a horizontal force P at the tip, which produces a constant normal force $N = -P$.

a) Set up the differential equation for the transverse displacement $w(x)$, and find the solution with four arbitrary constants.

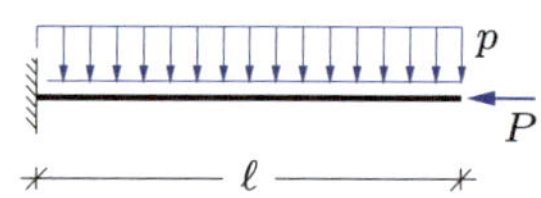

b) Use the four boundary conditions to determine the expression for the displacement $w(x)$.

c) Find an expression for the transverse tip displacement w_{tip} and the moment at the support M_{sup}.

d) Determine the critical load P_c corresponding to $w_{\text{tip}} \to \infty$.

Exercise 5.4. The figure shows a simply supported beam of length ℓ with constant bending stiffness EI_z. At the right support the beam is loaded by a local moment M_B and a horizontal force P. The latter produces the constant normal force $N = -P$.

a) Set up the differential equation for the transverse displacement $w(x)$, and find the solution with four arbitrary constants.

b) Use the four boundary conditions to determine the expression for the displacement $w(x)$.

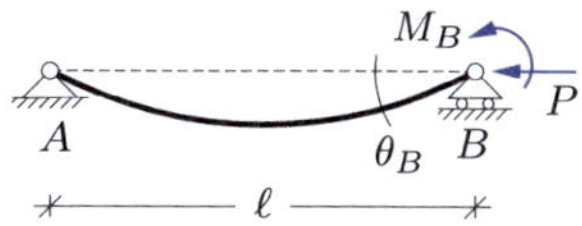

c) Find the expression for the rotation $\theta(x)$, and use this expression to show that the relation between the support rotation $\theta_B = \theta(\ell)$ and the applied moment M_B can be written as

$$M_B = \frac{(k\ell)^2}{1 - k\ell\cot(k\ell)}\,\frac{EI_z}{\ell}\,\theta_B\,.$$

d) The stiffness of the beam with respect to the rotation at the right support can be expressed in non-dimensional form as $M_B\ell/(3EI_z\theta_B)$, which is unity for $P = 0$. Use the solution in c) to find the expression for this rotational stiffness and explain what happens when $P \to P_E$.

Exercise 5.5. The figure shows a column of length ℓ with a fixed support in A and a fixed support with horizontal rollers at the top in B. The column is loaded by an axial compression force P, producing the constant normal force $N = -P$.

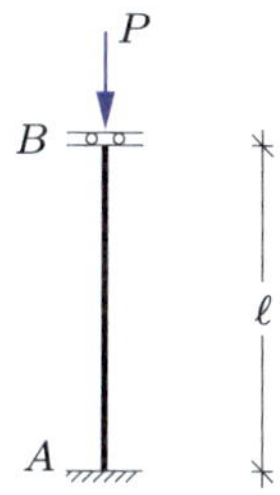

a) Sketch the critical buckling form of the column and estimate the equivalent column length ℓ_e and the critical load ratio P_c/P_E.

b) Use the general solution $w(x)$ with four arbitrary constants and the four boundary conditions to determine P_c/P_E and ℓ_e. Compare with the results in a).

Exercise 5.6. The figure shows a column of length ℓ with a simple support in A and a fixed support with horizontal rollers at the top in B. The column is loaded by an axial compression force P, producing the constant normal force $N = -P$.

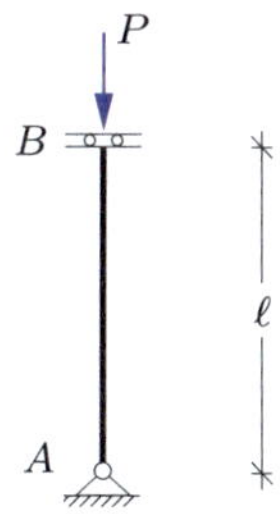

a) Sketch the critical buckling form of the column and estimate the equivalent column length ℓ_e and the critical load ratio P_c/P_E.

b) Use the general solution $w(x)$ with four arbitrary constant and the four boundary conditions to determine P_c/P_E and ℓ_e. Compare with the results in a).

Exercise 5.7. The figure shows ideal columns with intermediate supports. As discussed in Section 5.2.2 an approximative result for the equivalent column length ℓ_e is the longest distance between neighboring support or inflection points.

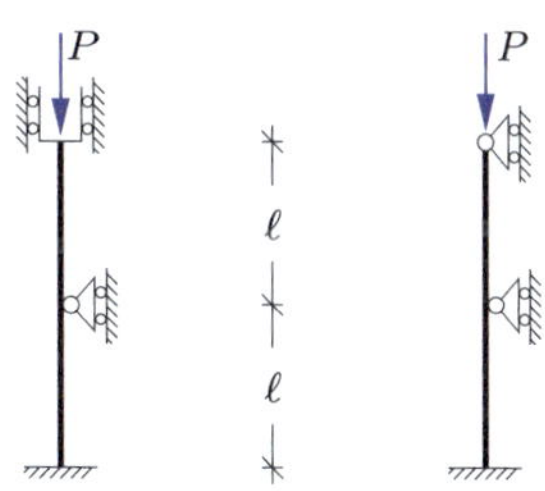

a) Sketch the buckling form for each of the columns and estimate the equivalent column length. Note that the inflection point for the second column is located below the intermediate support.

b) Determine the critical buckling load ratio P_c/P_E associated with the estimated column lengths. Compare with $P_c/P_E = 2.0$ and 1.3, respectively, obtained by numerical analysis.

Exercise 5.8. The figure shows an elastic column of length ℓ with rectangular cross-section with dimensions a and b. The column is part of a wall structure containing horizontal beams. This supports the motion in the plane of the wall, while motion is possible in the out-of-plane direction. The cross-section area is $A = ab$. The moment of inertia in the plane of the wall is $I_y = \frac{1}{12}Aa^2$, while it is $I_z = \frac{1}{12}Ab^2$ out of the plane. From a stability point of view it is important that the column has larger stiffness out of the plane direction than in the plane, which implies that $b > a$. The question is how much? The column is assumed to be fully fixed at both ends in both directions.

a) Determine the effective column length ℓ_e^y and the critical load P_c^y corresponding to buckling in the plane of the wall.

b) Determine the effective column length ℓ_e^z and the critical load P_c^z corresponding to buckling out of the plane of the wall.

c) Determine the ratio b/a such that $P_c^y = P_c^z$.

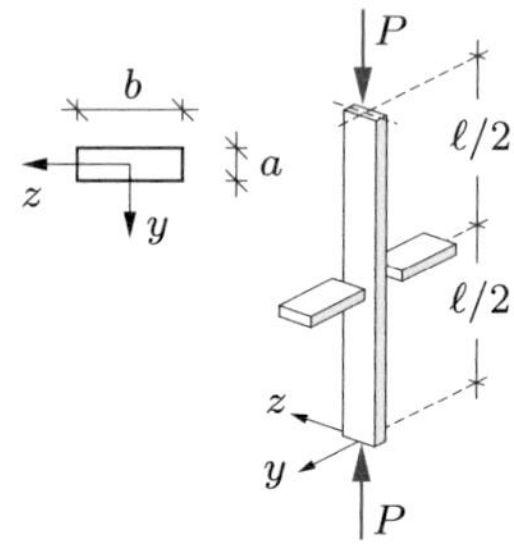

Exercise 5.9. Consider a linear elastic column of length ℓ, with simple supports at both ends and in both directions. The cross-section of the column is an I-profile as shown in the figure, with height h, width b and wall thickness t.

a) Determine the two moments of inertia I_y and I_z, neglecting the contribution from the web.

b) Find the ratio h/b so that the Euler load P_E is the same with respect to buckling in the two directions.

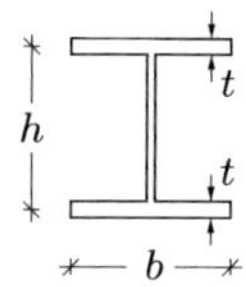

Exercise 5.10. The figure shows an I-profile with height h, width b and wall thickness t. The contribution of the web to the bending stiffness is omitted.

a) Make a sketch of the cross-section and indicate the points that limit the kernel with respect to bending in the two directions of symmetry.

b) Find the kernel area by superposition of the stress distributions from bending in the two directions of symmetry.

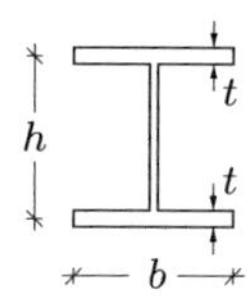

Exercise 5.11. Consider a simply supported column of length $\ell = 4.00\,\mathrm{m}$, and with quadratic cross-section of dimensions $50\,\mathrm{mm} \times 50\,\mathrm{mm}$. The material is steel with elastic modulus $E = 210\,\mathrm{GPa}$ and yield stress $\sigma_Y = 250\,\mathrm{MPa}$.

a) Find the critical stress σ_{el} corresponding to elastic buckling.

b) Determine the slenderness λ and the relative slenderness λ_r.

c) Find the length of the column $\ell = \ell_*$, which corresponds to the limit between a short and a long column.

Exercise 5.12. Consider a simply supported column of length $\ell = 4.00\,\mathrm{m}$, and with quadratic cross-section of dimensions $50\,\mathrm{mm} \times 50\,\mathrm{mm}$. The material is steel with elastic modulus $E = 210\,\mathrm{GPa}$ and yield stress $\sigma_Y = 250\,\mathrm{MPa}$. The imperfection of the column is represented by the parameter $\alpha = 0.22$.

a) Determine the relative slenderness λ_r.

b) Determine the ratio σ_c/σ_Y by Perry-Robertson's formula.

c) Determine the stress σ_{el} for the euler column and the ratio σ_c/σ_{el}. Comment on the influence of the imperfection.

Exercise 5.13. The figure shows an Euler column with a cruciform cross-section with width a and wall thickness t. The moment of inertia is $I_z = \frac{1}{12}a^3t$. The length of the column is $\ell = 25a$. The material is steel with elastic modulus $E = 210\,\text{GPa}$ for stresses below the yield stress $\sigma_Y = 235\,\text{MPa}$. The imperfection is given by the parameter $\alpha = 0.1$.

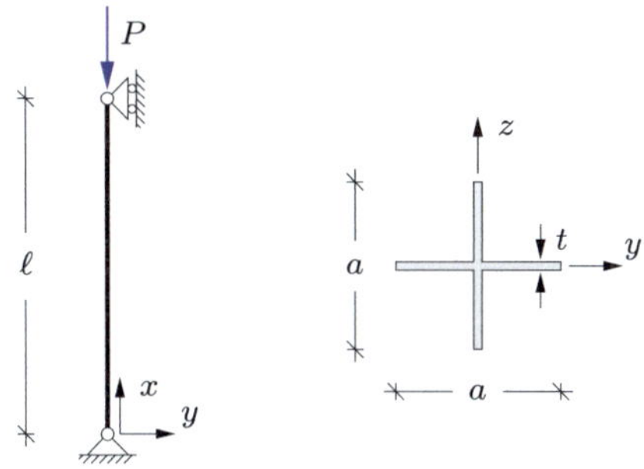

a) Determine the stress σ_{el} corresponding to the Euler load P_E.

b) Find the relative slenderness λ_r and determine if the column is long or short.

c) Determine σ_c/σ_Y by Perry-Robertson's formula, and the ratio σ_c/σ_{el}.

Exercise 5.14. The figure shows a column ACB of length 2ℓ. For deflection in the y-direction the column has a simple support in A and simple supports with vertical rollers in B and C. For deflection in the z-direction the column has a fixed support in A and a simple support with vertical rollers in B. The column is loaded by an axial compression force P in B, whereby the constant normal force is $N = -P$. As shown in the figure to the right the cross section of the column is quadratic with height a and width a. The column is elastic with elastic modulus E for stresses below the yield stress σ_Y. It is assumed that $E/\sigma_Y = 900$.

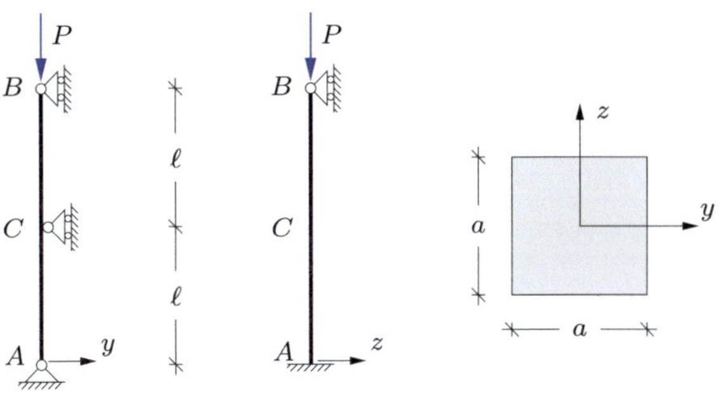

a) Sketch the buckling form for deflections in the y- and z-direction, respectively. Determine the corresponding equivalent column lengths ℓ_e^y and ℓ_e^z.

b) Determine the critical load P_{el} for the elastic column.

c) Determine the relative slenderness λ_r with respect to the critical buckling form, and find the value of ℓ/a that corresponds to the transition between a short and a long column.

6 The Force Method

Statically determinate structures are characterized by having the minimum number of supports and internal connections. While this permits direct analysis based on statics alone, it also entails some disadvantages such as vulnerability if any part of the structure is damaged. In addition statically determinate frames are prone to rather large moments e.g. at mid-span and at corners. This feature of the moment distribution typically leads to rather flexible structures. Thus, there are several reasons why most current structures are statically indeterminate. With proper care statically indeterminate structures can be designed such that the maximum section force is reduced, the structures are less flexible, and vulnerability to failure of individual structural elements may be reduced. A simple but typical example of moment redistribution and increased stiffness is a bridge with a main girder that is continuous over the supports.

The analysis of statically indeterminate structures requires information of the stiffness of the individual structural elements. There are two basically different approaches to the analysis of statically indeterminate structures. In the force method an equivalent statically determinate structure is introduced e.g. by introducing hinges at selected points and removing supports. In the name 'force method' the force refers to any component of the section forces and thereby also includes moments. The analysis then consists in determining the moment pairs necessary for restraining the relative rotation at the hinges

S. Krenk, J. Høgsberg, *Statics and Mechanics of Structures*,
DOI 10.1007/978-94-007-6113-1_6,

and prevent the motion at the temporarily removed supports. This method combines static analysis to determine the section forces in the equivalent statically determinate structure and calculation of the elastic displacements which must be compensated to reestablish the original structure. These subjects have been treated in some detail in Chapters 3 and 4. Thus, the main objective of the present chapter is to present and illustrate the systematic use of these principles to the analysis of statically indeterminate beam and frame structures. As the idea of the force method is to close the auxiliary kinematic discontinuities it is sometimes called the 'method of consistent deformations'.

The force method is rather straightforward to use in hand calculations of small structures with a few degrees of static indeterminacy. However, each step in the analysis involves determination of the internal forces in the full equivalent statically determinate structure. This aspect of the force method makes it less convenient for computer implementation. The stiffness method is based on a different way of thinking. In the stiffness method the structure is characterized by nodes, connected by structural elements. The properties of the structure are related to the structural elements between the nodes, and the unknown parameters of the analysis are the displacements of the nodes. An example of this approach was presented for truss structures in Section 2.5. The stiffness method for frames is presented in Chapter 7, first as a hand calculation method and subsequently generalized to Finite Element format.

6.1 Principle of the force method

The effect of making a structure statically indeterminate is here briefly indicated by the two simply supported two-span beams shown in Fig. 6.1. The beam in Fig. 6.1a has a hinge at the intermediate support. This permits a rotation discontinuity $\Delta\theta_C = \theta_C^+ - \theta_C^-$ between the two beams at C, whereby the bending moment in C vanishes. This provides an additional equilibrium condition $M_C = 0$ required to determine the four reaction forces by pure statics, as discussed in Chapter 1. Thus, the beam structure in Fig. 6.1a is statically determinate.

The beam shown in Fig. 6.1b is continuous at the support C thereby maintaining continuity of the beam rotation at the support ($\Delta\theta_C = 0$). The rotation continuity is the result of a non-vanishing moment at the support ($M_C \neq 0$). As indicated in the figure this leads to a redistribution of the bending moment in the beam, whereby both the maximum moments and the deformations are reduced. However, equilibrium of the beam gives only three equilibrium conditions, one short of the four reaction components, and the moment M_C can therefore not be determined by statics alone. Thus, the beam structure in Fig. 6.1b is statically indeterminate.

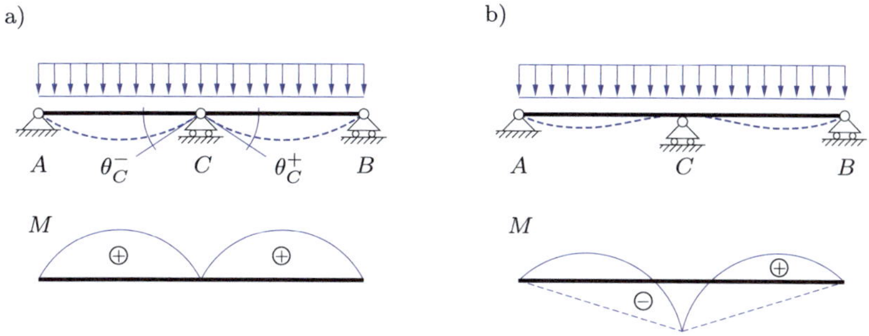

Fig. 6.1: Moment reduction in beam by additional rotation constraint in C.

Statically indeterminate two-span beam

The principle of the force method for solving statically indeterminate structures is demonstrated by the two-span beam shown in Fig. 6.2. The beam structure has four reactions, where the horizontal reaction in A can be found directly by horizontal equilibrium,

$$\rightarrow \qquad\qquad R'_A = 0\,.$$

This leaves the three vertical reactions R_A, R_B and R_C as unknown, and since only two equilibrium equations remain the structure is one time statically indeterminate, or it has a degree of indeterminacy equal to 1.

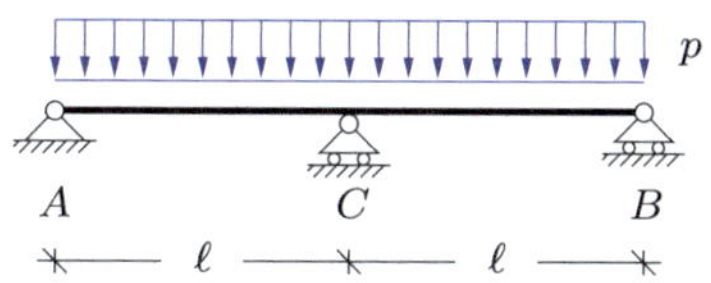

Fig. 6.2: Statically indeterminate two-span beam with distributed load.

The first step is to convert the structure to an equivalent statically determinate structure. This can be done in several ways, as discussed in Section 6.2.1. In the present example the vertical reaction R_C at the center of the beam is considered as an external force acting on the structure. This initially unknown force is denoted as X_1 instead of R_C. With X_1 as an external load the equivalent structure in Fig. 6.3 is statically determinate. The equivalent structure is now analyzed for two load cases: the actual distributed load p, and the new vertical force X_1 acting in C. As the structure is elastic, the principle of superposition applies: Thus the structure can be analyzed for each individual load case, and the result is obtained by superposition of the individual load cases, as explained previously in Section 3.4.1. The value of the auxiliary force X_1 is determined from a condition on the resulting displacement. In the present case this condition is that the structure has zero vertical displacement at C. Thus, the unknown external force X_1 must be determined to satisfy this condition. It is therefore necessary to determine the vertical displacement at C for both the two load cases. This is done by use of the principle of virtually work, developed in detail in Chapter 4.

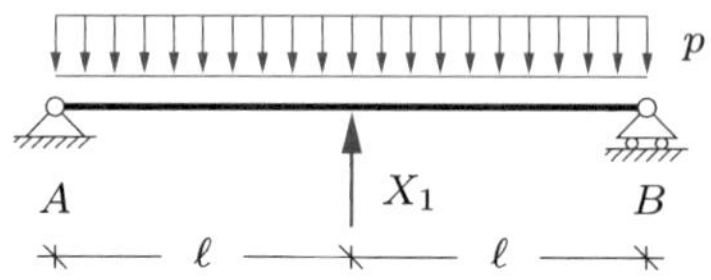

Fig. 6.3: Equivalent statically determinate structure with additional force X_1.

The auxiliary force X_1 represents the force needed to enforce the associated kinematic constraint in the statically determinate structure. As demonstrated in the following it is possible to have multiple auxiliary forces X_j's, and they can represent both external forces and moments as well as internal force and moment couples. The auxiliary forces X_j are denoted as redundant forces or redundant components.

First, the equivalent structure is analyzed for the actual distributed load without the redundant component, i.e. with $X_1 = 0$. This load case is shown in Fig. 6.4a. The reactions are

$$R_A^0 = R_B^0 = p\ell \, ,$$

and Fig. 6.4a shows the parabolic moment distribution M^0. The index 0 is used to refer to the load case with the actual external load, in the absence of any redundant force components. The displacement of C from the distributed load is determined by the virtual work equation described in Section 4.4 using the test load $X_1 = 1$. This load case and the corresponding moment distribution $M^1(x)$ are shown in Fig. 6.4b. The transverse displacement is typically dominated by the contribution from beam bending, and only this contribution is included in the following. The transverse displacement at C from the distributed load then follows from the virtual work equation (4.43) as

$$\xi_{10} = \int_0^\ell \frac{M^1 M^0}{EI} \, dx = \frac{2}{EI} \frac{\ell}{12} \left(\frac{1}{2} p\ell^2 \right) \left(-\frac{5}{2} \ell \right) = -\frac{5}{24} \frac{p\ell^4}{EI} \, .$$

The displacement is denoted ξ_{10}, where the first subscript indicates that this displacement corresponds to the redundant force X_1 in location and direction, while the second subscript 0 refers to the load as being the actual uniformly distributed load. The integral is evaluated by using row five in Table 4.1.

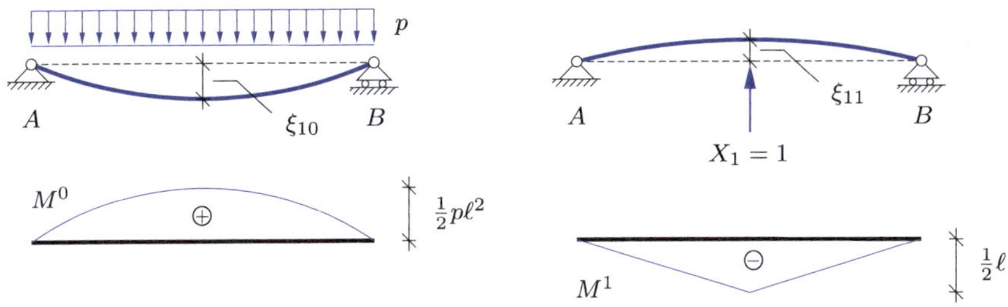

Fig. 6.4: Moment in statically determinate beam: a) Actual load, b) test load $X_1 = 1$.

The second load case corresponds to a unit magnitude of the redundant force, $X_1 = 1$. This load case is shown in Fig. 6.4b with reactions

$$R_A^1 = R_B^1 = -\tfrac{1}{2}$$

and moment distribution $M^1(x)$. The vertical displacement at C from this unit load is determined by the virtual work equation (4.43) as

$$\xi_{11} = \int_0^\ell \frac{M^1 M^1}{EI}\, dx = \frac{2}{EI}\frac{\ell}{3}\left(\frac{1}{2}\ell\right)^2 = \frac{1}{6}\frac{\ell^3}{EI}\,.$$

The first subscript in ξ_{11} refers to the resulting displacement component and is therefore unchanged, while the second subscript is now 1, referring to the load case $X_1 = 1$. The index system is explained in greater detail in Section 6.2.3, where the general procedure for the force method is presented.

The total displacement ξ_1 in C is determined by superposition of the displacements from the two load cases, giving the condition of zero displacement at C as

$$\xi_1 = \xi_{10} + \xi_{11} X_1 = 0\,.$$

The initially unknown redundant component X_1 follows from this condition as

$$X_1 = -\frac{\xi_{10}}{\xi_{11}} = \tfrac{5}{4}\, p\,\ell\,.$$

Note, that X_1 has the dimension of force. The absolute value EI of the beam stiffness does not enter into the expression, while the fact that the two parts of the beam has the same stiffness in this case has been accounted for in the calculation of the displacement ξ_{10} and the coefficient ξ_{11}.

It if often convenient to evaluate the reactions before proceeding to the calculation of the internal forces. In this example X_1 represents the vertical reaction R_C, which therefore is determined directly as

$$R_C = X_1 = \tfrac{5}{4}\,p\ell\,.$$

The two remaining reactions are determined subsequently by equilibrium or by superposition. The latter approach gives

$$R_A = R_A^0 + R_A^1 X_1 = p\ell + (-\tfrac{1}{2})\tfrac{5}{4}\,p\ell = \tfrac{3}{8}\,p\ell\,, \qquad R_B = R_A\,.$$

Vertical equilibrium confirms that $R_A + R_B + R_C = 2p\ell$.

The moment distributions of the individual load cases are always available from the analysis since they have been used in the virtual work equation to determine the equation coefficients. Thus, the resulting moment distribution can be determined by superposition. Alternatively, the moment distribution can be found by a standard section analysis, as the reactions have just been been obtained. In this example the moment distribution is determined by both methods.

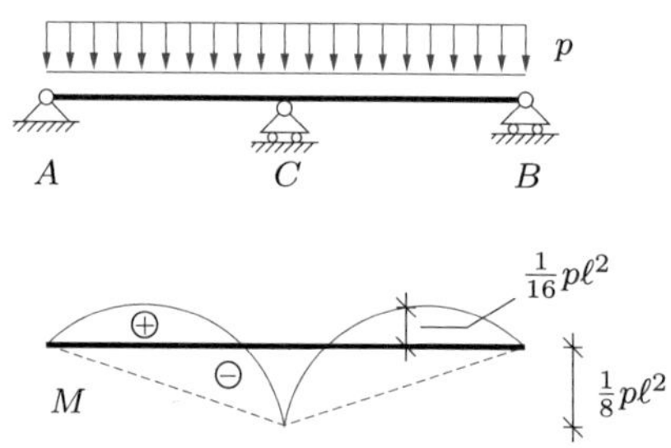

Fig. 6.5: Moment distribution.

The moment distributions M^0 and M^1 shown in Fig. 6.4 are parabolic and piecewise linear, respectively. With the origin of the x-coordinate at A the expressions in terms of x are

$$M^0(x) = \tfrac{1}{2}p\,x(2\ell - x)\,, \qquad M^1(x) = \begin{cases} -\tfrac{1}{2}\,x & \text{for } x < \ell\,, \\ -\tfrac{1}{2}(2\ell - x) & \text{for } x > \ell\,. \end{cases}$$

The expression for the resulting moment is then found by superposition as

$$M(x) = M^0(x) + M^1(x)X_1 = \begin{cases} \tfrac{1}{8}p\,x(3\ell - 4x) & \text{for } x < \ell\,, \\ \tfrac{1}{8}p(2\ell - x)(4x - 5\ell) & \text{for } x > \ell\,. \end{cases}$$

The total moment distribution is shown in Fig. 6.5. The numerically largest moment $M_{\max} = -M(\ell) = \frac{1}{8}p\ell$ is located in C, while the moment at the center of the two spans is $M(\frac{1}{2}\ell) = \frac{1}{16}p\ell$. The local maximum of the moment in each of the spans occurs where the shear force is zero: For the left span this gives

$$Q(x_{\max}) = R_A - p\,x_{\max} = 0 \qquad \Rightarrow \qquad x_{\max} = \tfrac{3}{8}\ell\,,$$

with local maximum $M(x_{\max}) = \frac{9}{128}p\ell^2 < M_{\max}$.

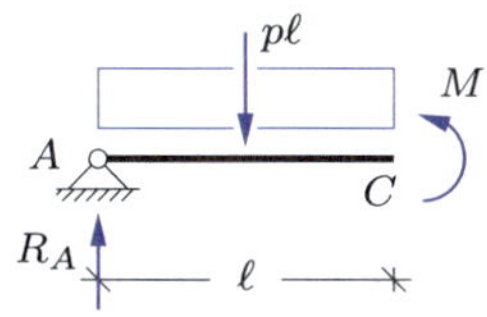

Fig. 6.6: Moment equilibrium in C.

The moment at C can also be determined by section force analysis, as shown in Fig. 6.6, where moment equilibrium for the section at C gives

$$\overset{\frown}{C} \qquad M(\ell) + \tfrac{1}{2}\ell\,p\ell - \ell\,R_A = 0 \quad \Rightarrow \quad M(\ell) = \tfrac{3}{8}\,p\ell^2 - \tfrac{1}{2}\,p\ell^2 = -\tfrac{1}{8}\,p\ell\,.$$

This corresponds to the previously found value and defines the dashed lines in the moment distribution in Fig. 6.5. The final moment distribution is then obtained by superimposing the parabola with maximum value $\frac{1}{8}p\ell^2$, following the procedure presented in Section 3.4.2.

6.2 The general force method

This section presents the general form and a step-by-step procedure of the force method. First the construction of an equivalent statically determinate structure is addressed, with emphasis on the relation between the released kinematic constraints and the corresponding redundant components. The principle of the method is then illustrated for a two times statically indeterminate structure, where coupling effects appear and the solution is obtained by matrix calculus. Finally, the general procedure of the force method for an n times statically indeterminate structure is summarized.

6.2.1 Released structure

For a structure that is n times statically indeterminate, the basis of the force method is the release of n kinematic constraints, whereby the structure becomes statically determinate with n conjugate redundant static components, generating n additional load cases. In the previous example a vertical reaction force was replaced by a vertical force, but in general the kinematic constraints can be of both external or internal type.

External releases

The degree of statically indeterminacy is often found by simply counting unknown reaction components and comparing this to the number of independent equilibrium equations. Therefore, it is often straight forward to simply replace the necessary number of reactions by redundant components X_j. Figure 6.7a shows a fixed support, which represents a reaction moment and two reaction forces. Figures 6.7b-d show the three situations in which the rotation, the vertical displacement and the horizontal displacement have been released, whereby the reaction moment, the vertical reaction force and the horizontal reaction force, respectively, are introduced as a redundant static component X_j.

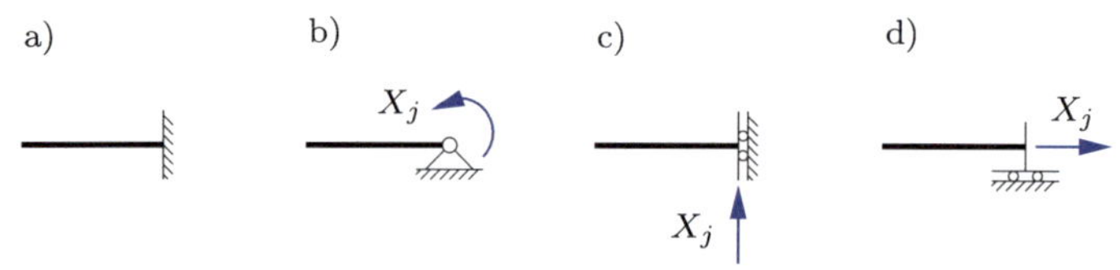

Fig. 6.7: Releasing support conditions.

Figure 6.8a shows a two-span beam with a fixed support in A and simple supports with horizontal rollers in B and C. There are five reactions and only three equilibrium equations, whereby the degree of indeterminacy is 2. The horizontal reaction can be obtained directly from horizontal equilibrium as $R'_A = 0$. Thus, the two redundant components can be chosen among the remaining four reactions. The force method is based on the determination of displacements (or rotations) by the virtual work equation, and therefore simple moment distributions are computationally advantageous. For the beam in Fig. 6.8a this can be obtained by replacing the reactions in B and C by redundant forces, whereby the structure is transformed into a cantilever beam, or by replacing the reaction moment in A and the reaction force in C, resulting in the simply supported beam shown in Fig. 6.8b.

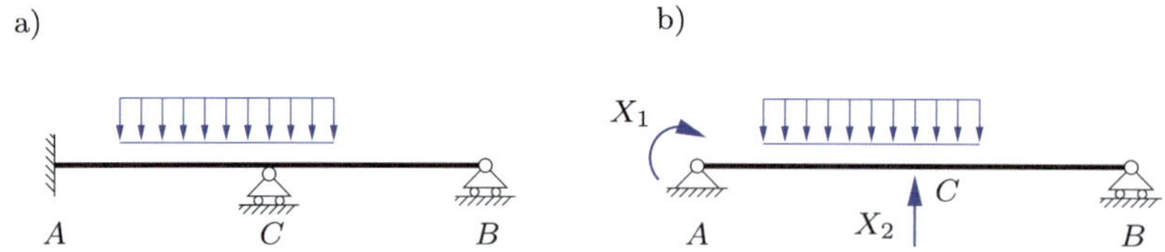

Fig. 6.8: Two-span beam with reactions as redundant components X_1 and X_2.

Internal releases

The equivalent statically determinate structure can also be constructed by releasing internal kinematic constraints and thereby introducing conjugate internal force or moment couples as redundant components. As illustrated in Fig. 6.9a the bending moment, normal force and shear force are the three available internal forces. Equilibrium in a section requires that internal forces always appear in opposing pairs, and thus redundant components following internal releases are introduced as moment and/or force couples. This is indicated in Figs. 6.9b–d, where X_j represents the moment, the shear force and the normal force, respectively.

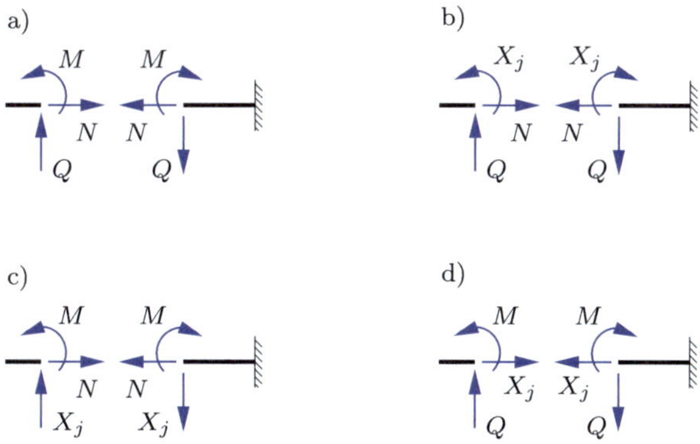

Fig. 6.9: Internal forces as redundant components.

The force method relies on the determination of displacements (or rotations) based on virtual work, where the influence of normal and shear forces is typically omitted. This implies that simple moment distributions often result in a simple analysis, and as demonstrated in the examples of this chapter the moment distributions can often be simplified by using internal moments as redundant components. Hereby, the moment distributions are often localized to limited regions with simple parabolic or linear shapes that are easily integrated in connection with the virtual work equation. Removing the internal bending moment capacity of the beam corresponds to introducing a hinge, as shown in Fig. 6.10, where the redundant component X_j is then added as a pair of opposing moments.

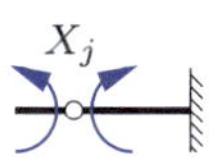

Fig. 6.10: Internal moment couple as redundant force.

Figure 6.11a shows a two span beam with both fixed and simple supports. Previously in Fig. 6.8 the equivalent statically determinant structure was constructed by replacing the reaction moment in A and the reaction force in C by redundant components X_1 and X_2. Instead of the reaction in C, the bending moment in C can be used as redundant component as illustrated in Fig. 6.11b. The hinge is placed in the beam in C and the redundant component X_2 is then introduced as a pair of opposing moments. It is important to note that the pair of moments X_2 cancel in the equilibrium of the full structure. However, the internal moment in C is now $M_C = X_2$, and thus X_2 contributes to the internal moment distribution.

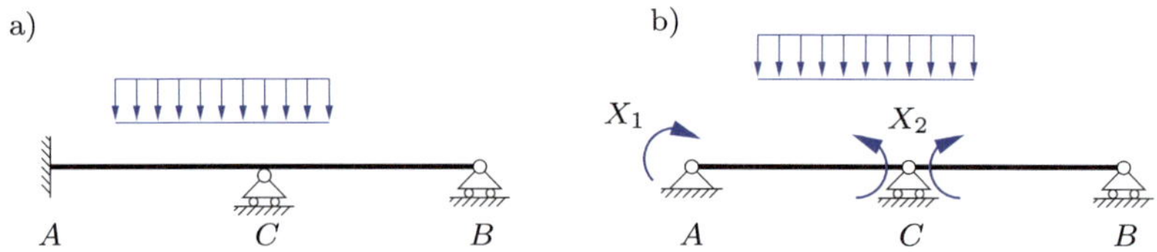

Fig. 6.11: Two-span beam with internal moment in C as redundant force.

Kinematic determinacy of released structure

As discussed above the first step in the force method is to construct an equivalent statically determinate structure and introduce the corresponding conjugate forces/moments as redundant components. In particular, when replacing internal force couples by redundant components it is important that the structure remains kinematically determinate in the sense that no mechanisms are created. For the two-span beam in Fig. 6.11 the reaction moment in A and the internal moment in C have been chosen as redundant components. This structure is statically and kinematically determinate and thus contain no kinematic mechanisms. Another hypothetical choice of the redundant components is shown in Fig. 6.12, where X_1 is the bending moment in C and X_2 is the reaction force in B. As indicated in the figure this choice creates a mechanism and can therefore not be used for the static analysis.

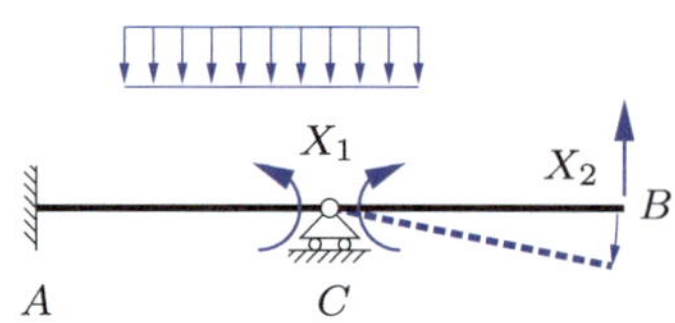

Fig. 6.12: Unstable two-span beam with mechanism in C.

6.2.2 The basic steps

The basic steps of the force method are illustrated here by the two-span beam in Fig. 6.13. This beam is two times statically indeterminate, and thus an equivalent statically determinate structure is obtained by releasing two suitable kinematic constraints.

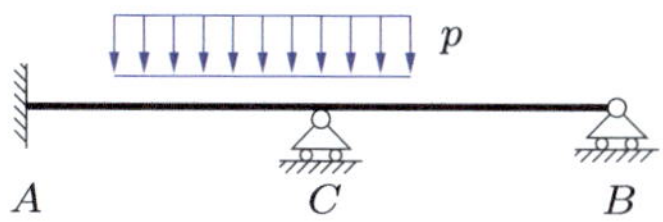

Fig. 6.13: Two-span beam with degree 2 of statically indeterminacy.

The steps of the force method for the two times statically indeterminate two-span beam are as follows:

1) Construct an equivalent statically determinate and stable structure with redundant static components X_1 and X_2.
2) Determine the displacements ξ_{10} and ξ_{20} in the equivalent structure from the external load.
3) Determine the displacements ξ_{11} and ξ_{21} from the load $X_1 = 1$, and the displacements ξ_{12} and ξ_{22} from the load $X_2 = 1$ on the equivalent structure.
4) Form the kinematic constraint equations and determine the redundant components X_1 and X_2.
5) Determine the reactions by superposition or by statics.
6) Determine the distribution of the internal forces by superposition or by equilibrium conditions.

The individual steps of the force method are explained in the following with emphasis on procedure and solution technique.

1) Statically determinate structure and redundant components. The kinematic constraints against rotation in A and vertical displacement in C are released, and the corresponding redundant components are $X_1 = M_A$ and $X_2 = R_C$, as illustrated in Fig. 6.14.

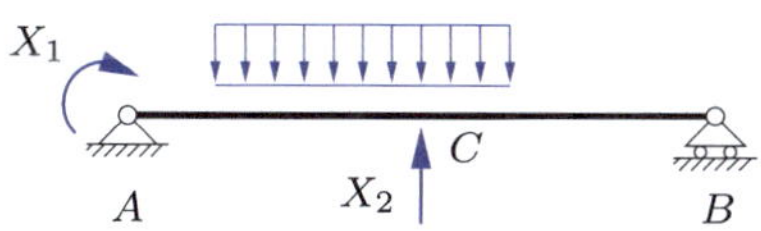

Fig. 6.14: Statically determinate structure with redundant components.

2) Displacements from external load. First the displacements of the equivalent structure with the external loading, but without redundant forces, are determined as illustrated in Fig. 6.15, showing the reactions R_A^0 and R_B^0 and the internal moment $M^0(x)$. The figure also shows the rotation ξ_{10} and the displacement ξ_{20} conjugate to the redundant static components X_1 and X_2. X_1 is a clockwise moment and thus ξ_{10} is positive as a clockwise rotation, while X_2 is an upward force, whereby ξ_{20} is positive if representing an upward displacement.

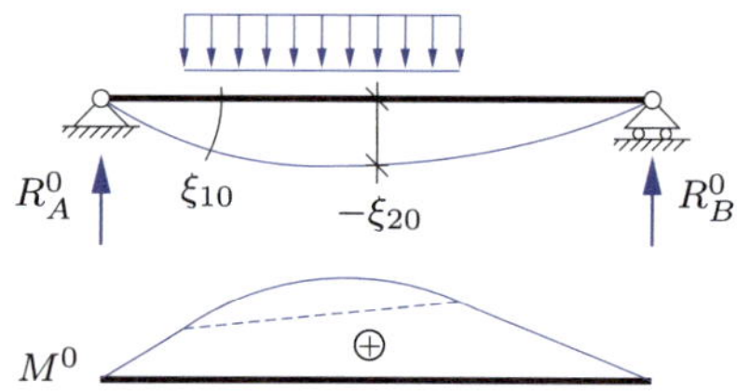

Fig. 6.15: Moment distribution $M^0(x)$ for external load.

The rotation ξ_{10} in A and the displacement ξ_{20} in C are determined by the virtual work equations (4.43) and (4.45), including only the bending moment contribution,

$$\xi_{10} = \int_0^\ell \frac{M^1 M^0}{EI}\, dx\,, \qquad \xi_{20} = \int_0^\ell \frac{M^2 M^0}{EI}\, dx\,.$$

The corresponding two test load cases are $X_1 = 1$ and $X_2 = 1$ with moment distributions $M^1(x)$ and $M^2(x)$, respectively, shown in Fig. 6.16.

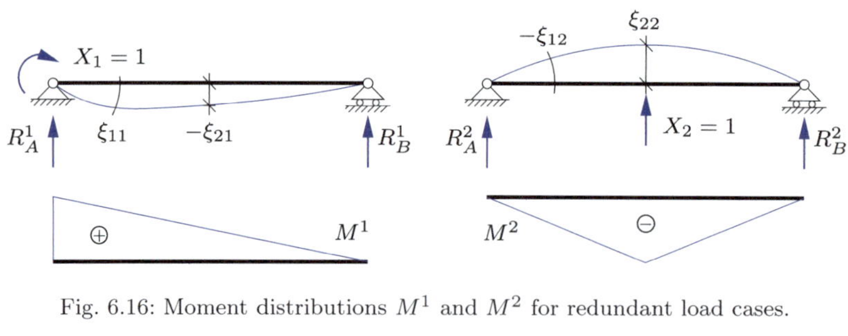

Fig. 6.16: Moment distributions M^1 and M^2 for redundant load cases.

3) Displacements from redundant components. The two remaining load case are for $p = 0$ and the combinations $X_1 = 1$, $X_2 = 0$ and $X_1 = 0$, $X_2 = 1$, respectively. They correspond to the test loads shown in Fig. 6.16, where the displacements and rotations ξ_{11}, ξ_{12}, ξ_{21} and ξ_{22} from the unit loads are also indicated.

In the load case $X_1 = 1$ (with $X_2 = 0$) shown in Fig. 6.16a the rotation ξ_{11} in A and the displacement ξ_{21} in C are determined by the virtual work equations as

$$\xi_{11} = \int_0^\ell \frac{M^1 M^1}{EI}\, dx\,, \qquad \xi_{21} = \int_0^\ell \frac{M^2 M^1}{EI}\, dx\,.$$

The first subscript denotes the number of the displacement/rotation, while the second subscript refers to the index of the redundant static component X_1. In the remaining load case $X_2 = 1$ the rotation ξ_{12} in A and the displacement ξ_{22} in C are

$$\xi_{12} = \int_0^\ell \frac{M^1 M^2}{EI}\, dx = \xi_{21}\,, \qquad \xi_{22} = \int_0^\ell \frac{M^2 M^2}{EI}\, dx\,.$$

It is important to note that because all the loads $X_j = 1$ are normalized to unity the dimensions of ξ_{ij} do not correspond to actual displacements or rotations. Instead they have dimensions of displacement or rotation per unit force or moment, corresponding to flexibility (inverse of stiffness). Thus, ξ_{ij}'s are referred to as *flexibility coefficients*, and the diagonal flexibility coefficients ξ_{jj} are positive, because they are given as the integral of $M^j M^j / EI \geq 0$.

4) Redundant components. The redundant components X_1 and X_2 can now be determined by the kinematic constraints that have been released in the original structure to obtain the equivalent statically determinate structure. In the present case the rotation ξ_1 in A and the transverse displacement ξ_2 in C can be determined by superposition of the three load cases,

$$\begin{aligned} \xi_1 &= \xi_{10} + \xi_{11} X_1 + \xi_{12} X_2\,, \\ \xi_2 &= \xi_{20} + \xi_{21} X_1 + \xi_{22} X_2\,, \end{aligned}$$

where the flexibility coefficients are multiplied by the respective redundant components because they have been computed for unit loads. In the case of multiple equations it is often convenient to express these in matrix form,

$$\begin{bmatrix} \xi_{11} & \xi_{12} \\ \xi_{21} & \xi_{22} \end{bmatrix} \begin{bmatrix} X_1 \\ X_2 \end{bmatrix} = \begin{bmatrix} \xi_1 \\ \xi_2 \end{bmatrix} - \begin{bmatrix} \xi_{10} \\ \xi_{20} \end{bmatrix},$$

where the flexibility coefficients are assembled in the flexibility matrix on the left side. For fixed supports $\xi_1 = 0$ and $\xi_2 = 0$, and the redundant components can be found by solving the matrix equation

$$\begin{bmatrix} X_1 \\ X_2 \end{bmatrix} = - \begin{bmatrix} \xi_{11} & \xi_{12} \\ \xi_{21} & \xi_{22} \end{bmatrix}^{-1} \begin{bmatrix} \xi_{10} \\ \xi_{20} \end{bmatrix}.$$

For a 2×2 matrix the inverse is given explicitly as

$$\begin{bmatrix} \xi_{11} & \xi_{12} \\ \xi_{21} & \xi_{22} \end{bmatrix}^{-1} = \frac{1}{\xi_{11}\xi_{22} - \xi_{21}\xi_{12}} \begin{bmatrix} \xi_{22} & -\xi_{12} \\ -\xi_{21} & \xi_{11} \end{bmatrix},$$

and the two redundant forces are given by the expression

$$\begin{bmatrix} X_1 \\ X_2 \end{bmatrix} = -\frac{1}{\xi_{11}\xi_{22} - \xi_{21}\xi_{12}} \begin{bmatrix} \xi_{22} & -\xi_{12} \\ -\xi_{21} & \xi_{11} \end{bmatrix} \begin{bmatrix} \xi_{10} \\ \xi_{20} \end{bmatrix}.$$

For systems with more than two equations it might be necessary to solve these numerically.

5) Reactions. In the present example the redundant components have replaced the reaction moment in A and the vertical reaction force in C, which are therefore directly given as

$$M_A = X_1 \quad , \quad R_C = X_2 \,.$$

As indicated in Figs. 6.15 and 6.16 it is convenient to determine the reactions for each load case during the analysis. Thus, the remaining reactions at this stage often can be found by superposition,

$$\begin{bmatrix} R_A \\ R_B \end{bmatrix} = \begin{bmatrix} R_A^0 \\ R_B^0 \end{bmatrix} + \begin{bmatrix} R_A^1 & R_A^2 \\ R_B^1 & R_B^2 \end{bmatrix} \begin{bmatrix} X_1 \\ X_2 \end{bmatrix}.$$

Alternatively, since X_1 and X_2 are now known the remaining reactions can also be found by statics. The choice of procedure depends on the specific type of problem.

6) Internal forces. The moment distributions of the individual load cases have already been determined and the resulting moment distribution can be found by superposition as

$$M(x) = M^0(x) + M^1(x)X_1 + M^2(x)X_2 \,,$$

or by a standard static analysis once the reactions have been determined. In the present case it might be easiest to use a direct static analysis, as the moment distribution is fully determined, once the section moments have been determined at C and at the two ends of the interval of the distributed load.

6.2.3 Summary of the force method

The following summary of the force method represents the general case for an n times statically indeterminate structure.

1) Statically determinate structure and redundant component(s). Release n kinematic constraints to obtain a statically and kinematically determinate structure. Introduce n redundant components X_j as conjugate forces to the released constraints ξ_j.

2) Displacements from external load. The displacements (or rotations) of the statically determinate structure at the released constraints from the external load are found by the virtual work equation

$$\xi_{j0} = \int_L \frac{M^j M^0}{EI} \, ds \,, \qquad j = 1 \ldots n \,,$$

where M^0 and M^j are the moment distributions for the external load and the test load $X_j = 1$, respectively.

3) Flexibility coefficients. For load case j the only load acting on the structure is the unit redundant component $X_j = 1$. The displacements/rotations at the released constraints are found for the individual load cases by the virtual work equation

$$\xi_{ij} = \int_L \frac{M^i M^j}{EI} \, ds \,, \qquad i, j = 1 \ldots n \,,$$

where M^i and M^j are the moment distributions for the test loads $X_i = 1$ and $X_j = 1$, respectively. The virtual work integral implies symmetry

$$\xi_{ij} = \xi_{ji} \,,$$

and thus only half of the coupling coefficients need to be computed directly. The displacements/rotations ξ_{ij} are determined for unit loads and thus represent *flexibility coefficients.*

4) Redundant components. The total displacements/rotations ξ_j can be found by superposition of the individual load cases as

$$\begin{aligned}
\xi_1 &= \xi_{10} + \xi_{11} X_1 + \cdots + \xi_{1n} X_n \,, \\
\xi_2 &= \xi_{20} + \xi_{21} X_1 + \cdots + \xi_{2n} X_n \,, \\
&\vdots \qquad\qquad\qquad \vdots \\
\xi_n &= \xi_{n0} + \xi_{11} X_1 + \cdots + \xi_{nn} X_n \,.
\end{aligned}$$

These n equations are solved with respect to the n redundant components $X_1 \ldots X_n$. When releasing kinematic constraints, they are reimposed by the conditions $\xi_j = 0$, while $\xi_j \neq 0$ can be used to impose a finite displacement/rotation.

5) Reactions. Reactions that have been replaced by redundant forces are directly available, while the remaining reactions can be determined by superposition,

$$R_A = R_A^0 + R_A^1 X_1 + \ldots + R_A^n X_n \,.$$

When $X_1 \ldots X_n$ have been determined, the remaining reactions can also be found by static equilibrium.

6) Internal forces. The internal forces can be found by superposition of the individual load cases, e.g. the moment distribution

$$M = M^0 + M^1 X_1 + \ldots + M^n X_n \,,$$

and similarly for the normal force N and shear force Q. As the reactions have been determined, the internal forces can also be determined by static analysis as described in Chapter 3.

6.3 Application of the Force Method

In this section the procedure of the force method is illustrated by simple examples involving statically indeterminate beam structures. The extension to analysis of frame structures is considered in Section 6.4.

Example 6.1. Two-span beam – revisited. The two-span beam in Fig. 6.17 corresponds to the introductory example in Section 6.1, and it is revisited here to illustrate the use of an internal moment as redundant component.

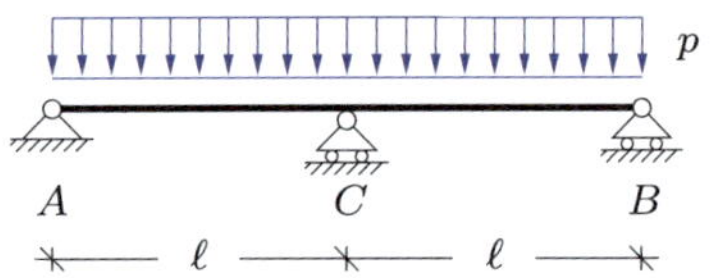

Fig. 6.17: Statically indeterminate two-span beam with distributed load.

The beam is one time statically indeterminate, and the rotation continuity in C is released by the introduction of an internal hinge as indicated in Fig. 6.18. The internal moment at C is the conjugate force to the released rotation, and M_C therefore acts as redundant component X_1. Note, that because the redundant component X_1 represents the internal moment it is introduced as a self-equilibrating moment pair in Fig. 6.18. The direction of X_1 is chosen in agreement with the sign convention for internal moments.

The actual load case with $X_1 = 0$ is shown in Fig. 6.19a. The corresponding moment M^0 vanishes in C due to the internal hinge, and the moment distribution is composed of two separate parabolic distributions with local maximum $\frac{1}{8}p\ell^2$. The rotation discontinuity ξ_{10} at C is determined by the virtual work equation, where the virtual moment distribution

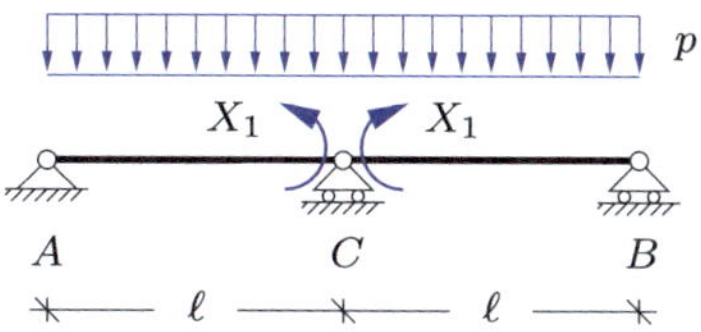

Fig. 6.18: Statically determinate beam with internal moment in C as redundant component.

$M^1(x)$ is obtained for $X_1 = 1$ as shown in Fig. 6.19b. It is seen that both $M^0(x)$ and $M^1(x)$ are symmetric with respect to the center of the beam, whereby

$$\xi_{10} = 2\int_0^\ell \frac{M^1 M^0}{EI}\,dx = \frac{2}{EI}\,\frac{\ell}{3}\Big(\frac{1}{8}\,p\ell^2\Big) = \frac{1}{12}\,\frac{p\ell^3}{EI}\,.$$

The integral is evaluated by row four in Table 4.1, and the above expression is non-dimensional, because ξ_{10} represents a rotation.

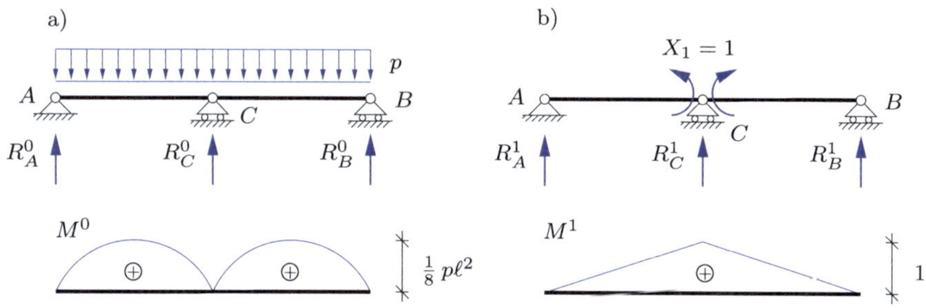

Fig. 6.19: Moment distribution.

The reactions are determined for each of the two load cases, and as shown in Fig. 6.20 the reaction in A is found by moment equilibrium about the section in C. For the actual load case p in Fig. 6.20a the moment in C vanishes due to the hinge, giving the reaction

$$\overset{\frown}{C} \qquad R_A^0\ell - \tfrac{1}{2}\ell\, p\ell = 0 \quad\Rightarrow\quad R_A^0 = \tfrac{1}{2}\,p\ell\,.$$

For the load case $X_1 = 1$ in Fig. 6.20b the internal moment in C is unity, which gives

$$\overset{\frown}{C} \qquad R_A^1\ell - 1 = 0 \quad\Rightarrow\quad R_A^1 = \frac{1}{\ell}\,.$$

The reactions in B can be found by the similar method for the right half of the beam, and the reactions in C are finally found by moment about A for the entire beam. This gives the following reaction components for the two load cases:

$$R_B^0 = R_A^0 = \tfrac{1}{2}\,p\ell\;,\; R_C^0 = p\ell\,,$$

$$R_B^1 = R_A^1 = \frac{1}{\ell}\;,\; R_C^1 = -\frac{2}{\ell}\,.$$

It is seen that vertical equilibrium is satisfied for both load cases: $R_A^0 + R_B^0 + R_C^0 = 2p\ell$ and $R_A^1 + R_B^1 + R_C^1 = 0$.

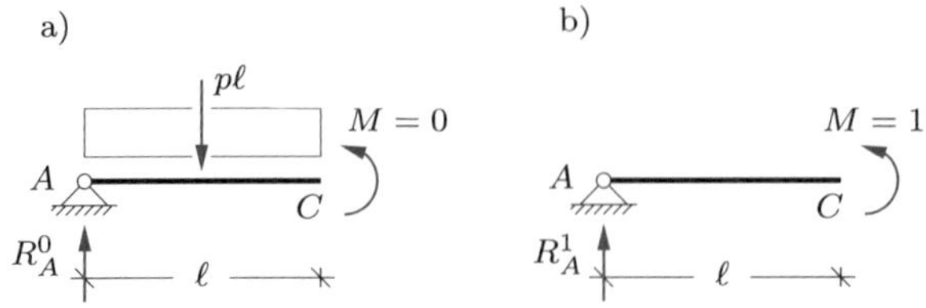

Fig. 6.20: Moment about section in C.

The flexibility coefficient ξ_{11} is governed by the load case in Fig. 6.19b, which only contains the unit load case $X_1 = 1$. The flexibility coefficient is found by the virtual work equation

$$\xi_{11} = 2\int_0^\ell \frac{M^1 M^1}{EI}\,dx = \frac{2}{3}\frac{\ell}{EI}.$$

It is seen that $\xi_{11} > 0$ and has dimensions $[\text{moment}]^{-1}$, which corresponds to [rotation/moment] and thereby rotation flexibility.

The resulting rotation discontinuity at C is found by superposition of the two load cases considered above,

$$\xi_1 = \xi_{10} + \xi_{11} X_1 .$$

The latter term is scaled by X_1 because ξ_{11} represents the moment flexibility determined for $X_1 = 1$. Continuity of the rotation at C is imposed by the condition $\xi_1 = 0$, and the redundant component is then determined from the above expression as

$$X_1 = -\frac{\xi_{10}}{\xi_{11}} = -\tfrac{1}{8}p\ell^2 .$$

This corresponds to the internal moment at C found previously in Fig. 6.5.

The reactions of the individual load cases are shown in Fig. 6.19, and superposition is therefore conveniently used in this example. For the reaction in A this gives

$$R_A = R_A^0 + R_A^1 X_1 = \tfrac{1}{2}p\ell - \frac{1}{\ell}\,\tfrac{1}{8}p\ell^2 = \tfrac{3}{8}p\ell ,$$

while the remaining reactions are determined similarly as

$$R_B = R_A = \tfrac{3}{8}p\ell , \qquad R_C = \tfrac{5}{4}p\ell .$$

It is seen that the reactions correspond to those obtained previously, satisfying vertical equilibrium: $R_A + R_B + R_C = 2p\ell$.

Finally, the section force distributions are determined, and the redundant component directly gives the internal moment $M_C = X_1$. Thus, the moment distribution is obtained by superimposing the moment parabola with local maximum $\frac{1}{8}p\ell^2$ to the linear curves connecting the moments in A, B and C as shown in Fig. 6.21a. The shear forces in A and B are in equilibrium with the corresponding reactions, giving

$$Q_A = R_A = \tfrac{3}{8}p\ell \quad , \qquad Q_B = -R_B = -\tfrac{3}{8}p\ell .$$

The shear force at C is found by placing a section at $C-$ immediately to the left of C, and then taking vertical equilibrium as illustrated in Fig. 6.21b,

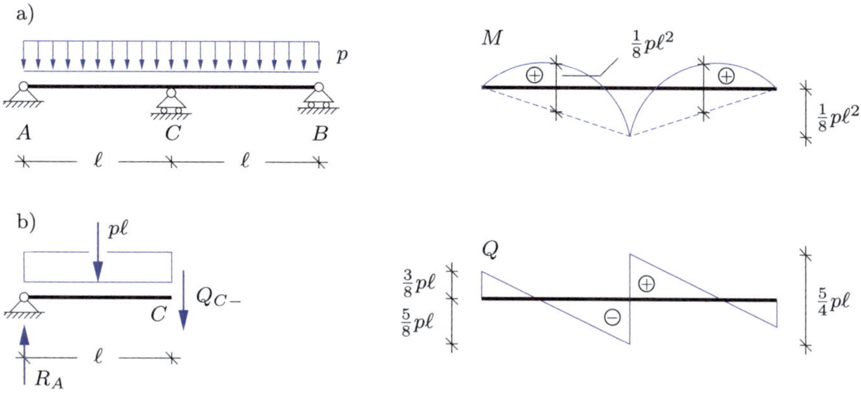

Fig. 6.21: Moment and shear force distribution.

$$\downarrow \qquad Q_{C-} + p\ell - R_A = 0 \qquad \Rightarrow \qquad Q_{C-} = -\tfrac{5}{8}\,p\ell\,.$$

Finally, the reaction R_C produces a discontinuity of $\frac{5}{4}p\ell$ in the shear force in C, and the resulting distribution of the shear force can be determined as shown in Fig. 6.21. □

Example 6.2. Two-span beam with fixed support. Figure 6.22 shows a two-span beam with a fixed support in A and simple supports in B and C. This structure was considered previously in Section 6.2.2 in a general context, while in this example the beam is loaded by a uniformly distributed load with intensity p.

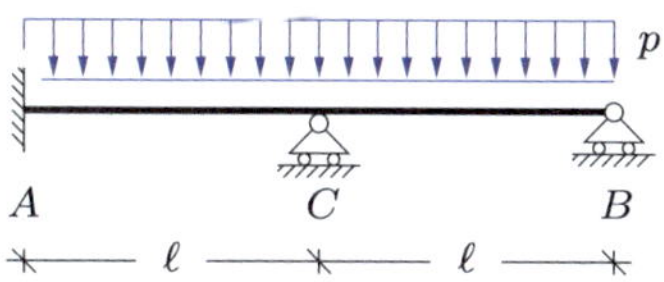

Fig. 6.22: Two-span beam with fixed support at A.

The structure is two times statically indeterminate and an equivalent statically determinate structure similar to that in Example 6.1 is obtained by releasing the rotation in A and the rotation continuity in C. Hereby, X_1 represents the reaction moment in A, while X_2 is the internal moment in C. Figure 6.23 shows the equivalent statically determinate beam structure with the redundant components.

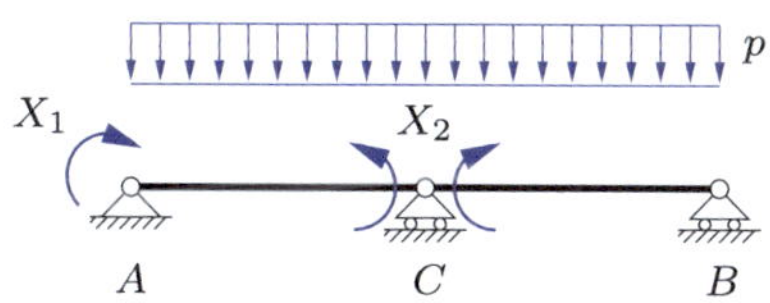

Fig. 6.23: Statically determinate structure and redundant components.

In the first load case only the distributed load is taken into account, while $X_1 = X_2 = 0$. Hereby, the system corresponds to the equivalent structure in Example 6.1, whereby moment distribution $M^0(x)$ and reactions R^0_A, R^0_B and R^0_C are directly available from Fig. 6.19a. Furthermore, the moment distribution $M^2(x)$ for the load case only containing $X_2 = 1$ corresponds to the moment distribution M^1 presented in Fig. 6.19b. This leaves $X_1 = 1$ as the only load case not already considered in Example 6.1. The moment distribution M^1 for the load case only containing $X_1 = 1$ is shown in Fig. 6.24, where the two remaining moment distributions $M^0(x)$ and $M^2(x)$ are also shown for convenience.

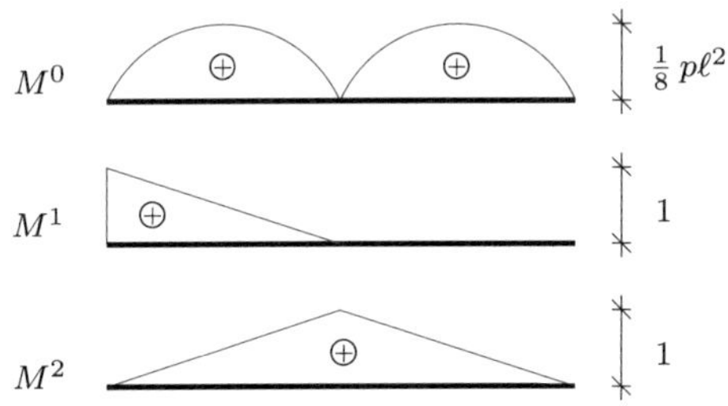

Fig. 6.24: Summary of moment distributions.

The rotation in A for the actual load case is found by the virtual work equation

$$\xi_{10} = \int_0^\ell \frac{M^1 M^0}{EI}\, dx = \frac{\ell}{3EI}\left(\frac{1}{8}p\ell^2\right) = \frac{1}{24}\frac{p\ell^3}{EI},$$

where only the left part of the beam contributes to the integral because $M^1(x) \equiv 0$ on the right part. The integral is solved by using row four of Table 4.1, where a linear curve is combined with a parabola. Similarly the change in rotation in C is found by virtual work, where a triangle and a parabola are combined for both parts of the two-span beam,

$$\xi_{20} = 2\,\xi_{10} = \frac{1}{12}\frac{p\ell^3}{EI}.$$

Figure 6.25 indicates the symmetric deformation form of the equivalent statically determinate structure, where the magnitude of the three rotations are equal, i.e. $\xi_{10} = \xi_{20}^- = \xi_{20}^+$. This verifies the relation $\xi_{20} = 2\xi_{10}$ used above.

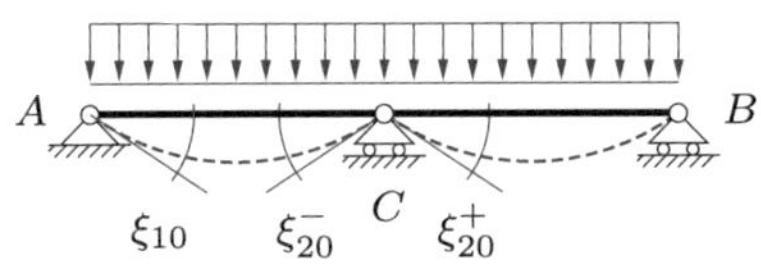

Fig. 6.25: Deformation of statically determinate structure.

The flexibility coefficients ξ_{11}, ξ_{21} and ξ_{12}, ξ_{22}, with $\xi_{12} = \xi_{21}$, represent the rotations in A and C for the two load cases $X_1 = 1$ and $X_2 = 1$, respectively. When combining the moment distributions $M^1(x)$ and $M^2(x)$ in the virtual work equation, the flexibility coefficients are determined as

$$\xi_{11} = \frac{1}{3}\frac{\ell}{EI}, \quad \xi_{22} = 2\xi_{11} = \frac{2}{3}\frac{\ell}{EI}, \quad \xi_{12} = \xi_{21} = \frac{1}{6}\frac{\ell}{EI}.$$

The dimension is the reciprocal of a moment, and thereby rotation flexibility.

Rotation at A and rotation discontinuity at C are constrained in the actual structure in Fig. 6.22, implying that $\xi_1 = 0$ and $\xi_2 = 0$. This gives two equations for X_1 and X_2. In matrix form these equations are

$$\begin{bmatrix} \xi_{11} & \xi_{12} \\ \xi_{21} & \xi_{22} \end{bmatrix} \begin{bmatrix} X_1 \\ X_2 \end{bmatrix} = - \begin{bmatrix} \xi_{10} \\ \xi_{20} \end{bmatrix},$$

and as demonstrated in Section 6.2.2 this determines the redundant components as

$$\begin{bmatrix} X_1 \\ X_2 \end{bmatrix} = - \frac{1}{\xi_{11}\xi_{22} - \xi_{21}\xi_{12}} \begin{bmatrix} \xi_{22} & -\xi_{12} \\ -\xi_{21} & \xi_{11} \end{bmatrix} \begin{bmatrix} \xi_{10} \\ \xi_{20} \end{bmatrix} = - \begin{bmatrix} 2 \\ 3 \end{bmatrix} \frac{p\ell^2}{28}.$$

It is seen that the magnitude of the moment is larger in C than in A, and that both moments are negative. This is in agreement with the sign convention chosen in Fig. 6.23, where X_1 and X_2 must be negative to prevent rotation in A and relative rotation at C.

The reactions are now determined, and horizontal equilibrium directly gives $R'_A = 0$. Furthermore, the reaction moment in A is represented directly by the first redundant component,

$$M_A = X_1 = -\tfrac{1}{14}\, p\ell^2 .$$

This leaves the vertical reactions in A, B and C. In this example the reactions from the individual load cases have not been determined, and the remaining reactions are therefore found by static equilibrium.

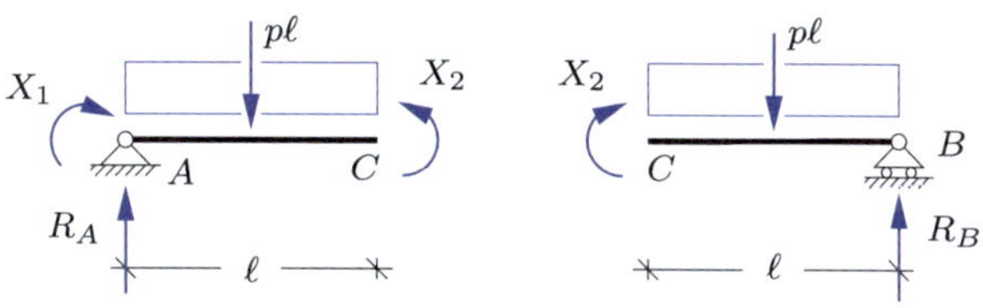

Fig. 6.26: Reactions by moment equilibrium.

Figure 6.26 shows the left and right part of the beam when placing a section in C. The purpose is to use moment equilibrium around C, and the shear forces at $C\pm$ are therefore not shown. For the left part AC moment about C gives the reaction in A,

$$\stackrel{\curvearrowright}{C} \qquad R_A\ell + X_1 - X_2 - \tfrac{1}{2}\ell\, p\ell = 0 \quad \Rightarrow \quad R_A = \tfrac{13}{28}\, p\ell .$$

Similarly, moment about C for the right part CB gives the reaction in B,

$$\stackrel{\curvearrowleft}{C} \qquad R_B\ell - X_2 - \tfrac{1}{2}\ell\, p\ell = 0 \quad \Rightarrow \quad R_B = \tfrac{11}{28}\, p\ell .$$

Finally, the vertical reaction in C is obtained by moment equilibrium for the entire structure with respect to B,

$$\stackrel{\curvearrowright}{B} \qquad R_C\ell + R_A 2\ell + X_1 - 2p\ell^2 = 0 \quad \Rightarrow \quad R_C = \tfrac{8}{7}\, p\ell .$$

Note, that X_2 represents a self-equilibrating internal moment pair without contribution to global equilibrium of the structure.

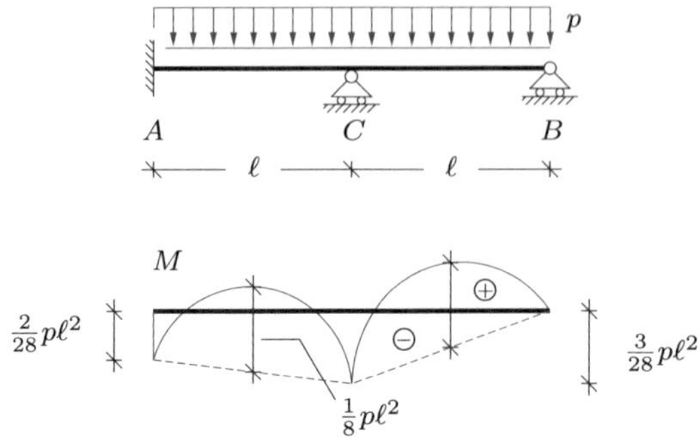

Fig. 6.27: Distribution of moment.

The moment distribution is governed by the internal moments $M_A = X_1$, $M_B = 0$ and $M_C = X_2$. These key values are plotted in the moment diagram in Fig. 6.27 and connected by (dashed) lines. The final moment distribution is then obtained by superimposing the parabolas from the distributed load. This example illustrates that the construction of the moment distribution is often simplified when the redundant components are chosen as suitable reaction moments and/or internal moments. □

Example 6.3. Two-span beam with local forces. Structures are often exposed to various types of loading conditions, e.g. wind excitation or snow loading, and the computational cost may be significantly reduced when results from previous static analyzes of the structure are reused in the analysis of a new load case. With respect to the force method the present example illustrates that only the displacements/rotations ξ_{j0} need to be re-calculated when changing the loading conditions, while the load-independent flexibility parameters ξ_{ij} may be carried on from a previous analysis.

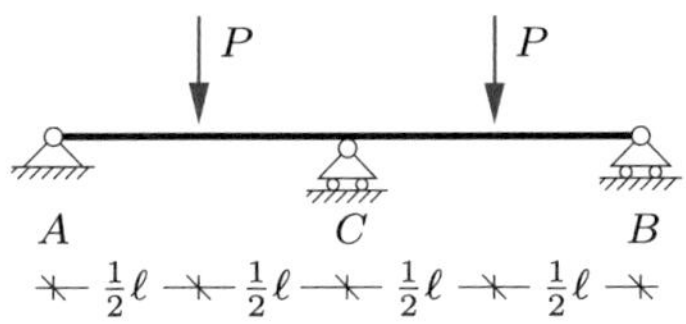

Fig. 6.28: Statically indeterminate two-span beam with local forces P.

Reconsider the simply supported two-span beam with distributed load in Example 6.1. Figure 6.28 shows the same beam structure, but loaded by two local forces P acting at the center of the two spans. The reuse of the previous results from Example 6.1 requires using the same equivalent statically determinate structure, and as shown in Fig. 6.29 the redundant component X_1 again represents the internal moment in C.

Figure 6.30 shows both the moment distribution $M^0(x)$ due to the actual forces P and the moment distribution $M^1(x)$ for $X_1 = 1$. Note, that $M^1(x)$ has previously been obtained in Example 6.1 and is therefore taken directly from Fig. 6.19. The rotation discontinuity at C is obtained by the virtual work equation,

$$\xi_{10} = \int_A^B \frac{M^1 M^0}{EI}\, dx = \frac{\ell}{3EI}\left(\frac{1}{4}P\ell\right)\frac{1}{2} + \frac{\ell}{6EI}\left(\frac{1}{4}P\ell\right)2 = \frac{1}{8}\frac{P\ell^2}{EI},$$

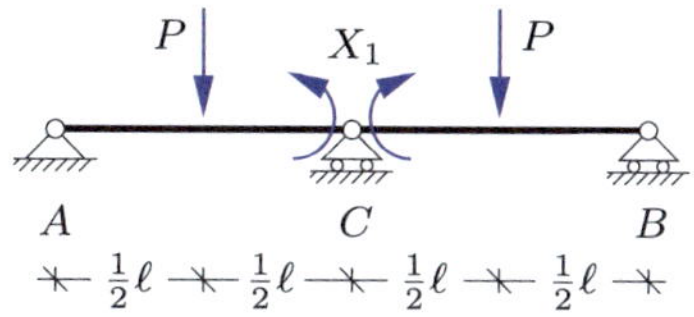

Fig. 6.29: Statically determinate structure and redundant component.

while the flexibility coefficient ξ_{11} can be taken directly from Example 6.1,

$$\xi_{11} = \frac{2}{3}\frac{\ell}{EI}.$$

In particular for structures that are several times statically indeterminate the reuse of the flexibility coefficients might save a significant amount of computational effort because the flexibility matrix need only be inverted once.

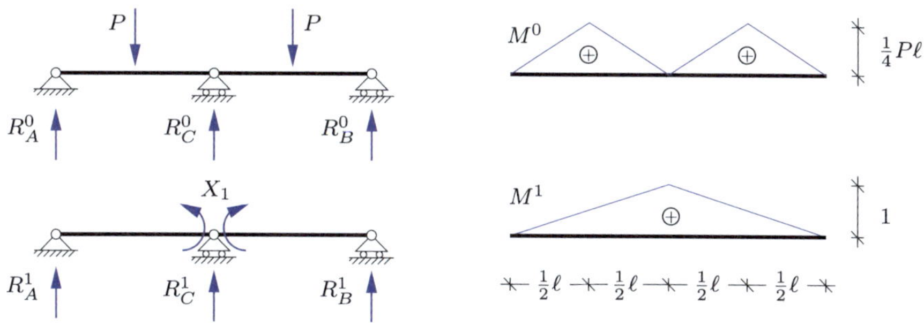

Fig. 6.30: Distribution of moment for external load and unit redundant moment.

In the original structure the rotation discontinuity at C is constrained, $\xi_1 = 0$, and this determines the redundant component

$$X_1 = -\frac{\xi_{10}}{\xi_{11}} = -\tfrac{3}{16}P\ell.$$

The reactions in Fig. 6.30a from the external loading are found as

$$R_A^0 = R_B^0 = \tfrac{1}{2}P, \qquad R_C^0 = P.$$

The reactions in Fig. 6.30b for the load case $X_1 = 1$ are taken from Example 6.1,

$$R_B^1 = R_A^1 = \frac{1}{\ell}, \qquad R_C^1 = -\frac{2}{\ell}.$$

The resulting reactions can now be determined by superposition,

$$R_B = R_A = \tfrac{5}{16}P, \qquad R_C = \tfrac{11}{8}P.$$

Note, that the reaction in C carries a larger part of the total vertical load compared to the case with the distributed load in Example 6.1.

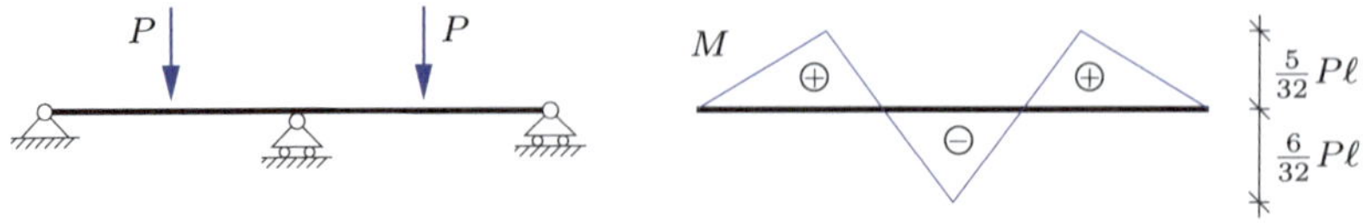

Fig. 6.31: Moment distribution.

The redundant component X_1 gives the internal moment in C directly, while the moment vanishes at the hinges in A and B. Thus, the moment distribution is fully determined once the moment M_P at the location of P is obtained. Superposition of the values obtained from Fig. 6.30 gives

$$M_P = M(\tfrac{1}{2}\ell) = M^0(\tfrac{1}{2}\ell) + M^1(\tfrac{1}{2}\ell)X_1 = \tfrac{1}{4}P\ell + \tfrac{1}{2}(-\tfrac{3}{16}P\ell) = \tfrac{5}{32}P\ell\,.$$

The final moment distribution is then determined by connecting the moment values at the points with transverse forces by straight lines as shown in Fig. 6.31. It is seen that the numerically largest moment $M_{\max} = \frac{6}{32}P\ell$ occurs in C. □

6.4 The force method for frame structures

Modern plane frame structures are typically designed with some kind of static indeterminacy arising from the use of more than the minimum three support components. The additional reaction components are deliberately introduced to obtain a stiffer structure, and to obtain some degree of balance between positive and negative moments, whereby the necessary moment capacity of the members of the frame is reduced. A simple illustration is provided by the rectangular frame shown in Fig. 6.32. The horizontal beam CD carries

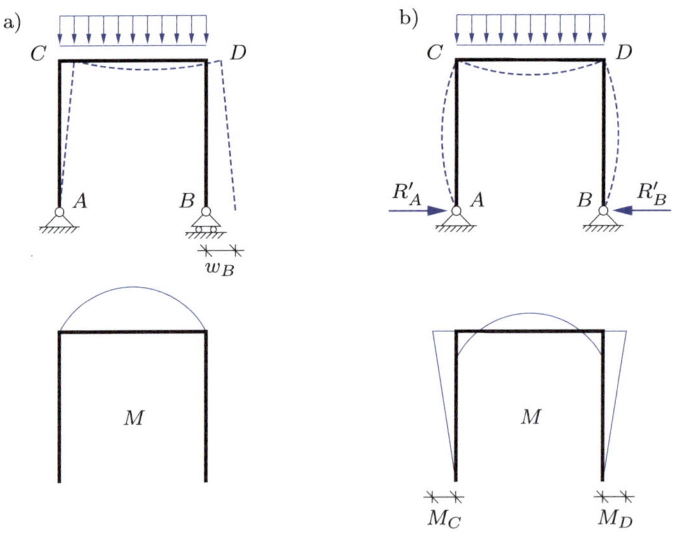

Fig. 6.32: Moment reduction in frame by additional support.

a uniformly distributed vertical load with intensity p. The figure shows two different sets of support conditions. In Fig. 6.32a the frame has a simple fixed support at A and a simple support on horizontal rollers at B. The reactions consist of two equal vertical forces at A and B, each carrying half of the load, while the horizontal reaction component at A vanishes. The horizontal beam CD curves upward, while the columns AC and BD remain straight. As a result there is outward motion w_B of the support B, and the moment distribution along CD is parabolic with zero moment at the frame corners C and D. Figure 6.32b shows a modified version of the frame, in which the support at B is replaced by a fixed simple support. This introduces horizontal reactions that prevent the outward motion of the supports and introduce a negative moment at the frame corners C and D, whereby the maximum moment in the horizontal beam CD is reduced. As a result this form of the frame is stiffer, and has a smaller maximum moment.

6.4.1 Simply supported frames

Simply supported plane frames are often one time statically indeterminate. It is possible to use one of the reactions as redundant component, but it may be advantageous to choose the internal moment at a suitable joint in the frame as X_1. This often leads to simple moment curves and reaction forces. Figure 6.33 shows some useful locations of X_1 for typical frame structures.

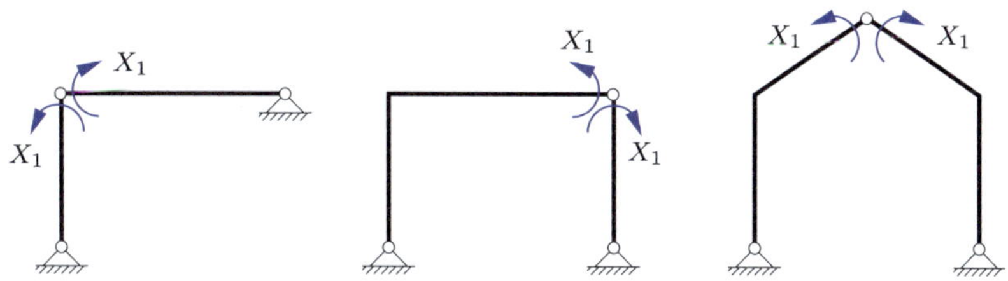

Fig. 6.33: Simply supported frames with the redundant component at a joint.

Example 6.4. Angle frame with distributed load. This example considers the simply supported angle frame ACB shown in Fig. 6.34a. A distributed vertical load with intensity p acts on the horizontal beam CB. The frame is one time statically indeterminate and as discussed above it is convenient to construct an equivalent statically determinate structure by releasing the rotation continuity at a suitable joint. Figure 6.34b shows a hinge introduced at the corner C, whereby the redundant component X_1 represents the internal moment M_C.

The moment distributions and the reactions are now determined for the actual load p without X_1, and for the unit load $X_1 = 1$ without p. Figure 6.35 shows these two load cases and the corresponding moment distributions. The rotation discontinuity at the joint C is found by the virtual work equation,

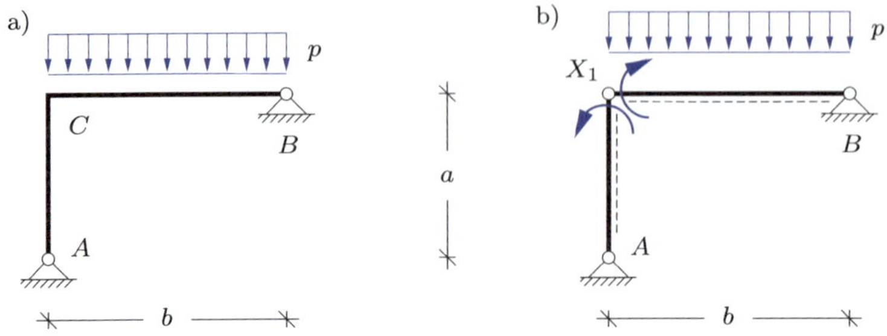

Fig. 6.34: Angle frame with distributed load.

$$\xi_{10} = \int_A^B \frac{M^0 M^1}{EI} \, ds = \frac{b}{3EI} \left(\frac{1}{8} pb^2 \right) = \frac{1}{24} \frac{pb^3}{EI},$$

where only the horizontal beam CB contributes to the integral, because $M^0(s) \equiv 0$ in the vertical beam AC. The integral is evaluated by row four in Table 4.1. It is observed that the rotation discontinuity ξ_{10} is positive, implying that the right angle of the joint in C will 'close' slightly. The corresponding flexibility coefficient is found by the integral

$$\xi_{11} = \int_A^B \frac{M_1 M_1}{EI} \, ds = \left(\frac{a}{3} + \frac{b}{3} \right) \frac{1}{EI} = \frac{1}{3} \frac{a+b}{EI},$$

following from the first row in Table 4.1.

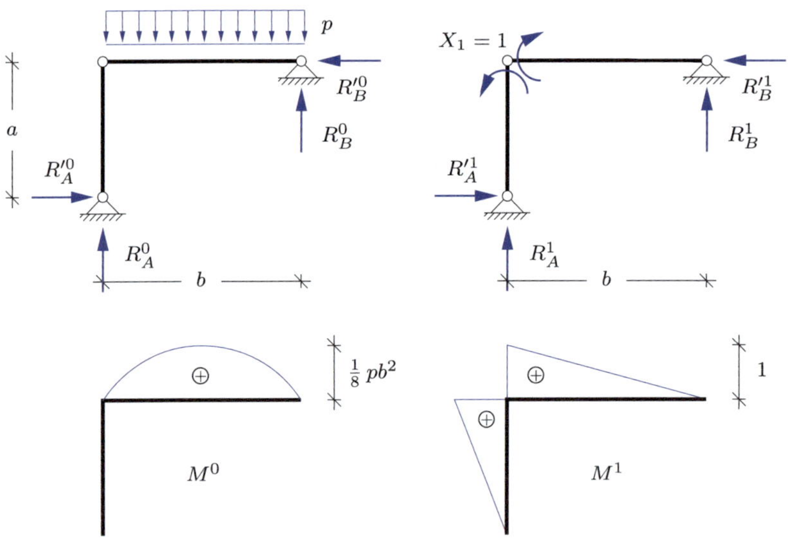

Fig. 6.35: Moment distribution and reactions for load cases.

In the actual frame structure the right angle of the corner at C is maintained during deformation of the frame, whereby $\xi_1 = \xi_{10} + \xi_{11} X_1 = 0$. This gives the redundant component

$$X_1 = -\frac{\xi_{10}}{\xi_{11}} = -\frac{1}{8}\frac{pb^3}{a+b},$$

representing the internal bending moment at C.

The reactions for the individual load cases are determined in accordance with the sign convention in Fig. 6.35. The reactions for the actual load p are

$$R_A^0 = R_B^0 = \tfrac{1}{2}\,pb\,, \qquad R_A'^0 = R_B'^0 = 0\,.$$

For the load case $X_1 = 1$ the reactions are

$$R_A^1 = -R_B^1 = -\frac{1}{b}, \qquad R_A'^1 = R_B'^1 = -\frac{1}{a},$$

and the final reactions can now be found by superposition of the individual load cases. For the vertical reaction in A this gives

$$R_A = R_A^0 + R_A^1 X_1 = \frac{pb}{2} + \Big(-\frac{1}{b}\Big)\Big(\frac{1}{8}\frac{pb^3}{a+b}\Big) = \frac{pb}{8}\frac{4a+5b}{a+b}\,.$$

The three remaining reaction forces can similarly be found by superposition as

$$R_B = \frac{pb}{8}\frac{4a+3b}{a+b}, \qquad R_A' = R_B' = \frac{1}{8}\frac{pb^3}{a(a+b)}\,.$$

It is seen that the horizontal and the vertical equilibrium are both satisfied.

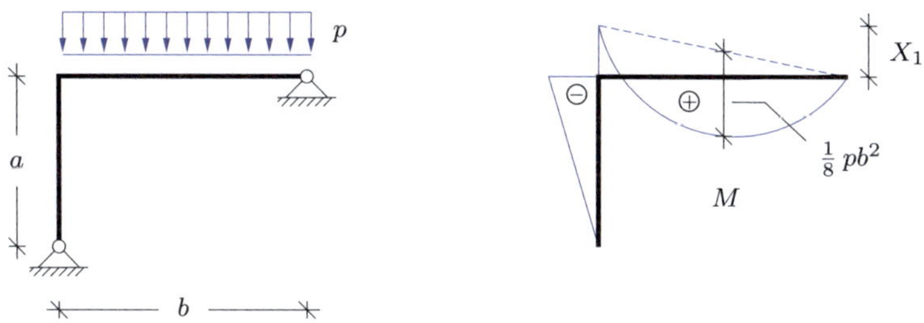

Fig. 6.36: Moment distribution.

The moment distribution is determined via the internal moment at the joint C,

$$M_C = X_1 = -\frac{1}{8}\frac{pb^3}{a+b}\,.$$

The moment distribution shown in Fig. 6.36 is linear in the vertical beam AC, while it is parabolic in the horizontal beam. The parabolic part with local maximum $\frac{1}{8}pb^2$ is superimposed on the (dashed) linear curve connecting the moment values in C and B. □

Example 6.5. Simply supported frame with combined load. Figure 6.37 shows a simply supported frame with distributed load p on CD and a vertical tip force P at E. The sign convention of the internal forces is defined by the dashed line, indicating the lower side of the individual beams. It is assumed that $P = \frac{1}{2}pa$, corresponding to half of the resultant of the distributed load.

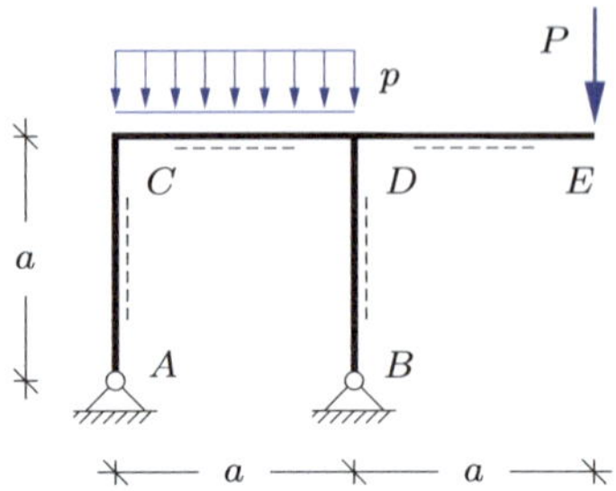

Fig. 6.37: Simply supported frame with distributed load p and local tip force P.

The frame has four reaction force components, while only three independent equilibrium equations are available. Thus, the frame structure is one time statically indeterminate. When choosing the redundant component as an internal moment, special care should be taken if placing a hinge in D, as the structure should remain kinematically determinate without mechanisms. Three possible choices are presented in Fig. 6.38. It is seen that the locations of the hinge in Figs. 6.38a,c imply that the horizontal reactions are zero in the load case $X_1 = 0$, whereby the corresponding moments in the vertical beams vanish. This might simplify the analysis and in the present example the internal moment pair at the joint C is chosen as the redundant component as shown in Fig. 6.39.

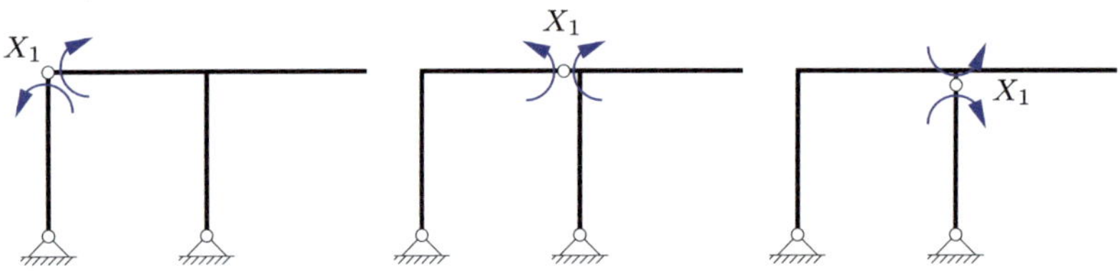

Fig. 6.38: Possible choices of the redundant component X_1.

The external load case is composed of two individual loads: the distributed load p on CD and the vertical tip force P at E. From a computational point of view it is convenient to determine the separate contributions to the virtual work equation from the two individual loads as shown in Fig. 6.40. The individual moment distributions for the two load components are also shown in the figure. The moment distribution M_a^0 due to the distributed load is parabolic on CD, while the moment distribution M_b^0 due to the tip force is piecewise linear. The rotation discontinuity at C is found by the equation of virtual work, where the contributions from the two load cases are added,

$$\xi_{10} = \int \frac{M_a^0 M^1}{EI}\, ds + \int \frac{M_b^0 M^1}{EI}\, ds\,.$$

The reactions for the load case $X_1 = 1$ are determined as follows. The horizontal reactions must be opposite and of equal magnitude to conserve horizontal equilibrium. Thus, they do not contribute to the global moment equilibrium about C, which determines $R_B = 0$. Vertical force equilibrium then gives $R_A = 0$. Finally, the horizontal component $R_A'^1$ is determined by moment equilibrium about the hinge at C of the left part of the frame,

$$\overset{\frown}{C} \qquad R_A'^1 a + 1 = 0 \quad \Rightarrow \quad R_A'^1 = -\frac{1}{a}\,.$$

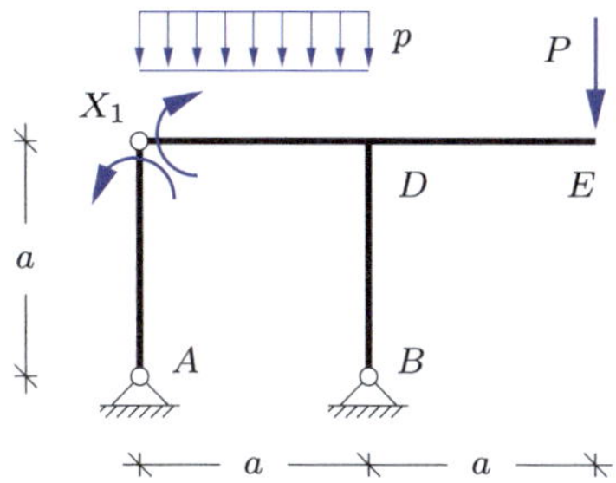

Fig. 6.39: Redundant component X_1 as moment in C.

The horizontal reaction in B then follows from horizontal equilibrium as $R_B'^1 = R_A'^1 = -1/a$. The moment distribution is linear over the individual beams and is determined from sections just below C and D. The moment distribution $M^1(s)$ is shown in Fig. 6.41.

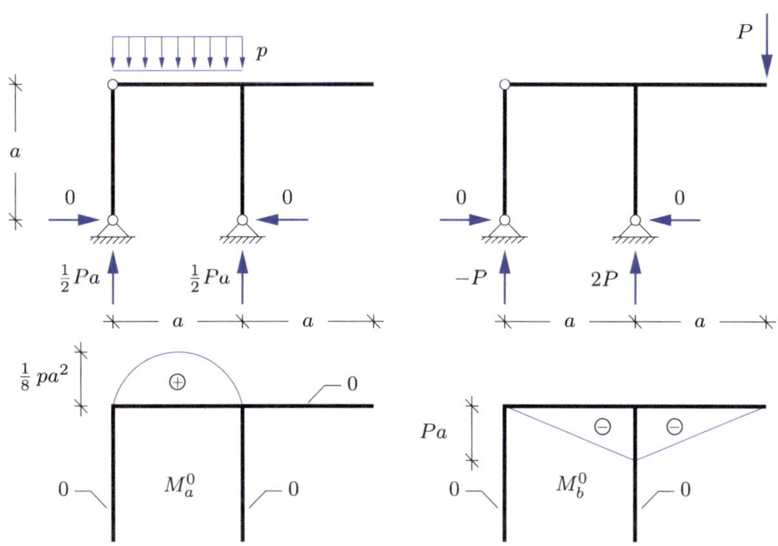

Fig. 6.40: Moment distributions from external load.

The two integrals in the virtual work equation are evaluated by the integral formulae in Table 4.1,

$$\xi_{10} = \frac{a}{3EI}\left(\tfrac{1}{8}\,pa^2\right)(1+1) + \frac{1}{EI}\,\tfrac{1}{2}\,a(-Pa) = \frac{1}{12}\,\frac{a^2}{EI}\,(PA - 6p)\,.$$

When introducing the relation $P = \frac{1}{2}pa$ the final expression becomes

$$\xi_{10} = -\,\frac{1}{6}\,\frac{pa^3}{EI}\,.$$

It is seen that the major part of the rotation discontinuity in the equivalent structure is due to the tip force P. The flexibility coefficient follows from the moment distribution in Fig. 6.41 as

$$\xi_{11} = \int_A^B \frac{M_1 M_1}{EI} ds = \frac{5}{3}\frac{a}{EI},$$

and the redundant component is then obtained as

$$X_1 = -\frac{\xi_{10}}{\xi_{11}} = \tfrac{1}{10}\, pa^2,$$

where the relation $P = \frac{1}{2}pa$ has been introduced.

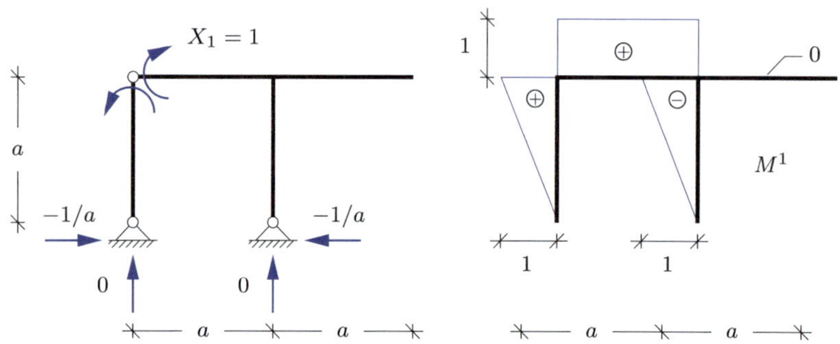

Fig. 6.41: Moment distributions from $X_1 = 1$.

The reactions have been determined for the three individual load cases in Figs. 6.40 and 6.41, and the resulting reactions can thus be determined by superposition. For $P = \frac{1}{2}pa$ the vertical reaction forces are

$$R_A = \tfrac{1}{2}pa - P = 0, \qquad R_B = \tfrac{1}{2}pa + 2P = \tfrac{3}{2}pa,$$

while the horizontal reactions are

$$R'_A = R'_B = \Big(-\frac{1}{a}\Big) X_1 = -\tfrac{1}{10}\, pa^2.$$

It is found that both horizontal and vertical force equilibrium are satisfied.

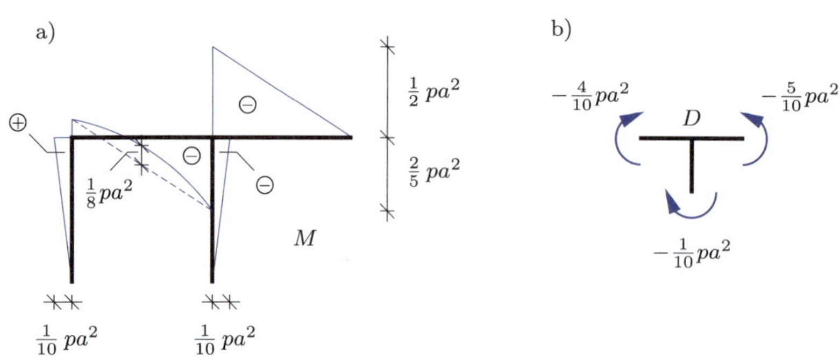

Fig. 6.42: a) Moment distribution, b) Equilibrium in D.

The moment distribution is obtained by determining the moment in C and D, whereafter the parabolic distribution from the distributed load is superimposed on CD. The moment at the left joint C is equal to the redundant component,

$$M_C = X_1 = \tfrac{1}{10}\,pa^2 .$$

At the joint D three internal moments occur – one for each beam attached to D. These moments are determined by superposition of the three load cases,

$$\begin{aligned} M_{DC} &= -Pa + X_1 = -\tfrac{2}{5}\,pa^2 ,\\ M_{DE} &= -Pa \qquad\quad = -\tfrac{1}{2}\,pa^2 ,\\ M_{DB} &= -X_1 \qquad\quad = -\tfrac{1}{10}\,pa^2 , \end{aligned}$$

where $P = \frac{1}{2}pa$ has again been introduced. The moment distribution is shown in Fig. 6.42a, where it is seen that the largest moment occurs in the cantilever beam DE at D,

$$M_{\max} = -M_{DE} = \tfrac{1}{2}\,pa^2 .$$

In joints connecting more than two elements it is a good idea to check moment equilibrium. In Fig. 6.42b the joint has been isolated and the three internal moments are applied. It is found that moment equilibrium is satisfied,

$$\overset{\frown}{C} \qquad -\tfrac{1}{10}\,pa^2 - \tfrac{4}{10}\,pa^2 - \left(-\tfrac{5}{10}\,pa^2\right) = 0 .$$

Similarly, horizontal and vertical force equilibrium could be checked at the joint. □

6.4.2 Frames with fixed supports

As demonstrated in the previous examples, simply supported frames are typically one time statically indeterminate and therefore conveniently analyzed by the force method. However, for frames with fixed supports, such as in Fig. 6.43, use of the force method requires three redundant components and thereby solution of three coupled equations. This is typically very cumbersome by analytical means, and some form of system reduction technique would therefore be of interest.

Symmetry conditions

Plane frame structures are often symmetric. This property can sometimes be used to separate the system into two parts, each with a reduced number of redundant components. Figure 6.43 shows a symmetric frame structure with a vertical force acting on the left inclined beam. Obviously the load is non-symmetric with respect to the line of symmetry of the frame. However, the load can be decomposed into a symmetric and an anti-symmetric part. These corresponding load cases are illustrated in Fig. 6.43 and summarized as follows:

- symmetric structure with symmetric load
- symmetric structure with anti-symmetric load

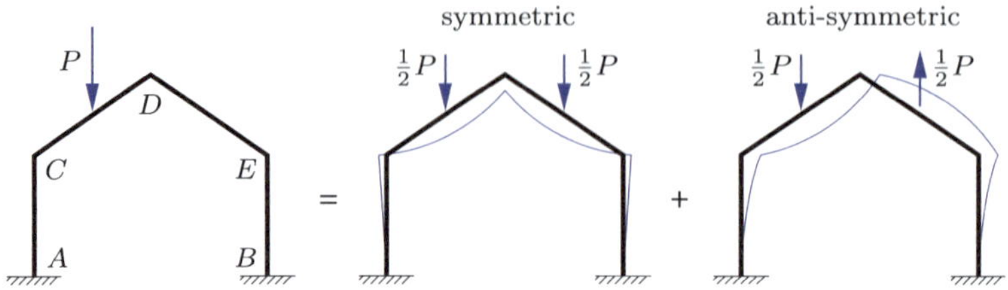

Fig. 6.43: Separation into symmetric and anti-symmetric load cases.

The deformation forms of the frame structure for each of the generic load cases are also indicated in Fig. 6.43. For the symmetric load the structure deforms symmetrically with respect to the line of symmetry. Thus, at the section in D along the line of symmetry of the structure the force component along the line of symmetry (in this case vertical) must vanish. The force component R'_D perpendicular to the line of symmetry (in this case horizontal) restrains D against motion in that direction, while the moment M_D restrains D against rotation. Typically, these two components shown in Fig. 6.44a do not vanish. For the ani-symmetric loading the frame deforms anti-symmetrically, and the force component perpendicular to the line of symmetry (in this case horizontal) and the moment therefore vanish. The remaining force component R_D along the line of symmetry (in this case vertical) restrains D against vertical motion and therefore does not vanish. This case is shown in Fig. 6.44b. These observations for symmetric plane frames can be generalized in the following symmetry conditions.

In a section located on the line of symmetry of a symmetric plane structure:

- the resulting internal force parallel to the line of symmetry vanishes for symmetric loading.
- the moment and the internal force perpendicular to the line of symmetry vanish for anti-symmetric loading.

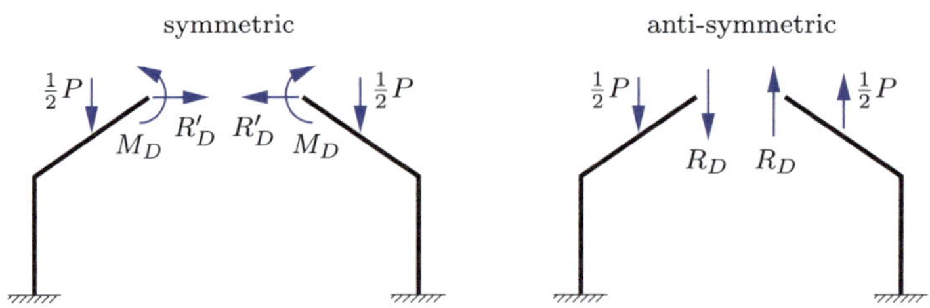

Fig. 6.44: Internal force components for symmetric and anti-symmetric loading.

It is observed that in Fig. 6.44 the moment M_D corresponds to the internal moment in D. However, due to the inclination of the beams CD and DE the resulting section forces R_D and R'_D are not equal to the shear force and the normal force.

When applying the symmetry conditions for the symmetric and the anti-symmetric load cases the degree of statically indeterminacy can be reduced. For the structure in Fig. 6.44 the left part of the symmetric load case contains five unknown reactions, three at A and two at D, and this part is two times statically indeterminant. For the anti-symmetric load case the degree of indeterminacy is reduced to one, with three unknown reactions at A and only a single unknown force component at D. Note, that similar symmetry conditions are also available for anti-symmetric structures.

Example 6.6. Symmetric frame with anti-symmetric load. In this example the use of symmetry conditions is illustrated for the symmetric rectangular frame shown in Fig. 6.45, loaded by horizontal forces P at the joints C and D. Thus, the loading is anti-symmetric with respect to the line of symmetry of the frame. The frame is symmetric and exposed to an anti-symmetric load. Therefore, the section at E only transmits a force component along the line of symmetry, and the left half of the frame is one time statically indeterminate. In this example the vertical force component in E is used as the redundant component X_1, as shown in Fig. 6.45b.

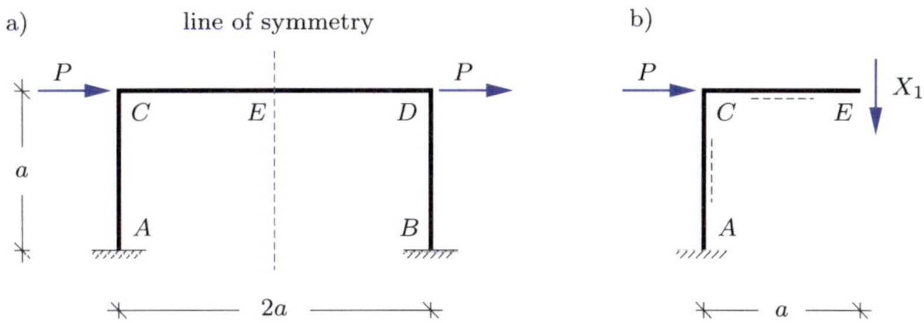

Fig. 6.45: Rectangular frame with fixed supports and anti-symmetric loading.

Figure 6.46 shows the left half of the frame with the actual load case P, but without X_1, and the unit load case with $X_1 = 1$ and without P. The figure also shows the corresponding moment distributions. The displacement in E from the external load P is obtained by the equation of virtual work as

$$\xi_{10} = \int_A^C \frac{M^0 M^1}{EI}\, ds = \frac{M^1}{EI}\int_A^C M^0\, ds = \frac{-a}{EI}\,\frac{1}{2}\,a\,(-Pa) = \frac{1}{2}\,\frac{Pa^3}{EI}\,,$$

where only the vertical beam AC contributes because $M^0(s) \equiv 0$ in the horizontal beam. Because $M^1(s)$ is constant on AC it can be taken outside the integral, which then represents the area $\frac{1}{2}a(-Pa)$ underneath the triangular M^1-curve. The flexibility coefficient is obtained by combining $M^1(s)$ with itself in the virtual work equation. This gives

$$\xi_{11} = \int_A^E \frac{M^1 M^1}{EI}\, ds = \frac{(-a)^2}{EI}\,a + \frac{a}{3EI}\,(-a)^2 = \frac{4}{3}\,\frac{a^3}{EI}\,.$$

The kinematic component $\xi_1 = \xi_{10} + \xi_{11} X_1$ associated with the vertical redundant force X_1 represents a discontinuity in the transverse displacement at E. In the actual structure in Fig. 6.45a the left and right part of the structure are rigidly joined at E, whereby the displacement discontinuity $\xi_1 = 0$. This gives a condition for determining the redundant

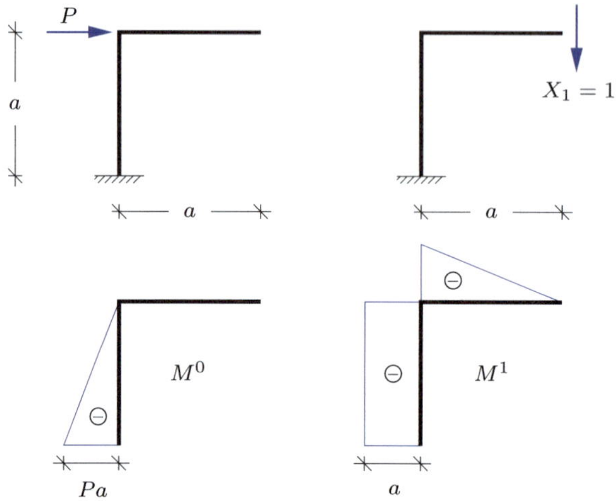

Fig. 6.46: Left half of frame with shear force in E as redundant component.

component,

$$X_1 = -\frac{\xi_{10}}{\xi_{11}} = -\tfrac{3}{8}\,P\,.$$

The horizontal beam CED is perpendicular to the line of symmetry, and the redundant component X_1 therefore represents the shear force at E.

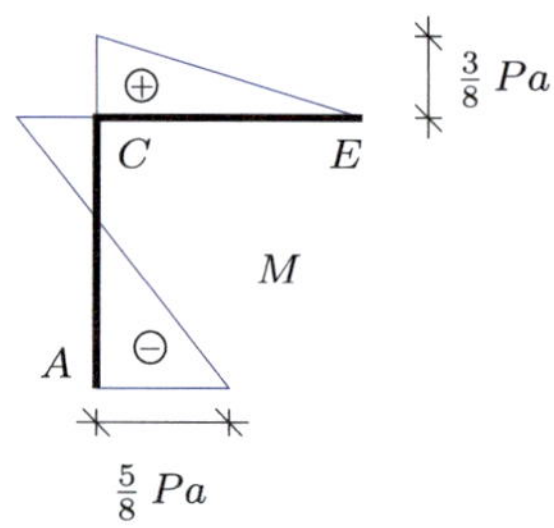

Fig. 6.47: Moment distribution for left half of frame.

The internal moment at A and C are obtained by superposition,

$$M_A = -Pa + (-a)X_1 = -\tfrac{5}{8}\,Pa\,, \qquad M_C = (-a)X_1 = \tfrac{3}{8}\,Pa\,.$$

Figure 6.47 shows the moment distribution, and it is seen that the maximum moment $M_{\max} = \frac{5}{8}Pa$ occurs at the fixed support at A – and because of symmetry also at B. □

6.5 Exercises

Exercise 6.1. The figure shows a beam of length ℓ with a fixed support in A and a simple support in B. A vertical force P is acting at the center of the beam in C.

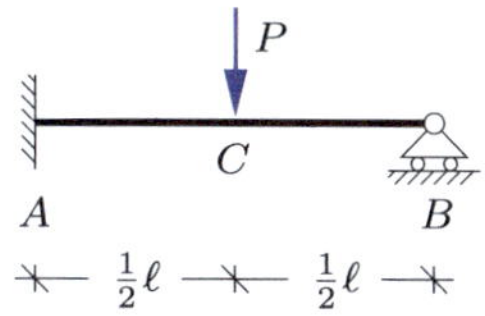

a) Determine the degree of statically indeterminacy.
b) Determine the reaction moment in A.
   Hint: Choose this moment as X_1.
c) Determine the moment distribution.
d) Determine the distribution of the shear force.

Exercise 6.2. The figure shows a beam of length ℓ with a fixed support in A and a simple support in B. The beam is loaded by a distributed load with constant intensity p. Note that the structure is similar to that in the previous exercise. Furthermore, the solutions to this problem can be found in Example 4.6, where the differential equation has been solved.

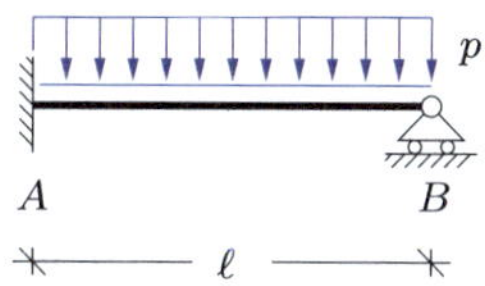

a) Use the force method to determine the reaction moment in A.
b) Find the remaining reactions.
c) Determine the moment distribution.
d) Determine the distribution of the shear force.

Exercise 6.3. The figure shows a beam of length ℓ with a fixed support in A and a simple support in B. A vertical force P acts at the tip of the beam in C.

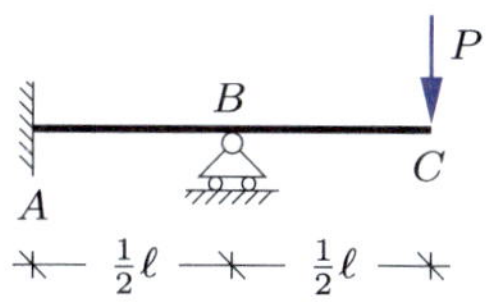

a) Use the force method to determine the reaction moment in A.
b) Find the remaining reactions.
c) Determine the distribution of the moment M and the shear force Q.

Exercise 6.4. The figure shows a two-span beam similar to that in Examples 6.1 and 6.3, but in this case only with distributed load on the left span AC.

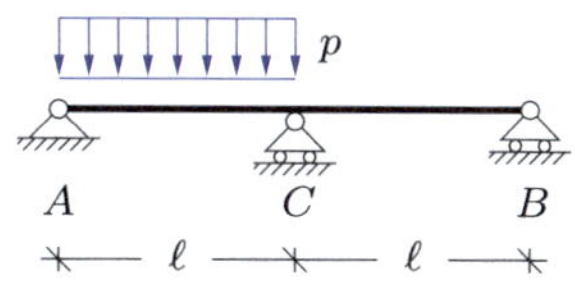

a) Use the force method to find the internal moment in C.
b) Find the reactions.
c) Determine the distribution of the moment M and the shear force Q.

Exercise 6.5. The figure shows a two-span beam loaded by a distributed load with constant intensity p. The left span AB has the length ℓ and bending stiffness EI. The right span has the double length 2ℓ and double bending stiffness $2EI$.

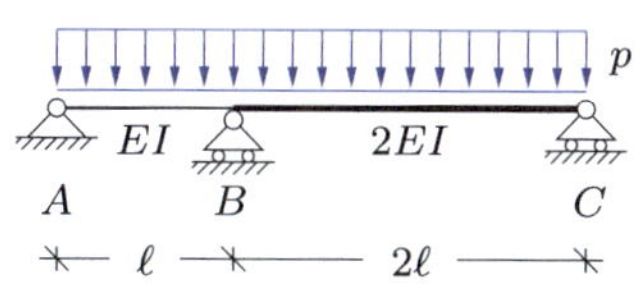

a) Use the force method to determine the bending moment in B.
b) Find the reactions.
c) Determine the distribution of the moment M and the shear force Q.

Exercise 6.6. The figure shows a beam of length ℓ with fixed supports in A and B. A vertical force P acts at the center of the beam at C.

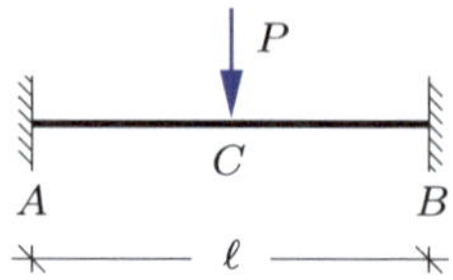

a) Apply symmetry conditions and determine the degree of indeterminacy.

b) Use the force method to determine the reaction moment in A.

c) Find the remaining reactions.

d) Determine the distribution of the moment M and the shear force Q.

e) Replace P by a distributed load with intensity p and repeat b)–d). Note that this problem has been solved in Example 4.8.

Exercise 6.7. The figure shows a three-span beam, where the length of each span is ℓ. The beam is loaded by a distributed load with intensity p on a) the center span, and b) the right span, respectively. Use the force method to analyze the structure for each load case following the outline presented below.

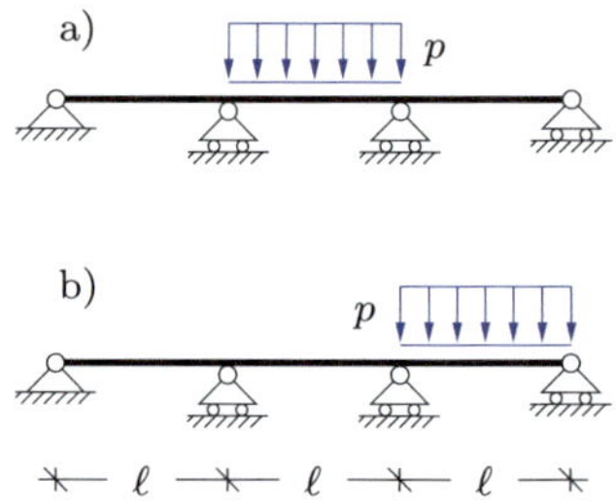

a) Determine internal moments at the two intermediate supports.

b) Find the reactions.

c) Determine the distribution of the moment M.

Exercise 6.8. The figure shows a five-span beam loaded by a distributed load with constant intensity p. The length of each span is ℓ.

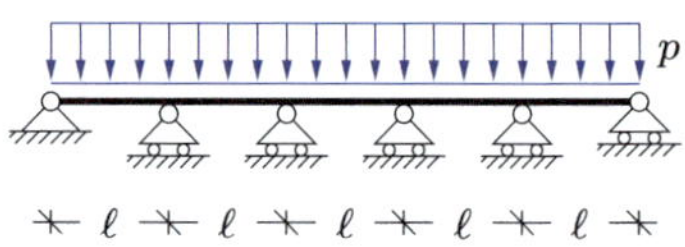

a) Determine the degree of indeterminacy before and after applying symmetry conditions.

b) Use the force method to find the reactions for the left half of the beam.

c) Determine the distribution of the moment M and the shear force Q in the left half of the beam.

Exercise 6.9. The figure shows a simply supported rectangular frame with a distributed load acting in the vertical direction on the horizontal beam CD with constant intensity p.

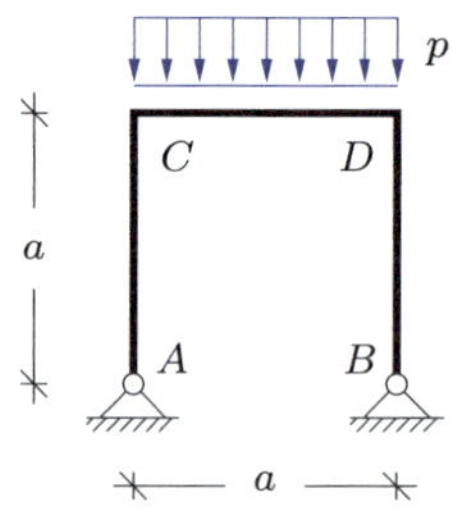

a) Use the force method to determine the internal moment in the upper left joint C.

b) Determine the reactions.

c) Determine the moment distribution and the magnitude and location of the maximum moment.

d) Determine the distribution of the normal and shear force.

Exercise 6.10. Reconsider the simple frame with a distributed load of Exercise 6.9. In this exercise a horizontal force P acts at the corner C in addition to the distributed load. Structure and loading are shown in the figure below.
It can be assumed that $P = \frac{1}{2}pa$.

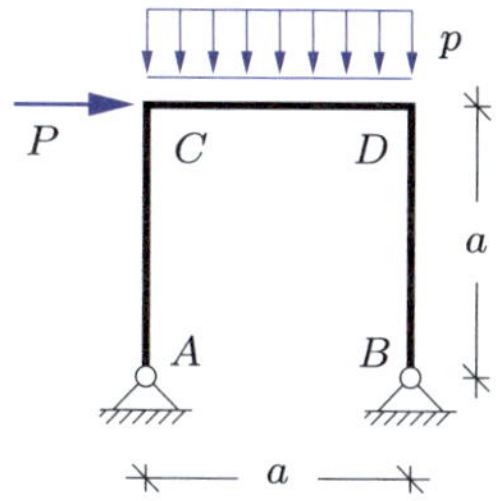

a) Determine the internal moment in the upper left corner C.

b) Determine the reactions.

c) Determine the moment distribution and the magnitude and location of the maximum moment.

d) Determine the distribution of the normal and shear force.

Exercise 6.11. The figure shows a simply supported rectangular frame with a distributed load p acting in the horizontal direction on AC. The length of the left vertical beam AC is $3a$, while the length of the two other beams CD and BD is $2a$.

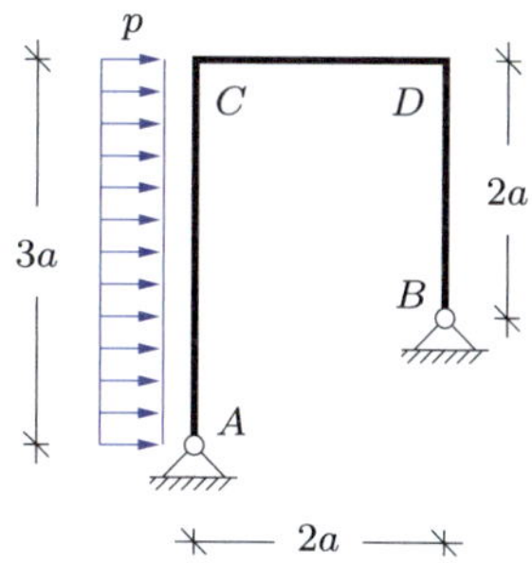

a) Use the force method to determine the internal moment in the upper left joint C.

b) Determine the reactions.

c) Determine the moment distribution and the magnitude and location of the maximum moment.

d) Determine the distribution of the normal and shear force.

Exercise 6.12. The figure shows a rectangular frame with a cantilever, similar to the frame structure in Example 6.5. In this exercise the loading of the example is replaced by a distributed load with intensity p, acting in the vertical direction on the horizontal cantilever DE. The dimensions of the frame are given in terms of a, as shown in the figure.

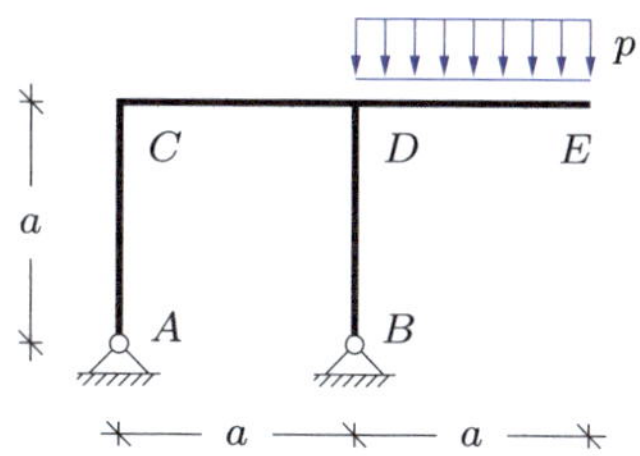

a) Use the force method to determine the internal moment in the upper left joint C.

b) Determine the reactions.

c) Determine the moment distribution and the magnitude and location of the maximum moment.

d) Determine the distribution of the normal and shear force.

e) Check equilibrium in the joint D.

Exercise 6.13. The figure shows a T-frame with simple fixed supports in A and B. The frame is loaded by a distributed load with intensity p, acting in the vertical direction on the horizontal cantilever CD. The dimensions of the frame are given in terms of a as shown in the figure.

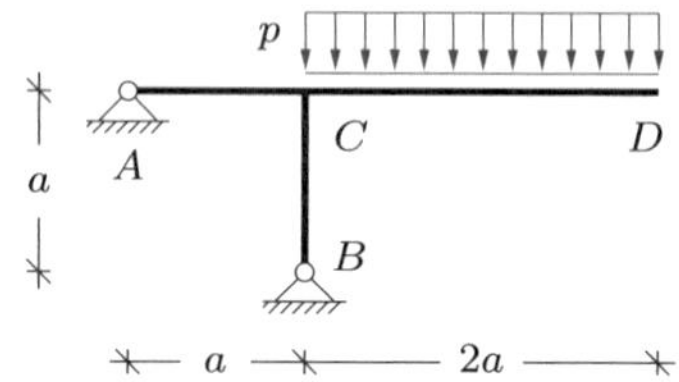

a) Find a statically determinate system with redundant component X_1.

b) Determine the reactions.

c) Determine the moment distribution and the magnitude and location of the maximum moment.

d) Check moment equilibrium in the joint C.

Exercise 6.14. The figure shows a simply supported beam AB with an additional vertical support in C by the bar element CD. The dimensions of the structure are given in terms of a, as shown in the figure. The elastic modulus is E for both beam and bar, while the moment of inertia for the beam is I and the cross sectional area of the bar is A.

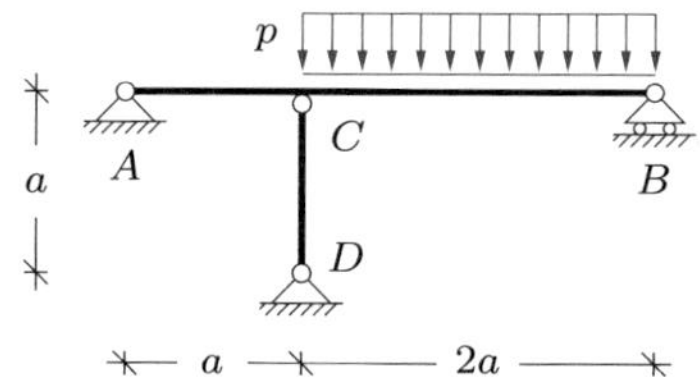

a) Use the force method to determine the bar force N_{CD}.

b) Determine the reactions.

c) Determine the moment distribution and the magnitude and location of the maximum moment.

d) Include the contribution of the normal force of the bar element in the virtual work equation. Redetermine the normal force N_{CD} and compare with the result in a). Assume that the cross sectional area A of the bar relates to the moment of inertia I of the beam by $A = 2000\, I/a^2$.

Exercise 6.15. The figure shows a simply supported beam AB with an additional vertical support in C by the inclined bar element CD. The dimensions of the structure are given in terms of a, as shown in the figure. The bending stiffness of the beam is EI, and contributions to the deformations from shear and normal forces are neglected.

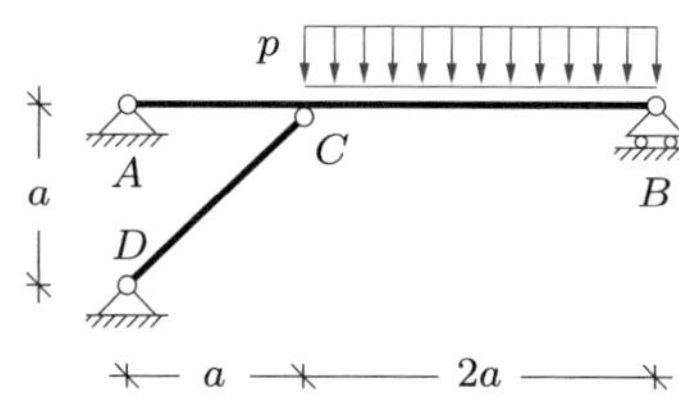

a) Use the force method to determine the normal force N_{CD} in the inclined bar CD.

b) Determine the reactions.

c) Determine the moment distribution and the magnitude and location of the maximum moment.

d) Determine the distribution of normal and shear force.

Exercise 6.16. The figure shows a typical geometry for frames used in barns and warehouses, with simple fixed supports in A and B. The dimensions of the frame are given in terms of a, as shown in the figure. A distributed load is acting in the vertical direction on CD.

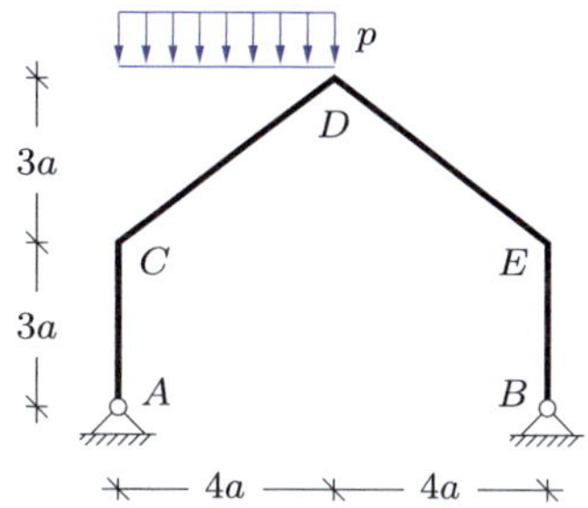

a) Use the force method to determine the internal moment in D.

b) Determine the reactions.

c) Determine the moment distribution and the magnitude and location of the maximum moment.

Deformation and Element Methods for Frames

7

In statically indeterminate structures the distribution of the internal forces depends on the stiffness of the individual parts of the structure. There are two basically different approaches to this problem. In the force method, described in the previous chapter, a sufficient number of connections in the structure are released and the corresponding local section forces necessary for closing the connection are determined. This method works well for simple structures, formed by linear elastic beams. However, the method requires the determination of the distribution of the internal forces in the statically determinate structure for each of the redundant components obtained by releasing connections, and for large structures this constitutes a considerable task. An alternative approach is to consider the structure as formed by individual beams, connected at nodes. Initially all nodes are considered fixed, and they are then released one at a time. The correct solution is obtained by finding the combination of node displacements, that do not require any additional constraining forces at the nodes. This approach has a number of advantages. First, the release of a constraint only affects beams directly connected to the constraint, and the equations of the method therefore only require a local analysis. Furthermore, this local analysis only involves individual beams, and it can therefore easily be given a general systematic form, suitable for computer analysis. This so-called finite element formulation is quite general and can be developed from the principle of virtual work for plates, shells and solid bodies as well. Here, the main focus is on frame structures, and the development will be based mainly on the equilibrium equations.

S. Krenk, J. Høgsberg, *Statics and Mechanics of Structures*,
DOI 10.1007/978-94-007-6113-1_7,

The chapter covers the classic deformation method of frames, intended for hand calculation, as well as the finite element method for frame structures. The deformation method is developed in Sections 7.1 and 7.2 for simple plane frames. The stiffness properties of a beam are developed by use of virtual work in the form of a set of basic deformation cases, including the effect of shear flexibility. These deformation cases are then used in the deformation method to determine the effect of sequentially imposing displacements of the constrained nodes. The procedure follows the classic deformation method, in which a manageable size of the problem is obtained by neglecting the effect of axial deformation of the individual beams.

The finite element formulation for elastic frames is obtained by rearranging the procedure of the deformation method into a systematic matrix format. The basic idea, already illustrated for truss structures in Section 2.5, is to represent each beam as an element with a stiffness matrix, including all displacement components at each of its nodes. These elements are then assembled into a frame structure, and the nodal displacements are determined by solution of the corresponding equation system. Two types of beam elements are developed here: beam elements with shear flexibility in Section 7.3.1, and beam-column elements in Section 7.3.2. The beam bending element is the typical element for frame analysis, while the beam-column element enables extension of the column analysis of Chapter 5 to a linearized stability analysis of frames. The finite element formulation has been implemented for plane frames in the MATLAB code MINIFRAME described in Section 7.4.

7.1 Stiffness of beams

The basic idea of the deformation method for frames is illustrated in Fig. 7.1. Figure 7.1a shows a T-frame with a distributed load on the part BC. The supports provide six reaction components. As only three reactions are needed for a structure without hinges, this implies that the frame is three times statically indeterminate. Thus, use of the force method would imply the release of three constraints and the introduction of three unknown force components. In the deformation method the point of view is changed. When neglecting axial deformation of the beams the joint B is fixed in space, but can rotate as determined by the loading of the frame. The rotation is indicated by the angle

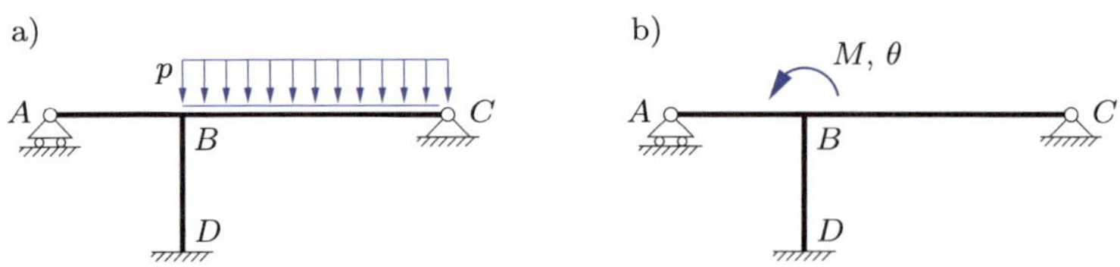

Fig. 7.1: T-frame, a) with distributed load, b) moment and rotation at joint.

θ in Fig. 7.1b. The idea of the deformation method is to consider the original problem as a superposition of two sub-problems. In the first sub-problem the node is prevented from rotating by imposing a concentrated moment M_0 at the node. Hereby the frame is converted to an assembly of independent beams, all with simple ideal support conditions, either fully fixed or with a simple support permitting rotation. When the moment distribution for the individual beams are known, the magnitude of the moment M_0 required to constrain the rotation of the joint B can be determined. The second sub-problem consists in imposing a unit rotation of the joint B. All three beams joined at B are subjected to the same unit rotation, and the total moment can therefore be obtained as the sum of the end-moments of the individual beams, when subjected to a unit rotation of the end-sections. The real situation is represented by a superposition of these two sub-problems, where the rotation of the node is determined to cancel the total external moment at the joint B. Thus, a central part of the deformation method is the development of a series of simple deformation load cases in which a unit deformation – here in the form of a rotation – is imposed at one end of a beam. This issue is addressed in the rest of this section, after which the deformation method is developed in Section 7.2.

7.1.1 Symmetric and anti-symmetric bending

The stiffness of a beam is characterized by the deformation generated by the application of end loads that are in equilibrium. For a plane beam there are three equilibrium states: constant normal force, constant bending moment, and constant shear force. The latter is accompanied by a bending moment of linear variation. The beam stiffness used in the deformation method for frames is due to bending and shear, while the effect of axial deformation and normal forces are typically neglected. Thus, the relevant stiffness characteristics are described by only two states of deformation. These are conveniently taken as symmetric bending, where there is no shear force, and anti-symmetric bending with a constant shear force. The various special cases needed for use in the deformation method or the finite element method for frames can then subsequently be obtained by linear combination of these two basic cases of deformation. An additional advantage of this simple approach is that it is straight-forward to incorporate the shear flexibility effect without introducing any additional complications in the derivation. However, it is important to note, that most frame structures consist of fairly slender elements with fairly low shear flexibility. In most structures analyzed by the deformation method using hand calculations it is therefore justified to neglect the effect of shear flexibility that will make the calculations more extensive. On the other hand, the finite element formulation for frame structures makes use of a beam element in which the inclusion of the shear flexibility simply consists in a set of appropriate coefficients. Therefore, the shear effect is included in the derivation of the stiffness properties of beams in the present section, as it serves as basis for both the deformation and the finite element methods.

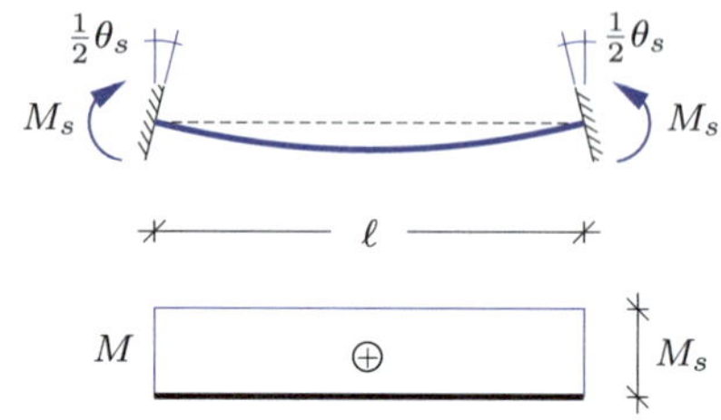

Fig. 7.2: Symmetric bending of beam.

The case of symmetric bending is illustrated in Fig. 7.2. Opposing bending moments of magnitude M_s are acting at the ends of the beam. The beam is assumed to be symmetric, and the moments then generate opposite rotations $\pm\frac{1}{2}\theta_a$ at the two end cross-sections of the beam. It is noted that in the present case there is no shear force, and thus the rotation of the cross-section is equal to the slope of the beam axis at the ends. The theory is fairly easily extended to non-symmetric beams, see e.g. Krenk (1994). The rotations are normalized via the factor $\frac{1}{2}$ in order for the external work to be represented as $\frac{1}{2}\theta_s M_s + \frac{1}{2}\theta_s M_s = \theta_s M_s$. When using the principle of virtual work discussed in Section 4.4 with the static field corresponding to the moment distribution $M(x)$ and the kinematic field consisting of the corresponding curvature distribution $\kappa(x) = M(x)/EI$, the virtual work equation takes the form

$$\theta_s M_s = \int_\ell \frac{M(x)M(x)}{EI}\, ds = \frac{\ell}{EI} M_s^2 \,. \tag{7.1}$$

This corresponds to the stiffness relation

$$M_s = \frac{EI}{\ell}\,\theta_s \tag{7.2}$$

for symmetric bending of the beam.

The case of anti-symmetric bending is illustrated in Fig. 7.3. Here, identical moments M_a are applied to the ends of the beam. This results in a total external moment of $2M_a$, that is counteracted by the shear force of magnitude $Q = 2M_a/\ell$. The beam ends do not translate, and the external work is therefore described entirely by the rotation of the to end moments as $\frac{1}{2}\theta_a M_a + \frac{1}{2}\theta_a M_a = \theta_a M_a$. In this case the internal work contains contributions from the shear force $Q(x)$ as well as from the moment $M(x)$, whereby the equality of external and internal work takes the form

$$\theta_a M_a = \int_\ell \Big\{ \frac{M(x)M(x)}{EI} + \frac{Q(x)Q(x)}{GA_s} \Big\} ds = \frac{\ell}{3EI} M_a^2 + \frac{\ell}{GA_s} Q^2 , \tag{7.3}$$

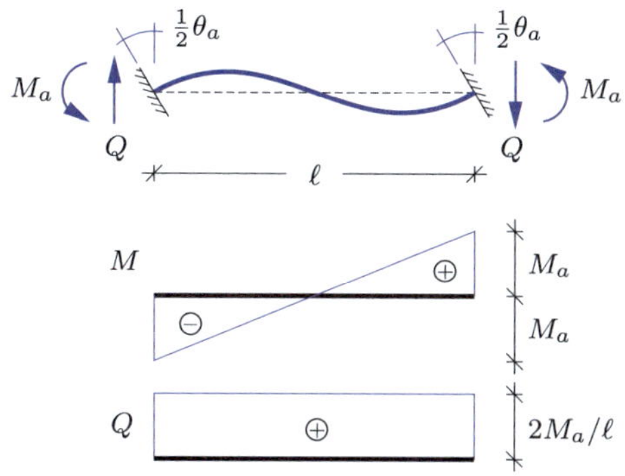

Fig. 7.3: Anti-symmetric bending of beam.

where A_s is the equivalent shear area of the cross-section. Upon substitution of the shear force Q in terms of the moment M_a this provides the anti-symmetric flexibility relation

$$\theta_a = \frac{\ell}{3EI} M_a + \frac{4\ell}{GA_s\ell^2} M_a = \frac{\ell}{3EI}\Big(1 + \frac{12EI}{GA_s\ell^2}\Big) M_a\,. \qquad (7.4)$$

The last term represents the additional flexibility introduced by the shear deformation. This shear flexibility effect is conveniently characterized by the shear flexibility parameter

$$\Phi = \frac{12EI}{GA_s\ell^2}\,. \qquad (7.5)$$

When introducing this notation, the stiffness relation becomes

$$M_a = \frac{3EI}{(1+\Phi)\ell}\,\theta_a \qquad (7.6)$$

for anti-symmetric bending of the beam. The corresponding shear force is

$$Q = \frac{6EI}{(1+\Phi)\ell^2}\,\theta_a. \qquad (7.7)$$

The subscript on the shear force is left out, as the symmetric part does not contain a shear force.

7.1.2 Basic cases of imposed deformation

There are two types of basic load cases for a homogeneous beam AB, those in which a unit rotation of an end cross-section is imposed shown in Figs. 7.4–7.5, and those in which a relative transverse translation of unit magnitude is imposed as shown in Figs. 7.6–7.7. These figures contain the reaction components, including the effect of shear flexibility. This effect is often included in the formulation of beam elements. However, in the deformation method this effect is often omitted, and the full set of results and their symmetric forms without the shear flexibility effect is given in Table 7.1 for easy reference.

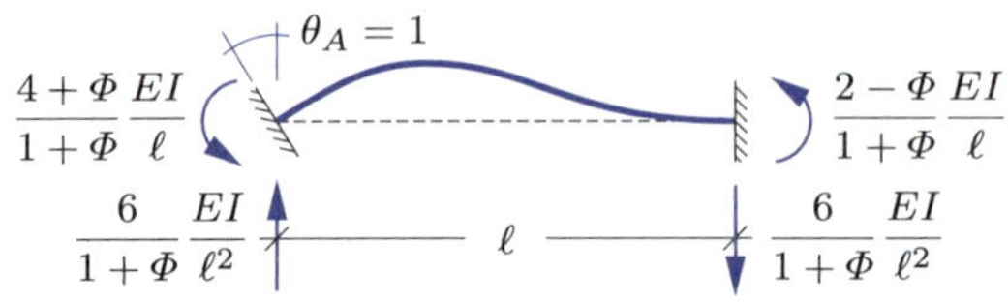

Fig. 7.4: End-section rotation for fixed support.

First the unit rotation load case of Fig. 7.4 is considered. With the sign convention used in the figures, the end moments are

$$M_A = -M_s + M_a\,, \qquad M_B = M_s + M_a\,. \tag{7.8}$$

The rotation of the end cross-sections are related to their symmetric and anti-symmetric parts as

$$\theta_A = -\tfrac{1}{2}\theta_s + \tfrac{1}{2}\theta_a\,, \qquad \theta_B = \tfrac{1}{2}\theta_s + \tfrac{1}{2}\theta_a = 0\,. \tag{7.9}$$

It follows from the sum and difference of these equations that the symmetric and anti-symmetric rotations are

$$\theta_a = -\theta_s = \theta_A = 1\,. \tag{7.10}$$

The end moments then follow by substitution of the symmetric and antisymmetric moments (7.2) and (7.6), respectively, into (7.8),

$$\begin{aligned} M_A &= \frac{EI}{\ell} + \frac{3EI}{(1+\Phi)\ell} = \frac{4+\Phi}{1+\Phi}\frac{EI}{\ell}\,, \\ M_B &= -\frac{EI}{\ell} + \frac{3EI}{(1+\Phi)\ell} = \frac{2-\Phi}{1+\Phi}\frac{EI}{\ell}\,. \end{aligned} \tag{7.11}$$

The shear force Q follows from (7.7) with $\theta_a = \theta_A = 1$,

$$Q = \frac{6}{1+\Phi}\frac{EI}{\ell^2}\,. \tag{7.12}$$

These results are shown in Fig. 7.4. Note, that the shear force follows directly from moment equilibrium as $Q = (M_A + M_B)/\ell$. The reactions of this load case and its symmetric counterpart are given in the first row of Table 7.1. The results of the symmetric load case follows from rotating the original load case by 180° in the plane.

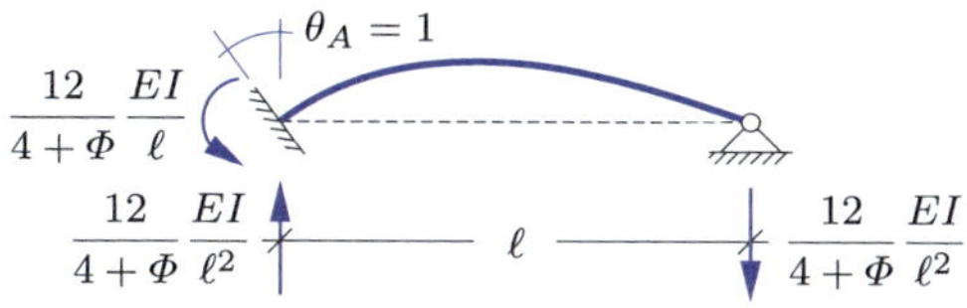

Fig. 7.5: End-section rotation for simple support.

In the practical application of the deformation method it is convenient also to have the corresponding load cases, in which a unit rotation is imposed at the cross-section at one end of the beam, while the other end has a simple support, permitting free rotation. This load cases is shown in Fig. 7.5. The moment vanishes at B, and thus the moment superposition relations (7.8) here take the form

$$M_A = -M_s + M_a\,, \qquad M_B = M_s + M_a = 0. \tag{7.13}$$

Thus, the symmetric and anti-symmetric parts of the moment are given in terms of M_A as

$$2M_a = -2M_s = M_A\,. \tag{7.14}$$

The rotation θ_a is now expressed in terms of the moment M_A by use of the relations (7.2) and (7.6),

$$\theta_A = -\tfrac{1}{2}\theta_s + \tfrac{1}{2}\theta_a = -\frac{\ell}{2EI}M_s + \frac{(1+\Phi)\ell}{6EI}M_a = \frac{4+\Phi}{12}\frac{\ell}{EI}M_A\,. \tag{7.15}$$

The stiffness is the inverse relation with $\theta_A = 1$,

$$M_A = \frac{12}{4+\Phi}\frac{EI}{\ell}\,, \tag{7.16}$$

and the shear force follows from moment equilibrium as $Q = M_A/\ell$, whereby

$$Q = \frac{12}{4+\Phi}\frac{EI}{\ell^2}\,. \tag{7.17}$$

These results are shown in Fig. 7.5. The results are included together with those of the symmetric load case in the second row in Table 7.1. The results of the symmetric case again follow from those already derived by a 180° rotation of the beam and its loads in the plane.

The nodes of a frame may rotate and translate. The cases involving imposed rotations have been covered in Figs. 7.4 and 7.5, and the similar cases involving an imposed translation of an end-section are now considered. The first of these, shown in Fig. 7.6, involves a beam AB in which a unit transverse translation is imposed on the cross-section at A. Within the degree of approximation involved in the theory of infinitesimal deformation, used here in

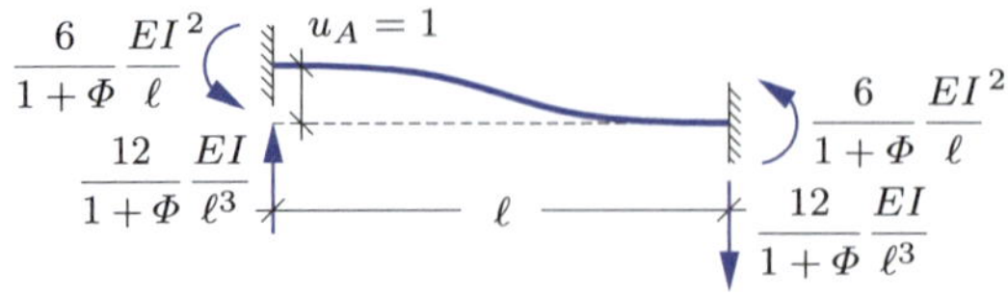

Fig. 7.6: Transverse translation for fixed support.

describing beam deformation, this case corresponds to imposing a clock-wise rotation of $\frac{1}{2}\theta_a = 1/\ell$ about B of the case of anti-symmetric bending, solved previously in connection with Fig. 7.3. The resulting end-moments and shear force have already been derived, and are given explicitly in Fig. 7.6a. The complementary case, in which the end-section at B is given a unit translation, follows from a simple change in sign and is included for the case without shear flexibility as the third row of Table 7.1. These load cases play an important role in the swaying of frames with fixed supports.

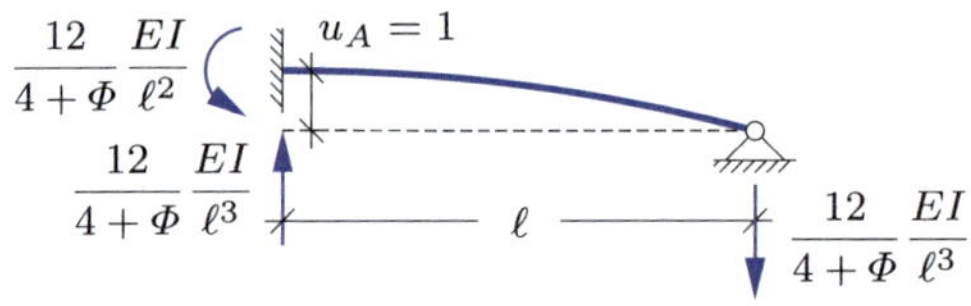

Fig. 7.7: Transverse translation for simple support.

Figure 7.7 shows the similar case of imposed translation, but now on a beam with a simple support at the other end. The results follow directly from those in Fig. 7.5, when the geometry is rotated and the imposed angle scaled by $\theta = \pm 1/\ell$. These results are included without shear flexibility as the last row of Table 7.1.

The load cases describing the internal forces generated by a unit displacement serve to determine how a statically indeterminate frame distributes the load to the supports. In most cases the shear flexibility effect, represented by Φ, can be neglected, and the formulae can be used with $\Phi = 0$. A simple example of load distribution is given below.

Example 7.1. Load distribution in angle frame.

Figure 7.8 shows an angle frame ABC in which the individual beam members AB and BC are of length a and with bending stiffness EI. The frame has a fixed support at A and a simple support at C. The load consists of an external moment M_0 applied at the corner B, and the issue is, how the moment is distributed to the two supports. The effect of shear deformation is neglected, corresponding to $\Phi = 0$ in both beams.

The problem is solved by imposing a rotation of magnitude θ_0 of the joint B, and determining the corresponding internal forces by use of the unit displacement load cases determined

Table 7.1: Constraining forces on deformed beams.

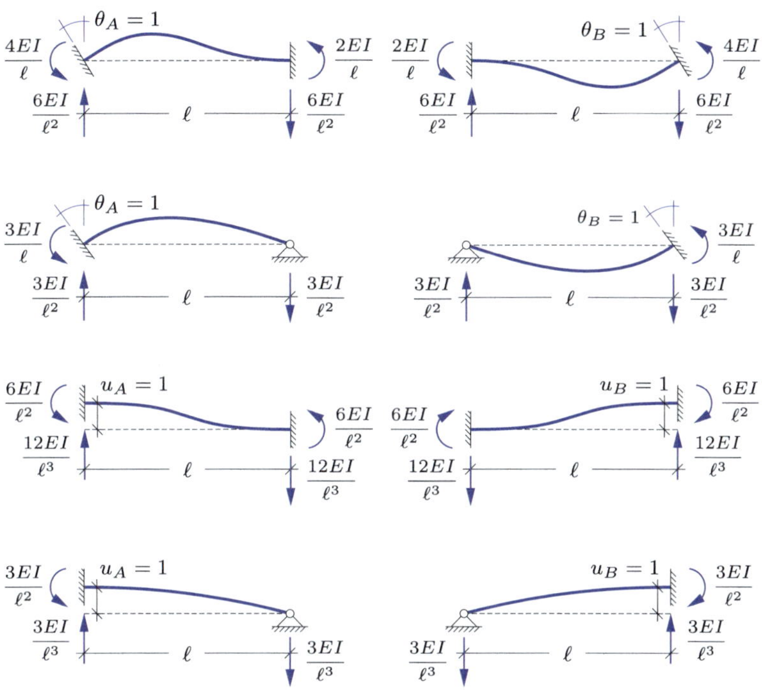

above. Rotation of the joint B by θ_0 corresponds to rotating one end cross-section of both the beams AB and BC. The beam AB then corresponds to the load case in Fig. 7.4, while the beam BC corresponds to the load case in Fig. 7.5. The moments and transverse forces corresponding to these load cases are shown in Fig. 7.9.

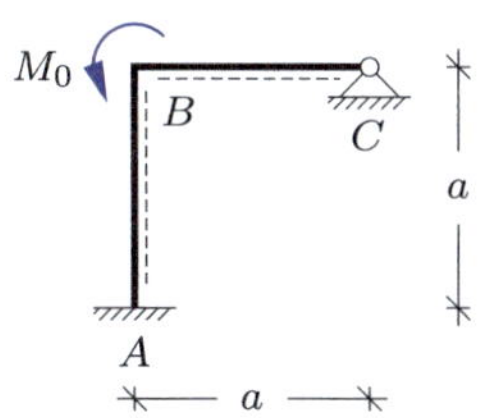

Fig. 7.8: Angle frame with external corner moment M_0.

The moment M_0 imposed at the joint B corresponds to the sum of the moments transferred to the two beams at B, and thus it follows from the figure that

$$M_0 = \frac{4EI}{a}\theta_0 + \frac{3EI}{a}\theta_0 = \frac{7EI}{a}\theta_0 ,$$

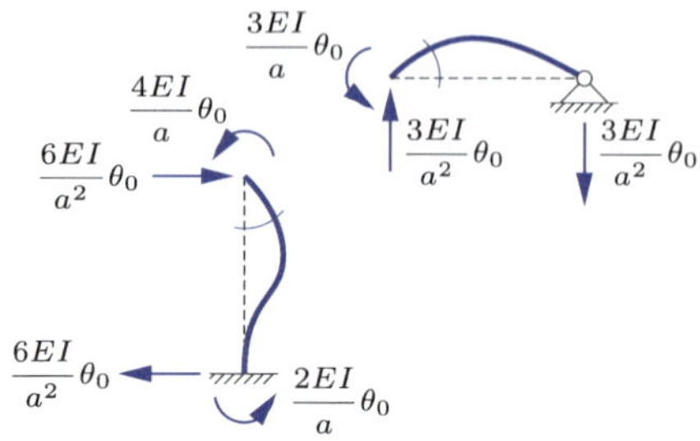

Fig. 7.9: Individual beams with common corner rotation θ_0.

where the smaller moment comes from the beam BC with the more flexibile support. This equation determines the rotation angle as

$$\theta_0 = \frac{a}{7EI}M_0 .$$

With this value of the rotation the normalized internal force diagram in Fig. 7.9 can be evaluated in terms of actual magnitudes, shown in Fig. 7.10a. The moments at B are

$$M_{BA} = \frac{4EI}{a}\theta_0 = \tfrac{4}{7}M_0 , \qquad M_{BC} = \frac{3EI}{a}\theta_0 = \tfrac{3}{7}M_0$$

where M_{BA} is the moment at B in the beam BA, and M_{BC} is the moment at B in the beam BC.

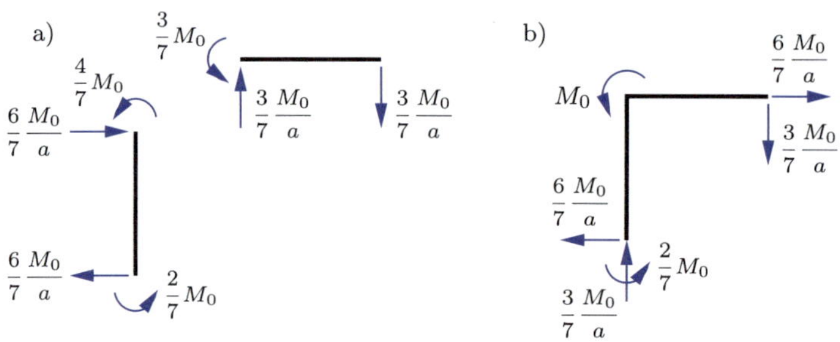

Fig. 7.10: a) End loads on individual beams, b) Assembled frame with reactions.

The internal forces in Fig. 7.10a are taken directly from the basic unit deformation load cases, and therefore do not contain normal forces components. The normal force in the beam AB must provide the upward transverse force $\frac{3}{7}M_0/a$, and thus must carry a compressive force of this magnitude. Similarly, the normal force in BC must produce the transverse force $\frac{6}{7}M_0/a$ in the vertical beam AB as indicated in the figure. After determination of the normal forces the reactions on the assembled frame can be shown in Fig. 7.10b.

It is seen that the total loads on the frame, including reactions, consist of the external moment M_0, a reaction moment $\frac{2}{7}M_0$, and two force couples of $\pm\frac{6}{7}M_0/a$ and $\pm\frac{3}{7}M_0/a$, respectively. A total moment balance gives

$$M_0 + \frac{2}{7}M_0 - \frac{6}{7}\frac{M_0}{a}a - \frac{3}{7}\frac{M_0}{a}a = 0,$$

thus demonstrating moment equilibrium.

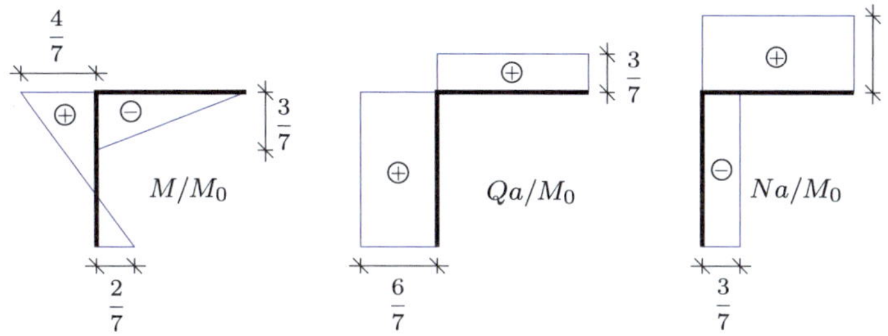

Fig. 7.11: Internal forces in angle frame.

Once the reactions have been determined, the internal force distributions can be generated as shown in Fig. 7.11. The moment and shear force distributions also follow directly from the diagram of the individual beams in Fig. 7.10a. It is seen that the larger part of the corner moment M_0 is taken in the beam AB with the fixed support. □

7.1.3 Loads on constrained beams

If all loads act directly on the nodes of the structure, the reactions and the distribution of the internal forces can be determined solely by use of the stiffness properties of the individual beam elements, derived above. However, most frame structures carry loads acting on the individual structural members, and therefore an additional step is needed, in which the local loads on the beams are distributed to the nodes. This distribution of the loads is accomplished by first considering the loaded beam as fully constrained. This requires suitable constraining forces and moments, that are later released in the form of loads at the nodes of the frame structure. Thus, the magnitude of the forces necessary to constrain the ends of a loaded beam must be known.

The constraining forces/moments are shown for several simple load cases in Table 7.2. The fully constrained beams are shown in the left column with the corresponding load cases for a beam with a simple support at the right end of the beam are shown in the right column. It is seen, that when the right support is changed from rigid to simple the moment and reaction force at the left end are increased. Conversely, at the end of the beam permitted to rotate, the reaction force is reduced.

The constraining forces shown in Table 7.2 have been calculated without including the effect of shear flexibility. The fully symmetric cases a) and c) are independent of the shear flexibility parameter Φ, and the other cases only exhibit a small redistribution of the reactions, resulting from the lack of symmetry. The approximate nature of the loads specified in most situations hardly justifies the additional complications of including the dependence of Φ in the following calculations.

Table 7.2: Constraining forces on loaded beams.

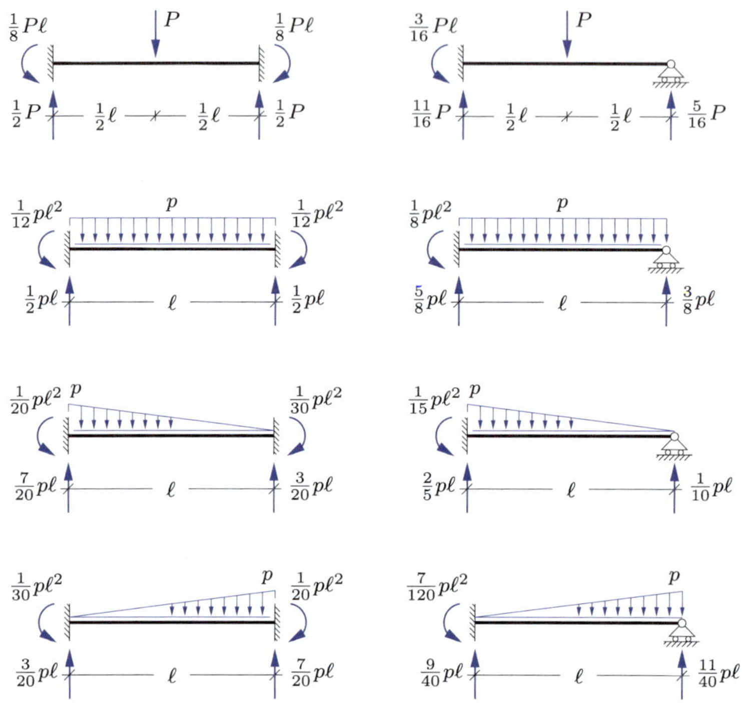

7.2 Deformation method for frames

In this section the deformation method for beams and frames is developed in a systematic way, first carefully considering two specific structures and then summarizing the general procedure. The first structure is a two-span beam, solved by introducing a single constraint. The second structure generalizes the procedure by considering a frame with two constraints. These two cases serve to introduce the procedure as well as the notation in a specific context, and subsequently the general procedure and notation are summarized in a concise form.

Two-span beam with a single constraint

The simplest case of the deformation method, in which only a single constraint is needed, is illustrated in Fig. 7.12 showing a homogeneous beam with bending stiffness EI that is continuous over the two spans AB and BC, each of length ℓ. The beam is fixed at A and supported by simple supports on horizontal rollers at B and C. The load consists of a vertical load P applied to the center of the span BC.

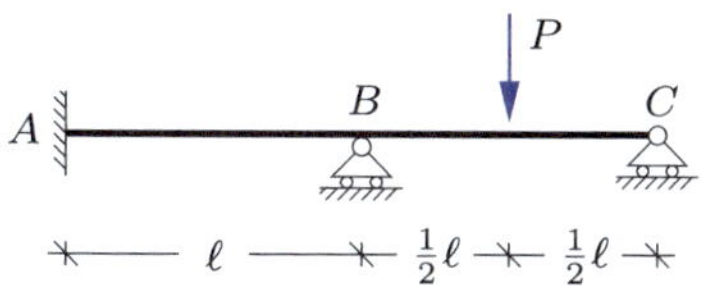

Fig. 7.12: Two-span beam with concentrated load.

The structure is twice statically indeterminate, and solution by the force method would then require the release of two constraints, e.g. the vertical reactions in B and C. In the deformation method the first step is to introduce constraints to reduce the full structure to a number of beams with fixed or simply supported ends. In the present case this is attained by constraining the rotation of node B. The rotation at B is prevented by introducing a constraining moment Z_{10} at B as illustrated in Fig. 7.13. Hereby the structure is reduced to a beam AB fixed at both ends, and a beam BC with a transverse force at the center, fixed at B and simply supported at C. The first subscript on Z_{10} identifies the degree of freedom, while the second subscript 0 identifies Z_{10} as the moment constraining the corresponding degree of freedom, when the structure is acted upon by the external load.

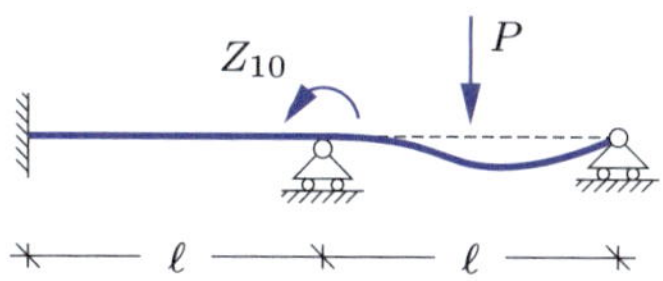

Fig. 7.13: Constraining moment on two-span beam.

The constraining moment Z_{10} must have a magnitude corresponding to constraining both the beam AB and the beam BC. In the present case the beam AB is unloaded, and thus requires no constraining moment, while the constraining moment of the beam BC follows from the figure to the right in the first row of Table. 7.2,

$$Z_{10} = M_{BA}^0 + M_{BC}^0 = 0 + \tfrac{3}{16} P\ell = \tfrac{3}{16} P\ell.$$

In the present notation the superscript 0 on M_{BA}^0 indicates a constraining moment, and the subscript BA identifies the location as node B of the beam BA. The constraining moment Z_{10}, as well as its contributing parts, are considered positive when acting in the counter-clockwise direction.

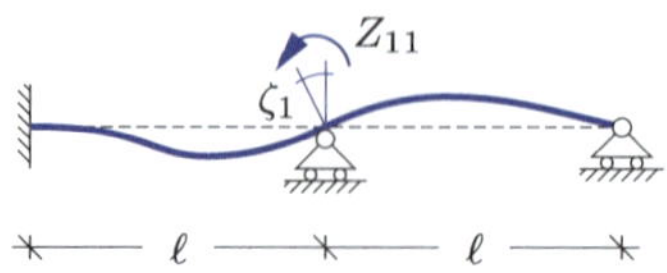

Fig. 7.14: Imposed unit rotation on two-span beam.

In the actual structure the node B can rotate, and this rotation is denoted by ζ_1. Figure 7.14 illustrates an imposed unit rotation $\zeta_1 = 1$ at node B and the corresponding moment Z_{11}. The first subscript 1 indicates that Z_{11} is a moment contribution, while the second subscript defines this as a contribution from a unit rotation associated with ζ_1. The forces and moments in the beams AB and BC associated with a unit rotation at node B have been derived as deformation load cases in Section 7.1.2, and are listed to the right in the first row and the to the left in the second row of Table 7.1. When using the results for the end moments, the moment necessary to impose a unit rotation at B is found as

$$Z_{11} = M_{BA}^1 + M_{BC}^1 = 4\frac{EI}{\ell} + 3\frac{EI}{\ell} = 7\frac{EI}{\ell}.$$

In this relation the superscript is changed to 1 to indicate that this moment is associated with a unit deformation of ζ_1.

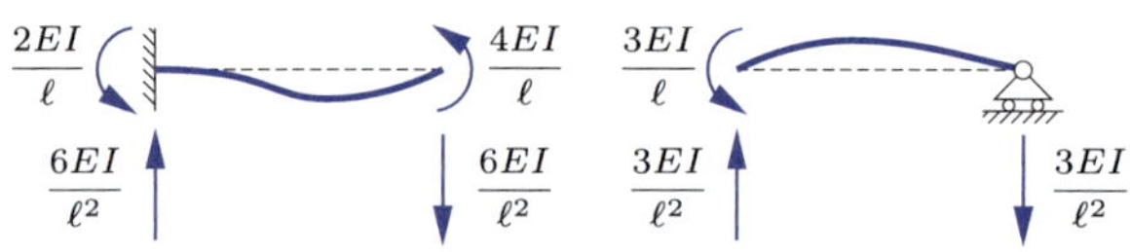

Fig. 7.15: Constraining forces and moments for imposed unit rotation $\zeta_1 = 1$.

In the actual structure there is no imposed external moment at B. This provides the following equation,

$$Z_1 = Z_{10} + Z_{11}\zeta_1 = 0.$$

This is an equation for the initially unknown rotation ζ_1 with the solution

$$\zeta_1 = -\frac{Z_{10}}{Z_{11}} = -\frac{3}{112}\frac{P\ell^2}{EI}.$$

The rotation ζ_1 of node B is negative, indicating a clockwise rotation.

Internal forces and reactions can now be evaluated by considering the full solution as the superposition of the constrained case of Fig. 7.13 and the unit rotation case from Fig. 7.15, multiplied by the parameter ζ_1. In practice, it is often most convenient to calculate the reactions from this superposition principle, and then to evaluate the section force distributions from an ordinary static analysis, based directly on the loads and reactions. This is the procedure shown here.

First, it follows directly from the considered load cases that the horizontal reaction force at A vanishes, $R'_A = 0$. The vertical reaction forces are evaluated by the superposition principle. For the vertical reaction in A

$$R_A = R_A^0 + R_A^1\zeta_1 = 0 + 6\frac{EI}{\ell^2}\zeta_1 = -\tfrac{9}{56}P,$$

where all components are positive in the upward direction. It is seen that the reaction R_A is negative, and thereby downward. It receives no contribution from the constrained load case, because the beam AB carries no external load. The vertical reaction at B is

$$R_B = R_B^0 + R_B^1\zeta_1 = \tfrac{11}{16}P - 3\frac{EI}{\ell^2}\zeta_1 = \tfrac{43}{56}P.$$

Here the reaction R_B^1 corresponding to a unit rotation is a combination of of a downward component from AB and an upward component from BC. Finally, the reaction at C follows as

$$R_C = R_C^0 + R_C^1\zeta_1 = \tfrac{5}{16}P - 3\frac{EI}{\ell^2}\zeta_1 = \tfrac{22}{56}P.$$

The reaction forces are shown in Fig. 7.16, where the arrows indicate the corresponding positive direction.

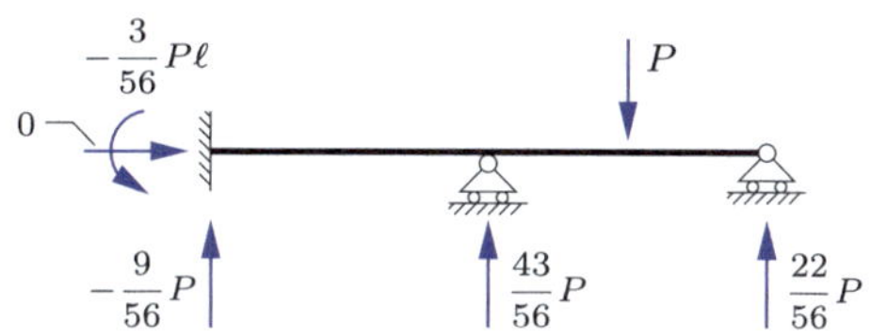

Fig. 7.16: Load and reactions on two-span beam.

It is seen that the sum of the reactions give $R_A + R_B + R_C = P$, corresponding to vertical projection equilibrium. However, it follows from the expression for the reactions in terms of ζ_1 that the sum of the vertical reactions is independent of the value of ζ_1, and thus the check does not verify the correctness of the solution, but merely constitutes a useful consistency check.

The reaction moment M_A can now be determined, either by taking moment equilibrium of the structure including load and reaction forces as shown in Fig. 7.16, or by use of the superposition procedure. The latter gives

$$M_A = M_A^0 + M_A^1 \zeta_1 = 0 + 2\frac{EI}{\ell}\zeta_1 = -\tfrac{3}{56}P\ell.$$

It is easily verified that this moment satisfies global moment equilibrium of the structure.

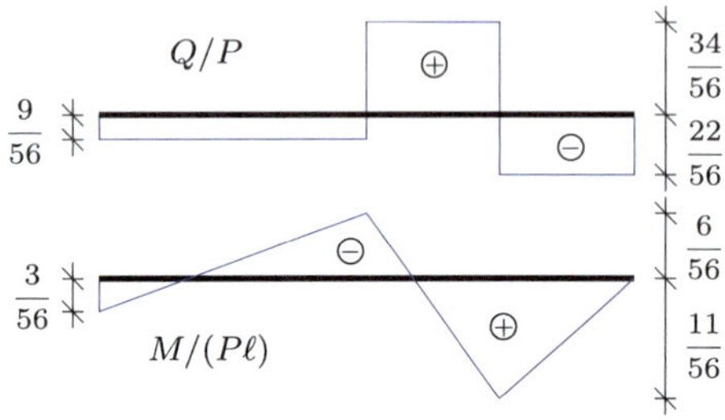

Fig. 7.17: Shear force and moment distributions for on two-span beam.

The section forces follow directly from the structure with the load and reaction components indicated in Fig. 7.16. When the reactions have been determined, it is immaterial for this part of the analysis that the structure is statically indeterminate. The moment and shear force distributions are shown in Fig. 7.17. It is seen that the maximum moment $M_{\max} = \frac{11}{56}P\ell$ is found under the load. This is a reduction relative to if the load had been carried only by the beam BC with simple supports, in which case $M_{\max} = \frac{14}{56}P\ell$.

Frame with two constraints

The procedure and notation of the deformation method is now extended to structures with two constraints by considering the simple frame shown in Fig. 7.18. The frame consists of a horizontal continuous beam ABC, supported by vertical beams BD and CE. For simplicity of the presentation the bending stiffness of all members is EI and all lengths are a as indicated in the figure, but these features are not important for the principles of the method.

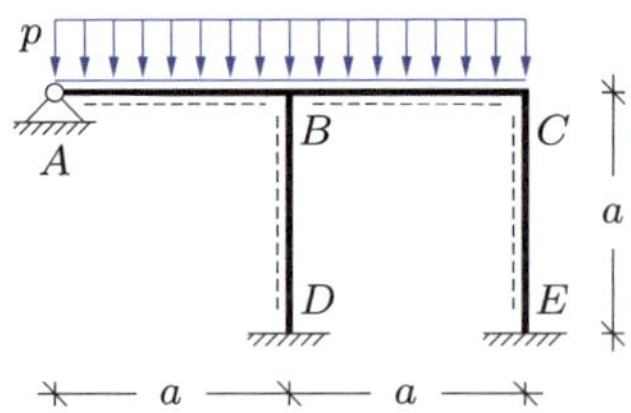

Fig. 7.18: Frame with distributed load.

First, the nodes that should be constrained against motion – rotation and translation – are identified. In the present case the location of all nodes are fixed in space, and only the rotation of the internal nodes B and C should be constrained against rotation. Thus, the problem has two degrees of freedom, and the rotations and constraining moments are identified by subscript(s) $j = 1, 2$, where $j = 1$ refers to B, while $j = 2$ refers to C. External moments Z_{10} and Z_{20} are then introduced to simultaneously constrain the rotation of the nodes B and C, respectively, as shown in Fig. 7.19. The first subscript identifies the node, at which the moment is acting, while the second subscript indicates that the moments are constraining motion from external loads.

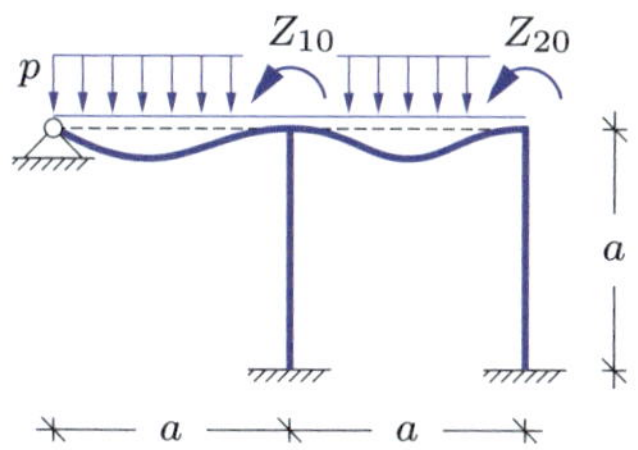

Fig. 7.19: Constraining moments on frame.

The magnitude of the constraining moments Z_{10} and Z_{20} are determined from the second row in Table 7.2. For simplicity all external moments introduced in the present analysis are considered positive, when acting in the counter-clockwise direction. The beam AB is simply supported at A, and therefore constraining rotation at B requires the moment $M^0_{BA} = -\frac{1}{8}pa^2$, as shown in the right figure of the second row in Table 7.2. The beam BC is constrained against rotation at D, and thus constraining B in this beam requires the moment $M^0_{BC} = \frac{1}{12}pa^2$, as shown in corresponding right figure in Table 7.2. Note, that the moment M^0_{BA} acts in the clockwise direction, and the moment M^0_{BC} in the counter-clockwise direction. The constraining moment Z_{10} is the sum of these two moments,

$$Z_{10} = M^0_{BA} + M^0_{BC} = -\tfrac{1}{8}pa^2 + \tfrac{1}{12}pa^2 = -\tfrac{1}{24}pa^2.$$

The moment needed to prevent rotation at C is given at the right of the second row in Table 7.2 as $M^0_{CB} = -\frac{1}{12}pa^2$, whereby the constraining moment at C is

$$Z_{20} = M^0_{CB} = -\tfrac{1}{12}pa^2.$$

With these constraining moments all internal nodes are fixed against translation and rotation, and only nodes with simple support can rotate.

Next, the displacement components ζ_1 and ζ_2 associated with the constraints Z_{10} and Z_{20} are given unit magnitude, one at a time. In the present case ζ_1 is the rotation of node B, while ζ_2 is the rotation of node C. Figure 7.20a

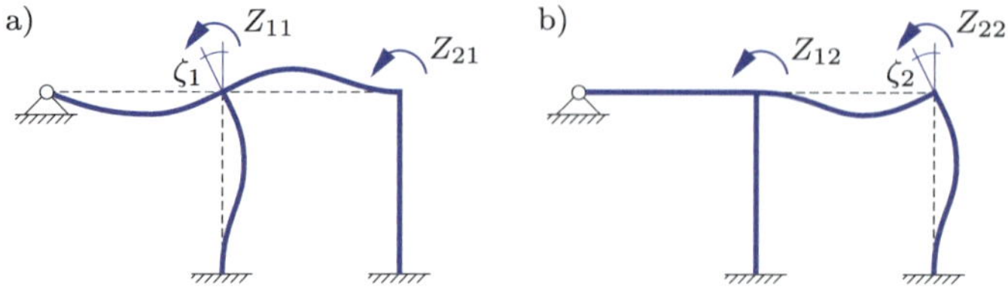

Fig. 7.20: Imposed unit rotations on frame.

shows a unit rotation of node B corresponding to $\zeta_1 = 1$, while the other nodes with the exception of simple supports are constrained. This requires a moment Z_{11} at node B and a moment Z_{21} at node D. Similarly, a unit rotation of node D corresponding to $\zeta_2 = 1$ with the other nodes constrained requires a moment Z_{22} at node D and a moment Z_{21} at node C, as shown in Fig. 7.20. The moments in the individual beams generated by these imposed rotations are found from the basic cases of imposed deformation presented in Table 7.1.

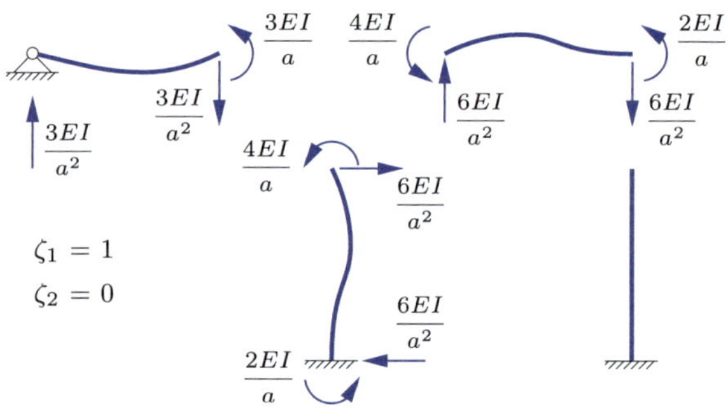

Fig. 7.21: Constraining forces and moments for imposed unit rotation $\zeta_1 = 1$.

The end-moments and transverse forces introduced in the individual beams by an imposed unit rotation $\zeta_1 = 1$ are shown in Fig. 7.21. The beam AB has an imposed rotation and a simply supported end corresponding to the second row in Table 7.1, while both of the beams BC and BD have an imposed rotation and a fully constrained end corresponding to the first row in Table 7.1. The imposed moment Z_{11} is the sum of all end moments at node B,

$$Z_{11} = M_{BA}^1 + M_{BC}^1 + M_{BD}^1 = 3\frac{EI}{a} + 4\frac{EI}{a} + 4\frac{EI}{a} = 11\frac{EI}{a}.$$

The superscript 1 indicates that the moments correspond to the imposed deformation $\zeta_1 = 1$. Note, that in Z_{11} each of the contributing moments is positive, as each beam produces resistance to the imposed rotation. The constraining moment Z_{21} at node D comes from the beam BC,

$$Z_{21} = M_{CB}^{1} = 2\frac{EI}{a}.$$

The remaining end forces and moments are used to determine the reactions and the distribution of internal forces, once ζ_1 and ζ_2 have been determined.

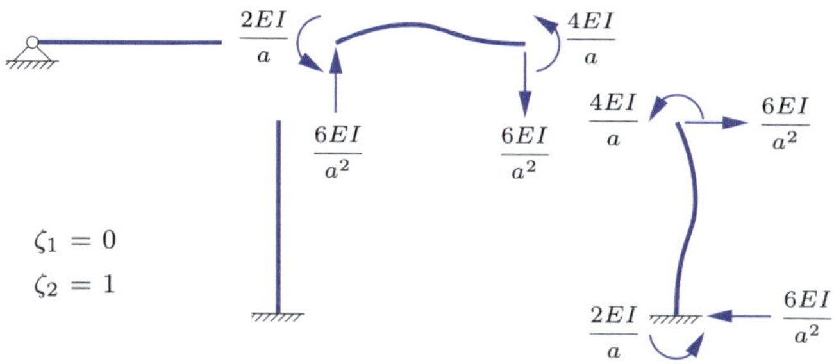

Fig. 7.22: Constraining forces and moments for imposed unit rotation $\zeta_2 = 1$.

Figure 7.22 shows the end moments and transverse forces introduced in the individual beams by imposing the unit rotation $\zeta_2 = 1$ at node C. The imposed moment Z_{22} is the sum of all end-moments of node C,

$$Z_{22} = M_{CB}^{2} + M_{CE}^{2} = 4\frac{EI}{a} + 4\frac{EI}{a} = 8\frac{EI}{a}.$$

Finally, the constraining moment Z_{12} at node C comes from the beam BC,

$$Z_{12} = M_{BC}^{2} = 2\frac{EI}{a}.$$

It is observed that $Z_{12} = Z_{21}$, corresponding to a symmetric coefficient matrix Z_{ij}. This is a general property, following from the principle of virtual work as discussed later.

The initially unknown rotations ζ_1 and ζ_2 are now determined by considering the actual deformation and internal forces as a superposition of the case of constrained loads, shown in Fig. 7.19, and the cases of imposed rotations ζ_1 and ζ_2, respectively, shown in Fig. 7.20. In the actual state of the frame structure the constraining moments Z_1 and Z_2 at the nodes B and C vanish, thus providing the equations

$$Z_1 = Z_{10} + Z_{11}\zeta_1 + Z_{12}\zeta_2 = 0\,,$$

$$Z_2 = Z_{20} + Z_{21}\zeta_1 + Z_{22}\zeta_2 = 0\,.$$

This is an equation system of the form

$$Z_{11}\zeta_1 + Z_{12}\zeta_2 = -Z_{10}\,,$$

$$Z_{21}\zeta_1 + Z_{22}\zeta_2 = -Z_{20}\,,$$

in which the coefficients Z_{ij} have jut been calculated. Thus, in the present problem the equations are

$$\frac{EI}{a}\begin{bmatrix} 11 & 2 \\ 2 & 8 \end{bmatrix}\begin{bmatrix} \zeta_1 \\ \zeta_2 \end{bmatrix} = \frac{pa^2}{24}\begin{bmatrix} 1 \\ 2 \end{bmatrix}.$$

The solution is obtained by pre-multiplication with the inverse matrix,

$$\begin{bmatrix} \zeta_1 \\ \zeta_2 \end{bmatrix} = \frac{pa^3}{24EI}\,\frac{1}{84}\begin{bmatrix} 8 & -2 \\ -2 & 11 \end{bmatrix}\begin{bmatrix} 1 \\ 2 \end{bmatrix} = \frac{1}{504}\frac{pa^3}{EI}\begin{bmatrix} 1 \\ 5 \end{bmatrix}.$$

Thus, both nodes rotate counter-clockwise, and $\zeta_2 = 5\zeta_1$.

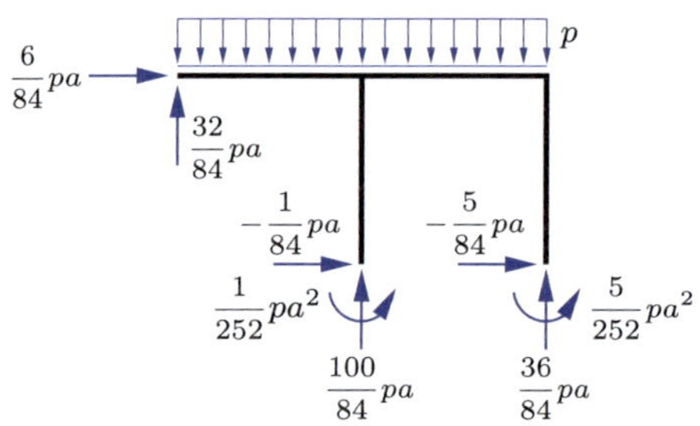

Fig. 7.23: Loads and reactions on frame.

Figure 7.23 shows the loads and reactions on the frame. The reactions are determined as follows. First, the reaction moments are evaluated, as they are determined directly by superposition of the corresponding load cases. Reaction moments are positive in the counter-clockwise direction. At A there is a simple support, whereby $M_A = 0$. At D the reaction moment is

$$M_D = M_D^0 + M_D^1\zeta_1 + M_D^2\zeta_2 = 0 + 2\frac{EI}{a}\zeta_1 + 0 = \tfrac{1}{252}pa^2,$$

where the unit rotation moment M_D^1 is given in Fig. 7.21. Similarly, at the support E

$$M_E = M_E^0 + M_E^1\zeta_1 + M_E^2\zeta_2 = 0 + 0 + 2\frac{EI}{a}\zeta_2 = \tfrac{5}{252}pa^2,$$

where the unit rotation moment M_E^2 is given in Fig. 7.22. Neither of the reaction moments contain a contribution from the constrained frame with external loads, as neither of the adjoining beams are loaded directly.

The reaction force in the direction transverse to the beam also follows directly from the reaction components of the load cases. Horizontal reactions are positive towards the right, and vertical reactions are positive upwards. At the simple support A the vertical component of the reaction force is

$$R_A = R_A^0 + R_A^1\zeta_1 + R_A^2\zeta_2 = \tfrac{3}{8}pa + 3\frac{EI}{a^2}\zeta_1 + 0 = \tfrac{32}{84}pa.$$

The horizontal reaction at D is determined by the transverse force component in the beam DB as

$$R_D' = R_D'^0 + R_D'^1\zeta_1 + R_D'^2\zeta_2 = 0 - 6\frac{EI}{a^2}\zeta_1 + 0 = -\tfrac{1}{84}pa.$$

This component is given entirely in terms of ζ_1 and the transverse force component for the corresponding unit deformation given in Fig. 7.21. Similarly, the horizontal reaction component at E is given by the transverse force in the beam EC as

$$R_E' = R_E'^0 + R_E'^1\zeta_1 + R_E'^2\zeta_2 = 0 + 0 - 6\frac{EI}{a^2}\zeta_2 = -\tfrac{5}{84}pa.$$

This component is given entirely in terms of ζ_2 and the transverse force component for the corresponding unit deformation given in Fig. 7.22.

The remaining reactions are normal forces in the corresponding beams and they are therefore not represented explicitly by the load cases. The horizontal reaction component in A is most easily found by horizontal equilibrium of the full loaded frame,

$$R_A' = -R_D' - R_E' = \tfrac{6}{84}pa.$$

The vertical reaction R_D is determined from the transverse forces in ABC at B,

$$R_D = R_D^0 + R_D^1\zeta_1 + R_D^2\zeta_2 = \big(\tfrac{5}{8} + \tfrac{1}{2}\big)pa + (6-3)\frac{EI}{a^2}\zeta_1 + 6\frac{EI}{a^2}\zeta_2 = \tfrac{100}{84}pa.$$

Similarly, the vertical reaction R_E is determined from the transverse forces in ABC at C,

$$R_E = R_E^0 + R_E^1\zeta_1 + R_E^2\zeta_2 = \tfrac{1}{2}pa - 6\frac{EI}{a^2}\zeta_1 - 6\frac{EI}{a^2}\zeta_2 = \tfrac{36}{84}pa.$$

It is easily verified that the sum of the vertical reactions equal the load.

Once the reactions have been determined as illustrated in Fig. 7.23 it is a standard procedure to calculate the associated internal force distributions. The moment distribution in the frame is shown in Fig. 7.24. The basic behavior of the frame is illustrated by the moment distribution over the beam

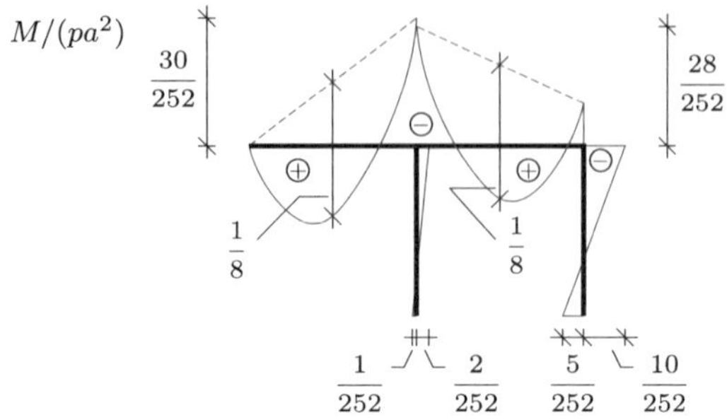

Fig. 7.24: Moment distribution in frame.

ABC. If the nodes B and C were fully constrained the moment curve over BC would have been symmetric. However, while the node B has a much smaller rotation than node C it appears as nearly fixed and thereby retains the rather large moment, while node C appears as a flexible support and thereby reduces the moment at C.

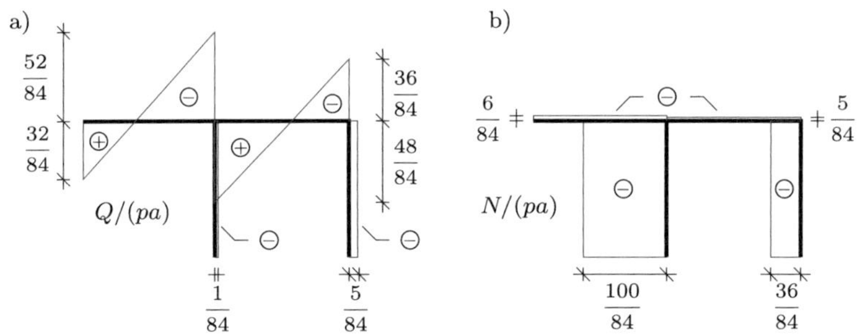

Fig. 7.25: (a) Shear and (b) normal force distribution in frame.

The shear force and and normal force distributions are shown in Fig. 7.25a and b, respectively. The figures clearly illustrate that the two legs of the frame primarily act as columns in compression.

General procedure of the deformation method

On the basis of the two examples the deformation method for analysis of beam and frame structures can now be described in concise terms. A number of points of the structure are designated as nodes. The initially unconstrained degrees of freedom of these nodes are denoted $\zeta_1, \cdots, \zeta_n$. These degrees of freedom may include displacements and rotations. The loading is now considered to be applied to a structure in which the motion of the nodes has been constrained by imposing forces and moments $Z_{10}, \cdots, Z_{n0}$ corresponding to the degrees of freedom $\zeta_1, \cdots, \zeta_n$. The forces/moments $Z_{10}, \cdots, Z_{n0}$ needed to constrain the nodes are obtained by summation of the forces/moments needed to constrain the individual beams, joined at the corresponding node. The constraining forces/moments in the loaded beams are typically obtained from pre-calculated load cases, as e.g. those of Table 7.2.

The constraints are now released *one at a time* and given a unit displacement/rotation, $\zeta_j = 1$, while the other displacement components are still constrained, $\zeta_k = 0$ for $k \neq j$. This requires that concentrated loads $Z_{1j}, \cdot, Z_{nj}$ are imposed, corresponding to the degrees of freedom $\zeta_1, \cdots, \zeta_n$. The concentrated loads corresponding to an imposed unit deformation $\zeta_j = 1$, consists of an imposed load Z_{jj}, and a set of restraining loads Z_{jk}, $k \neq j$. The imposed load Z_{jj} is the sum of loads needed to enforce the unit displacement in all connected beams, while the constraints Z_{jk} are found directly from the beam in question. Only nodes directly connected with the node with degree of freedom j need to be restrained, and thus the analysis has a local character, in contrast to the force method. The coefficients Z_{ij} are formed from individual beams with an imposed end displacement. The basic end displacement load cases are given in Figs. 7.4 to 7.7. The reactions are given in a general form including the dependence of shear flexibility. In many cases the shear flexibility effect is modest, and the shear flexibility effect is omitted in most hand calculations.

The total imposed loads Z_i are found by superposition of the contribution from the external load with constrained nodes plus a contribution from each of the displacements $\zeta_1, \cdots, \zeta_n$,

$$Z_i = Z_{i0} + Z_{i1}\zeta_1 + \cdots + Z_{in}\zeta_n\,, \qquad i = 1, \cdots, n. \tag{7.18}$$

In the real structure there are no imposed loads, and thus the displacements $\zeta_1, \cdots, \zeta_n$ are determined by the condition that the sum of all imposed loads vanish at each node,

$$\begin{aligned} Z_1 &= Z_{10} + Z_{11}\zeta_1 + \cdots + Z_{1n}\zeta_n = 0 \\ Z_2 &= Z_{20} + Z_{21}\zeta_1 + \cdots + Z_{2n}\zeta_n = 0 \\ &\vdots \qquad\qquad \vdots \qquad\qquad\quad \vdots \\ Z_n &= Z_{n0} + Z_{n1}\zeta_1 + \cdots + Z_{nn}\zeta_n = 0. \end{aligned} \tag{7.19}$$

This condition constitutes n equations for the initially unknown displacement components $\zeta_1, \cdots, \zeta_n$.

When the displacement components $\zeta_1, \cdots, \zeta_n$ have been determined, reactions and internal forces can be calculated. There are basically two approaches. In the first reactions and internal force distributions are determined for the constrained loaded structure and for each imposed deformation case. The result is then obtained by superposition. It is often simpler to use the superposition approach only for the reactions, for example

$$R = R_{i0} + R_{i1}\zeta_1 + \cdots + R_{in}\zeta_n\,, \qquad i = 1, \cdots, n, \tag{7.20}$$

and then to determine the internal force distributions by a static analysis of the structure with the full load and reactions. In this way superposition of internal force distributions over the beams is avoided.

Although in practice the constraint loads Z_{i0} and Z_{ij} are often obtained from a table of simple load cases for a single beam, it is instructive to review a procedure for calculation of the coefficients Z_{ij} by use of the principle of virtual work as described in Section 4.4.2. Consider the case with all nodes constrained, except $\zeta_i = 1$. The corresponding loads are $Z_{i1}, \cdots, Z_{in}$ and the internal force distributions are $M_i(s)$, $Q_i(s)$ and $N_i(s)$. This combination of loads and internal forces is now used as the static field in the principle of virtual work together with the virtual displacement field corresponding to the unit motion $\zeta_j = 1$. The corresponding virtual strain field is $\kappa_j = M_j(s)/EI$, $\gamma_j = Q_j(s)/GA_s$ and $\varepsilon_j = N_j(s)/EA$. There are no distributed loads or discontinuities in the virtual displacement field, and it therefore follows from the principle of virtual work in the form (4.43) and (4.45) that the external work $Z_{ij}\zeta_j = Z_{ij}$ is given as

$$Z_{ij} = \int \Big\{ \frac{M_i(s)M_j(s)}{EI} + \frac{Q_i(s)Q_j(s)}{GA_s} + \frac{N_i(s)N_j(s)}{EA} \Big\} ds\,, \tag{7.21}$$

where the internal force fields correspond to the imposed isolated unit displacements $\zeta_i = 1$ and $\zeta_j = 1$, respectively. The integral relation (7.21) implies, that the stiffness coefficients Z_{ij} satisfy the symmetry relations

$$Z_{ji} = Z_{ij}\,, \qquad i, j = 1, \cdots, n. \tag{7.22}$$

Thus, the equation system (7.19) of the deformation method is symmetric. The contribution from the normal force is usually negligible, and is therefore omitted. For slender beams of isotropic material the shear contribution can also often be neglected, while it may be important for short beams and for composite beams with stiff webs and a core of a more flexible material.

Example 7.2. Influence of relative stiffness in symmetric frame. Figure 7.26 shows a symmetric frame $ABCD$ with fixed supports at A and D. The frame supports a uniformly distributed load of intensity p over BC. Both frame and load are symmetric, and thus the deformation and the internal forces have symmetry properties. The columns AB and DC have bending stiffness EI_1, while the horizontal beam BC has bending stiffness EI_2. The reactions and internal force distribution are analyzed, using a slight modification of the deformation method as described above, and the influence of the relative stiffness of the frame members is illustrated.

The displacements are symmetric, and thus the displacement components to be constrained are the rotation of the corner nodes B and C. Due to symmetry the constraining moments are of equal magnitude but opposite orientation, and they can therefore both be represented by the constraining moment Z_{10} as shown in Fig. 7.27. The magnitude of the constraining

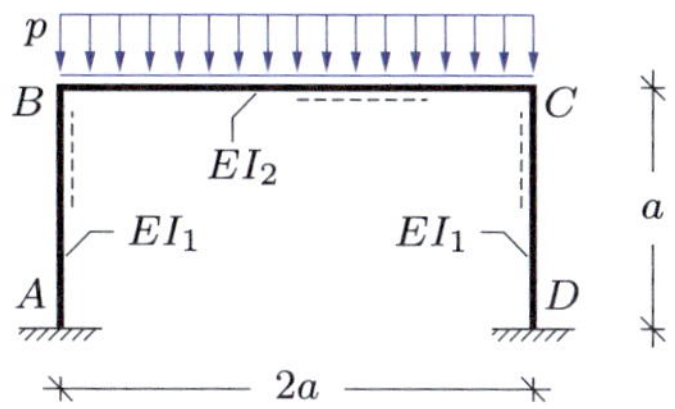

Fig. 7.26: Symmetric frame with column stiffness EI_1 and beam stiffness EI_2.

moments follow from Table 7.2 as

$$Z_{10} = \tfrac{1}{12}p(2a)^2 = \tfrac{1}{3}pa^2.$$

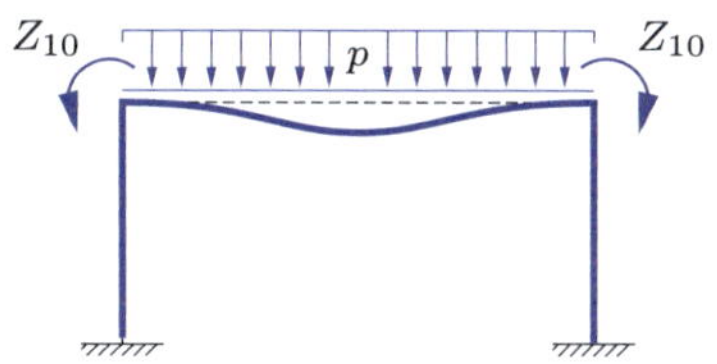

Fig. 7.27: Constraining moments Z_{10} at corners.

The rotation of the corners are of the same magnitude but opposite orientation, and the unit deformation case is therefore defined as a deformation where both corners rotate simultaneously as illustrated in Fig. 7.28. The magnitude of the rotation is denoted ζ_1, and the magnitude of the moments needed to produce a unit rotation is denoted Z_{11}.

The moments and shear forces in the individual members of the frame corresponding to a unit rotation are shown in Fig. 7.29. The moments and shear forces in the columns correspond to the unit end-rotation case illustrated in Fig. 7.4, while the beam BC experiences uniform bending as illustrated in Fig. 7.2. The moment Z_{11} to be applied at both the corners follows from adding the contributions from the column and the beam,

$$Z_{11} = \frac{4EI_1}{a} + \frac{EI_2}{a}.$$

The corner rotation follows from the condition that the total moment applied to each corner must vanish,

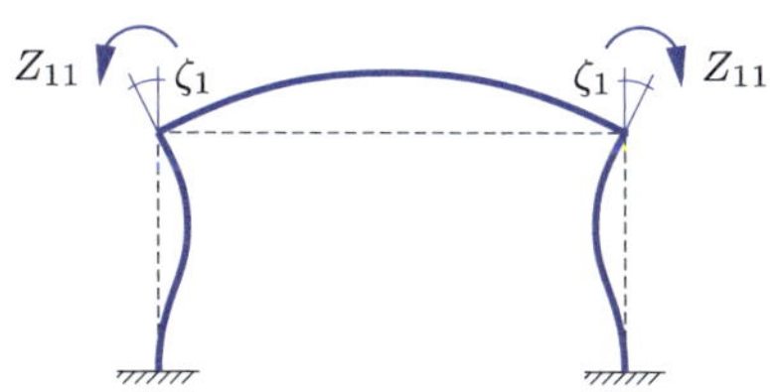

Fig. 7.28: Imposed unit rotation $\zeta_1 = 1$.

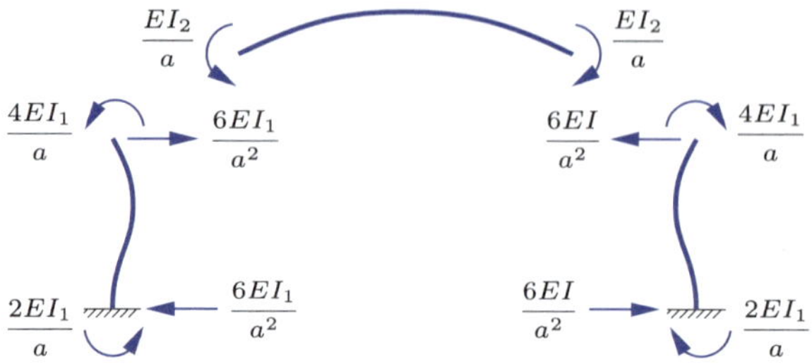

Fig. 7.29: Beam moments and shear forces for $\zeta_1 = 1$.

$$Z_1 = Z_{10} + Z_{11}\zeta_1 = 0,$$

whereby

$$\zeta_1 = -\frac{Z_{10}}{Z_{11}} = -\frac{\frac{1}{3}pa^3}{4EI_1 + EI_2}.$$

The transverse reaction forces and the reaction moments now follow from their normalized values given in Fig. 7.29, when multiplied by ζ_1. This product involves the relative stiffness of the columns and the beam, conveniently expressed in terms of the parameter

$$\alpha = \frac{4EI_1}{4EI_1 + EI_2}.$$

It is seen that this parameter is the ratio of the moment rotating the top of one of the columns, to the moment needed to rotate the frame corner. Clearly, this ratio has a lower limit of zero for extremely flexible columns and increases towards unity for very stiff columns. The reactions are given in terms of the parameter α in Fig. 7.30.

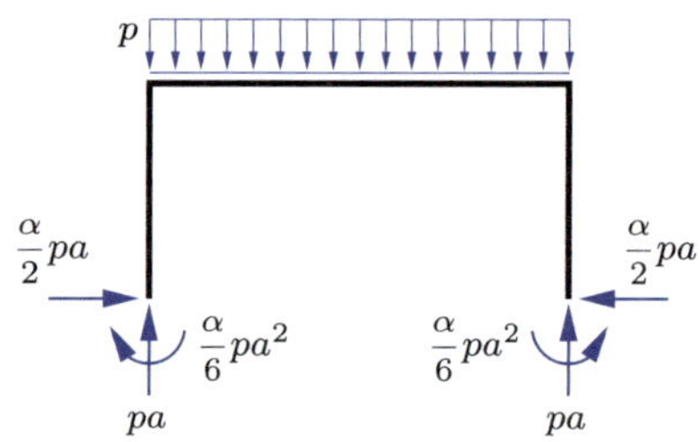

Fig. 7.30: Loads and reactions on frame.

With the load and the reactions given, the moment distribution follows from simple statics. The support moments have the magnitude $\frac{1}{6}\alpha pa^2$, while the corner moments are $-\frac{1}{3}\alpha pa^2$. The moment distribution and the transverse force in the two columns are proportional to α, implying that moments in the columns decrease with decreasing relative stiffness of the columns.

In the special case where the bending stiffness of the beam and the columns are identical $\alpha = \frac{4}{5}$. In this case the corner moments are $-\frac{4}{15}pa^2$, while the moment at the center of the beam is $\frac{7}{30}pa^2$, i.e. roughly the same magnitude. □

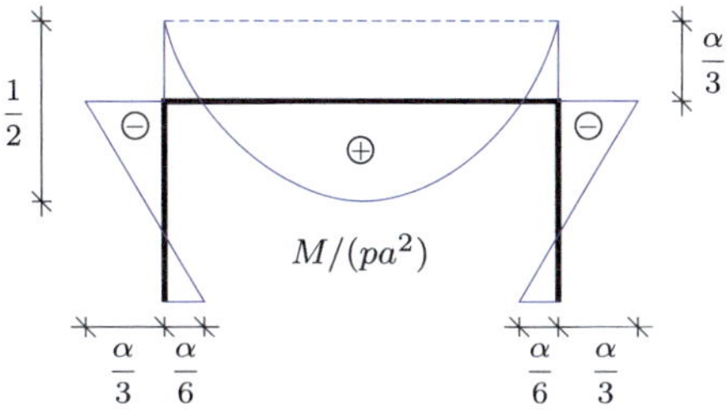

Fig. 7.31: Moment distribution in frame.

Example 7.3. Swaying of symmetric frame with anti-symmetric load. Figure 7.32 shows a symmetric frame $ABCD$ loaded by a horizontal force $2P$ at the corner B. The beam BC is considered inextensible, and thus the load can be considered as distributed equally between the nodes B and C. All members have bending stiffness EI.

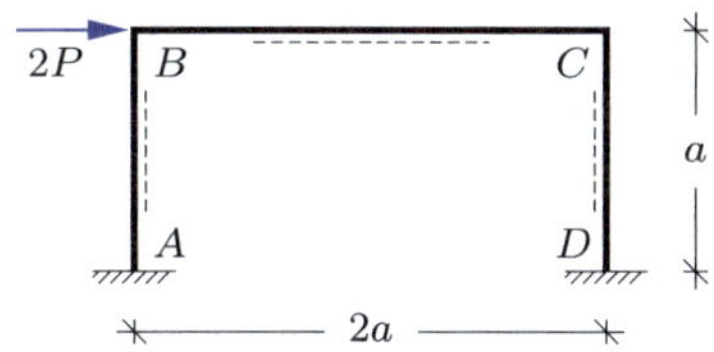

Fig. 7.32: Symmetric frame with horizontal force.

In the present case full constraint of the corner nodes B and C requires both a pair of equal horizontal forces Z_{10} and a pair of moments Z_{20} as shown in Fig. 7.33. It is clear from the figure, that if the horizontal constraining forces balance the external load, there will be no need for constraining moments at the corners, and thus

$$Z_{10} = -P, \qquad Z_{20} = 0.$$

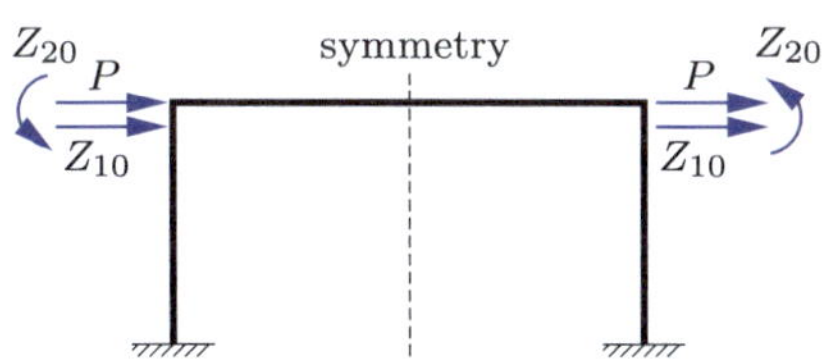

Fig. 7.33: Constraining forces Z_{10} and moments Z_{20} at corners.

The deformation modes corresponding to the constraining forces Z_{10} and moments Z_{20} are represented by a horizontal translation ζ_1 and a counter-clockwise rotation ζ_2 of both the corner nodes B and C as shown in Fig. 7.34.

The deformation modes are anti-symmetric, and therefore only the left half of the frame need to be considered when calculating the stiffness coefficients Z_{ij}. The coefficients Z_{11} and Z_{21} corresponding $\zeta_1 = 1$ can be identified from Fig. 7.35a as the horizontal force and the moment at node B. The beam BC is not deformed in this case, and thus these coefficients follow directly from the column AB as

$$Z_{11} = 12\frac{EI}{a^3}\,, \qquad Z_{21} = 6\frac{EI}{a^2}\,.$$

Similarly, the coefficients Z_{12} and Z_{22} corresponding $\zeta_2 = 1$ can be identified from Fig. 7.35b as the horizontal force and the total moment at node B. Note, that the beam BC is here deformed in anti-symmetric bending, corresponding to a hinge at its center on the line of symmetry and no axial force.

$$Z_{12} = 6\frac{EI}{a^2}\,, \qquad Z_{22} = 4\frac{EI}{a} + 3\frac{EI}{a} = 7\frac{EI}{a}\,.$$

Note, that by symmetry $Z_{12} = Z_{21}$.

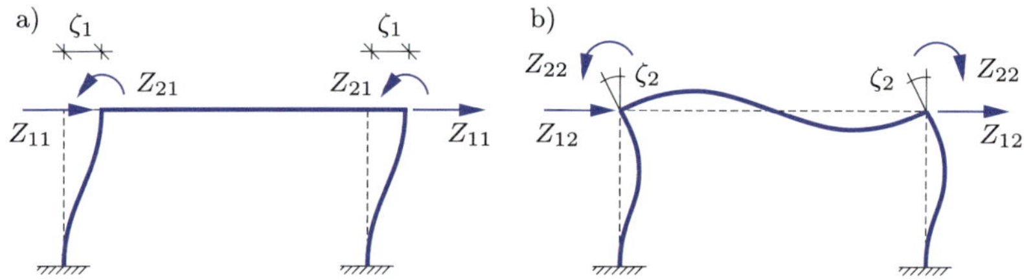

Fig. 7.34: Imposed unit displacements. (a) corner translation ζ_1, (b) corner rotation ζ_2.

The total imposed horizontal force and moment at each of the corner nodes must vanish, whereby

$$Z_1 = Z_{10} + Z_{11}\zeta_1 + Z_{12}\zeta_2 = 0\,,$$
$$Z_2 = Z_{20} + Z_{21}\zeta_1 + Z_{22}\zeta_2 = 0\,.$$

When inserting the Z_{ij} coefficients just calculated the following equations are obtained,

$$\frac{EI}{a^3}\begin{bmatrix} 12 & 6a \\ 6a & 7a^2 \end{bmatrix}\begin{bmatrix} \zeta_1 \\ \zeta_2 \end{bmatrix} = \begin{bmatrix} P \\ 0 \end{bmatrix}.$$

The rotation ζ_2 follows e.g. by subtracting $2/a$ times the second equation from the first, and ζ_1 then follows immediately in terms of ζ_2 from the second equation. The result is

$$\zeta_1 = \frac{7}{48}\frac{Pa^3}{EI}\,, \qquad \zeta_2 = -\frac{1}{8}\frac{Pa^2}{EI}\,.$$

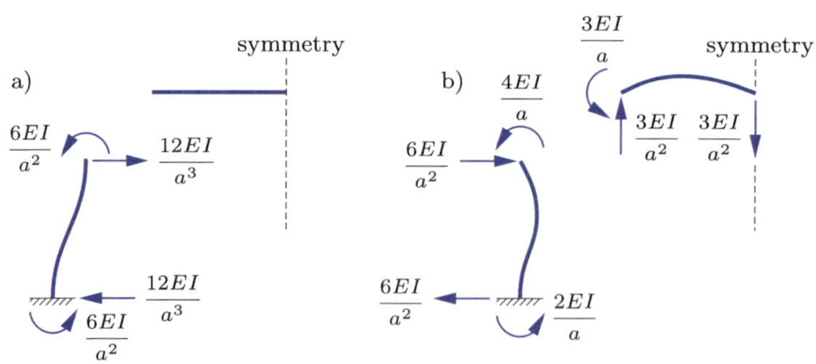

Fig. 7.35: Beam forces and moments for unit displacements: a) $\zeta_1 = 1$, b) $\zeta_2 = 1$.

The horizontal reactions $R'_A = R'_D$ can be calculated from the transverse forces at A in Fig. 7.35. When considered positive towards the left

$$\leftarrow \qquad R'_A = R'_D = 12\frac{EI}{a^3}\zeta_1 + 6\frac{EI}{a^2}\zeta_2 = P\,.$$

Alternatively, this result could have been determined by horizontal projection of forces. Now, this projection serves as a check. The reaction moments $M_A = M_D$, considered positive in the counter-clockwise direction, follow from weighted superposition of the moments in Fig. 7.35 as

$$\curvearrowleft \qquad M_A = M_D = 6\frac{EI}{a^2}\zeta_1 + 6\frac{EI}{a}\zeta_2 = \tfrac{1}{8}Pa\,.$$

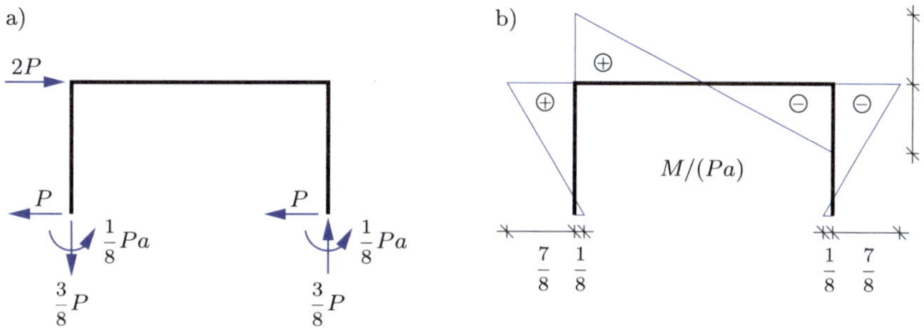

Fig. 7.36: a) Load and reactions, b) Moment distribution.

The vertical reactions $R_A = -R_D$ can be determined from the transverse force at the center of the beam BC,

$$\downarrow \qquad R_A = -R_D = -3\frac{EI}{a^3}\zeta_2 = \tfrac{3}{8}P\,.$$

Alternatively, the vertical reactions can be determined from moment equilibrium of the complete frame, including both the reaction moments $M_A = M_D$. The load and the reactions on the full frame are shown in Fig. 7.36a, and the corresponding moment distribution in Fig. 7.36b. If the frame had fixed simple supports at A and D the horizontal reactions would still be $R'_A = R'_D = P$, and the moment at the corners would then be $M_B = -M_C = Pa$. The constraint of the supports is seen to reduce the corner moments by $\frac{1}{8}Pa$. □

The examples have demonstrated, how the deformation method can be used to determine the internal forces and the displacements in statically indeterminate beam and frame structures. The analysis is systematic and quite straightforward in principle. However, once the constrained degrees of freedom have been found by solving an equation system, the remaining determination of the internal forces is complicated by the fact that axial deformation has been omitted from the analysis, and therefore the normal forces must be determined via a separate static analysis of the structure. In most cases the influence of the axial deformation on the resulting distribution of the internal forces is small, and thus this approximation appears attractive, as it reduces the number of degrees of freedom that needs to be constrained, and thereby reduces the size of the equation system. When solving the problem numerically the balance changes, and it is advantageous to include the axial deformation, as it leads to a more systematic method for determining the internal forces and displacements of the structure. In that case it is advantageous to formulate the problem within the format of the finite element method, illustrated for truss structures in Section 2.5. The corresponding finite element formulation of frame structures is described in the following two sections – first considering the individual beam element in Section 7.3, and then assembling the elements into a model of a frame structure in Section 7.4.

7.3 Beam elements

The typical beam element treated here consists of a straight beam connecting the two nodes A and B. The properties of the element are first expressed in a local frame of reference $\{x', y'\}$ with the element placed along the x'-axis as shown in Fig. 7.37. The two nodes have three generalized force components,

$$\mathbf{f}'_A = [\, f^A_{x'}, f^A_{y'}, m^A \,]^T\,, \quad \mathbf{f}'_B = [\, f^B_{x'}, f^B_{y'}, m^B \,]^T\,, \tag{7.23}$$

where $f_{x'}$ is the axial force component, $f_{y'}$ is the transverse force component, and m is the moment in the counter-clockwise direction. The corresponding generalized displacement components at A and B are shown in Fig. 7.37b,

$$\mathbf{u}'_A = [\, u^A_{x'}, u^A_{y'}, \theta^A \,]^T\,, \quad \mathbf{u}'_B = [\, u^B_{x'}, u^B_{y'}, \theta^B \,]^T\,, \tag{7.24}$$

where $u_{x'}$ is the axial translation component, $u_{y'}$ is the transverse translation component, and θ is the counter-clockwise rotation.

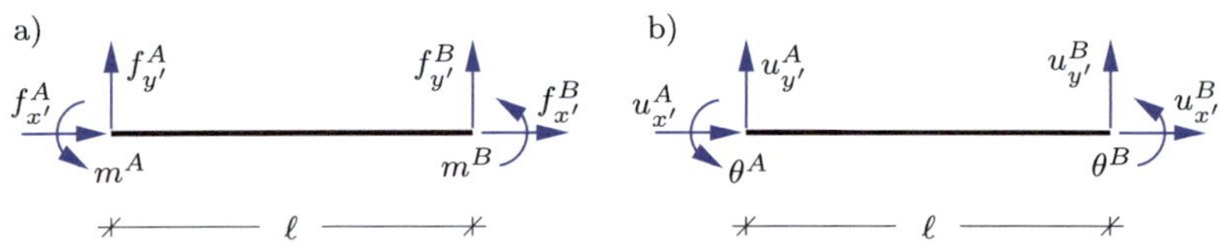

Fig. 7.37: Two-dimensional beam element. a) node forces, b) node displacements.

Each of the generalized displacement components are associated with a displacement mode as illustrated in Fig. 7.38, and the displacement of the beam element is obtained by superposition of these displacement modes.

Traditionally, the nodal forces generated by the nodal displacements are obtained by solving the differential equations for the displacements of the beam for each of the six unit displacement cases shown Fig. 7.38. However, greater generality as well as simplicity is gained by using the flexibility of the deformation modes, illustrated for the two bending modes in Section 7.1.1. When using the flexibility formulation, variation of cross-section properties as well as curvature can be accounted for explicitly without obtaining the displacement field, see e.g. Krenk (1994). Here, the presentation is limited to two particular types of straight homogeneous beam elements: the bending element with shear flexibility, and the beam-column element.

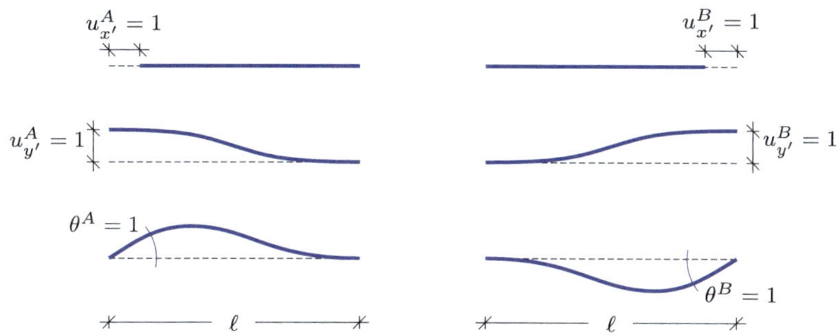

Fig. 7.38: Displacement modes of two-dimensional beam element.

7.3.1 Beam bending element

The beam bending element gives a relation between the generalized forces and the generalized moments shown in Fig. 7.37. For the case of a straight homogeneous beam the necessary expressions have already been obtained, when neglecting a possible column effect from the normal force – the normal force relation in connection with bars, and the unit displacement bending cases in Section 7.1. The results are expressed in the form of a relation between the generalized forces of the element, contained in the six-component vector

$$[\,\mathbf{f}'^T_A, \mathbf{f}'^T_B\,] = [\,f^A_{x'}, f^A_{y'}, m^A, f^B_{x'}, f^B_{y'}, m^B\,], \tag{7.25}$$

and the six-component generalized displacement vector

$$[\,\mathbf{u}'^T_A, \mathbf{u}'^T_B\,] = [\,u^A_{x'}, u^A_{y'}, \theta^A, u^B_{x'}, u^B_{y'}, \theta^B\,]. \tag{7.26}$$

The relation is linear and can be written in the generic block matrix format

$$\begin{bmatrix}\mathbf{f}'_A\\ \mathbf{f}'_B\end{bmatrix} = \underbrace{\begin{bmatrix}\mathbf{K}'_{AA} & \mathbf{K}'_{AB}\\ \mathbf{K}'_{BA} & \mathbf{K}'_{BB}\end{bmatrix}}_{\mathbf{K}'_{\text{beam}}} \begin{bmatrix}\mathbf{u}'_A\\ \mathbf{u}'_B\end{bmatrix}, \tag{7.27}$$

where $\mathbf{K}'_{\text{beam}}$ is the stiffness matrix of the beam element, when located in a local frame of reference along the x'-axis. In compact form the element stiffness matrix relation is

$$\mathbf{f}'_e = \mathbf{K}'_{\text{beam}}\mathbf{u}'_e \tag{7.28}$$

where $\mathbf{f}'_e = [\mathbf{f}'^T_A, \mathbf{f}'^T_B]^T$ and $\mathbf{u}'_e = [\mathbf{u}'^T_A, \mathbf{u}'^T_B]^T$ contain all generalized force and displacement components of the element.

For unit displacements at node A the sub-matrix $\mathbf{K}'_{AA}$ represents the generalized forces at node A, while the sub-matrix $\mathbf{K}'_{BA}$ represents the generalized forces at node B. These are the generalized forces at A and B, respectively, corresponding to the three unit deformation cases illustrated in the first column of Fig. 7.38. The first column in these matrices corresponds to the unit deformation $u^A_{x'} = 1$, which generates a normal force of magnitude $-EA/\ell$. The second column corresponds to the unit transverse displacement $u^A_{y'} = 1$. This case corresponds to Fig. 7.6, where the end forces are indicated. Finally, the generalized forces corresponding to a unit rotation $\theta^A = 1$ follows from Fig. 7.4. The sub-matrices $\mathbf{K}'_{AB}$ and $\mathbf{K}'_{BB}$ similarly represent the generalized forces at A and B, respectively, from unit displacements at B. The stiffness matrix $\mathbf{K}'_{\text{beam}}$ follows from collecting these generalized forces in matrix format,

$$\mathbf{K}'_{\text{beam}} = \begin{bmatrix} \frac{EA}{\ell} & 0 & 0 & -\frac{EA}{\ell} & 0 & 0 \\ 0 & \frac{12}{1+\Phi}\frac{EI}{\ell^3} & \frac{6}{1+\Phi}\frac{EI}{\ell^2} & 0 & -\frac{12}{1+\Phi}\frac{EI}{\ell^3} & \frac{6}{1+\Phi}\frac{EI}{\ell^2} \\ 0 & \frac{6}{1+\Phi}\frac{EI}{\ell^2} & \frac{4+\Phi}{1+\Phi}\frac{EI}{\ell} & 0 & -\frac{6}{1+\Phi}\frac{EI}{\ell^2} & \frac{2-\Phi}{1+\Phi}\frac{EI}{\ell} \\ -\frac{EA}{\ell} & 0 & 0 & \frac{EA}{\ell} & 0 & 0 \\ 0 & -\frac{12}{1+\Phi}\frac{EI}{\ell^3} & -\frac{6}{1+\Phi}\frac{EI}{\ell^2} & 0 & \frac{12}{1+\Phi}\frac{EI}{\ell^3} & -\frac{6}{1+\Phi}\frac{EI}{\ell^2} \\ 0 & \frac{6}{1+\Phi}\frac{EI}{\ell^2} & \frac{2-\Phi}{1+\Phi}\frac{EI}{\ell} & 0 & -\frac{6}{1+\Phi}\frac{EI}{\ell^2} & \frac{4+\Phi}{1+\Phi}\frac{EI}{\ell} \end{bmatrix}. \tag{7.29}$$

The shear flexibility is represented via the non-dimensional parameter

$$\Phi = \frac{12EI}{GA_s\ell^2} \tag{7.30}$$

introduced in (7.5) in connection with the anti-symmetric bending problem with constant shear force distribution. The classic result for the so-called Euler beam without shear flexibility is obtained as the special case of $\Phi = 0$, corresponding to infinite shear stiffness. Clearly, in a computer program it is advantageous to implement the full expression, and obtain the classic theory as a special case.

7.3.2 Beam-column element

For frames with slender members it is important to be able to include the effect of reduced flexibility due to normal compression forces. In slender beams the shear flexibility effect is most often negligible, and the beam-column element is therefore derived for a beam without shear flexibility. The basic equations were derived in Section 5.1, and only the main points needed for development of the beam-column element are summarized here. The differential equation for the transverse displacement of a beam in the presence of a normal tension force N was derived as (5.9). In the present formulation the element is located along the x'-axis, and the transverse direction is y' with displacement $u_{y'}$. In this notation the bending of a beam without transverse load is governed by the differential equation

$$\frac{d^2}{dx'^2}\Big(EI\,\frac{d^2u_{y'}}{dx'^2}\Big) - \frac{d}{dx'}\Big(N\,\frac{du_{y'}}{dx'}\Big) = 0\,. \tag{7.31}$$

The typical application involves a compressive normal force, denoted by $P = -N$. For a homogeneous beam this equation can be expressed as

$$\frac{d^4u_{y'}}{dx'^4} + k^2\,\frac{d^2u_{y'}}{dx'^2} = 0\,, \tag{7.32}$$

where the parameter k has been introduced by the definition

$$k^2 = \frac{P}{EI}\,. \tag{7.33}$$

Real-valued parameters k correspond to compression, and the corresponding expressions for a tension force can be obtained by using complex notation, whereby trigonometric functions translate into their corresponding hyperbolic counterpart. However, the linearized form of the present theory only contains k^2, and these results can therefore be expressed directly in terms of the normal force N.

The general solution to the homogeneous 4'th order beam-column equation (7.33) is

$$u_{y'}(x') = C_1 + C_2\,kx' + C_3\cos(kx') + C_4\sin(kx')\,. \tag{7.34}$$

In the present notation with transverse displacement $u_{y'}(x')$ the moment follows from (7.34) as

$$\frac{M(x')}{EI} = \frac{d^2u_{y'}}{dx'^2} = -C_3\,k^2\cos(kx') - C_4\,k^2\sin(kx')\,. \tag{7.35}$$

The shear force at the end sections will be determined directly from equilibrium of the element, when needed.

Symmetric bending

As in the case of the beam bending element it is convenient to construct the element stiffness matrix from the symmetric and anti-symmetric deformation modes. The symmetric bending mode is shown in Fig. 7.39 with rotations $\pm\frac{1}{2}\theta_s$ of the end sections.

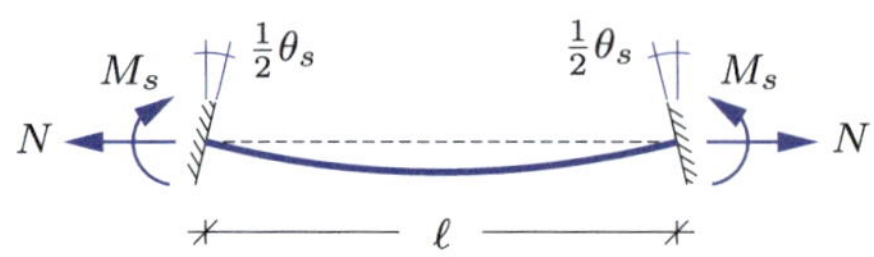

Fig. 7.39: Symmetric bending with normal force.

The y'-axis of the coordinate system is placed in the central line of symmetry, and thus the solution will be in the form of the symmetric part of (7.34),

$$u_{y'}^s(x') = C_1 + C_3 \cos(kx'). \tag{7.36}$$

The boundary conditions consist of vanishing displacement and prescribed rotation at $\pm\frac{1}{2}\ell$,

$$\begin{aligned} u_{y'}^s(\tfrac{1}{2}\ell) &= C_1 + \cos(\tfrac{1}{2}k\ell)\, C_3 = 0\,, \\ u_{y'}^{s\prime}(\tfrac{1}{2}\ell) &= \quad - k\sin(\tfrac{1}{2}k\ell)\, C_3 = \tfrac{1}{2}\theta_s\,. \end{aligned} \tag{7.37}$$

The constant C_3 follows from the second equation as

$$C_3 = -\frac{\frac{1}{2}\theta_s}{k\sin(\frac{1}{2}k\ell)}\,. \tag{7.38}$$

The moment at the end-sections then follows from (7.35) in the form

$$M_s = \varphi\,\frac{EI}{\ell}\,\theta_s \tag{7.39}$$

with the stiffness coefficient for symmetric bending

$$\varphi(k\ell) = (\tfrac{1}{2}k\ell)\cot(\tfrac{1}{2}k\ell)\,. \tag{7.40}$$

The case of zero normal force is represented by the limit $k\ell = 0$, for which $\varphi(0) = 1$. This case recovers the previous result (7.2) for symmetric bending without axial force. The variation of the stiffness coefficient φ is shown as a function of the normal compressive force P in Fig. 7.40, normalized with respect to the Euler load P_E. It is seen that the bending stiffness decreases with increasing compression, and turns negative when exceeding the Euler load.

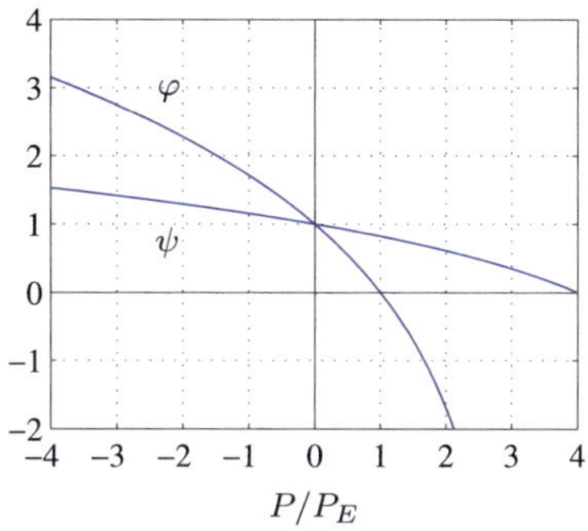

Fig. 7.40: Beam-column bending coefficients.

The shear force in the beam is determined by equilibrium of the full element. The normal forces are co-linear and therefore do not contribute to the moment equilibrium. Thus the shear force vanishes in the case of symmetric bending, $Q = 0$.

Anti-symmetric bending

The case of anti-symmetric bending in the presence of a normal force is shown in Fig. 7.41 with rotation $\frac{1}{2}\theta_a$ of both end sections.

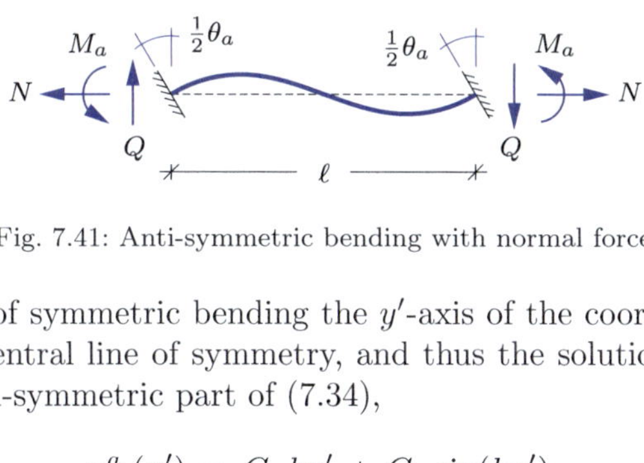

Fig. 7.41: Anti-symmetric bending with normal force.

As in the case of symmetric bending the y'-axis of the coordinate system is placed in the central line of symmetry, and thus the solution will be in the form of the anti-symmetric part of (7.34),

$$u^a_{y'}(x') = C_2\, kx' + C_4 \sin(kx')\,. \tag{7.41}$$

Also in this case the boundary conditions consist of vanishing displacement and prescribed rotation at $\pm\frac{1}{2}\ell$,

$$\begin{aligned} u^a_{y'}(\tfrac{1}{2}\ell) &= \tfrac{1}{2}\ell\, C_2 + \sin(\tfrac{1}{2}k\ell)\, C_4 = 0\,, \\ u^{a\prime}_{y'}(\tfrac{1}{2}\ell) &= k\, C_2 - k\sin(\tfrac{1}{2}k\ell)\, C_3 = \tfrac{1}{2}\theta_a\,. \end{aligned} \tag{7.42}$$

The constant C_4 follows from these equations as

$$C_3 = -\frac{\frac{1}{4}\ell\theta_a}{\sin(\frac{1}{2}k\ell) - (\frac{1}{2}k\ell)\cos(\frac{1}{2}k\ell)}. \tag{7.43}$$

The moment at the end sections then follows from (7.35) in the form

$$M_a = \psi\,\frac{3EI}{\ell}\,\theta_a \tag{7.44}$$

with the stiffness coefficient for anti-symmetric bending

$$\psi(k\ell) = \frac{\frac{1}{12}(k\ell)^2}{1 - (\frac{1}{2}k\ell)\cot(\frac{1}{2}k\ell)} = \frac{\frac{1}{12}(k\ell)^2}{1-\varphi}. \tag{7.45}$$

The coefficient is normalized such that the limit of vanishing normal force corresponds to $\psi(0) = 1$. This case recovers the previous result (7.6) for anti-symmetric bending without axial force in the case of vanishing shear flexibility, $\Phi = 0$. The variation of the coefficient ψ is shown as a function of the normal compressive force P in Fig. 7.40. Also in this case the bending stiffness decreases with increasing compression. In this case the stiffness turns negative when exceeding the load $4P_E$, corresponding to the stability load of an Euler column with a node in the middle.

The shear force in the beam is determined by equilibrium of the full element. Also in this case the normal forces are co-linear and therefore do not contribute to the moment equilibrium. The shear force then is determined as, $Q = 2M_a/\ell$.

Translation of end-section

The deformation mode for translation of one of the end-sections is illustrated in Fig. 7.42. The deformation mode is generated by equal moments M_t, applied at both end-sections. This is similar to the antisymmetric bending shown in Fig. 7.41, but in the present case the normal force N and the shear force Q are along a different set of axes. The displacements are assumed 'small' and thus the effect of the normal force can be represented via the parameter $k^2 = -N/EI$, using the normal force N shown in Fig. 7.42.

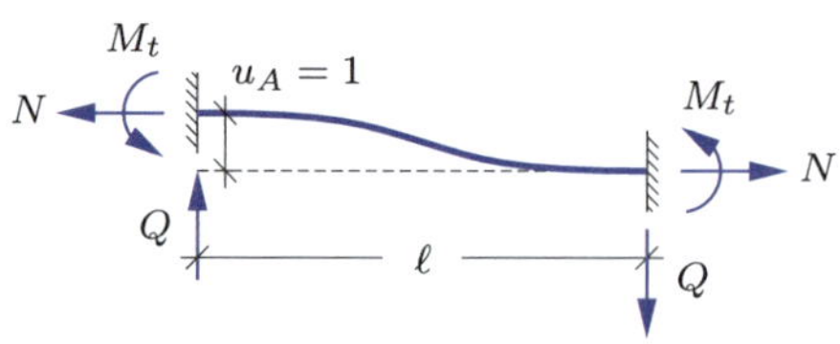

Fig. 7.42: Translation of end section.

The equivalent anti-symmetric rotation angle $\frac{1}{2}\theta_a$ is identified by considering a counter-clockwise rotation of magnitude $1/\ell = \frac{1}{2}\theta_a$. Thus, the moment M_t needed for the translation $u_{y'}^A = 1$ follows from (7.44) as

$$M_t = \psi \frac{6EI}{\ell^2} . \tag{7.46}$$

The normal force N is considered given, and the shear force Q then follows from moment equilibrium of the element as

$$Q = \frac{2}{\ell} M_t + \frac{1}{\ell} N = 12\psi \frac{EI}{\ell^3} + k^2 \frac{EI}{\ell} = \big(12\psi - (k\ell)^2 \big) \frac{EI}{\ell^3} . \tag{7.47}$$

When substituting ψ from (7.45) the expression for the shear force takes the compact form

$$Q = 12 \varphi \psi \frac{EI}{\ell^3} . \tag{7.48}$$

Thus, the contribution of the normal force to the moment equilibrium introduces an extra factor φ into the expression for the shear force.

Beam-column stiffness matrix

The 6×6 stiffness matrix contains the constraining forces/moments corresponding to the six unit deformation cases shown in Fig. 7.38. Extension generates a normal force of magnitude $N = EA(u_{x'}^B - u_{x'}^A)$, and the corresponding nodal forces appear in the first and fourth column and row in the stiffness matrix (7.49). The forces and moments corresponding to the case of a unit translation of node A with the other degrees of freedom constrained appear as the second column of the stiffness matrix. This case of deformation was illustrated in Fig. 7.42, and the moment and transverse force given in (7.46) and (7.48), respectively. The deformation in which node A is given a unit rotation, $\theta^A = 1$, while the other degrees of freedom are constrained, is obtained by superimposing the symmetric and anti-symmetric cases of deformation with $\theta_a = -\theta_s = 1$. The corresponding constraint forces and moments appear as the third column in the stiffness matrix. The last two columns follow from considering the symmetric load cases. In total these six cases of unit deformation defines the stiffness matrix

$$\mathbf{K}'_{\rm beam} = \begin{bmatrix} \frac{EA}{\ell} & 0 & 0 & -\frac{EA}{\ell} & 0 & 0 \\ 0 & 12\varphi\psi\frac{EI}{\ell^3} & 6\psi\frac{EI}{\ell^2} & 0 & -12\varphi\psi\frac{EI}{\ell^3} & 6\psi\frac{EI}{\ell^2} \\ 0 & 6\psi\frac{EI}{\ell^2} & (3\psi+\varphi)\frac{EI}{\ell} & 0 & -6\psi\frac{EI}{\ell^2} & (3\psi-\varphi)\frac{EI}{\ell} \\ -\frac{EA}{\ell} & 0 & 0 & \frac{EA}{\ell} & 0 & 0 \\ 0 & -12\varphi\psi\frac{EI}{\ell^3} & -6\psi\frac{EI}{\ell^2} & 0 & 12\varphi\psi\frac{EI}{\ell^3} & -6\psi\frac{EI}{\ell^2} \\ 0 & 6\psi\frac{EI}{\ell^2} & (3\psi-\varphi)\frac{EI}{\ell} & 0 & -6\psi\frac{EI}{\ell^2} & (3\psi+\varphi)\frac{EI}{\ell} \end{bmatrix}. \qquad (7.49)$$

In the stiffness matrix the entries corresponding to the various forms of bending include the coefficients ϕ and ψ that depend on the normal force N as given by (7.40) and (7.45) and illustrated in Fig. 7.40. It is seen from the figure that within the range $|N| \lesssim P_E$, the coefficients φ and ψ are nearly linear functions of the normal force. Thus, there is a substantial range of normal forces, for which a linear approximation constitutes a fair representation of the effect of the normal force. If needed, the linearized approximation can be improved by sub-dividing the elements, whereby the element length ℓ decreases. Hereby the linearized form becomes an attractive option for performing a stability analysis as described in Section 7.4.2.

The linearized form of the stiffness matrix (7.49) is obtained by using a Taylor series expansion of the two functions $\varphi(k\ell)$ and $\psi(k\ell)$. It follows directly from the Taylor expansion of the cot-function that

$$\varphi(k\ell) = \tfrac{1}{2}k\ell \cot(\tfrac{1}{2}k\ell) \simeq 1 - \tfrac{1}{3}(\tfrac{1}{2}k\ell)^2 - \tfrac{1}{45}(\tfrac{1}{2}k\ell)^4 , \qquad (7.50)$$

and the ψ-function expansion then follows from the last expression in (7.45) as

$$\psi(k\ell) = \frac{\tfrac{1}{3}(\tfrac{1}{2}k\ell)^2}{1-\varphi} \simeq 1 - \tfrac{1}{15}(\tfrac{1}{2}k\ell)^2 . \qquad (7.51)$$

In these relations $(k\ell)^2 = -N\ell^2/EI$, whereby the linearized expressions in the normal force N take the form

$$\varphi \simeq 1 + \frac{1}{12}\frac{N\ell^2}{EI}, \qquad \psi \simeq 1 + \frac{1}{60}\frac{N\ell^2}{EI}. \qquad (7.52)$$

When using these linearized expressions, the stiffness matrix (7.49) can be written as the sum of two matrices,

$$\mathbf{K}'_{\rm beam} \simeq \mathbf{K}^{c\prime}_{\rm beam} + \mathbf{K}^{g\prime}_{\rm beam} , \qquad (7.53)$$

where $\mathbf{K}^{c\prime}_{\rm beam}$ is the constitutive stiffness matrix corresponding to $\varphi = \psi = 1$,

$$
\mathbf{K}^{c\,\prime}_{\rm beam} = \begin{bmatrix}
\dfrac{EA}{\ell} & 0 & 0 & -\dfrac{EA}{\ell} & 0 & 0 \\
0 & 12\dfrac{EI}{\ell^3} & 6\dfrac{EI}{\ell^2} & 0 & -12\dfrac{EI}{\ell^3} & 6\dfrac{EI}{\ell^2} \\
0 & 6\dfrac{EI}{\ell^2} & 4\dfrac{EI}{\ell} & 0 & -6\dfrac{EI}{\ell^2} & 2\dfrac{EI}{\ell} \\
-\dfrac{EA}{\ell} & 0 & 0 & \dfrac{EA}{\ell} & 0 & 0 \\
0 & -12\dfrac{EI}{\ell^3} & -6\dfrac{EI}{\ell^2} & 0 & 12\dfrac{EI}{\ell^3} & -6\dfrac{EI}{\ell^2} \\
0 & 6\dfrac{EI}{\ell^2} & 2\dfrac{EI}{\ell} & 0 & -6\dfrac{EI}{\ell^2} & 4\dfrac{EI}{\ell}
\end{bmatrix} \tag{7.54}
$$

and $\mathbf{K}^{g\,\prime}_{\rm beam}$ is the so-called geometric stiffness matrix, corresponding to the linear terms in the normal force N,

$$
\mathbf{K}^{g\,\prime}_{\rm beam} = \frac{N}{30\ell} \begin{bmatrix}
0 & 0 & 0 & 0 & 0 & 0 \\
0 & 36 & 3\ell & 0 & -36 & 3\ell \\
0 & 3\ell & 4\ell^2 & 0 & -3\ell & -\ell^2 \\
0 & 0 & 0 & 0 & 0 & 0 \\
0 & -36 & -3\ell & 0 & 36 & -3\ell \\
0 & 3\ell & -\ell^2 & 0 & -3\ell & 4\ell^2
\end{bmatrix} . \tag{7.55}
$$

It is observed that the constitutive stiffness matrix (7.54) is a special case of the form (7.29) in which the shear flexibility has been omitted. In practice, the geometric stiffness is less sensitive to the details of the shape functions than the constitutive stiffness, as it depends on the first derivatives while the latter depends on the second derivatives. Thus, a convenient option is to use the form (7.29) including shear flexibility for the constitutive stiffness, while retaining the simple form (7.55) for the geometric stiffness.

7.3.3 Transformation to global form

The beam elements treated above are located in a local frame of reference $\{x', y'\}$ with the beam axis along the x'-axis. In order to use these elements in a model of a structure the corresponding generalized displacement and force components must be transformed into a common global frame of reference $\{x, y\}$ as illustrated in Fig. 7.43. The generalized displacement components $\mathbf{u}' = [u_{x'}, u_{y'}, \theta]^T$ in the local frame are related to the corresponding components $\mathbf{u} = [u_x, u_y, \theta]^T$ in the global frame by the transformation

$$
\mathbf{u}' = \mathbf{A}\,\mathbf{u}\,, \qquad \mathbf{u} = \mathbf{A}^T\mathbf{u}'\,, \tag{7.56}
$$

with the component transformation matrix for the generalized displacements at a node given by

$$\mathbf{A} = \begin{bmatrix} \cos\alpha & \sin\alpha & 0 \\ -\sin\alpha & \cos\alpha & 0 \\ 0 & 0 & 1 \end{bmatrix}. \tag{7.57}$$

In this transformation the displacement components $[u_x, u_y]$ transform as a vector, while the rotation θ is the same in both frames of reference. The transformation between local and global components is identical for the generalized forces.

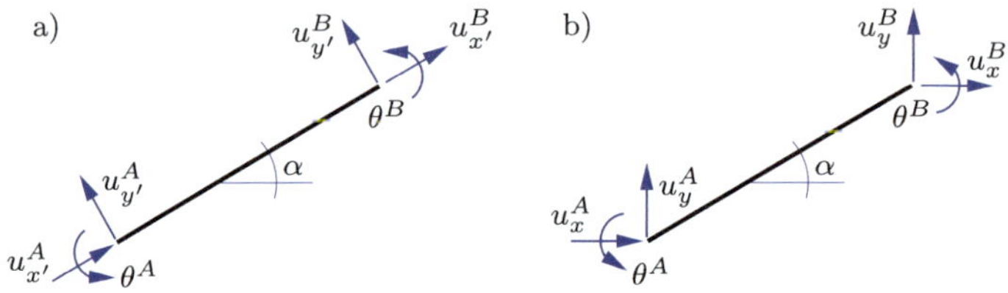

Fig. 7.43: Beam element: a) Local components, b) Global components.

The beam element has two nodes, and the components at each of the nodes must be transformed according to (7.56). It is convenient to combine the transform matrix for the components at a single node into a diagonal block matrix

$$\mathbf{A}_e = \begin{bmatrix} \mathbf{A} & \mathbf{0} \\ \mathbf{0} & \mathbf{A} \end{bmatrix} \tag{7.58}$$

that transforms the components at all element nodes at the same time. In terms of this element transformation matrix the six-components relations for the generalized displacements are

$$\mathbf{u}'_e = \mathbf{A}_e \mathbf{u}_e\,, \qquad \mathbf{u}_e = \mathbf{A}_e^T \mathbf{u}'_e\,. \tag{7.59}$$

The local form of the element stiffness relation (7.28) is transformed into global component form by pre-multiplication with $\mathbf{A}_e^T$, whereby

$$\mathbf{A}_e^T \mathbf{f}'_e = \mathbf{A}_e^T \mathbf{K}'_{\text{beam}} \mathbf{u}'_e = \mathbf{A}_e^T \mathbf{K}'_{\text{beam}} \mathbf{A}_e \mathbf{u}_e\,. \tag{7.60}$$

The left side is recognized as the global components of the generalized forces via the transform (7.59), whereby the global form of the element stiffness relation takes the form

$$\mathbf{f}_e = \mathbf{K}_{\text{beam}} \mathbf{u}_e \tag{7.61}$$

with the global element stiffness matrix given by

$$\mathbf{K}_{\text{beam}} = \mathbf{A}_e^T \mathbf{K}'_{\text{beam}} \mathbf{A}_e\,. \tag{7.62}$$

The global form of the stiffness matrix can be assembled into a stiffness matrix for a frame structure as explained in the following section.

7.4 Finite element method for frames

The finite element analysis of a frame structure is organized in a way similar to that of a truss structure, described in Section 2.5. Therefore only the main points are discussed here with reference to the frame shown in Fig. 7.44. When using one beam element for each member of the frame, the structural model has five nodes, here ordered sequentially as A, B, C, D, E. The frame consists of the four elements AB, BC, BD and CE with global element stiffness matrices $\mathbf{K}_{AB}$ etc., obtained as described in the previous section.

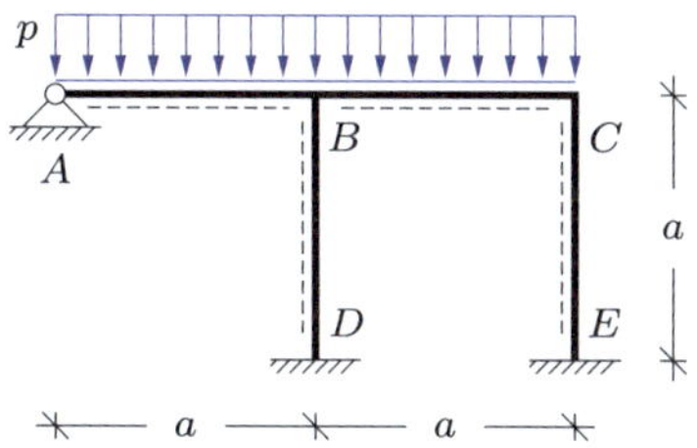

Fig. 7.44: Frame with distributed load.

The stiffness matrix of the structure is formed by including the stiffness contributions from each of the elements. In order to do this all element matrices must first be transformed to a common global frame of reference, and the parts of the stiffness matrix must be associated with the degrees of freedom as organized for the total structure. This procedure, usually called assembling the stiffness matrix, is illustrated in (7.63). Here $\mathbf{K}^1$ is the stiffness matrix of element AB denoted as element No. 1. The stiffness matrix consists of four sub-matrices $\mathbf{K}_{AA}$, $\mathbf{K}_{AB}$, $\mathbf{K}_{BA}$ and $\mathbf{K}_{BB}$ as indicated in (7.28). The first subscript indicates the node of the resulting generalized force, when unit displacements are imposed at the node indicated by the second subscript. The corresponding sub-matrices of element No. 1 are located at the first and second row and column, associated with the element nodes A and B. Element No. 2 is associated with nodes B and C and the sub-matrices therefore enter the rows and columns 2 and 3.

$$
\mathbf{K}^1 = \begin{bmatrix} \mathbf{K}_{AA} & \mathbf{K}_{AB} & - & - & - \\ \mathbf{K}_{BA} & \mathbf{K}_{BB} & - & - & - \\ - & - & - & - & - \\ - & - & - & - & - \\ - & - & - & - & - \end{bmatrix}, \quad \mathbf{K}^2 = \begin{bmatrix} - & - & - & - & - \\ - & \mathbf{K}_{BB} & \mathbf{K}_{BC} & - & - \\ - & \mathbf{K}_{CB} & \mathbf{K}_{CC} & - & - \\ - & - & - & - & - \\ - & - & - & - & - \end{bmatrix},
$$

$$
\mathbf{K}^3 = \begin{bmatrix} - & - & - & - & - \\ - & \mathbf{K}_{BB} & - & \mathbf{K}_{BD} & - \\ - & - & - & - & - \\ - & \mathbf{K}_{DB} & - & \mathbf{K}_{DD} & - \\ - & - & - & - & - \end{bmatrix}, \quad \mathbf{K}^4 = \begin{bmatrix} - & - & - & - & - \\ - & - & - & - & - \\ - & - & \mathbf{K}_{CC} & - & \mathbf{K}_{CE} \\ - & - & - & - & - \\ - & - & \mathbf{K}_{EC} & - & \mathbf{K}_{EE} \end{bmatrix}. \tag{7.63}
$$

Element No. 3 is the beam/column connecting nodes B and D, and in this case the sub-matrices are therefore located in the rows and columns 2 and 4. Similarly for element No. 4 connecting nodes C and E. The actual assembling process starts with an empty structure stiffness matrix $\mathbf{K}$, and then loops over all beam elements, adding each of the four element sub-matrices directly into its correct global position. There is no need to form the element matrices explicitly in the global format. The association of the two nodes of a beam element with global node numbers is described in the topology matrix $\mathbf{T}$.

The resulting equilibrium equations are of the form

$$\mathbf{K}\,\mathbf{u} = \mathbf{f}\,, \tag{7.64}$$

where $\mathbf{K}$ is the global stiffness matrix, $\mathbf{u}$ is the global displacement vector, and $\mathbf{f}$ is the global load vector. The loads can either be associated with a node or an element. While nodal loads are entered directly into the global equations, element based loads must be defined in connection with an element, and then translated into equivalent nodal loads. This procedure makes use of the constraint forces illustrated in Table 7.2. Later, when calculating the section forces the contribution from element based loads also need representation of the local variation corresponding to the load distribution within the element.

The support conditions, constraining the displacement at nodes, are implemented as for the case of truss structures in Section 2.5. A simple, but approximate procedure, is to retain the full equation system and then introduce a stiff spring as a diagonal term for each constrained degree of freedom. Alternatively, the constrained degrees of freedom can be eliminated by removing the corresponding rows and columns from the equation system. The reactions can then be recovered from the force components generated by the removed rows of the stiffness matrix, as explained in detail in Section 2.5.2.

7.4.1 The *MiniFrame* program

The principles described in the previous sections have been implemented in a small Finite Element program MiniFrame using the high level programming language Matlab. The structure of the program is similar to that of the MiniTruss program described in Section 2.5.3, while element details, external loads and internal forces are new here. The main features of the program and its data structure are explained in relation to the specific frame shown in Fig. 7.44 and already analyzed by the deformation method in Section 7.2.

The program is built as a script file `MiniFrame.m` that serves as a driver that reads a data file and activates subroutines that set up the model, form the global stiffness matrix, apply the load, and solve the constrained equations for the displacement of all nodes of the supported structure. The structure

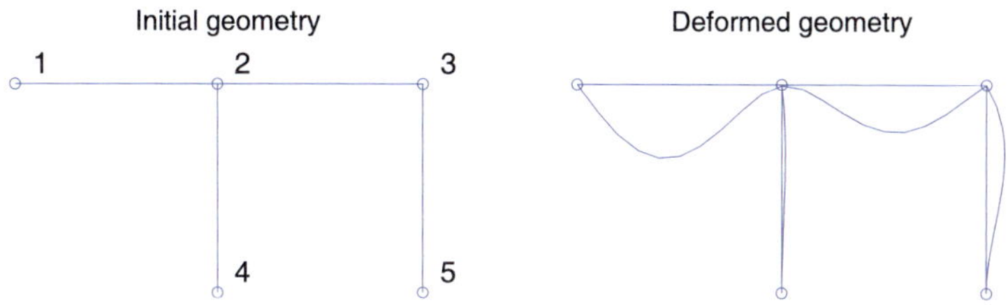

Fig. 7.45: Frame geometry plots: a) Initial, b) Deformed.

and content of the data file corresponding to the frame in Figs. 7.44–7.46 are described in the following.

Node data. The frame structure is described in an xy-coordinate system with the horizontal x-axis through the column support points D and E and the vertical y-axis vertical through the left support at A. For the purpose of illustration the horizontal distances are given in terms of a, and the vertical distances in terms of b. The node coordinates are given in the array `X`, with each node corresponding to one row. The first part of the data file then is

```
% Width 'a' and height 'b' of truss
a = 4.0; b =  4.0;

% Coordinates of nodes X = [x y],
X = [  0.0      b
        a       b
       2*a      b
        a      0.0
       2*a     0.0  ];
```

The node coordinates `[x,y]` are given in the order of the node number, starting with node 1. Thus, the node number is not given explicitly, but implied by the row number in the node coordinate matrix `X`.

Element data. The beam elements are defined in the topology matrix `T`. Each row of this matrix defines an element, by listing its two nodes by their node number, and by giving a third number identifying a set of element properties, given as a row of parameter values in the element properties matrix `H`.

```
% Topology matrix T = [node1 node2 propno],
T = [ 1  2   1
      2  3   1
      2  4   2
      3  5   2 ];
```

```
% Element property matrix H = [ E A I G As ],
H = [ 2.1e11  1.00e-3  1.00e-6  8.1e10  0.80e-3
      2.1e11  1.00e-3  1.00e-6  -1.00   -1.00   ];
```

The element properties consist of the elastic modulus `E` and the section area `A` and moment of inertia `I`. If the element includes shear flexibility, the shear modulus `G` and the shear area `As` must also be included in the element data. In classic analysis the shear flexibility effect is often neglected. This corresponds to the condition of infinitely large shear stiffness `G*As`. In the data file this condition is identified by a non-positive shear parameter, i.e. by `G` ≤ 0 or `As` ≤ 0 as illustrated for material No. 2 used for the columns in the sample file. The shear flexibility parameter Φ is calculated in the function `kebeam`, when forming the beam element stiffness matrix.

Loads. The loads can be given as concentrated loads at the nodes or as distributed transverse loads with uniform distribution within the length of a beam element. The concentrated loads are specified in the load matrix `P`. This matrix contains a row for each loaded node. The data row specifies the node number and the generalized force components. In the present example there are no concentrated loads. Thus the following line of code describing a concentrated downward vertical force at node No. 2 is just an illustration.

```
% Prescribed loads P = [ node Px Py M ]
P = [ 2   0.000 -1.00e4 0.000 ];
```

The elements can support a uniform transverse load specified by the load intensity p. The transverse downward load on the frame in Fig. 7.44 is generated by the input array

```
% Prescribed loads p = [ elno p ]
p = [ 1  2.0e4
      2  2.0e4 ];
```

The load data arrays are processed by the function `febeam` and the specific load components are entered into the global load vector `f`. The program MINIFRAME accepts imposed nodal displacements as input, and thus it may happen that there are no loads in the form of generalized forces. The code therefore checks for the existence of data arrays `P` and `p`.

Support conditions. The support conditions are given in the constraint matrix `C`. The constraint matrix contains a row for each constrained generalized displacement component. In the present example there are 2 constrained displacement components at node 1, and 3 generalized displacement constraints at the nodes 4 and 5.

```
% Constraints C = [ node 'dof' (uc) ]
C = [ 1  1
      1  2
      4  1
      4  2
      4  3
      5  1
      5  2
      5  3 ];
```

The last column is optional. In the case of an imposed displacement it contains the magnitude of the constrained displacement uc.

Graphics. The MiniFrame program produces two plots of the structure: a plot of the initial geometry without deformation including node numbers, and another plot of the deformed structure after application of the load, Fig. 7.45. The displacements are scaled to give a visual impression of the deformation, that would typically not be directly visible. The coordinate window used for the plots is controlled via definition of the plot axes, specified in the array

```
% Axes used for geometry plots [Xmin Xmax Ymin Ymax]
PlotAxes = [-0.40*a 2.50*a  -0.25*a 1.40*a];
```

The deformed geometry is plotted using the computed nodal displacements, accounting for the difference between cross-section and center line rotation at the element ends by using the shear force and the equivalent shear strain.

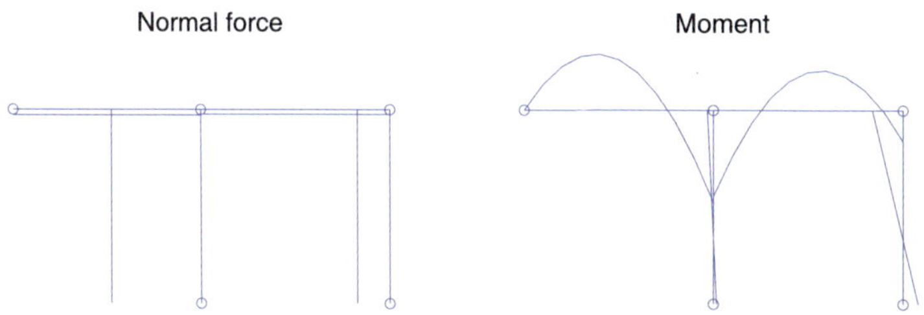

Fig. 7.46: Internal force distributions: a) normal force, b) internal moment.

The MiniFrame program also provides plots of the internal force and moment distributions within the elements as shown in Fig. 7.46. The internal forces at the ends of the element are calculated from the element stiffness matrix, and the moment is corrected with a parabolic variation in the case of a transverse load on the element.

Analysis process. The analysis procedure using the MiniFrame program is quite similar to that of MiniTruss. The first step is to read the appropriate data file into memory, either by writing the data file name DoubleFrame in the command window, or by uploading `DoubleFrame.m` to the MATLAB editor and pressing the `F5` key from the editor. The data is now available in active memory and the analysis is carried out by activating the script file `MiniFrame.m` from the command window or by pressing the `F5` key with `MiniFrame.m` in the editor. A sample of the file `MiniFrame.m` is shown below.

```
% Nodal loads into load vector
if exist('P','var')
    f = loadnode(f,P,dof);
end

% Element loads into load vector
if exist('p','var')
    f = loadelem(f,p,T,X,dof);
end

% Global stiffness matrix
K = kbeam(T,X,H,dof);

% Solve stiffness equation
[u,r,ic] = solveeq(K,f,C,dof);

% Nodal displacements
Un = reshape(u,dof,size(X,1))';

% Element section forces
Se = sbeam(T,X,H,Un,11);
if exist('p','var')
    Se = sebeam(Se,p,T,X);
end

% Element displacements
Ue = ubeam(T,X,H,Un,Se,11);
if exist('p','var')
    Ue = uebeam(Ue,p,T,X,H);
end
```

The program activates the following processes: i) builds up the full load vector `f` from nodal loads `P` and transverse element loads `p`, ii) generates the structure stiffness matrix `K`, iii) solves the constrained equation system including the support conditions, iv) presents the generalized node displacements, in the column format `u` and in matrix format `Un`, and finally v) computes the internal forces `Se` and the displacements `Ue` in the beam elements.

7.4.2 Stability analysis of frames

A compressive normal force in a beam reduces its stiffness as discussed in Chapter 5 in connection with column stability. The same effect may be important in a frame structure, where the stiffness reduction in symmetric and in anti-symmetric bending of an individual member are expressed in terms of the coefficients φ and ψ, illustrated in Fig. 7.40. A direct way to include this stiffness reduction is to use the beam-column element stiffness matrix (7.49), whereby the global equilibrium equations take the form

$$\mathbf{K}(N_e)\,\mathbf{u} = \mathbf{f}\,, \tag{7.65}$$

where the notation $\mathbf{K}(N_e)$ indicates that the stiffness matrix depends on the normal forces in the individual elements. In a statically indeterminate structure the internal forces, including the normal force, depend on the stiffness of the elements and thereby the equilibrium equation (7.65) becomes non-linear.

In practice, the problem is often solved without use of non-linear analysis by introducing the linearized form (7.53), in which the stiffness matrix is represented as the sum of the constitutive stiffness matrix $\mathbf{K}^c$ and the geometric stiffness matrix $\mathbf{K}^g$. This gives the global equilibrium equations as

$$[\,\mathbf{K}^c + \mathbf{K}^g\,]\,\mathbf{u} = \mathbf{f}\,. \tag{7.66}$$

This set of equations is then solved approximately by assuming that the redistribution of the normal forces within the structure is only moderate. The load is then represented in the form $\alpha\mathbf{f}_0$, where the scalar variable α acts as a load factor. The analysis then proceeds in three steps. First the displacements $\mathbf{u}_0$ are calculated without accounting for geometric stiffness effects by solving the equilibrium equations when including only the constitutive stiffness,

$$\mathbf{K}^c\,\mathbf{u}_0 = \mathbf{f}_0\,. \tag{7.67}$$

Then the normal forces in the elements are calculated from the constitutive element stiffness matrices for the displacement field $\mathbf{u}_0$. Finally, the critical value of the load factor α, at which the structure looses its stiffness and buckles is obtained from the eigenvalue problem

$$[\,\mathbf{K}^c + \alpha\mathbf{K}^g\,]\,\mathbf{u} = \mathbf{0}\,. \tag{7.68}$$

The smallest eigenvalue α determines the lowest ideal stability load $\alpha\mathbf{f}_0$ and the corresponding buckling mode $\mathbf{u}$, when neglecting deformation before buckling. This is equivalent to the elastic stability load introduced in the design procedure for columns in Section 5.3.

The linearized stability problem is implemented as an extension of the MiniFrame program called MiniFrameS. This program first solves the linear initial problem (7.67) with a reference load $\mathbf{f}_0$ using the procedure described in connection with the MiniFrame program in the previous section. This solution determines the normal force in each of the elements, and these normal forces are used to form the global geometric stiffness matrix $\mathbf{K}^g$. The eigenvalue problem (7.68) is then solved for the load factor α, which appears as output together with a plot of the corresponding buckling mode. The linearized stability analysis is illustrated in the following two examples.

Example 7.4. Number of elements in beam-columns. The linear eigenvalue problem 7.68 is based on a linearized form of the stiffness matrix. This introduces an approximation regarding the shape of the buckling modes, and it is important to determine the appropriate number of elements in a finite element model for a buckling problem. This is illustrated with reference to a column with a fixed support at one and and a simple support at the other. Shear flexibility is neglected in the present example. The solution of this problem was treated in Example 5.2, where the buckling load was found by an iterative procedure as $P_c = 2.0457\, P_E$, with P_E as the Euler buckling load for the similar simply supported column.

Fig. 7.47: Ideal column with fixed/simple support: a) initial geometry, b) buckling mode.

The Finite Element model is illustrated in Figure 7.47 with $n_{el} = 4$ elements, showing graphs from MINIFRAMES. The left end is fixed, while the right end has a simple support permitting motion in the axial direction as well as rotation. It is seen that the shape of the buckling mode has an inflexion point and this places extra demands on the shape functions in the element model.

n_{el}	1	2	3	4
P/P_c	1.486	1.026	1.006	1.002
ℓ_e/a_{el}	0.574	1.381	2.091	2.794

The results are summarized in the table for $n_{el} = 1, 2, 3, 4$ elements. The second row gives the buckling load P obtained by the Finite Element model, normalized with the analytically determined critical load P_c. The third row gives the ratio of the effective column length ℓ_e to the element length a_{el}. This number indicates the number of elements per effective column length and thus has relevance for general frame structures, where the effective column length of a member is simply the distance between inflexion points of the buckling mode. The results indicate that use of a single element is clearly insufficient, providing less than one element per effective column length and leading to an overestimation of the critical load by about 50 pct. The results indicate that an accurate determination of the buckling load requires 3 elements in the present case, corresponding to two or more elements per effective column length. For a column with two fixed supports this corresponds to a minimum of four elements. Thus, the simplicity of the linearized analysis is attained at the cost of a more detailed element model. □

Example 7.5. Buckling of angle-frame. The finite element model of an angle frame with fixed simple supports is shown in Fig. 7.48a. The horizontal and vertical dimension is a. The load consists of a uniformly distributed downward load of intensity p on the horizontal part of the frame, i.e. a total load of pa. The load is characterized by the non-dimensional load factor $\alpha = \frac{1}{2}pa/P_E$, where P_E is the Euler load of a simply supported column with the same properties as the vertical part of the fame. Shear flexibility is neglected, and the effect of axial strain is negligible. The results therefore only depend on the load factor α.

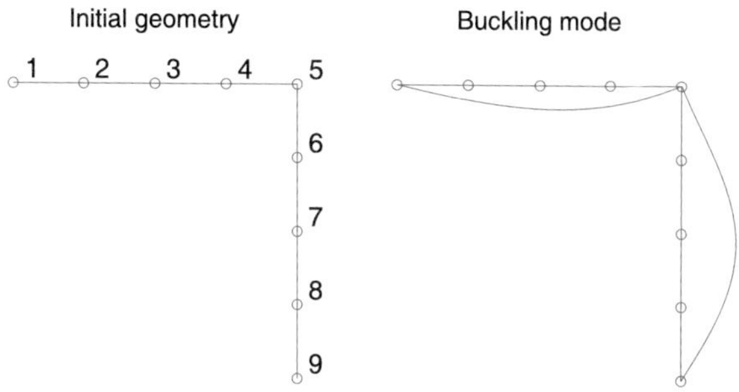

Fig. 7.48: Angle frame with simple supports and uniformly distributed load.

The effect of the number of elements is illustrated via the buckling mode in Fig. 7.48b and the buckling load shown in tabular form for $n_{el} = 1, 2, 3, 4$ together with the buckling load at convergence, obtained for $n_{el} \geq 15$. In principle only the vertical member needs to be subdivided, as the horizontal member is nearly without normal load and the stiffness is therefore correctly represented by the constitutive stiffness matrix.

n_{el}	1	2	3	4	≥ 15
P/P_E	1.617	1.238	1.229	1.227	1.226

The first part of the analysis consists in determination of displacements and internal forces, when neglecting the flexibility effect of the normal force. The moment distribution is shown in Fig. 7.49a. It consists of a linear variation vanishing at the supports and with corner value $-M_0$, supplemented by a parabolic variation along the horizontal member. When neglecting the effect of axial strain the corner moment is easily determined by the deformation method as $M_0 = \frac{1}{16}pa^2$.

When increasing the load the vertical member becomes increasingly flexible, whereby the magnitude of the corner moment decreases. The rotation of the corner node is constrained by the horizontal member, and thus the vertical member can support more than the Euler load P_E. When passing P_E the rotation stiffness becomes negative, and the corner moment changes sign. Thus, the corner moment becomes positive before the buckling load p_c is reached. The development of the corner moment with the normalized load p/p_c can be

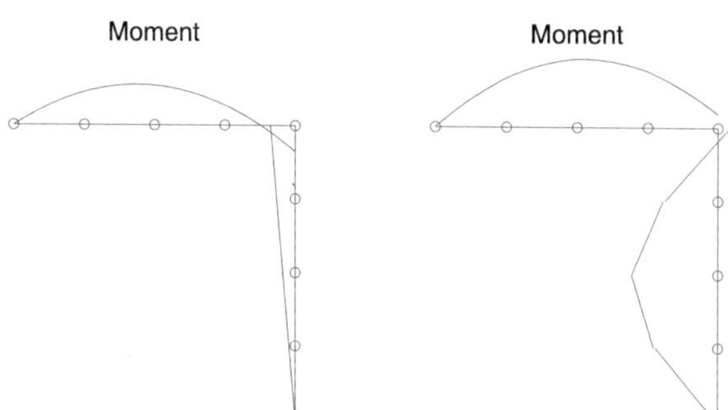

Fig. 7.49: Normalized moment distribution: a) $p/p_c \simeq 0$, b) $p/p_c = 0.8$.

calculated by using the full stiffness matrix $\mathbf{K}^c + \alpha\mathbf{K}^g$ and is shown in the following table.

p/p_c	0.1	0.2	0.3	0.4	0.5	0.6	0.7	0.8	0.9
M/M_0	0.99	0.95	0.90	0.75	0.62	0.43	0.12	−0.52	−2.43

The gradual transition of the corner moment from a negative value, constraining the horizontal beam, to a positive value, constraining the vertical column, is an example of redistribution of the internal forces due to change in stiffness of the various parts of the structure with increasing normal load. This is a non-linear effect, and a better representation can be obtained by recalculating the normal force from the full stiffness matrix in an iterative procedure. □

7.5 Exercises

Exercise 7.1. The figure shows a statically indeterminate frame structure.

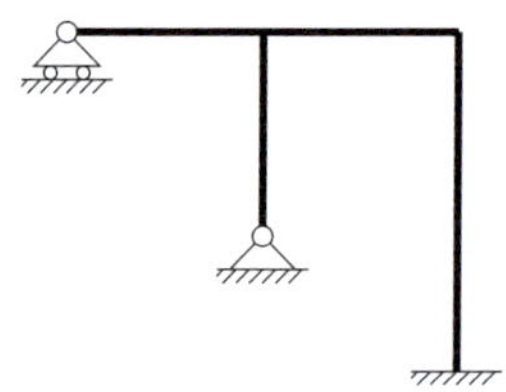

a) Identify by how many components the structure is statically indeterminate, and show a set of redundant forces for use with the force method.

b) Show the kinematic degrees of freedom to be used with the deformation method.

c) Is it advantageous to use the deformation method for this frame?

d) Introduce an additional horizontal constraint at the left support, and repeat a)–c).

Exercise 7.2. The figure shows four statically indeterminate beam and frame structures, and the following questions are answered for each of the structures.

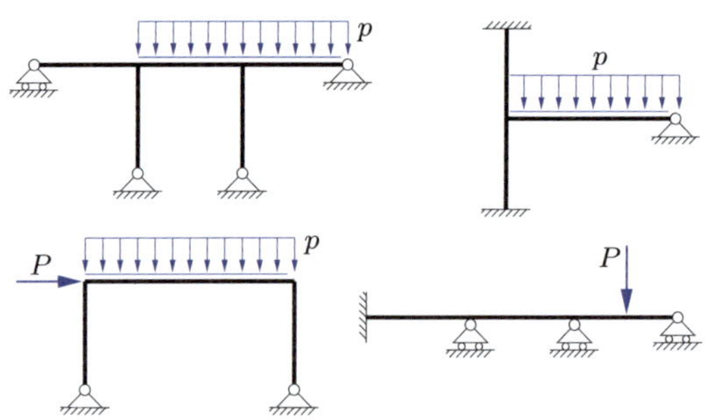

a) Identify by how many components the structure is statically indeterminate, and show a possible set of generalized forces for use with the force method.

b) Show the kinematic degrees of freedom to be used with the deformation method.

c) Is it advantageous to use the deformation method for this frame.

Exercise 7.3. Consider the load case Fig. 7.5, in which the cross-section A of a beam AB with a simple support at B is rotated $\theta_A = 1$. Determine the rotation θ_B at the simple support, including the effect of shear flexibility.

Exercise 7.4. The figure shows a continuous beam over two spans with a fixed support in A and simple supports with horizontal rollers in B and C. The length of AB is ℓ, while the length of BC is $\alpha\ell$. The beam is loaded by a uniformly distributed load with vertical intensity p. The bending stiffness for all beams is EI and the influence of shear flexibility is neglected ($\Phi = 0$).

a) Use the deformation method to determine the rotation at B.

b) Determine the reactions and the moment distribution in the frame for the special case $\alpha = 1$.

c) The rotation at B can be clockwise or counterclockwise, depending on the magnitude of α. Determine the sign conditions.

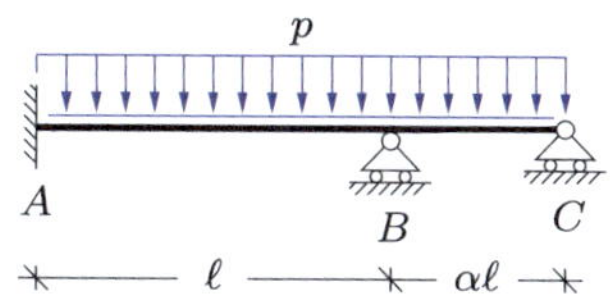

Exercise 7.5. The figure shows a T-frame. The three beams of length a and bending stiffness EI are joined rigidly at B and have simple supports at A, C and D, respectively. The shear flexibility effect is neglected ($\Phi = 0$).

a) Use the deformation method to determine the rotation at B.

b) Determine all reactions on the frame.

c) Determine the distribution of moment, shear force and normal force in the frame.

d) Replace the simple supports in A and D by fixed supports and repeat the questions in a)–c) for the modified structure.

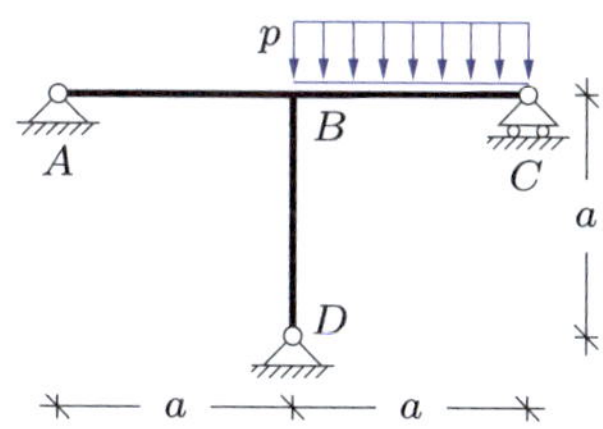

Exercise 7.6. The figure shows an angle frame similar to that in Example 7.1, but now loaded by a distributed load p on the horizontal beam BC. Both beams have length a and bending stiffness EI.

a) Use the deformation method to determine the rotation at B, when assuming $\Phi = 0$.

b) Determine the reactions and the moment distribution in the frame.

c) Include the effect of shear flexibility, $\Phi > 0$, and repeat the analysis in a) and b).

d) Replace the simple support in C by a fixed support and repeat the analysis in a) and b).

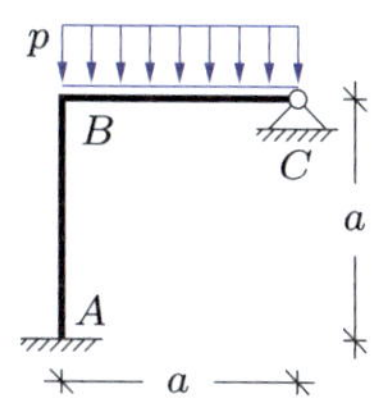

Exercise 7.7. Change the load in the angle frame in Exercise 7.6 to a concentrated vertical downward force P acting at the center of BC, and repeat the analysis in a) and b) without the influence of shear flexibility.

Exercise 7.8. The figure shows an angle frame with a fixed support in A and a simple support in C. Length and bending stiffness are $2a$ and EI for the vertical beam AB and a and αEI for the horizontal beam BC. The frame is loaded by a distributed load with horizontal intensity p on the vertical beam AB. The influence of shear flexibility is neglected in the analysis ($\Phi = 0$).

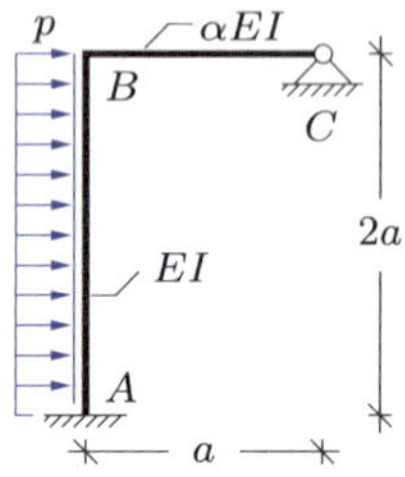

a) Use the deformation method to determine the rotation at B for $\alpha = \frac{1}{2}$.

b) Determine the reactions and the moment distribution in the frame.

c) Determine the distribution of the shear force and the normal force in the frame.

d) Determine the rotation at B for $\alpha \to 0$, and comment on the associated reduction in rotational stiffness.

Exercise 7.9. The figure shows a T-frame with a fixed support in A and simple supports in D and E. The elements AB, BC, CD and CE all have length a and bending stiffness EI. The frame is loaded by a vertical force P at B. The influence of shear flexibility is neglected ($\Phi = 0$).

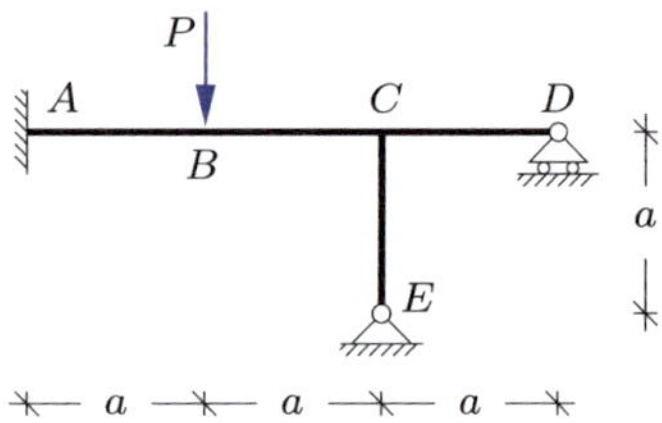

a) Use the deformation method to determine the rotation of the joint C.

b) Determine all reactions on the frame.

c) Determine the moment distribution.

d) Determine the distribution of the shear and normal force.

Exercise 7.10. The figure shows a frame consisting of a column ABC supporting a horizontal beam DB loaded by a uniformly distributed load with intensity p. The frame has a fixed support in A and fixed simple support in D and a simple support with vertical rollers in C. The beam elements have the lengths $|AB| = |BC| = a$ and $|DB| = 2a$, and bending stiffness EI. The influence of shear flexibility is neglected in the analysis.

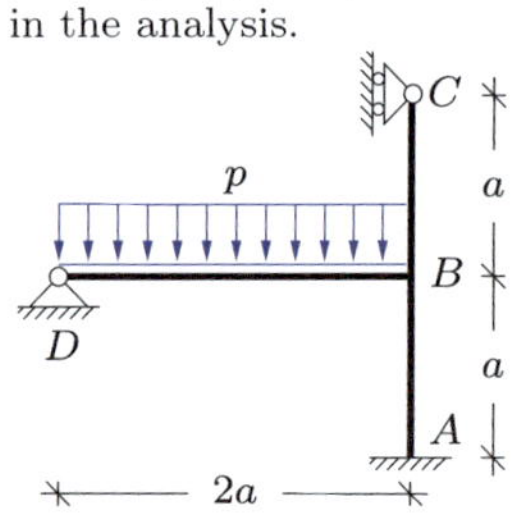

a) Use the deformation method to determine the rotation of the joint B.

b) Determine all reactions on the frame.

c) Determine the moment distribution in the frame.

d) Determine the distribution of shear force and normal force in the frame.

Exercise 7.11. The figure shows a beam structure with a fixed support in A, intermediate transverse supports in B and C, and a free end in D with a transverse tip load P. The length of each span is a and the bending stiffness is EI for the entire beam. The effect of shear flexibility is neglected ($\Phi = 0$).

a) Determine the moment M_C at the intermediate support in C.

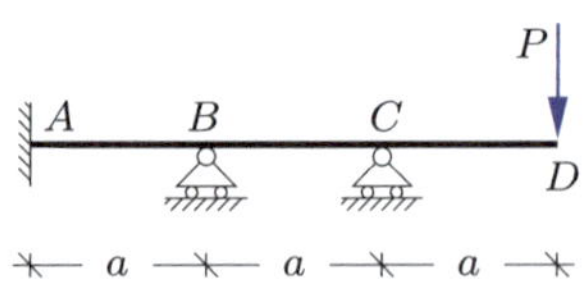

b) Use the deformation method to determine the rotation at the intermediate support in B.

c) Determine the reactions.

d) Determine the moment and shear force distribution in the beam.

Exercise 7.12. The figure shows a beam structure with a fixed support in A and simple supports with horizontal rollers in B, C and D. The beam is loaded by a uniformly distributed load p on the two outer spans BCD. The length of each span is a and the bending stiffness is EI for the entire beam. The effect of shear flexibility is neglected ($\Phi = 0$).

a) Use the deformation method to determine the rotations at the intermediate supports in B and C.

b) Determine the reactions and the moment distribution in the beam.

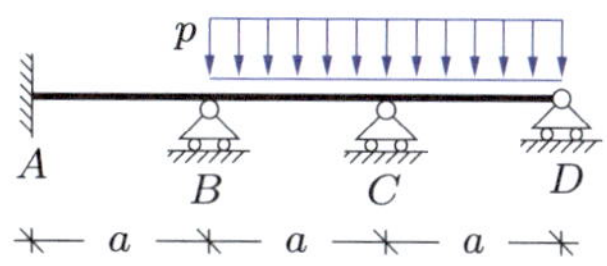

Exercise 7.13. The figure shows a T-frame similar to that in Exercise 7.5, where the left simple support now permits horizontal motion.

a) Use the deformation method to determine the rotation and horizontal displacement at B.

b) Determine all reactions on the frame.

c) Determine the moment distribution in the frame.

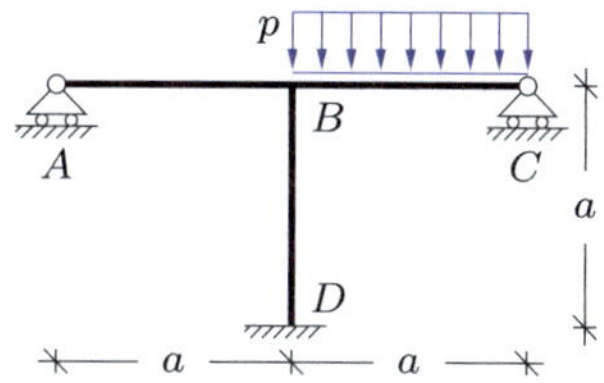

Exercise 7.14. The figure shows a frame similar to that in Exercise 7.10, where the rollers at the support in D now permits horizontal motion of beam DB.

a) Use the deformation method to determine the rotation and the horizontal displacement at B.

b) Determine all reactions on the frame.

c) Determine the moment distribution in the frame.

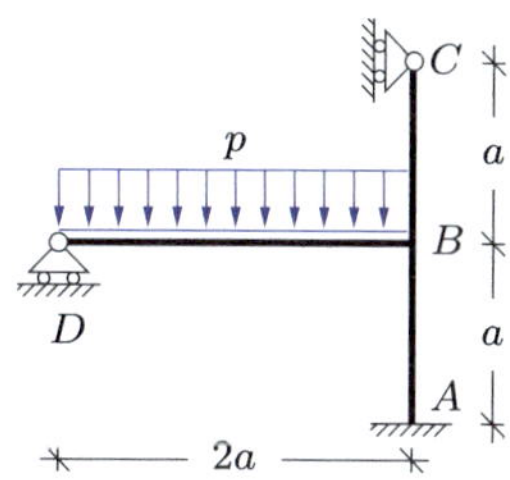

Exercise 7.15. Consider the angle frame in Example 7.1 with $E = 210 \cdot 10^9$, $A = 10^{-3}$, $I = 10^{-6}$, $a = 10$ and $M_0 = 10^3$.

a) Create a data file `AngleFrameM0.m` to be used in `MiniFrame`.

b) Determine the rotation at the corner of the frame and compare with θ_0 in Example 7.1.

c) Determine the reactions and the internal moments at the corner, and compare with the results obtained in Example 7.1.

d) Plot the distribution of the internal forces M, Q and N and compare with the diagrams in Fig. 7.11.

Exercise 7.16. Consider the angle frame with distributed load in Exercise 7.6. Use the following parameters: $E = 210 \cdot 10^9$, $A = 10^{-3}$, $I = 10^{-6}$, $a = 10$ and $p = 100$.

a) Create a data file `AngleFramep.m` to be used in `MiniFrame`.

b) Find the magnitude and location of the maximum transverse displacement and the maximum moment.

c) Take the influence of shear flexibility into account with $G = E/2.6$ and $A_s = 0.8A$. Compare with the results in b).

Exercise 7.17. Consider the frame in Exercise 7.10 and use the following parameters: $E = 210 \cdot 10^9$, $A = 10^{-3}$, $I = 10^{-6}$, $a = 10$ and $p = 100$.

a) Create a data file `FrameStorey.m` to be used in `MiniFrame`.

b) Find the reactions on the frame.

c) Find the magnitude and location of the maximum transverse displacement and the maximum moment.

d) Permit the simple support to move horizontally and compare with the results in Exercise 7.14 and in c).

Exercise 7.18. Consider the T-frame in Exercise 7.5 and use the following parameters: $E = 210 \cdot 10^9$, $A = 10^{-3}$, $I = 10^{-6}$, $a = 10$ and $p = 100$.

a) Create a data file `TFrame.m` to be used in `MiniFrame`.

b) Find the reactions on the frame.

c) Plot the distribution of the section forces and find the maximum moment.

Exercise 7.19. Consider the T-frame in Exercise 7.13 and use the following parameters: $E = 210 \cdot 10^9$, $A = 10^{-3}$, $I = 10^{-6}$, $a = 10$ and $p = 100$.

a) Create a data file `TFrameSway.m` to be used in `MiniFrame`.

b) Find the reactions on the frame.

c) Plot the distribution of the section forces and find the maximum moment.

d) Take the influence of shear flexibility into account with $G = E/2.6$ and $A_s = 0.8A$. Compare with the results in b).

Stresses and Strains | 8

In structural elements like beams and frames, statics and equilibrium are formulated in terms of section forces and moments. The section forces and moments represent the accumulated effect of local force distributions expressed in terms of so-called stresses. Similarly the deformation of structural elements is expressed in terms of e.g. extension and curvature. These deformations lead to deformation at the local level inside the structural element, and these deformations are expressed in terms of strain. The mechanics of deformable bodies is called continuum mechanics and is based on the notion of a *state* of the material at each individual point of the body expressed in terms of stresses and strains. Some simple states of stresses and strains have been introduced for bars, beams and columns already. This chapter gives a general presentation of stresses and strains within a material body. The relation between the stresses and strains depends on the particular material under consideration and is governed by the so-called constitutive relations of the material, e.g. linear elasticity. While the present chapter lays the ground by defining general properties of stresses and strains, the following chapter gives a brief description of linear elastic material behavior, and presents some commonly used criteria for material failure. The presentation is deliberately kept at the level of a complement to the theory of structures. More detailed accounts of continuum mechanics can be found e.g. in Mase and Mase (1999) and Fung (1965).

S. Krenk, J. Høgsberg, *Statics and Mechanics of Structures*,
DOI 10.1007/978-94-007-6113-1_8,

Section 8.1 gives a precise definition of stress, that is valid for general loading conditions, and discusses the conditions on the stresses necessary for equilibrium of the body. Deformation and its precise description in terms of strains is the subject of Section 8.2. Here it is also explained, how a general but small displacement of a deformable body can be resolved as the sum of translation, rotation and strain. The equilibrium equations constitute a set of balance equations for the external load and the internal stresses. These equations can be rewritten as a statement about balance of external and internal work by multiplication with a displacement field. The resulting principle of virtual work is derived in Section 8.3. The principle of virtual work plays a central role in the formulation of numerical theories for analysis of continuous bodies, such as the finite element method. Special states of stress and strain find wide application in engineering practice. The states of plane stress and plane strain are discussed in detail in Section 8.4, that also describes the representation of a general state of stress or strain in terms of its principal components and the associated graphical representation by the Mohr circle diagram.

8.1 Stress

The simple case of normal stress on a cross-section with area A of a homogeneous bar was discussed in Section 2.4.1. The idea was to consider a subdivision in the form of parallel bars each carrying their part of the force N, and this lead to the idea of a normal stress $\sigma = N/A$. This is a special instance of the concept of stress, treated in detail in the following. The idea of similarity, when joining similar bodies or changing their dimensions by scaling was originally introduced by GALILEI (1564–1642), while the introduction of the general concept of stress is due to CAUCHY (1789–1857), see e.g. Timoshenko (1983).

8.1.1 The stress vector

Figure 8.1a shows a continuous body, here in the form of a beam, that is subjected to external loads. In order to characterize the distribution of the load through the body, the body is separated into two parts by a section as indicated in the figure. Figure 8.1b shows one of the parts created by the section. The figure shows a small area dA in the section together with the force $d\mathbf{T}$ acting on this area. The stress vector at a point within this area of the section is defined as the limit of the ratio of the force $d\mathbf{T}$ to the area dA on which it acts,

$$\mathbf{t} = \lim_{dA\to 0} \frac{d\mathbf{T}}{dA} = \frac{\text{Force}}{\text{Area}}, \qquad \Big[\frac{\text{N}}{\text{m}^2}\Big] = \big[\,\text{Pa}\,\big]. \tag{8.1}$$

The units are shown in square brackets. When the magnitude of the force is expressed in Newton [N] and the area in square meters [m^2] the corresponding unit for the magnitude of the stress is Pascal [Pa]. The force and stress vectors have the same units as the corresponding magnitude of the vector.

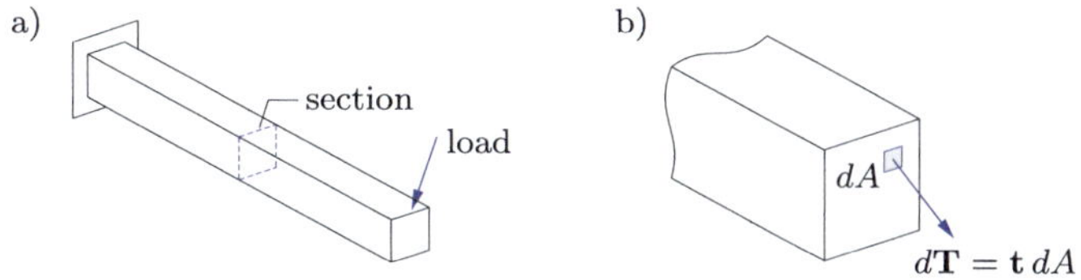

Fig. 8.1: Stress vector $\mathbf{t}$ as normalized force per area.

By the rule of equality of action and reaction the two parts of the body must act upon each other by force distributions over the section that are equal in magnitude but opposite in direction. Thus, if the selected part of the body is acted upon by a stress vector $\mathbf{t}$ at a point of the section, the other part of the body must be acted upon by the stress vector $-\mathbf{t}$. A particular instance is the case where the section is part of the surface of the body, in which case the stress vector $\mathbf{t}$ at a particular point of the surface has the character of an external surface load, such as for example pressure created on a structure by wind or water.

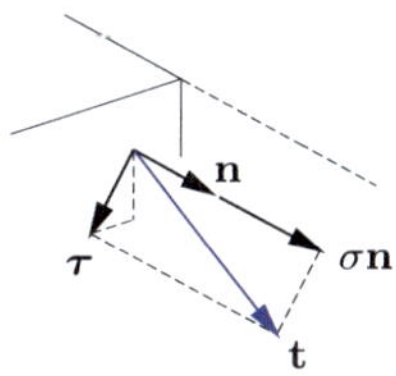

Fig. 8.2: Normal stress σ and shear stress vector $\boldsymbol{\tau}$.

In many problems the direction of the stress vector acting on a section is of importance. Figure 8.2 shows a small part of a section with outward unit normal vector $\mathbf{n}$. The normal stress is defined as the projection of the stress vector on the normal. Thus, the normal stress has the magnitude

$$\sigma = \mathbf{n}^T\mathbf{t}\,. \tag{8.2}$$

By the definition of the normal vector $\mathbf{n}$ as pointing outward it is seen that positive normal stress corresponds to tension. The remaining part of the stress vector, lying in the surface, is called the shear stress. This part is found by subtracting the normal stress vector $\sigma\mathbf{n}$ from the total stress vector $\mathbf{t}$ as shown in the figure. Thus, the shear stress vector is

$$\boldsymbol{\tau} = \mathbf{t} - \mathbf{n}\sigma\,. \tag{8.3}$$

The resolution into normal and shear stress components depends on the orientation of the section as illustrated in the following example.

Example 8.1. Stress on inclined section. Figure 8.3a shows a tension test specimen with tension force P and cross-section area A. The figure shows a section inclined by the angle θ, and thus the area of this section is $A_\theta = A/\cos\theta$.

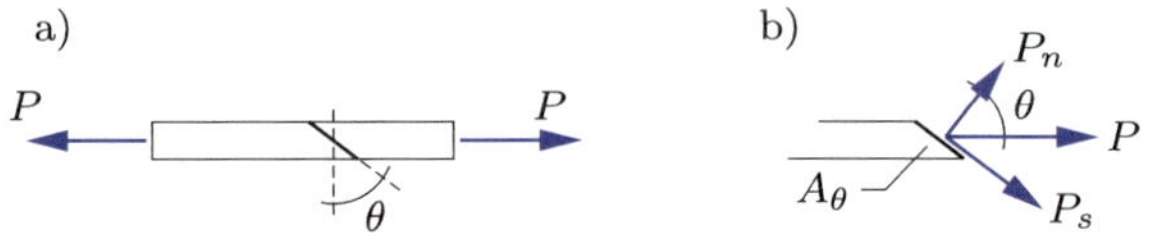

Fig. 8.3: Stress vector on inclined section in tension bar.

Resolution of the force P in the normal and in-plane directions gives the force components

$$P_n = P\cos\theta\,, \qquad P_s = P\sin\theta$$

as shown in Fig. 8.3b. The normal and shear stresses are obtained by normalizing these force components with respect to the area of the inclined section A_θ,

$$\sigma = \frac{P_n}{A_\theta} = \frac{P}{A}\cos^2\theta\,, \qquad \tau = \frac{P_s}{A_\theta} = \frac{P}{A}\cos\theta\,\sin\theta\,.$$

Note, that the stress components σ and τ are not found by the ordinary rule of vector projection, but include products of trigonometric functions. □

8.1.2 General stress components

The full three-dimensional state of stress at a point $\mathbf{x} = [x, y, z]^T$ inside the body is described in terms of components by considering a small cube with sides $d\mathbf{x} = [dx, dy, dz]^T$ as illustrated in Fig. 8.4. A complete description of the state of stress at the point $\mathbf{x}$ requires the stress vectors on each of the sides of the cube. In a homogeneous state of stress this amounts to the stress vectors $\mathbf{t}_x$, $\mathbf{t}_y$ and $\mathbf{t}_z$ acting on the planes orthogonal to the coordinate directions x, y and z through the point $\mathbf{x}$. These three stress vectors are shown on the corresponding 'front' faces of the cube in Fig. 8.5a.

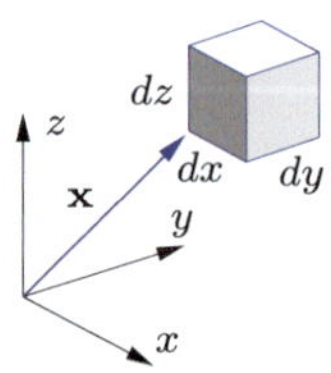

Fig. 8.4: Infinitesimal cube $d\mathbf{x}$ located at $\mathbf{x}$.

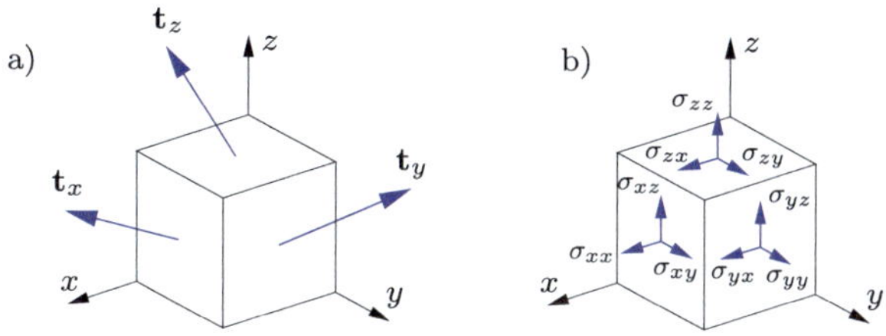

Fig. 8.5: a) Stress vectors $\mathbf{t}_x, \mathbf{t}_y, \mathbf{t}_z$, b) Stress components $\sigma_{xx}, \sigma_{xx}, \cdots$.

Each of the three stress vectors $\mathbf{t}_x$, $\mathbf{t}_y$ and $\mathbf{t}_z$ are represented by three components as illustrated in Fig. 8.5b. Thus, for instance the components of the stress vector $\mathbf{t}_x$ are $[\sigma_{xx}, \sigma_{xy}, \sigma_{xz}]$, where the first subscript refers to the coordinate direction defining the section, while the second subscript indicates the direction of the component. The components of the three stress vectors $\mathbf{t}_x$, $\mathbf{t}_y$ and $\mathbf{t}_z$ are traditionally arranged as the rows of the 3×3 stress component matrix

$$\boldsymbol{\sigma} = \begin{bmatrix} \sigma_{xx} & \sigma_{xy} & \sigma_{xz} \\ \sigma_{yx} & \sigma_{yy} & \sigma_{yz} \\ \sigma_{zx} & \sigma_{zy} & \sigma_{zz} \end{bmatrix} . \tag{8.4}$$

It is seen from Fig. 2.5b that the diagonal elements σ_{xx}, σ_{yy}, σ_{zz} of the stress component matrix (8.4) are the normal stresses on three coordinate planes through the point $\mathbf{x}$, while the off-diagonal stress components with different subscripts are the shear stresses on these planes. The mean value of the normal stresses is called the mean stress,

$$\sigma_m = \tfrac{1}{3}(\sigma_{xx} + \sigma_{yy} + \sigma_{zz}) . \tag{8.5}$$

The mean stress is closely related to the concept of pressure as illustrated in the following example.

Example 8.2. Static fluid pressure. In a fluid in static equilibrium under the action of gravity there are no shear stress components, and the stress state is completely determined by the hydrostatic pressure p, determined from the weight of the overlying fluid column. The hydrostatic pressure increases with depth proportional to the specific gravity γ. If z denotes a vertical coordinate with $z = 0$ at the surface of the fluid, the hydrostatic pressure is

$$p = -\gamma z .$$

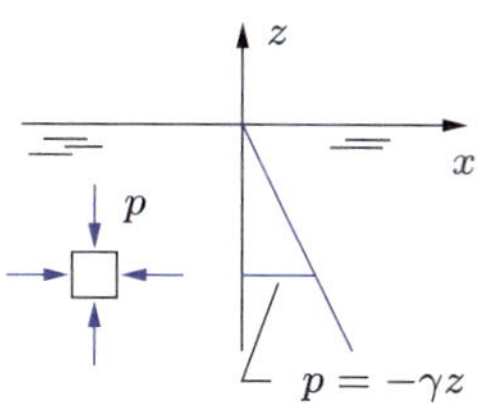

Fig. 8.6: Hydrostatic stress in fluid with specific gravity γ.

The stress component matrix of this hydrodrostatic stress state corresponds to equal normal stress $-p$ on all three coordinate surfaces, and vanishing shear stresses, whereby

$$\boldsymbol{\sigma} = \begin{bmatrix} -p & 0 & 0 \\ 0 & -p & 0 \\ 0 & 0 & -p \end{bmatrix} = -p \begin{bmatrix} 1 & 0 & 0 \\ 0 & 1 & 0 \\ 0 & 0 & 1 \end{bmatrix}.$$

This stress state is fully characterized by the mean stress $\sigma_m = -p$. □

The stress component matrix is (8.4) symmetric. This follows from moment equilibrium about each of the three lines parallel with with the axes through the center of a small cube. The principle is illustrated for the stress components σ_{xz} and σ_{zx} in Fig. 8.7. Moment equilibrium about the line through the center of the cube parallel with the y-axis consists of two force couples that must balance: a force couple of magnitude $\sigma_{zx}dxdy$ acting with moment arm dz, and a force couple of magnitude $\sigma_{xz}dydz$ acting with moment arm dx. Balance of the moments of these two force couples implies that

$$(\sigma_{zx}\, dx\, dy)\, dz = (\sigma_{xz}\, dy\, dz)\, dx\,. \tag{8.6}$$

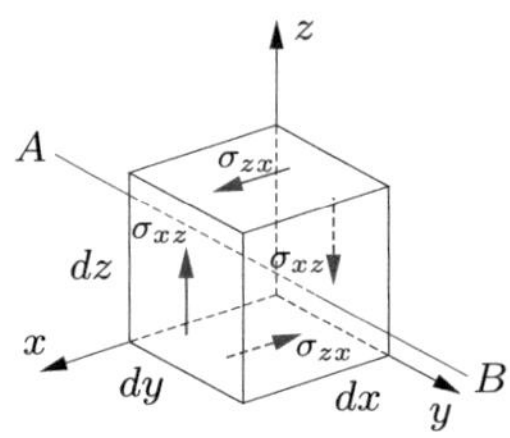

Fig. 8.7: Stress components for moment equilibrium about y-axis.

The factors representing the infinitesimal side lengths of the cube are identical on both sides. From this equation $\sigma_{zx} = \sigma_{xz}$, and by interchange of indices it follows that similar symmetry relations apply to the other off-diagonal stress components, whereby

$$\boldsymbol{\sigma} = \begin{bmatrix} \sigma_{xx} & \sigma_{xy} & \sigma_{xz} \\ & \sigma_{yy} & \sigma_{yz} \\ \text{Sym.} & & \sigma_{zz} \end{bmatrix}. \tag{8.7}$$

This implies that there are only 6, and not 9, independent stress components at a point. A special notation for the stress components, generally used in numerical computations such as the finite element method, using only the 6 independent components is discussed in Section 8.3.2.

Stress on an arbitrary section

The individual components of the stress component matrix $\boldsymbol{\sigma}$ refer to stresses on planes parallel to the coordinate planes of a particular $\{x, y, z\}$ coordinate system. In many cases it is of interest to express the stress vector on a section defined by the normal unit vector $\mathbf{n} = [n_x, n_y, n_z]^T$ in terms of the stress components of the stress matrix $\boldsymbol{\sigma}$ with reference to the coordinate axes. A special case was already considered in Example 8.1, and here the general theory is presented. The original theory for the components of the stress vector on an inclined section due to CAUCHY (1789–1857) remains virtually unchanged.

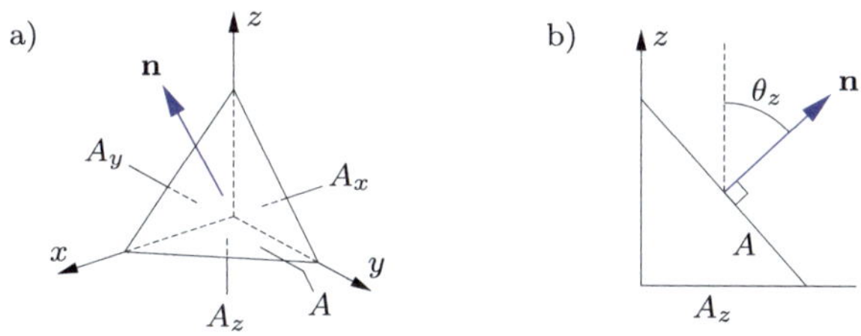

Fig. 8.8: a) Tetrahedron with normal $\mathbf{n}$ on inclined surface, b) Area A_z in the xy-coordinate plane by projection of the inclined surface area A.

Figure 8.8a shows a tetrahedron defined by the three coordinate planes and a plane with normal unit vector $\mathbf{n} = [n_x, n_y, n_z]^T$. The area of the inclined triangle, defined by the intersecting coordinate planes, is denoted A. The triangular areas in the coordinate planes are denoted A_x, A_y and A_z as indicated in the figure. Figure 8.8b shows a section through the triangle A containing the z-axis. The area A_z follows by projection of the area A through the angle θ_z as shown in the figure. The general result is

$$\begin{bmatrix} A_x \\ A_y \\ A_z \end{bmatrix} = \begin{bmatrix} A\cos\theta_x \\ A\cos\theta_y \\ A\cos\theta_z \end{bmatrix} = A \begin{bmatrix} n_x \\ n_y \\ n_z \end{bmatrix}, \tag{8.8}$$

where the last relation follows from the fact that the components of the unit vector are the direction cosines, $\mathbf{n} = [\cos\theta_x, \cos\theta_y, \cos\theta_z]^T$.

The stress vector $\mathbf{t}^n$ on the inclined plane with normal vector $\mathbf{n}$ is now found by setting up the equilibrium equations for the tetrahedron. The stress vectors on the triangles in the coordinate planes are $-\mathbf{t}_x$, $-\mathbf{t}_y$ and $-\mathbf{t}_z$, respectively. The minus sign is because the outward normal on each of these triangles is in the negative coordinate direction. A force balance is set up by multiplication of each of the stress vectors with the appropriate triangle area,

$$A\,\mathbf{t}^n = A_x\mathbf{t}_x + A_y\mathbf{t}_y + A_z\mathbf{t}_z\,. \tag{8.9}$$

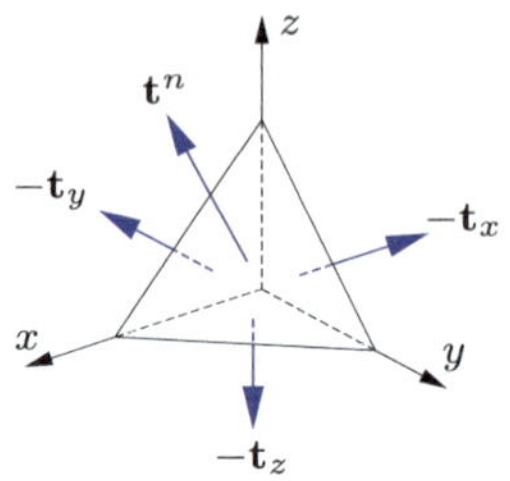

Fig. 8.9: Stress vectors $\mathbf{t}^n$, and $-\mathbf{t}_x$, $-\mathbf{t}_y$, $-\mathbf{t}_z$ on sides of tetrahedron.

In this equation a volume force term has been omitted, because it is proportional to the volume of the tetrahedron, and therefore is of higher order in the limit of vanishing size of the tetrahedron. As shown in (8.4) the components of the stress vectors are the rows of the stress component matrix $\boldsymbol{\sigma}$, and the projected areas are expressed in terms of the components of the normal vector $\mathbf{n}$ by (8.8). After division of the relation (8.9) by the area A it therefore takes the compact form

$$\mathbf{t}^n = \boldsymbol{\sigma}^T \mathbf{n} = \boldsymbol{\sigma}\, \mathbf{n}\,, \tag{8.10}$$

where the last relation follows from the symmetry of the stress component matrix. The component form of this relation is

$$\begin{bmatrix} t_x^n \\ t_y^n \\ t_z^n \end{bmatrix} = \begin{bmatrix} \sigma_{xx} & \sigma_{xy} & \sigma_{xz} \\ \sigma_{yx} & \sigma_{yy} & \sigma_{yz} \\ \sigma_{zx} & \sigma_{zy} & \sigma_{zz} \end{bmatrix} \begin{bmatrix} n_x \\ n_y \\ n_z \end{bmatrix}. \tag{8.11}$$

This relation enables calculation of the stress vector components on inclined sections, and thereby the normal and shear stresses as shown in Example 8.1.

The stress vector relation (8.11) can also be used to obtain the stress component transformation rule between the original $\{xyz\}$-coordinate system and a new $\{x'y'z'\}$-coordinate system, illustrated in Fig. 8.10. Let the direction of the new x'-axis be defined by a unit vector $\mathbf{n}^{x'}$ with the components $[n_x^{x'}, n_y^{x'}, n_z^{x'}]$ in the original xyz-coordinate system. The stress vector on a section defined by the $y'z'$-plane orthogonal to the x'-axis is therefore defined

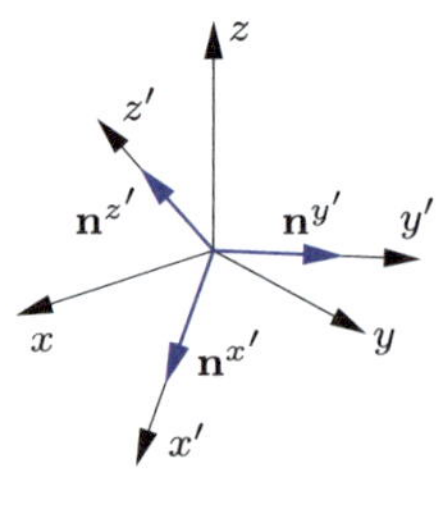

Fig. 8.10: Transformation from xyz-coordinate system to new $x'y'z'$-coordinate system.

by (8.10) as

$$\mathbf{t}^{x'} = \boldsymbol{\sigma}\,\mathbf{n}^{x'}. \tag{8.12}$$

The y'- and z'-axis are defined by unit vectors $\mathbf{n}^{y'} = [n_x^{y'}, n_y^{y'}, n_z^{y'}]^T$ and $\mathbf{n}^{z'} = [n_x^{z'}, n_y^{z'}, n_z^{z'}]^T$, respectively. The stress vectors on the three new coordinate planes are therefore given by an extension of (8.12) as

$$[\,\mathbf{t}^{x'}, \mathbf{t}^{y'}, \mathbf{t}^{z'}\,] = \boldsymbol{\sigma}\,[\,\mathbf{n}^{x'}, \mathbf{n}^{y'}, \mathbf{n}^{z'}\,]. \tag{8.13}$$

The transformation is described by the 3×3 matrix formed by the components of the new unit base vectors as columns,

$$\mathbf{A}^T = \begin{bmatrix} n_x^{x'} & n_x^{y'} & n_x^{z'} \\ n_y^{x'} & n_y^{y'} & n_y^{z'} \\ n_z^{x'} & n_z^{y'} & n_z^{z'} \end{bmatrix}. \tag{8.14}$$

By introducing this matrix as $\mathbf{A}^T$ uniformity of notation is obtained with component transformations used e.g. in (7.56) and (10.46). The new stress components are formed by projection of the stress vectors on the new $x'y'z'$-axes. These projections are obtained from the stress vector relation (8.13) by pre-multiplication with $\mathbf{A}$, whereby the stress transformation relation takes the form

$$\boldsymbol{\sigma}' = \mathbf{A}\,\boldsymbol{\sigma}\,\mathbf{A}^T. \tag{8.15}$$

It is seen that this transformation involves pre- as well as post-multiplication with the coordinate transformation matrix $\mathbf{A}$. This implies that stress components do not transform via simple products with trigonometric functions like vectors, but via a transform involving products of trigonometric functions. Coordinate transformations and special properties associated with particular coordinate systems are discussed in Section 8.4.

8.1.3 Equilibrium

Figure 8.11 shows a continuous body with volume V and surface S. The body is acted upon by distributed volume forces $\mathbf{p}(\mathbf{x})$ and surface forces represented by the stress vector $\mathbf{t}(\mathbf{x})$. The inclusion of the argument $\mathbf{x}$ in the volume and surface force distributions indicates that they may depend on the location described by the coordinates $[x, y, z]$. The loads generate a stress state $\boldsymbol{\sigma}(\mathbf{x})$ in the body. In order for the body to remain at rest, each part of the body must be in equilibrium. This imposes a set of conditions on the stress distribution in the body, the so-called equilibrium conditions. The equilibrium conditions are independent of the particular material properties of the body under consideration, and they are therefore often considered as the most important conditions for the determination of the state of stress in a body at rest, while the material properties can be described in an approximate way, if necessary.

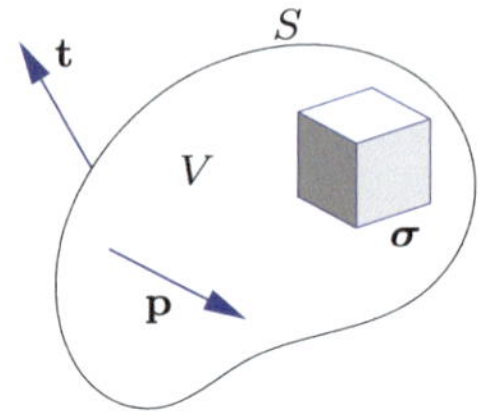

Fig. 8.11: Continuous body with volume V and surface S.

A direct way to establish the equilibrium conditions at a point $\mathbf{x} = [x, y, z]^T$ of a continuous body is to consider a small cube with sides $d\mathbf{x} = [dx, dy, dz]^T$ as illustrated in Fig. 8.4. The sides of the cube are parallel with the coordinate planes of the $\{xyz\}$-coordinate system, and a set of local axes through the point $\mathbf{x}$ can then be introduced as shown in Fig. 8.5. The sides of the cube coinciding with the local coordinate planes are acted upon by the stress vectors $-\mathbf{t}_x$, $-\mathbf{t}_y$ and $-\mathbf{t}_z$, respectively. The minus sign is due to the fact that these 'back' surfaces have normals in the negative coordinate directions. The similar 'front' surfaces have stresses that may be different from the back surfaces. For a small cube the change in stress across the cube can be represented by a Taylor series expansion. Consider the stress vector $\mathbf{t}_x$ on a surface parallel to the yz-coordinate plane. A Taylor expansion around the point $\mathbf{x}$ is

$$\mathbf{t}_x(\mathbf{x} + d\mathbf{x}) \;=\; \mathbf{t}_x(\mathbf{x}) + \frac{\partial \mathbf{t}_x(\mathbf{x})}{\partial x}\, dx + \cdots , \tag{8.16}$$

where the dots represent higher order terms in the components of $d\mathbf{x}$. At the end of the argument a limit operation in which $d\mathbf{x} \to \mathbf{0}$ is considered, and only the first derivatives give finite contributions. The higher order terms can therefore be omitted already at this stage. Figure 8.12 shows the first order Taylor representation of the stress vectors acting on the front surfaces of the cube.

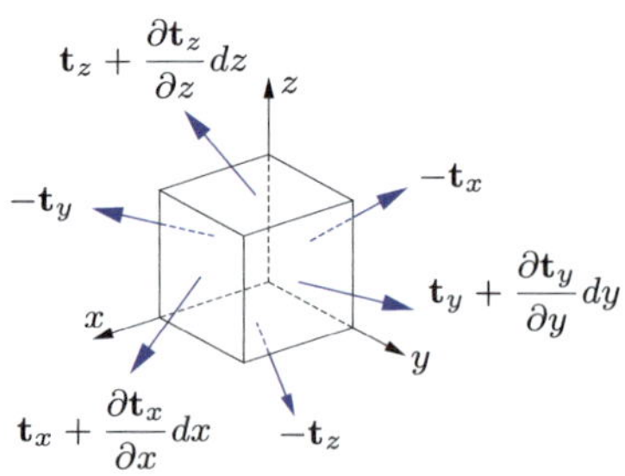

Fig. 8.12: Small cube with stress vectors $\mathbf{t}_x, \mathbf{t}_y, \mathbf{t}_z$ on the sides.

When adding the force contributions from opposing surfaces orthogonal to the x-axis it is observed that the force $-\mathbf{t}_x dydz$ on the 'back' surface cancels the force from the first Taylor term on the front surface, leaving the first order contribution $(\partial\mathbf{t}_x/\partial x)dxdydz$. Adding the surface contributions from the three opposing sets of sides and the volume force contribution $\mathbf{p}\,dxdydz$ gives the equilibrium condition

$$\Big(\frac{\partial\mathbf{t}_x}{\partial x}dx\Big)dydz + \Big(\frac{\partial\mathbf{t}_y}{\partial y}dy\Big)dzdx + \Big(\frac{\partial\mathbf{t}_z}{\partial z}dz\Big)dxdy \;+\; \mathbf{p}\,dxdydz \;=\; \mathbf{0}\,. \tag{8.17}$$

It is seen that all terms contain the common factor $dV = dxdydz$, which can therefore be removed from the equation by division by dV, leaving the equilibrium condition in the form

$$\frac{\partial\mathbf{t}_x}{\partial x} + \frac{\partial\mathbf{t}_y}{\partial y} + \frac{\partial\mathbf{t}_z}{\partial z} + \mathbf{p} = \mathbf{0}\,. \tag{8.18}$$

The component form of this vector equation is obtained by using that the components of the stress vectors are stored as the rows of the stress component matrix, which by symmetry correspond to the columns. Thus, the first column of the stress component matrix is differentiated with respect to x, the second with respect to y, and the third with respect to z. This gives the following three equilibrium equations in terms of the stress components,

$$\begin{aligned}
\frac{\partial\sigma_{xx}}{\partial x} + \frac{\partial\sigma_{yx}}{\partial y} + \frac{\partial\sigma_{zx}}{\partial z} + p_x &= 0\,,\\
\frac{\partial\sigma_{xy}}{\partial x} + \frac{\partial\sigma_{yy}}{\partial y} + \frac{\partial\sigma_{zy}}{\partial z} + p_y &= 0\,,\\
\frac{\partial\sigma_{xz}}{\partial x} + \frac{\partial\sigma_{yz}}{\partial y} + \frac{\partial\sigma_{zz}}{\partial z} + p_z &= 0\,.
\end{aligned} \tag{8.19}$$

These equations must be satisfied by any static stress field, irrespective of the material behavior.

When using the three-dimensional equilibrium equations, e.g. to derive the principle of virtual work or in connection with a finite element formulation of numerical procedures, it is convenient to use a more compact notation. The differential operators appear in a systematic way that can be captured by introducing the gradient operator

$$\boldsymbol{\nabla} = \Big[\frac{\partial}{\partial x}, \frac{\partial}{\partial y}, \frac{\partial}{\partial z}\Big]. \tag{8.20}$$

The equilibrium equations (8.19) can then be expressed in the compact form

$$(\boldsymbol{\nabla}\,\boldsymbol{\sigma})^T + \mathbf{p} = \mathbf{0}\,. \tag{8.21}$$

In this formula the components are combined according to the normal rules of matrix multiplication, whereby the differential operator acts on each of the columns of the stress component matrix $\boldsymbol{\sigma}$ in turn. The transposition is introduced to present the result in column format corresponding to the component equations (8.19).

8.2 Deformation and strain

A solid body is typically deformed, when acted upon by loads. The motion of the body is described in terms of the displacement field $\mathbf{u}(\mathbf{x})$, and it is important to extract a precise measure of the deformation, e.g. in the form of extension of bars and curvature of beams. For a deformable solid body the description of deformation must apply to an arbitrarily small region around each point of the body, and the local deformation is described in terms of strains associated with each point of the body. In the following the theory of strain is developed – first in general form, and then linearized corresponding to small displacements. The general non-linear formulation is beyond the scope of the present text, but is increasingly used in connection with finite element analysis of structures, see e.g. Zienkiewicz and Taylor (2000) or Krenk (2009).

8.2.1 Strain

The theory of deformation of a continuous body is conveniently developed by considering the relative elongation in a selected direction AB at a selected point A. The problem is illustrated in Fig. 8.13, showing a continuous body before and after deformation. A point initially located at $\mathbf{x}$ moves to its current position $\bar{\mathbf{x}} = \mathbf{x} + \mathbf{u}$. The problem is posed as follows. Find the relative elongation of an infinitesimal line element $d\mathbf{x}$ of initial length ds. In the displacement process the line increment $d\mathbf{x}$ changes into $d\bar{\mathbf{x}} = d\mathbf{x} + d\mathbf{u}$. It is convenient to characterize the elongation of the line increment $d\mathbf{x}$ by the so-called Green strain, defined as

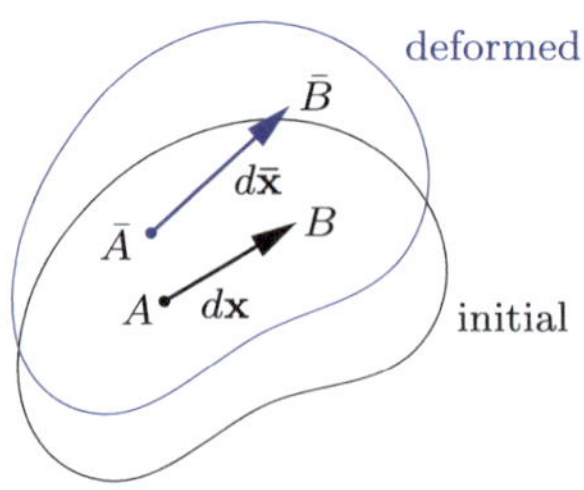

Fig. 8.13: Continuous body before and after deformation.

$$\varepsilon_G = \frac{d\bar{s}^2 - ds^2}{2\,ds^2} = \frac{d\bar{s} + ds}{2\,ds}\,\frac{d\bar{s} - ds}{ds}. \tag{8.22}$$

The idea behind this strain definition is that it is formed by the squares of the length of the line increments $d\mathbf{x}$ and $d\bar{\mathbf{x}}$, and this enables simple calculation via the vector components, while at the same time corresponding closely to the simple linear axial strain (2.4) for small displacements. This follows from the last form in which the first factor is a normalized form of the mean value of the initial length and the final length, and thus is close to unity for the moderate elongations often encountered in engineering practice. Under this condition the Green strain closely approximates the relative elongation, represented by the second factor. The original definition in terms of the square of the involved lengths, permits the Green strain to be expressed directly by the scalar products of the vectors $d\mathbf{x}$ and $d\bar{\mathbf{x}}$,

$$\varepsilon_G = \frac{d\bar{\mathbf{x}}^T d\bar{\mathbf{x}} - d\mathbf{x}^T d\mathbf{x}}{2\,ds^2} = \frac{(d\mathbf{x} + d\mathbf{u})^T(d\mathbf{x} + d\mathbf{u}) - d\mathbf{x}^T d\mathbf{x}}{2\,ds^2}. \tag{8.23}$$

When the scalar products are carried out on the individual terms, the resulting expression is

$$\varepsilon_G = \frac{d\mathbf{x}^T}{ds}\frac{d\mathbf{u}}{ds} + \frac{1}{2}\frac{d\mathbf{u}^T}{ds}\frac{d\mathbf{u}}{ds} \simeq \frac{d\mathbf{x}^T}{ds}\Big[\frac{\partial\mathbf{u}}{\partial\mathbf{x}}\Big]\frac{d\mathbf{x}}{ds}. \tag{8.24}$$

The second term is quadratic in the displacement derivative, and has been omitted as a higher order term. The linear first term has been reformulated to a differentiation 'through' $\mathbf{x}$ by using the chain rule. In this formula $d\mathbf{x}/ds = \mathbf{n}$ is a unit vector in the direction AB. The term $[\partial\mathbf{u}/\partial\mathbf{x}]$ is the displacement gradient matrix

$$\frac{\partial\mathbf{u}}{\partial\mathbf{x}} = \begin{bmatrix} \partial u/\partial x & \partial u/\partial y & \partial u/\partial z \\ \partial v/\partial x & \partial v/\partial y & \partial v/\partial z \\ \partial w/\partial x & \partial w/\partial y & \partial w/\partial z \end{bmatrix}. \tag{8.25}$$

In the approximate linear expression for the axial Green strain, the displacement gradient matrix is pre- and post-multiplied by the components of the unit vector $d\mathbf{x}/ds = \mathbf{n}$. Thus, only the symmetric part of the displacement gradient matrix $[\partial\mathbf{u}/\partial\mathbf{x}]$ contributes to the strain. This symmetric part is defined as the linear strain matrix

$$\boldsymbol{\varepsilon} = \frac{1}{2}\Big(\Big[\frac{\partial\mathbf{u}}{\partial\mathbf{x}}\Big] + \Big[\frac{\partial\mathbf{u}}{\partial\mathbf{x}}\Big]^T\Big). \tag{8.26}$$

The linear strain components are arranged in the symmetric component matrix

$$\boldsymbol{\varepsilon} = \begin{bmatrix} \varepsilon_{xx} & \varepsilon_{xy} & \varepsilon_{xz} \\ \varepsilon_{yx} & \varepsilon_{yy} & \varepsilon_{yz} \\ \varepsilon_{zx} & \varepsilon_{zy} & \varepsilon_{zz} \end{bmatrix}, \tag{8.27}$$

with components following from (8.25) and (8.26) in the form

$$
\begin{aligned}
\varepsilon_{xx} &= \frac{\partial u}{\partial x}, & \varepsilon_{yz} &= \varepsilon_{zy} = \frac{1}{2}\Big(\frac{\partial v}{\partial z} + \frac{\partial w}{\partial y}\Big),\\
\varepsilon_{yy} &= \frac{\partial v}{\partial y}, & \varepsilon_{zx} &= \varepsilon_{xz} = \frac{1}{2}\Big(\frac{\partial w}{\partial x} + \frac{\partial u}{\partial z}\Big),\\
\varepsilon_{zz} &= \frac{\partial w}{\partial z}, & \varepsilon_{xy} &= \varepsilon_{yx} = \frac{1}{2}\Big(\frac{\partial u}{\partial y} + \frac{\partial v}{\partial x}\Big).
\end{aligned}
\tag{8.28}
$$

There are 9 strain components, of which only 6 are independent due to symmetry of the strain component matrix (8.27). This is similar to the 9 components of the symmetric stress component matrix (8.4).

Each of the linearized strain components in the matrix $\boldsymbol{\varepsilon}$ can be given a simple and direct physical interpretation. It follows from the linearized form of (8.24) that the axial strain in a direction described by the unit vector $\mathbf{n}$ is given by

$$
\varepsilon^{n} = \mathbf{n}^T \boldsymbol{\varepsilon}\, \mathbf{n}\,. \tag{8.29}
$$

In particular this implies, that the relative elongation in the direction of the x-axis is given by

$$
\varepsilon_{xx} = [1,0,0]\begin{bmatrix} \varepsilon_{xx} & \varepsilon_{xy} & \varepsilon_{xz}\\ \varepsilon_{yx} & \varepsilon_{yy} & \varepsilon_{yz}\\ \varepsilon_{zx} & \varepsilon_{zy} & \varepsilon_{zz}\end{bmatrix}\begin{bmatrix}1\\0\\0\end{bmatrix} = \frac{\partial u}{\partial x}\,. \tag{8.30}
$$

Similar relations hold for the other coordinate directions, and thus the state of strain given by

$$
\boldsymbol{\varepsilon} = \begin{bmatrix} \varepsilon_{xx} & 0 & 0\\ 0 & \varepsilon_{yy} & 0\\ 0 & 0 & \varepsilon_{zz}\end{bmatrix}. \tag{8.31}
$$

corresponds to relative extension of the axes by ε_{xx}, ε_{yy} and ε_{zz}, respectively. This state of strain is illustrated in Fig. 8.14, showing a unit cube before and after deformation. The strain components ε_{xx}, ε_{yy} and ε_{zz} are called normal strains or axial strains.

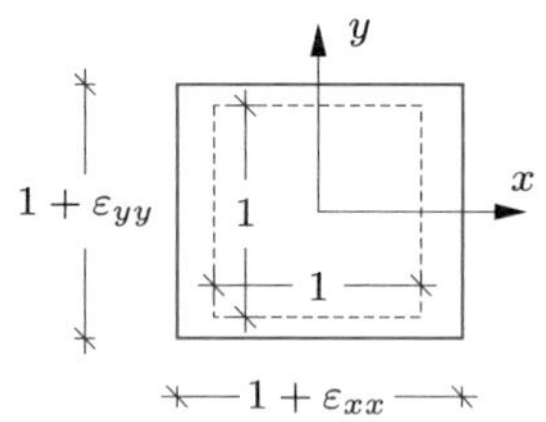

Fig. 8.14: Unit square before and after axial straining $\varepsilon_{xx}, \varepsilon_{yy}$.

The initial volume of the unit cube shown in Fig. 8.14 is 1, while the volume after deformation is found as the product of the side lengths $(1+\varepsilon_{xx})$, $(1+\varepsilon_{yy})$

and $(1 + \varepsilon_{zz})$. The relative change in volume is

$$\frac{\bar{V} - V}{V} = (1 + \varepsilon_{xx})(1 + \varepsilon_{yy})(1 + \varepsilon_{zz}) - 1 \tag{8.32}$$

$$= \varepsilon_{xx} + \varepsilon_{yy} + \varepsilon_{zz} + \varepsilon_{yy}\varepsilon_{zz} + \varepsilon_{xx}\varepsilon_{zz} + \varepsilon_{xx}\varepsilon_{yy} + \varepsilon_{xx}\varepsilon_{yy}\varepsilon_{zz} \,.$$

The last four terms are of higher order, and will be negligible for small strains. Thus, the linearized volume strain is defined by

$$e = \varepsilon_{xx} + \varepsilon_{yy} + \varepsilon_{zz} \,. \tag{8.33}$$

It can be shown that the volume strain e defined by the sum of the diagonal components is independent of the particular coordinate system used. This invariance is an important characteristic e.g. in connection with the development of relations between stresses and strains, see Chapter 9 and a general discussion e.g. by Mase and Mase (1999) and Ottosen and Ristinmaa (2005).

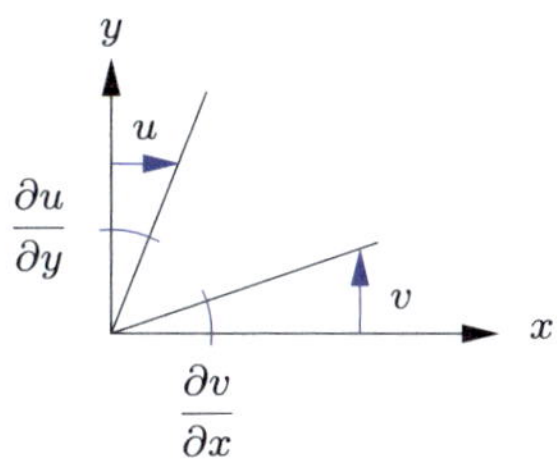

Fig. 8.15: Shear strain components and change of angle between axes.

The physical interpretation of the off-diagonal strain component ε_{xy} is illustrated in Fig. 8.15. The figure shows the gradients of the displacement components u and v, changing the orientation of the y- and x-axis, respectively. The two contributions *diminish* the angle between the x- and the y-axis by

$$\gamma_{xy} = \frac{\partial v}{\partial x} + \frac{\partial u}{\partial y} = 2\varepsilon_{xy} \,. \tag{8.34}$$

The off-diagonal components of $\boldsymbol{\varepsilon}$ are called shear strains, while the change in angles like $\gamma_{xy} = 2\varepsilon_{xy}$ are called angle strain due to their direct interpretation as a change of the angle between the axes.

Transformation of strain components

If the strain component matrix $\boldsymbol{\varepsilon}$ is known in a given $\{xyz\}$-coordinate system, the strain components $\boldsymbol{\varepsilon}'$ in another $\{x'y'z'\}$-coordinate system can be found by a simple matrix transformation, similar to the stress component transformation (8.15). The two coordinate systems are shown in Fig. 8.10, and the relation between the coordinates is then given by

$$\mathbf{x}' = \mathbf{A}\,\mathbf{x} \tag{8.35}$$

in terms of the matrix $\mathbf{A}$, already introduced in (8.14). The matrix $\mathbf{A}$ is orthogonal, and thus its inverse is given by its transpose. The inverse coordinate transformation therefore is

$$\mathbf{x} = \mathbf{A}^T\mathbf{x}'\,. \tag{8.36}$$

When the original coordinates $\mathbf{x}$ are considered as functions of the new coordinates $\mathbf{x}'$, differentiation of this relation gives

$$\frac{\partial \mathbf{x}}{\partial \mathbf{x}'} = \mathbf{A}^T. \tag{8.37}$$

The new displacement components $\mathbf{u}'$ are given in terms of the original components $\mathbf{u}$ by a relation similar to the coordinate relation (8.35),

$$\mathbf{u}' = \mathbf{A}\,\mathbf{u}\,. \tag{8.38}$$

The strain component matrix ε' in the new coordinate system is defined in terms of the displacement gradient as

$$\varepsilon' = \frac{1}{2}\Big(\Big[\frac{\partial \mathbf{u}'}{\partial \mathbf{x}'}\Big] + \Big[\frac{\partial \mathbf{u}'}{\partial \mathbf{x}'}\Big]^T\Big). \tag{8.39}$$

The new displacement gradient is expressed in terms of the original by using the displacement component transformation (8.38) and the chain rule for differentiation 'through' the coordinates $\mathbf{x}$,

$$\frac{\partial \mathbf{u}'}{\partial \mathbf{x}'} = \mathbf{A}\,\frac{\partial \mathbf{u}}{\partial \mathbf{x}}\frac{\partial \mathbf{x}}{\partial \mathbf{x}'} = \mathbf{A}\Big[\frac{\partial \mathbf{u}}{\partial \mathbf{x}}\Big]\mathbf{A}^T\,. \tag{8.40}$$

Substitution of this result into the strain definition (8.39) then gives

$$\varepsilon' = \mathbf{A}\frac{1}{2}\Big(\Big[\frac{\partial \mathbf{u}}{\partial \mathbf{x}}\Big] + \Big[\frac{\partial \mathbf{u}}{\partial \mathbf{x}}\Big]^T\Big)\mathbf{A}^T\,, \tag{8.41}$$

where the last term is obtained by matrix transposition. The central part of this formula is recognized as the strain matrix ε in the original coordinate system, whereby the transformation takes the form

$$\varepsilon' = \mathbf{A}\,\varepsilon\,\mathbf{A}^T\,. \tag{8.42}$$

This transformation is identical to the transformation rule (8.15) for the stress component matrix.

8.2.2 Rotation at a point

It was demonstrated in the previous section that within the small deformation assumption only the six components in the symmetric part of the displacement gradient matrix $[\partial\mathbf{u}/\partial\mathbf{x}]$ contributes to the linear strain components ε. The remaining three components describe a rotation of the material about the point. This is seen from Fig. 8.15 by observing that the derivative $\partial u/\partial y$ describes a clockwise rotation of the y-axis about the z-axis, while the derivative $\partial v/\partial x$ similarly describes a counter-clockwise rotation of the x-axis. Thus, the mean rotation of the x- and y-axis about the z-axis is given by

$$\omega_z = \frac{1}{2}\Big(\frac{\partial v}{\partial x} - \frac{\partial u}{\partial y}\Big). \tag{8.43}$$

Similar relations exist for the small rotation about the other two axes,

$$\omega_x = \frac{1}{2}\Big(\frac{\partial w}{\partial y} - \frac{\partial v}{\partial z}\Big), \qquad \omega_y = \frac{1}{2}\Big(\frac{\partial u}{\partial z} - \frac{\partial w}{\partial x}\Big). \tag{8.44}$$

These components define the small rotation vector $\boldsymbol{\omega} = [\omega_x, \omega_y, \omega_z]^T$. The components may also be arranged as the skew-symmetric part of the rotation matrix

$$\boldsymbol{\Omega} = \begin{bmatrix} 0 & -\omega_z & \omega_y \\ \omega_z & 0 & -\omega_x \\ -\omega_y & \omega_x & 0 \end{bmatrix}. \tag{8.45}$$

This form is often used in continuum mechanics, as it permits writing the small rotation of a vector $\mathbf{a}$ by the rotation vector $\boldsymbol{\omega}$ in the form of a direct matrix product, $\boldsymbol{\Omega}\mathbf{a} = \boldsymbol{\omega} \times \mathbf{a}$, thus avoiding the use of the vector cross-product.

8.2.3 Displacement decomposition

The previous two sections have described the deformation (strain) and rotation, that can be associated with a given displacement field $\mathbf{u}(\mathbf{x})$ under the 'small displacement' assumption. This can be viewed as a decomposition of the displacement field into the sum of a translation $\mathbf{u}$, a rotation $\boldsymbol{\omega}$, and a state of deformation (strain) ε, associated with the material at each point of the body. The first observation is that the displacement gradient matrix is the sum of the strain and rotation component matrices,

$$\frac{\partial\mathbf{u}}{\partial\mathbf{x}} = \varepsilon + \boldsymbol{\Omega}\,. \tag{8.46}$$

Thus, the displacement at a point defines the translation, while the displacement gradient components define the state of deformation in a neighborhood around the point in terms of a rotation and a local state of strain.

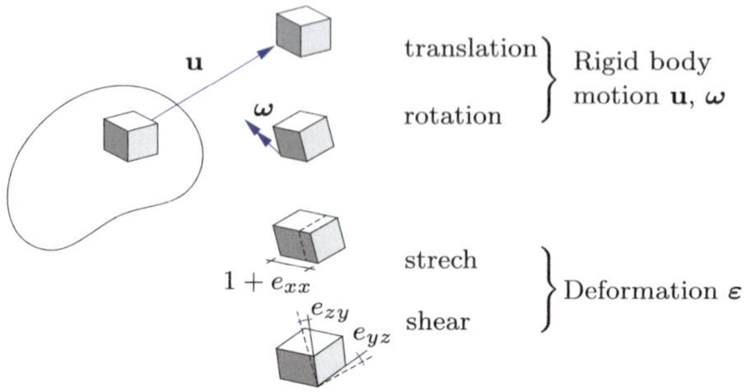

Fig. 8.16: Decomposition of displacement field into translation, rotation, and deformation.

The general effect of a displacement field on a small cube located at a point $\mathbf{x}$ is illustrated in Fig. 8.16. First the cube is subjected to a local rigid body motion consisting of a translation by the vector $\mathbf{u}$, followed by a rotation by the rotation vector $\boldsymbol{\omega}$. The cube is then deformed, by stretching $\varepsilon_{xx}, \varepsilon_{yy}, \varepsilon_{zz}$ along the axes, and by changes $2\varepsilon_{yz}, 2\varepsilon_{zx}, 2\varepsilon_{xy}$ of the angles between the axes.

8.3 Virtual work

The statics of a solid body is described by the distribution of stresses $\boldsymbol{\sigma}(\mathbf{x})$ and the associated volume and surface forces $\mathbf{p}(\mathbf{x})$ and $\mathbf{t}(\mathbf{x})$, while the kinematics is described by the displacement field $\mathbf{u}(\mathbf{x})$ and the associated strain field $\boldsymbol{\varepsilon}(\mathbf{x})$. When considering a static (stress) field that satisfies the equilibrium conditions and an independent kinematic (displacement) field the work of the static field through the displacements of the kinematic field satisfy a virtual work equation. This equation and the associated principle of virtual work play an important role in the development of discretized formulations for solid and structural mechanics problems, e.g. by the finite element method, see e.g. Zienkiewicz and Taylor (2000) and Krenk (2009). Special cases of the virtual work principle have been developed for bars an truss structures in Section 2.4.3 and for beams and frames in Section 4.4. The present section contains a derivation of the virtual work equation for a continuous body and a discussion of the boldface matrix notation for stresses and strains often used in computational methods, inspired by the virtual work equation.

8.3.1 Equation of virtual work

The stresses field $\boldsymbol{\sigma}(\mathbf{x})$ in a body in equilibrium must satisfy the equilibrium equations in terms of the volume forces $\mathbf{p}(\mathbf{x})$ distributed within the body, and generate a stress vector distribution $\mathbf{t}(\mathbf{x})$ on the surface of the body matching any imposed surface forces. This is illustrated in Fig. 8.17. The

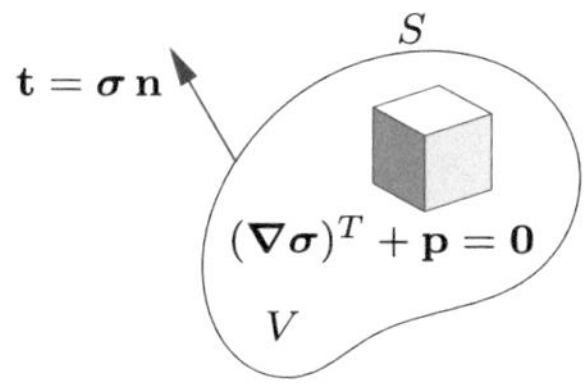

Fig. 8.17: Equilibrium of stress $\boldsymbol{\sigma}$, volume force $\mathbf{p}$ and surface traction $\mathbf{t}$.

stress field $\boldsymbol{\sigma}(\mathbf{x})$ satisfies the equilibrium equations (8.19). These equations express equilibrium of the stress and volume force components acting on a unit cube. The virtual work of these components is obtained by multiplication of each of the three equilibrium equations with the corresponding component of a virtual displacement field $\delta\mathbf{u}(\mathbf{x})$. As previously, the notation $\delta\mathbf{u}$ is used to indicate an infinitesimally small displacement field. The virtual work of a body with volume V is then obtained by integration over the volume in the form

$$\int_V \Big\{ \Big(\frac{\partial \sigma_{xx}}{\partial x} + \frac{\partial \sigma_{yx}}{\partial y} + \frac{\partial \sigma_{zx}}{\partial z} + p_x \Big) \delta u + \Big(\frac{\partial \sigma_{xy}}{\partial x} + \frac{\partial \sigma_{yy}}{\partial y} + \frac{\partial \sigma_{zy}}{\partial z} + p_y \Big) \delta v + \Big(\frac{\partial \sigma_{xz}}{\partial x} + \frac{\partial \sigma_{yz}}{\partial y} + \frac{\partial \sigma_{zz}}{\partial z} + p_z \Big) \delta w \Big\} dV = 0 . \tag{8.47}$$

The equality with zero, irrespective of the virtual displacement field, follows from the fact that the stress field satisfies the equilibrium equations.

This equation is now reformulated to a relation between internal virtual work performed by the stresses $\boldsymbol{\sigma}(\mathbf{x})$ and the external work performed by the stress vector distribution $\mathbf{t}(\mathbf{x})$ acting on the body surface S and the volume forces $\mathbf{p}(\mathbf{x})$ distributed inside the body. In the case of bars and beams the body is one-dimensional and the result was obtained via integration by parts. The three-dimensional equivalent to this procedure is the use of the divergence theorem. The divergence theorem states that the integral of the divergence of a vector field $\mathbf{f} = [f_x, f_y, f_z]^T$ over a volume V is equal to the flux of the vector field through the surface S enclosing the volume. In precise terms this amounts to the equality

$$\int_V \Big(\frac{\partial f_x}{\partial x} + \frac{\partial f_y}{\partial y} + \frac{\partial f_z}{\partial z} \Big) dV = \int_S (f_x n_x + f_y n_y + f_z n_z)\, dS . \tag{8.48}$$

The integrand on the left side is the divergence, while the integrand on the right side is the projection of the vector field on the outward normal to the surface, $\mathbf{f}^T\mathbf{n}$.

In the virtual work integral (8.47) the stresses only appear via their derivatives, while these are multiplied by the virtual work component. Application of the divergence theorem assumes that the individual terms of the integrand are partial derivatives. Thus, the virtual displacement components are placed within the differentiation, and the additional terms generated by this are subtracted subsequently. This gives the following equation,

$$\begin{aligned}\int_V \Big\{ & \frac{\partial(\sigma_{xx}\delta u)}{\partial x} + \frac{\partial(\sigma_{yx}\delta u)}{\partial y} + \frac{\partial(\sigma_{zx}\delta u)}{\partial z} + \frac{\partial(\sigma_{xy}\delta v)}{\partial x} + \frac{\partial(\sigma_{yy}\delta v)}{\partial y} \\ & + \frac{\partial(\sigma_{zy}\delta v)}{\partial z} + \frac{\partial(\sigma_{xz}\delta w)}{\partial x} + \frac{\partial(\sigma_{yz}\delta w)}{\partial y} + \frac{\partial(\sigma_{zz}\delta w)}{\partial z} - \Big(\sigma_{xx}\frac{\partial\delta u}{\partial x} \\ & + \sigma_{yx}\frac{\partial\delta u}{\partial y} + \sigma_{zx}\frac{\partial\delta u}{\partial z} + \sigma_{xy}\frac{\partial\delta v}{\partial x} + \sigma_{yy}\frac{\partial\delta v}{\partial y} + \sigma_{zy}\frac{\partial\delta v}{\partial z} + \sigma_{xz}\frac{\partial\delta w}{\partial x} \\ & + \sigma_{yz}\frac{\partial\delta w}{\partial y} + \sigma_{zz}\frac{\partial\delta w}{\partial z}\Big) + p_x\delta u + p_y\delta v + p_z\delta w\Big\} dV = 0\,. \end{aligned} \tag{8.49}$$

It is seen that the virtual displacement components appear within the parentheses defining the differentiation in the first nine terms, and that the additional terms containing derivatives of the virtual displacement components are subtracted in the following nine terms. In the next step the first nine terms are rearranged in divergence form, and the symmetry of the stress components is used to regroup the following set of terms,

$$\begin{aligned}\int_V \Big\{ & \frac{\partial}{\partial x}\Big(\sigma_{xx}\delta u + \sigma_{xy}\delta v + \sigma_{xz}\delta w\Big) + \frac{\partial}{\partial y}\Big(\sigma_{yx}\delta u + \sigma_{yy}\delta v + \sigma_{yz}\delta w\Big) \\ & + \frac{\partial}{\partial z}\Big(\sigma_{zx}\delta u + \sigma_{zy}\delta v + \sigma_{zz}\delta w\Big) - \Big(\sigma_{xx}\frac{\partial\delta u}{\partial x} + \sigma_{yy}\frac{\partial\delta v}{\partial y} + \sigma_{zz}\frac{\partial\delta w}{\partial z} \\ & + \sigma_{yz}\frac{1}{2}\Big(\frac{\partial\delta w}{\partial y} + \frac{\partial\delta v}{\partial z}\Big) + \sigma_{zx}\frac{1}{2}\Big(\frac{\partial\delta w}{\partial y} + \frac{\partial\delta v}{\partial z}\Big) + \sigma_{xy}\frac{1}{2}\Big(\frac{\partial\delta w}{\partial y} + \frac{\partial\delta v}{\partial z}\Big)\Big) \\ & + p_x\,\delta u + p_y\,\delta v + p_z\,\delta w\Big\} dV = 0\,. \end{aligned} \tag{8.50}$$

The first group of terms is seen to be of divergence form, and they are therefore equal to the surface integral of $\mathbf{n}^T(\boldsymbol{\sigma}\delta\mathbf{u})$. When changing the order of the factors and using the formula (8.10) to express the surface stress vector $\mathbf{t} = \boldsymbol{\sigma}\,\mathbf{n}$, the surface integrand takes the form $\delta\mathbf{u}^T\mathbf{t}$. The kinematic factors in the following six terms are recognized as virtual strains by use of (8.28). Thus, regrouping of the integrals leads to the following final form of the equation of virtual work,

$$\int_V \big(\delta\varepsilon_{xx}\,\sigma_{xx} + \delta\varepsilon_{yy}\,\sigma_{yy} + \delta\varepsilon_{zz}\,\sigma_{zz} + 2\delta\varepsilon_{yz}\,\sigma_{yz} + 2\delta\varepsilon_{zx}\,\sigma_{zx} + 2\delta\varepsilon_{xy}\,\sigma_{xy}\big)dV$$
$$= \int_S \big(\delta u\,t_x + \delta v\,t_y + \delta w\,t_z\big)dS + \int_V \big(\delta u\,p_x + \delta v\,p_y + \delta w\,p_z\big)dV\,. \tag{8.51}$$

The volume integral on the left side is the internal work of the stresses through the virtual strains, while the right side is the external work, consisting of the virtual work of surface stresses and volume forces through the virtual displacements.

8.3.2 Matrix and tensor notation

It was demonstrated in Sections 8.1.2 and 8.2.1 that the stress component matrix $\boldsymbol{\sigma}$ and the strain component matrix $\boldsymbol{\varepsilon}$ are symmetric. Thus, they only contain 6 independent components each, and not the 9 components contained in the double indexed matrix format. While the double indexed format accurately reflects the role of the components – e.g. in producing the stress vector by a product with the components of the normal vector – it is uneconomical in numerical computations. A simple alternative in the form of a one-dimensional array is suggested by the principle of virtual work. It is seen from the first integral in (8.51), that the internal virtual work can be expressed in terms of six components, provided a factor 2 is included on the shear components. It is customary to include this factor in the shear strains, that thereby become the angle strains γ_{yz}, γ_{zx} and γ_{xy} already introduced in (8.34). The stress and strain components are therefore often combined into the six-component arrays

$$\boldsymbol{\sigma} = \big[\,\sigma_{xx}, \sigma_{yy}, \sigma_{zz}, \sigma_{yz}, \sigma_{zx}, \sigma_{xy}\,\big]^T \tag{8.52}$$

and

$$\begin{aligned}\boldsymbol{\varepsilon} &= \big[\,\varepsilon_{xx}, \varepsilon_{yy}, \varepsilon_{zz}, 2\varepsilon_{yz}, 2\varepsilon_{zx}, 2\varepsilon_{xy}\,\big]^T \\ &= \big[\,\varepsilon_{xx}, \varepsilon_{yy}, \varepsilon_{zz}, \gamma_{yz}, \gamma_{zx}, \gamma_{xy}\,\big]^T,\end{aligned} \tag{8.53}$$

where the same boldface symbols are used as for the corresponding component matrices. With a bit of care the context defines the specific notation, and use of the same boldface symbols adds an intuitive aspect to the formulae. In the present context the internal work appears as a scalar product of the virtual strain array and the stress array in the form

$$\begin{aligned}\delta\boldsymbol{\varepsilon}^T\boldsymbol{\sigma} &= \delta\varepsilon_{xx}\,\sigma_{xx} + \delta\varepsilon_{yy}\,\sigma_{yy} + \delta\varepsilon_{zz}\,\sigma_{zz} \\ &\quad + 2\delta\varepsilon_{yz}\,\sigma_{yz} + 2\delta\varepsilon_{zx}\,\sigma_{zx} + 2\delta\varepsilon_{xy}\,\sigma_{xy}\,.\end{aligned} \tag{8.54}$$

In this notation the virtual work equation (8.51) takes the compact form

$$\int_V \delta\boldsymbol{\varepsilon}^T \boldsymbol{\sigma}\, dV = \int_V \delta\mathbf{u}^T \mathbf{p}\, dV + \int_S \delta\mathbf{u}^T \mathbf{t}\, dS\,. \tag{8.55}$$

This is the form used in the finite element literature.

8.4 Special states of stress and strain

Special two-dimensional states of stress and strain play important roles in practice. The special states of plane stress and plane strain are introduced in Section 8.4.1, and the corresponding two-dimensional component transformations are treated in detail in Section 8.4.2. An important property of a state of stress or strain is that a particular coordinate system can be found in which all shear components vanish. This coordinate system is called the principal coordinate system, and the normal stress/strain components in this coordinate system are called the principal stresses/strains. The identification of the principal coordinate system, and the corresponding principal stress/strain components is treated for the two-dimensional case in Section 8.4.3, where the graphical illustration of stress states by the Mohr circle construction is presented. The theory is generalized to three dimensions in Section 8.4.4. The principal stress and strain components play an important role in the description of material properties such as elastic deformation, yielding and failure, dealt with in Chapter 9.

8.4.1 Plane stress and plane strain

Many engineering problems are solved as two-dimensional problems. Figure 8.18a shows a typical example in which a plate is loaded by forces in the plane of the plate. The xy-plane is taken as the symmetry plane of the plate. For thin plates the stress component σ_{zz}, referring to transverse normal stress, is usually small and is therefore neglected. Similarly the average value of the stress components σ_{zx} and σ_{yz} vanish, and the corresponding stress components are therefore assumed negligible. This leads to the assumption

$$\sigma_{zz} = \sigma_{zx} = \sigma_{xz} = \sigma_{yz} = \sigma_{zy} = 0 \tag{8.56}$$

for the so-called condition of plane stress. The only non-vanishing stress components in plane stress are

$$\boldsymbol{\sigma} = \begin{bmatrix} \sigma_{xx} & \sigma_{xy} \\ \sigma_{yx} & \sigma_{yy} \end{bmatrix}. \tag{8.57}$$

The assumption of plane stress implies that the stress distribution in thin plates loaded in their own plane can be treated by only three stress components, σ_{xx}, σ_{yy} and $\sigma_{xy} = \sigma_{yx}$. This reduction simplifies the analysis considerably, and the assumption of plane stress is often used in practice.

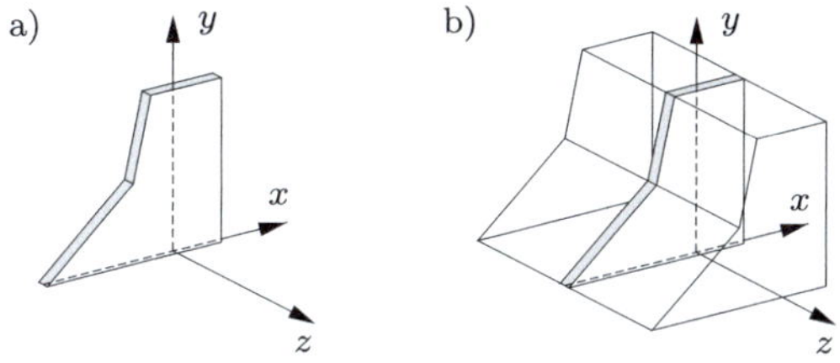

Fig. 8.18: a) Plane stress, b) plane strain.

Another important two-dimensional problem is illustrated in Fig. 8.18b, showing a slice of a dam. The displacements are constrained such that there is negligible displacements in the transverse direction, and thus $w = 0$. This condition can also be expressed as $\varepsilon_{zz} = \partial w/\partial z = 0$. In cases where the slice represents a plane of symmetry the strain components ε_{zx} and ε_{zy} also vanish. This leads to the conditions

$$\varepsilon_{zz} = \varepsilon_{xz} = \varepsilon_{zx} = \varepsilon_{yz} = \varepsilon_{zy} = 0\,, \tag{8.58}$$

defining the so-called state of plane strain. The non-vanishing strain components in the state of plane strain are

$$\boldsymbol{\varepsilon} = \begin{bmatrix} \varepsilon_{xx} & \varepsilon_{xy} \\ \varepsilon_{yx} & \varepsilon_{yy} \end{bmatrix}. \tag{8.59}$$

Due to symmetry there are only three independent strain components, giving a reduction of the problem size comparable to that of the state of plane stress.

8.4.2 Stress and strain transformations

Coordinate axes are often selected with view to a simple representation of the geometry of the problem at hand, and it is of interest to be able to calculate the stress and strain components relative to a different set of coordinate axes. The stress component transformation was briefly described in Section 8.1 and is here covered in more detail for the two-dimensional case. It turns out that the transformation of the strain components is the same as the transformation for the stress components, and the following derivations are therefore limited to the stress components, while the transformation for the strain components is summarized subsequently.

Transformation of plane stress components

Figure 8.19a shows the original set of axes $\{x, y\}$, and a new set of axes $\{x', y'\}$, obtained by rotating the original axes about the z-axis by the angle θ. The new axes are described by the two unit vectors

$$\mathbf{n}^{x'} = [\cos\theta, \sin\theta]^T, \qquad \mathbf{n}^{y'} = [-\sin\theta, \cos\theta]^T. \tag{8.60}$$

The stress components with respect to the original and the new axes are illustrated in Fig. 8.19.

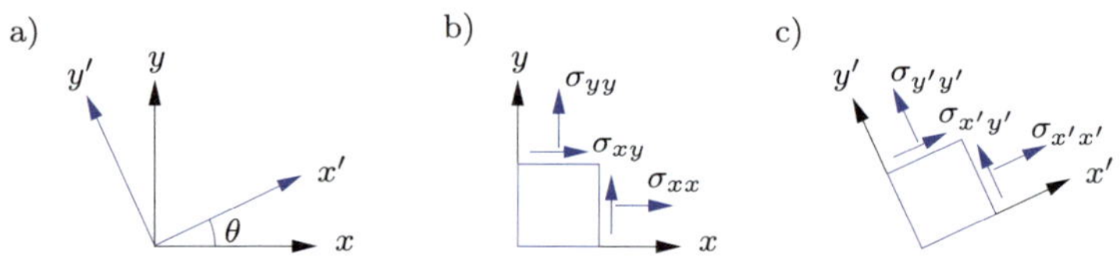

Fig. 8.19: 'Old' and 'new' stress components $\boldsymbol{\sigma}$ and $\boldsymbol{\sigma}'$.

The stress vector acting on sections with $\mathbf{n}^{x'}$ and $\mathbf{n}^{y'}$ as normal vectors are defined in terms of the original stress component matrix $\boldsymbol{\sigma}$ as

$$\mathbf{t}^{x'} = \boldsymbol{\sigma}\,\mathbf{n}^{x'}, \qquad \mathbf{t}^{y'} = \boldsymbol{\sigma}\,\mathbf{n}^{y'}. \tag{8.61}$$

These relations are conveniently combined into a single relation,

$$[\,\mathbf{t}^{x'}, \mathbf{t}^{y'}\,] = \boldsymbol{\sigma}\,[\,\mathbf{n}^{x'}, \mathbf{n}^{y'}\,]. \tag{8.62}$$

In this relation the left side is a 2×2 matrix containing the components of the two stress vectors as columns, while the last factor on the right side is the coordinate transformation matrix

$$\mathbf{A}^T = [\,\mathbf{n}^{x'}, \mathbf{n}^{y'}\,] = \begin{bmatrix} \cos\theta & -\sin\theta \\ \sin\theta & \cos\theta \end{bmatrix}. \tag{8.63}$$

The stress components $\boldsymbol{\sigma}'$ in the new coordinate system are obtained by projecting the stress vectors $\mathbf{t}^{x'}$ and $\mathbf{t}^{y'}$ on to the new axes,

$$\boldsymbol{\sigma}' = [\,\mathbf{n}^{x'}, \mathbf{n}^{y'}\,]^T\,[\,\mathbf{t}^{x'}, \mathbf{t}^{y'}\,]. \tag{8.64}$$

The first factor is recognized as $\mathbf{A}$, and when the stress vectors are substituted from (8.62) the following stress component transformation relation is obtained,

$$\boldsymbol{\sigma}' = \mathbf{A}\,\boldsymbol{\sigma}\,\mathbf{A}^T. \tag{8.65}$$

This is the special two-dimensional case of the general stress component transformation (8.15), here with the coordinate transformation matrix $\mathbf{A}$ given in terms of the angle θ in (8.63).

When introducing components in (8.65) the following component transformation relations are obtained

$$\begin{aligned}\sigma_{x'x'} &= \sigma_{xx}\cos^2\theta + \sigma_{yy}\sin^2\theta + 2\sigma_{xy}\sin\theta\cos\theta\,,\\ \sigma_{y'y'} &= \sigma_{xx}\sin^2\theta + \sigma_{yy}\cos^2\theta - 2\sigma_{xy}\sin\theta\cos\theta\,,\\ \sigma_{x'y'} &= \sigma_{xy}\big(\cos^2\theta - \sin^2\theta\big) + \big(\sigma_{yy}-\sigma_{xx}\big)\sin\theta\cos\theta\,.\end{aligned} \tag{8.66}$$

The transformation of stress components involves products of the trigonometric functions $\cos\theta$ and $\sin\theta$. These products are conveniently combined by use of the formulae for trigonometric functions of double angle,

$$\cos 2\theta = \cos^2\theta - \sin^2\theta\,, \qquad \sin 2\theta = 2\sin\theta\cos\theta\,. \tag{8.67}$$

When these formulae are used to express the products in the component transformation formula (8.66) the following stress component transformation formulae in terms of the double angle are obtained,

$$\begin{aligned}\sigma_{x'x'} &= \tfrac{1}{2}\big(\sigma_{xx}+\sigma_{yy}\big) + \tfrac{1}{2}\big(\sigma_{xx}-\sigma_{yy}\big)\cos 2\theta + \sigma_{xy}\sin 2\theta\,,\\ \sigma_{y'y'} &= \tfrac{1}{2}\big(\sigma_{xx}+\sigma_{yy}\big) - \tfrac{1}{2}\big(\sigma_{xx}-\sigma_{yy}\big)\cos 2\theta - \sigma_{xy}\sin 2\theta\,,\\ \sigma_{x'y'} &= \qquad\qquad\qquad\; - \tfrac{1}{2}\big(\sigma_{xx}-\sigma_{yy}\big)\sin 2\theta + \sigma_{xy}\cos 2\theta\,.\end{aligned} \tag{8.68}$$

It is readily confirmed that the transformation reduces to an identity for $\theta = 0$. For $\theta = 90°$ the transformation gives $\sigma_{x'x'} = \sigma_{yy}$, $\sigma_{y'y'} = \sigma_{xx}$ and $\sigma_{x'y'} = -\sigma_{xy}$.

The form of the transformation formulae (8.68) suggests calculating the sum and the difference of the normal stress components. This yields the following form of the stress component transformation,

$$\begin{aligned}\tfrac{1}{2}\big(\sigma_{x'x'}+\sigma_{y'y'}\big) &= \tfrac{1}{2}\big(\sigma_{xx}+\sigma_{yy}\big)\,,\\ \tfrac{1}{2}\big(\sigma_{x'x'}-\sigma_{y'y'}\big) &= \tfrac{1}{2}\big(\sigma_{xx}-\sigma_{yy}\big)\cos 2\theta + \sigma_{xy}\sin 2\theta\,,\\ \sigma_{x'y'} &= -\tfrac{1}{2}\big(\sigma_{xx}-\sigma_{yy}\big)\sin 2\theta + \sigma_{xy}\cos 2\theta\,.\end{aligned} \tag{8.69}$$

It is seen that the mean stress $\sigma_{x'x'}+\sigma_{y'y'}$ remains the same as in the original coordinate system, independent of the angle θ. It is therefore called *invariant*. The two 'components' $\frac{1}{2}(\sigma_{xx}-\sigma_{yy})$ and σ_{xy} transform like a vector, but with the double angle 2θ. This transformation rule is illustrated graphically by Mohr's circle in Section 8.4.3.

Example 8.3. Biaxial stress. A common condition in a test specimen consists of an axial stress σ_{xx} in combination with a transverse stress σ_{yy}, a so-called biaxial state of stress. This state of plane stress is illustrated in Fig. 8.20a. The transformation formula (8.69) enables direct determination of the stress components acting on a section inclined by the angle θ as shown in Fig. 8.20b.

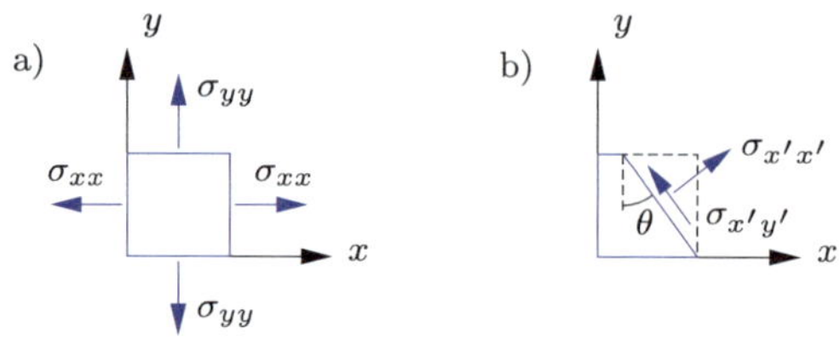

Fig. 8.20: Test specimen with normal stresses σ_{xx} and σ_{yy}.

Straightforward application of (8.69) gives the normal stress

$$\sigma_{x'x'} = \tfrac{1}{2}\big(\sigma_{xx} + \sigma_{yy}\big) + \tfrac{1}{2}\big(\sigma_{xx} - \sigma_{yy}\big)\cos 2\theta$$

and the shear stress

$$\sigma_{x'y'} = -\tfrac{1}{2}\big(\sigma_{xx} - \sigma_{yy}\big)\sin 2\theta\,.$$

This generalizes the result of Example 8.1 for uniaxial tension, $\sigma_{xx} = T/A$. Note, that if the two original stress components are equal, $\sigma_{xx} = \sigma_{yy}$, there is no shear stress on any section in the body. □

Example 8.4. Equal tension-compression. A special case of the biaxial stress state of Example 8.3 is the combination of compression and tension of equal magnitude.

$$-\sigma_{xx} = \sigma_{yy} = \sigma\,, \qquad \sigma_{xy} = 0\,.$$

This state of stress is illustrated in Fig. 8.21a. It is interesting to find the stress components on axes rotated $\theta = 45°$ relative to the original axes. In this case $\cos 2\theta = 0$ and $\sin 2\theta = 1$. The stress components on the rotated axes then follows from (8.69) as

$$\sigma_{x'x'} = \sigma_{y'y'} = 0\,, \qquad \sigma_{x'y'} = \sigma\,.$$

Thus, a combination of compression and tension of equal magnitude corresponds to a state of shear stress on axes rotated 45° relative to the original axes. □

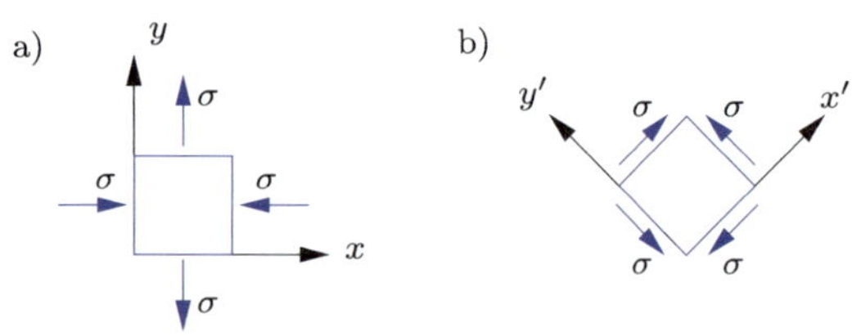

Fig. 8.21: Equal compression $\sigma_{xx} = -\sigma$ and tension $\sigma_{yy} = \sigma$.

Transformation of plane strain components

According to the general three-dimensional formula (8.42) the strain components transform in the same way as the stress components. The special two-dimensional transformation is therefore identical to that of stresses, whereby

$$\begin{aligned} \varepsilon_{x'x'} &= \tfrac{1}{2}\big(\varepsilon_{xx} + \varepsilon_{yy}\big) + \tfrac{1}{2}\big(\varepsilon_{xx} - \varepsilon_{yy}\big)\cos 2\theta + \varepsilon_{xy}\sin 2\theta\,, \\ \varepsilon_{y'y'} &= \tfrac{1}{2}\big(\varepsilon_{xx} + \varepsilon_{yy}\big) - \tfrac{1}{2}\big(\varepsilon_{xx} - \varepsilon_{yy}\big)\cos 2\theta - \varepsilon_{xy}\sin 2\theta\,, \\ \varepsilon_{x'y'} &= \qquad\qquad\qquad\quad - \tfrac{1}{2}\big(\varepsilon_{xx} - \varepsilon_{yy}\big)\sin 2\theta + \varepsilon_{xy}\cos 2\theta\,. \end{aligned} \tag{8.70}$$

The use of the strain transformation is illustrated by the following example.

Example 8.5. Strain gauge rosette. The strain components in the surface of a continuous body can be measured by so-called strain gauges. A strain gauge is a small length of conducting wire or foil with an electric resistance that depends on the elongation of the wire. By measuring the resistance the axial strain in the direction of the wire can be determined.

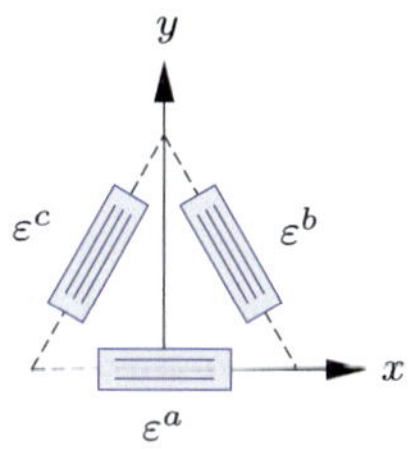

Fig. 8.22: Strain gauge delta rosette with equal angles 60°.

Figure 8.22 shows a delta type strain gauge rosette consisting of three strain gauges forming an isosceles triangle. The axial strains in the strain gauges are denoted ε^a, ε^b and ε^c, and a coordinate system $\{x, y\}$ is introduced as indicated. The strains ε^a, ε^b and ε^c in the strain gauges can now be expressed in terms of the three strain components ε_{xx}, ε_{yy} and ε_{xy} by use of (8.70a) with $\theta = 0°, 120°, -120°$, respectively. When using that

$$\cos(\pm 240°) = -\tfrac{1}{2}\,, \qquad \sin(\pm 240°) = \mp\tfrac{1}{2}\sqrt{3}\,,$$

the strain in the individual strain gauges are found to be

$$\varepsilon^a = \varepsilon_{xx}\,, \quad \varepsilon^b = \tfrac{1}{4}\varepsilon_{xx} + \tfrac{3}{4}\varepsilon_{yy} - \tfrac{1}{2}\sqrt{3}\,\varepsilon_{xy}\,, \quad \varepsilon^c = \tfrac{1}{4}\varepsilon_{xx} + \tfrac{3}{4}\varepsilon_{yy} + \tfrac{1}{2}\sqrt{3}\,\varepsilon_{xy}.$$

These relations are easily inverted to give the strain components ε_{xx}, ε_{yy} and ε_{xy} in terms of the measured axial strains ε^a, ε^b and ε^c,

$$\varepsilon_{xx} = \varepsilon^a\,, \quad \varepsilon_{yy} = \tfrac{2}{3}\big(\varepsilon^c + \varepsilon^b - \tfrac{1}{2}\varepsilon^a\big)\,, \quad \varepsilon_{xy} = \big(\varepsilon^c - \varepsilon^b\big)/\sqrt{3}.$$

The mean strain $\varepsilon_{xx} + \varepsilon_{yy}$ in the plane takes a particularly simple form,

$$\varepsilon_{xx} + \varepsilon_{yy} = \tfrac{2}{3}\big(\varepsilon^a + \varepsilon^b + \varepsilon^c\big).$$

It is seen that each of the strain gauges contribute equally to the mean strain. The factor $\frac{2}{3}$ can be determined directly be considering a state of equal extension $\varepsilon_{xx} = \varepsilon_{yy}$ in all directions. In that case it follows immediately that $\varepsilon^a = \varepsilon^b = \varepsilon^c = \varepsilon_{xx} = \varepsilon_{yy}$. □

Direct stress transformation by equilibrium

When a stress component on an inclined section is needed, it is expedient simply to set up the appropriate equilibrium equation. The method is illustrated with reference to Fig. 8.23, showing a section with normal inclined by the angle θ to the x-axis.

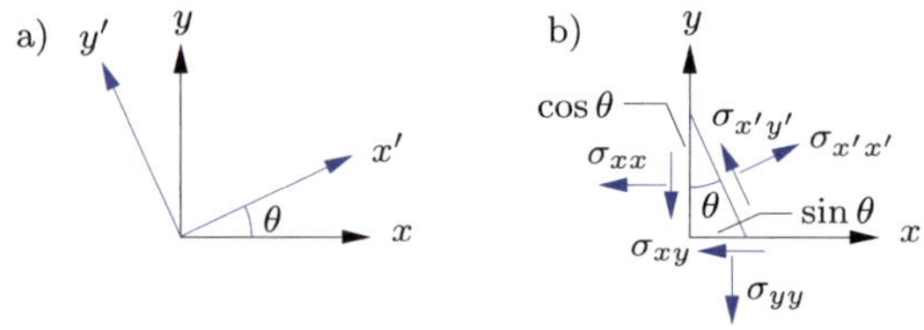

Fig. 8.23: 'New' stress components $\sigma_{x'x'}$ and $\sigma_{x'y'}$ by equilibrium.

The stress components $\sigma_{x'x'}$ and $\sigma_{x'y'}$ acting on the inclined section are determined by considering equilibrium of the triangle shown in Fig. 8.23b. Let the length of the inclined side be unity, whereby the side along the x-axis is $\sin\theta$ and the length of the side along the y-axis is $\cos\theta$. The normal stress $\sigma_{x'x'}$ then follows directly from force equilibrium in the direction of the normal to the inclined section,

$$\begin{aligned} \sigma_{x'x'} &= \big(\sigma_{xx}\cos\theta + \sigma_{xy}\sin\theta\big)\cos\theta \,+\, \big(\sigma_{yx}\cos\theta + \sigma_{yy}\sin\theta\big)\sin\theta \\ &= \sigma_{xx}\cos^2\theta \,+\, \sigma_{yy}\sin^2\theta \,+\, 2\,\sigma_{xy}\sin\theta\cos\theta\,. \end{aligned} \tag{8.71}$$

Similarly, force equilibrium along the direction of the tangent of the inclined section gives the shear stress component $\sigma_{x'y'}$,

$$\begin{aligned} \sigma_{x'y'} &= -\big(\sigma_{xx}\cos\theta + \sigma_{xy}\sin\theta\big)\sin\theta \,+\, \big(\sigma_{yx}\cos\theta + \sigma_{yy}\sin\theta\big)\cos\theta \\ &= -\big(\sigma_{xx} - \sigma_{yy}\big)\sin\theta\cos\theta \,+\, \sigma_{xy}\big(\cos^2\theta - \sin^2\theta\big). \end{aligned} \tag{8.72}$$

Clearly, these results are identical to (8.66a) and (8.66c).

Example 8.6. Inclined shear by equilibrium. The problem of shear produced by equal compression and tension considered in Example 8.4 has a very simple solution in terms of equilibrium of an appropriate triangle as shown in Fig. 8.24. When the length of the inclined section in Fig. 8.24b is taken as unity, force projection on the tangent gives

$$\tau = \big(\tfrac{1}{2}\sqrt{2}\,\sigma + \tfrac{1}{2}\sqrt{2}\,\sigma\big)/\sqrt{2} = \sigma\,,$$

where the factors inside the parenthesis represent the area of the sides with normal stress, while the last factor represents the projection on the inclined direction.

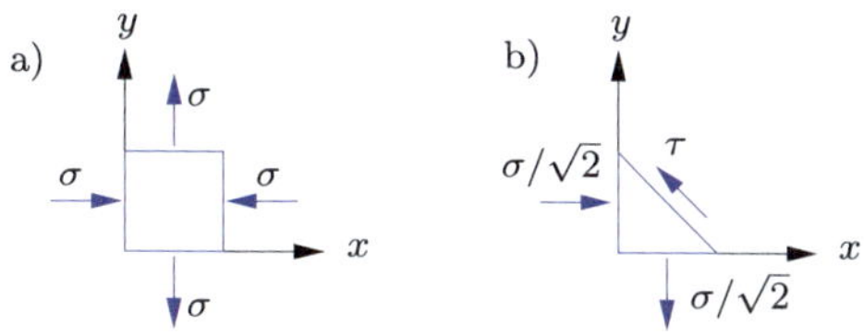

Fig. 8.24: Equilibrium for the shear problem, $-\sigma_{xx} = \sigma_{yy} = \sigma$.

It follows immediately from projection on the normal, that the normal stress on the inclined section vanishes, thus establishing a state of pure shear in the rotated system. □

8.4.3 Principal stresses and strains in a plane

Transformation of stress components provides the answer to the following important question. Is there for an arbitrary state of stress a particular coordinate system, in which all the shear stress components vanish? If such a system exists, the implication is that any stress state can be characterized by three normal stress components, acting on opposite sides of a small cube oriented appropriately in the body. It turns out that the answer is affirmative, and these particular normal stress components are called principal stresses, while the corresponding coordinate system is called the principal coordinate system. The principal stresses play a central role in the development of theories for material behavior as discussed in Chapter 9. The concept of principal stresses and the associated theory is first introduced for plane problems and then generalized to three-dimensional problems.

Principal stresses in plane stress

The concept of principal stresses in a two-dimensional state of stress is illustrated in Fig. 8.25. A stress state is given in terms of its components σ_{xx}, σ_{yy} and σ_{xy} in a coordinate system $\{x, y\}$. The problem is to determine if there is a coordinate system $\{x', y'\}$ in which the shear stress component $\sigma_{x'y'}$ vanishes. If this is the case, the small square shown in Fig. 8.25b will only have normal stress components.

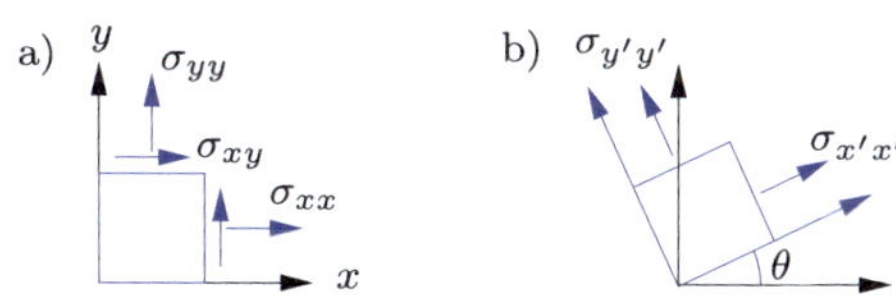

Fig. 8.25: Principal stresses σ_x, σ_y and rotation angle θ.

The shear stress component $\sigma_{x'y'}$ in a coordinate system that is rotated by the angle θ relative to the original system is given by (8.69c). The shear stress will vanish for an angle θ, determined by

$$\sigma_{x'y'} = -\tfrac{1}{2}(\sigma_{xx} - \sigma_{yy}) \sin 2\theta + \sigma_{xy} \cos 2\theta = 0 . \tag{8.73}$$

Thus, the angle defining the principal coordinate system is determined by

$$\tan 2\theta = \frac{2\sigma_{xy}}{\sigma_{xx} - \sigma_{yy}} . \tag{8.74}$$

In the principal coordinate system there are only two stress components, and they are therefore denoted by one subscript only, according to the definition

$$\sigma_{x'x'} = \sigma_x , \qquad \sigma_{y'y'} = \sigma_y , \qquad \sigma_{x'y'} = 0 . \tag{8.75}$$

The stress components σ_x and σ_y are the principal stresses.

In a rotated coordinate system the normal stress components are given by (8.69a) and (8.69b). The principal stresses σ_x and σ_y can be determined by combining these formulae with (8.74) for the angle θ. First it is observed that by (8.69a) the sum of the normal stresses is independent of the angle θ,

$$\tfrac{1}{2}(\sigma_x + \sigma_y) = \tfrac{1}{2}(\sigma_{xx} + \sigma_{yy}). \tag{8.76}$$

Similarly, the difference between the normal stresses follows as

$$\tfrac{1}{2}(\sigma_x - \sigma_y) = \tfrac{1}{2}(\sigma_{xx} - \sigma_{yy}) \cos 2\theta + \sigma_{xy} \sin 2\theta . \tag{8.77}$$

When adding the square of this equation to the square of the equation (8.73) the angle θ is eliminated, leaving the equation

$$\tfrac{1}{4}(\sigma_x - \sigma_y)^2 = \tfrac{1}{4}(\sigma_{xx} - \sigma_{yy})^2 + \sigma_{xy}^2 , \tag{8.78}$$

from which

$$\tfrac{1}{2}(\sigma_x - \sigma_y) = \pm\sqrt{\tfrac{1}{4}(\sigma_{xx} - \sigma_{yy})^2 + \sigma_{xy}^2} . \tag{8.79}$$

The principal stresses σ_x and σ_y are now obtained as the sum and difference of (8.76) and (8.79),

$$\left.\begin{matrix} \sigma_x \\ \sigma_y \end{matrix}\right\} = \frac{\sigma_{xx} + \sigma_{yy}}{2} \pm \sqrt{\left(\frac{\sigma_{xx} - \sigma_{yy}}{2}\right)^2 + \sigma_{xy}^2} . \tag{8.80}$$

This formula is illustrated in the so-called Mohr's circle diagram after a brief discussion of the corresponding result for strains.

Example 8.7. Tension-torsion of thin-walled cylinder. Some tests are carried out by subjecting a thin-walled cylinder to tension N and torsion M. This produces an axial tension stress σ and a shear stress τ in the cylinder wall as illustrated in Fig. 8.26.

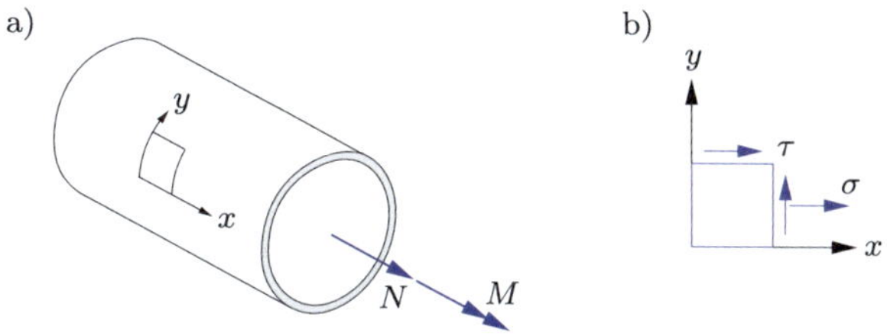

Fig. 8.26: Combined tension σ and shear τ stresses in thin-walled cylinder.

For a thin-walled cylinder the variation of the stresses over the wall thickness is negligible, and the stress state is therefore described by the components

$$\sigma_{xx} = \sigma\,, \quad \sigma_{yy} = 0\,, \quad \sigma_{xy} = \tau\,.$$

In this state of stress the principal stress components are

$$\left.\begin{matrix}\sigma_x\\ \sigma_y\end{matrix}\right\} = \tfrac{1}{2}\sigma \pm \sqrt{\tfrac{1}{4}\sigma^2 + \tau^2}\,,$$

and the angle θ follows from (8.74) as

$$\tan 2\theta = \frac{2\tau}{\sigma}\,.$$

If σ decreases, the principal stresses approach $\sigma_1 \simeq -\sigma_2 \simeq \tau$, and the angle approaches $\theta = 45°$. This state of stress corresponds to pure torsion with the principal stresses of equal magnitude and opposite sign, inclined by 45° to the axis of the cylinder. □

Principal strains in plane strain

The principal coordinate system with respect to strains is defined by the property that in this coordinate system the shear strain component(s) vanish. This property is illustrated in Fig. 8.27. In plane strain a square with sides parallel to the principal coordinate axes deforms into a rectangle, retaining right angles at the corners. The corresponding strains are called the principal strains.

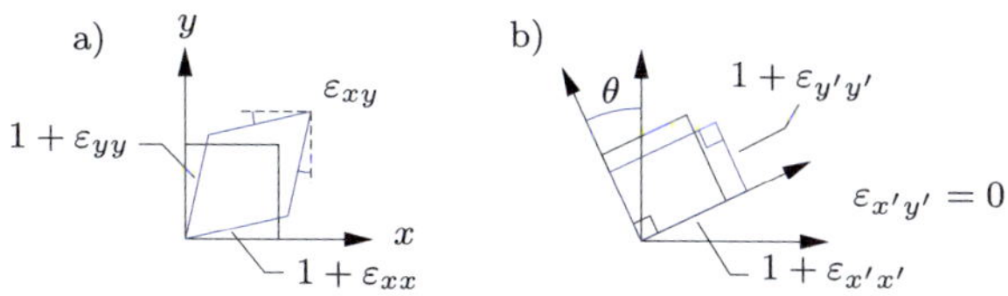

Fig. 8.27: Principal strains $\varepsilon_x, \varepsilon_y$ and angle θ.

The angle θ defining the principal strain coordinate system is determined from the shear strain formula (8.70c),

$$\tan 2\theta = \frac{2\varepsilon_{xy}}{\varepsilon_{xx} - \varepsilon_{yy}}. \tag{8.81}$$

The corresponding principal strains ε_1 and ε_2 follow from (8.70a) and (8.70b) in exactly the same way as for the corresponding principal stresses. Thus the result is

$$\left.\begin{array}{l}\varepsilon_x\\ \varepsilon_y\end{array}\right\} = \frac{\varepsilon_{xx} + \varepsilon_{yy}}{2} \pm \sqrt{\left(\frac{\varepsilon_{xx} - \varepsilon_{yy}}{2}\right)^2 + \varepsilon_{xy}^2}\,. \tag{8.82}$$

This formula can also be illustrated by Mohr's circle diagram, shown for stress components in the next section.

Mohr's circle for plane stress

The stress state at a point can be illustrated by a geometrical diagram due to MOHR (1835–1918). It is convenient to consider the principal coordinate system as the starting point. In this coordinate system the stress state is given by the principal stresses as $\sigma_{xx} = \sigma_x, \sigma_{yy} = \sigma_y, \sigma_{xy} = 0$. In the following the principal axes are chosen such that $\sigma_x \geq \sigma_y$. The stress state $\sigma_{x'x'}, \sigma_{y'y'}, \sigma_{x'y'}$ in a coordinate system rotated by the angle θ then follows from (8.68) as

$$\begin{aligned}
\sigma_{x'x'} &= \frac{\sigma_x + \sigma_y}{2} + \frac{\sigma_x - \sigma_y}{2}\cos 2\theta\,,\\
\sigma_{y'y'} &= \frac{\sigma_x + \sigma_y}{2} - \frac{\sigma_x - \sigma_y}{2}\cos 2\theta\,,\\
\sigma_{x'y'} &= \qquad\qquad\;\; - \frac{\sigma_x - \sigma_y}{2}\sin 2\theta\,,
\end{aligned} \tag{8.83}$$

where it has been used that the shear stress vanishes in the principal coordinate system. The stress state $\sigma_{x'x'}, \sigma_{y'y'}, \sigma_{x'y'}$ can be illustrated in the

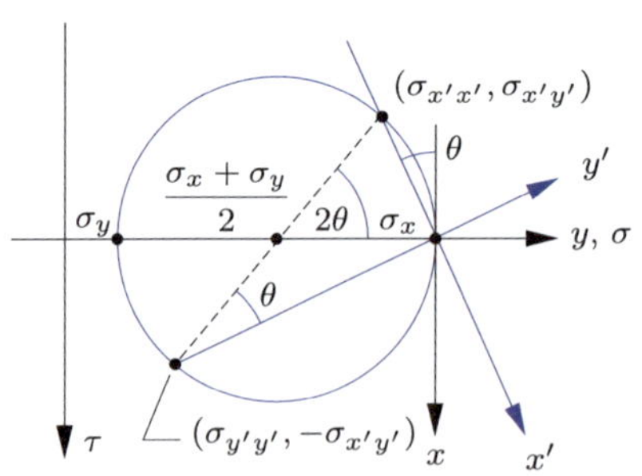

Fig. 8.28: Mohr's circle for plane stress components $\sigma_{x'x'}, \sigma_{y'y'}, \sigma_{x'y'}$.

$\{\sigma, \tau\}$-coordinate system shown in Fig. 8.28. In this coordinate system the stresses $[\sigma_{x'x'}, \sigma_{x'y'}]$ and $[\sigma_{y'y'}, -\sigma_{x'y'}]$ are located at diametrically opposite points of a circle centered at $\left[\frac{1}{2}(\sigma_x + \sigma_y), 0\right]$ with radius $\frac{1}{2}(\sigma_x - \sigma_y)$. The rotated $\{x'y'\}$-coordinate system is shown to illustrate the angle θ. By a geometric argument this angle appears at the point $[\sigma_{y'y'}, -\sigma_{x'y'}]$ as a periphery angle, and therefore appears with 2θ as center angle. With the center angle 2θ established the coordinates of the points $[\sigma_{x'x'}, \sigma_{x'y'}]$ and $[\sigma_{y'y'}, -\sigma_{x'y'}]$ are seen to reproduce the transformation formulae (8.83). As the angle θ increases, the points move counter-clockwise around the circle.

The circle is characterized by the parameters

$$\text{Center:}\ \frac{\sigma_x + \sigma_y}{2}, \qquad \text{Radius:}\ \frac{\sigma_x - \sigma_y}{2}, \qquad \tau_{\max} = \frac{\sigma_x - \sigma_y}{2}, \tag{8.84}$$

where the last formula states that the maximal shear stress on any section is given by the radius of the circle $\tau_{\max}$. It also follows directly from the figure that the maximum normal stress is the principal stress σ_x, and the minimum normal stress is the principal stress σ_y.

Example 8.8. Mohr's circle for tension-shear. The stress state in the wall of the thin-walled cylinder in tension and torsion considered in Example 8.18 was

$$\sigma_{xx} = \sigma, \quad \sigma_{yy} = 0, \quad \sigma_{xy} = \tau.$$

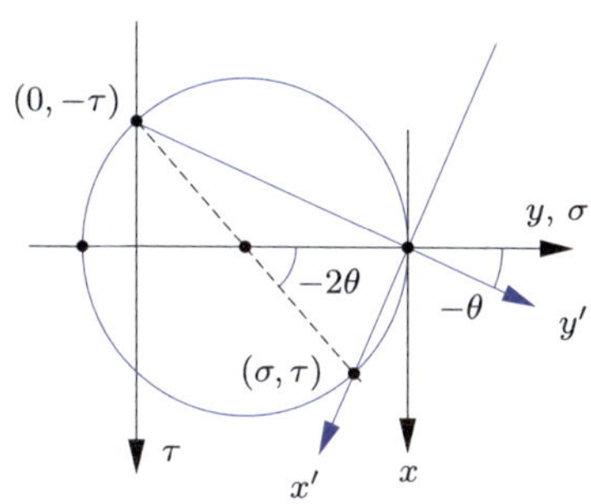

Fig. 8.29: Mohr's circle for combined tension and shear.

This stress state is shown in a Mohr's circle diagram in Fig. 8.29. Note, that the angle θ is negative in the diagram, because it denotes the angle from the principal axes to the axes of the stress state under consideration. □

The transformation formulae (8.70) for the strain components are identical to (8.68) for the stress components, and it is then clear that the Mohr's circle diagram can also be used to illustrate the strain components in a coordinate system that is rotated θ relative to the principal strain coordinate system.

8.4.4 Principal stresses in three dimensions

In three-dimensional problems the concept of principal stress is conveniently defined in terms of vectors. Figure 8.30 illustrates that when $\mathbf{n}$ is a unit vector in a principal direction, the corresponding stress vector is proportional to $\mathbf{n}$,

$$\mathbf{t}^n = \sigma\,\mathbf{n}\,, \tag{8.85}$$

and σ is the magnitude of the corresponding principal stress. In two-dimensional problems the direction is determined by a single angle θ, while in three dimensions the direction involves all three components of the vector $\mathbf{n}$. The technique for three-dimensional problems is therefore somewhat different from the simple two-dimensional problem considered above.

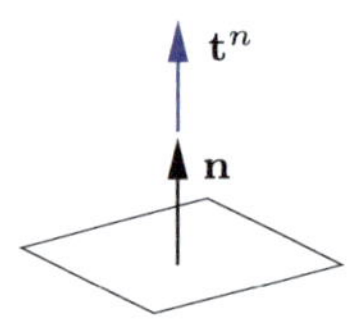

Fig. 8.30: Principal stress normal to the surface.

Analytical determination from invariants

In order to identify the principal directions the defining vector relation (8.85) is expressed by the stress component matrix $\boldsymbol{\sigma}$ as

$$\boldsymbol{\sigma}\,\mathbf{n} = \sigma\,\mathbf{n}\,. \tag{8.86}$$

When introducing the unit matrix $\mathbf{I}$, this relation can be written as

$$\big[\boldsymbol{\sigma} - \sigma\mathbf{I}\big]\,\mathbf{n} = \mathbf{0}\,. \tag{8.87}$$

This is an eigenvalue problem consisting of a set of homogeneous equations that will only have a non-trivial solution $\mathbf{n} = [n_x, n_y, n_z]^T$ for particular values of the parameter σ, the so-called eigenvalues. In component form the eigenvalue problem is

$$\begin{bmatrix} \sigma_{xx}-\sigma & \sigma_{xy} & \sigma_{xz} \\ \sigma_{yx} & \sigma_{yy}-\sigma & \sigma_{yz} \\ \sigma_{zx} & \sigma_{zy} & \sigma_{zz}-\sigma \end{bmatrix} \begin{bmatrix} n_x \\ n_y \\ n_z \end{bmatrix} = \begin{bmatrix} 0 \\ 0 \\ 0 \end{bmatrix}. \tag{8.88}$$

Non-trivial solutions to homogeneous equations can only be obtained, when the determinant of the coefficient matrix vanishes. Thus, the condition for a non-trivial solution is the determinant equation

$$\det\left[\boldsymbol{\sigma} - \sigma\,\mathbf{I}\right] = \begin{vmatrix} \sigma_{xx} - \sigma & \sigma_{xy} & \sigma_{xz} \\ \sigma_{yx} & \sigma_{yy} - \sigma & \sigma_{yz} \\ \sigma_{zx} & \sigma_{zy} & \sigma_{zz} - \sigma \end{vmatrix} = 0\,, \tag{8.89}$$

with the initially unknown parameter σ.

The determinant can be expanded into a cubic polynomial, whereby the equation takes the form

$$\sigma^3 - I_1\,\sigma^2 + I_2\,\sigma - I_3 = 0\,. \tag{8.90}$$

The three roots of this cubic equation are the three principal stresses. The value of the principal stresses are independent of the original coordinate system, and the coefficients I_1, I_2 and I_3 of the cubic equation must therefore also be independent of the particular choice of the initial coordinate system. The coefficients I_1, I_2 and I_3 are therefore called the (stress) invariants. They are evaluated by expanding the determinant, whereby

$$\begin{aligned}
I_1 &= \sigma_{xx} + \sigma_{yy} + \sigma_{zz}\,,\\
I_2 &= \begin{vmatrix} \sigma_{yy} & \sigma_{yz} \\ \sigma_{zy} & \sigma_{zz} \end{vmatrix} + \begin{vmatrix} \sigma_{zz} & \sigma_{xz} \\ \sigma_{zx} & \sigma_{xx} \end{vmatrix} + \begin{vmatrix} \sigma_{xx} & \sigma_{xy} \\ \sigma_{yx} & \sigma_{yy} \end{vmatrix}\\
&= \sigma_{yy}\sigma_{zz} + \sigma_{zz}\sigma_{xx} + \sigma_{xx}\sigma_{yy} - \sigma_{yz}^2 - \sigma_{xz}^2 - \sigma_{xy}^2\,, \qquad (8.91)\\
I_3 &= \begin{vmatrix} \sigma_{xx} & \sigma_{xy} & \sigma_{xz} \\ \sigma_{yx} & \sigma_{yy} & \sigma_{yz} \\ \sigma_{zx} & \sigma_{zy} & \sigma_{zz} \end{vmatrix} = \sigma_{xx}\,\sigma_{yy}\,\sigma_{zz} + 2\sigma_{yz}\,\sigma_{xz}\,\sigma_{xy}\\
&\qquad - \sigma_{xx}\,\sigma_{yz}^2 - \sigma_{yy}\,\sigma_{xz}^2 - \sigma_{zz}\,\sigma_{xy}^2\,.
\end{aligned}$$

It is observed that the mean stress is $\frac{1}{3}I_1$, and thereby invariant. This was already found for two-dimensional stress states in (8.69a).

Stress invariants are used to find the principal stresses by solving the cubic equation (8.90). They are also useful in theories of material behavior. When used in this connection they are often expressed in terms of the principal stresses. These expressions are found by removing the shear components from the general formulae (8.91), whereby

$$\begin{aligned}
I_1 &= \sigma_x + \sigma_y + \sigma_z\,,\\
I_2 &= \sigma_y\,\sigma_z + \sigma_x\,\sigma_z + \sigma_x\,\sigma_y\,, \qquad (8.92)\\
I_3 &= \sigma_x\,\sigma_y\,\sigma_z\,.
\end{aligned}$$

The transformation rules are identical for stress and strain components, and the present theory of principal stresses in terms of stress invariants therefore has an exact counterpart for principal strains in terms of strain invariants. The use of invariants in models of material behavior is discussed in Chapter 9.

Mohr's circles for triaxial stress

There is also a Mohr's circle diagram for a general three-dimensional stress state. However, it is much more complex than the two-dimensional case, and while the two-dimensional Mohr's circle can often be used in practical analysis of stress states, the three-dimensional Mohr's circle mainly serves as an illustration of the properties of a general state of stress.

Like in the two-dimensional case it is convenient to start in the principal coordinate system, and to assume the principal stresses to be ordered with respect to magnitude. Thus, in the principal coordinate system the stress state is given by

$$\boldsymbol{\sigma} = \begin{bmatrix} \sigma_x & & \\ & \sigma_y & \\ & & \sigma_z \end{bmatrix}, \qquad \sigma_x \geq \sigma_y \geq \sigma_z\,. \tag{8.93}$$

In this coordinate system the stresses on a section defined by the unit normal vector $\mathbf{n} = [n_x, n_y, n_z]$ is to be investigated. The stress vector on this section follows from (8.10) as

$$\mathbf{t}^n = \begin{bmatrix} \sigma_x & & \\ & \sigma_y & \\ & & \sigma_z \end{bmatrix} \begin{bmatrix} n_x \\ n_y \\ n_z \end{bmatrix} = \begin{bmatrix} \sigma_x n_x \\ \sigma_y n_y \\ \sigma_z n_z \end{bmatrix}. \tag{8.94}$$

The stress vector $\mathbf{t}^n$ has a normal component σ in the direction of $\mathbf{n}$ and a shear component τ in the section as illustrated in Fig. 8.2.

The objective now is to express the stress components σ and τ in terms of the components $[n_x, n_y, n_z]$ of the unit normal. The first step is to form three equations: one for the length of the normal vector, one for the length of the normal component, and one for the length of the stress vector. These three equations are

$$\begin{aligned} 1 &= |\mathbf{n}|^2 = n_x^2 + n_y^2 + n_z^2\,, \\ \sigma &= \mathbf{n}^T\mathbf{t}^n = \sigma_x\, n_x^2 + \sigma_y\, n_y^2 + \sigma_z\, n_z^2\,, \\ \sigma^2 + \tau^2 &= |\mathbf{t}^n|^2 = \sigma_x^2\, n_x^2 + \sigma_y^2\, n_y^2 + \sigma_z^2\, n_z^2\,. \end{aligned} \tag{8.95}$$

These three equations can be solved for n_x^2, n_y^2 and n_z^2. The solution is

$$n_x^2 = \frac{\tau^2 + (\sigma - \sigma_y)(\sigma - \sigma_z)}{(\sigma_x - \sigma_y)(\sigma_x - \sigma_z)}, \quad n_y^2 = \cdots \quad , \quad n_z^2 = \cdots\,, \tag{8.96}$$

where the expressions for n_y^2 and n_z^2 are found by cyclic permutation of the subscripts.

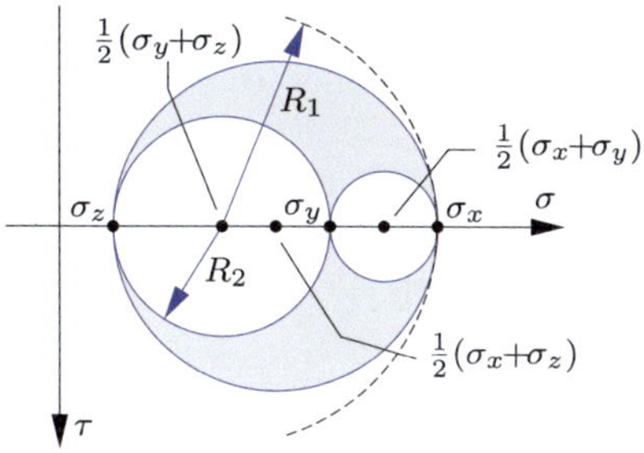

Fig. 8.31: Mohr's circles for a triaxial stress state.

The expression for n_x^2 can be rewritten as a relation between the stress components σ and τ in the form

$$(\sigma - \sigma_y)(\sigma - \sigma_z) + \tau^2 = n_x^2 (\sigma_x - \sigma_y)(\sigma_x - \sigma_z). \tag{8.97}$$

By rearranging the terms it is seen that this is the equation for a circle in the $\sigma\tau$-plane,

$$\Big(\sigma - \frac{\sigma_y + \sigma_z}{2}\Big)^2 + \tau^2 = \Big(\frac{\sigma_y - \sigma_z}{2}\Big)^2 + n_x^2 (\sigma_x - \sigma_y)(\sigma_x - \sigma_z). \tag{8.98}$$

This is the equation of a circle with center at $\big[\frac{1}{2}(\sigma_y + \sigma_z), 0\big]$ and radius R_x determined by

$$R_x^2 = \Big(\frac{\sigma_y - \sigma_z}{2}\Big)^2 + n_x^2 (\sigma_x - \sigma_y)(\sigma_x - \sigma_z). \tag{8.99}$$

The center is independent of the normal vector component n_x, while the radius attains a minimum for $n_x = 0$ and a maximum for $n_x = \pm 1$. It follows from (8.99) that

$$\Big(\frac{\sigma_y - \sigma_z}{2}\Big)^2 \le R_x^2 \le \Big(\frac{\sigma_y + \sigma_z}{2}\Big)^2 + \sigma_x^2 - \sigma_x\big(\sigma_y + \sigma_z\big). \tag{8.100}$$

By grouping the terms of the upper limit as a square the following limits on the radius are obtained,

$$\tfrac{1}{2}\big(\sigma_y + \sigma_z\big) - \sigma_z \le R_x \le \sigma_x - \tfrac{1}{2}\big(\sigma_y + \sigma_z\big). \tag{8.101}$$

It is seen that with the center at $[\frac{1}{2}(\sigma_y + \sigma_z), 0]$ the largest circle with radius R_1 passes through the point $[\sigma_x, 0]$, while the smallest circle with radius R_2 passes through the point $[\sigma_z, 0]$. These circles are shown in Fig. 8.31.

On the basis of the above analysis it can be concluded that the equation (8.96a) for the component n_x imposes the restriction that the stress state $[\sigma, \tau]$ must lie in the domain between two circles with center at $\big[\frac{1}{2}(\sigma_y+\sigma_z), 0\big]$, passing through all the three points $[\sigma_x, 0]$, $[\sigma_y, 0]$ and $[\sigma_z, 0]$. In a similar way the equations (8.96b) and (8.96c) generate equations for circles in terms of the components n_y and n_z. The stress state (σ, τ) must satisfy the conditions imposed by all three equations, each defining a domain between two circles. The common domain is shown as the hatched area in Fig. 8.31. It is seen from the figure that the maximum shear stress is given by

$$\tau_{\max} = \frac{\sigma_x - \sigma_z}{2}, \tag{8.102}$$

while the maximum and minimum normal stress is σ_y and σ_z, respectively. A more detailed understanding of the three-dimensional stress state at a point can be obtained by rotating the axes associated with any of the planes defined the principal directions, see e.g. Hibbeler (2005).

8.5 Exercises

Exercise 8.1. The figure shows a column loaded by a central compression force P. Four different cross section are considered in this exercise, and the dimensions are shown in the figure to the right, where the dimensions are $h = 60$ mm, $b = 75$ mm and $t = 10$ mm.

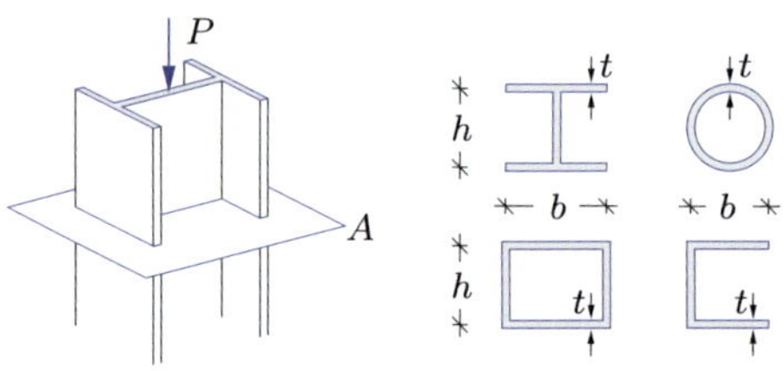

a) Determine the mean axial stress σ in the section A for each of the four cross sections.

b) Determine the magnitude of P that corresponds to $\sigma = 200$ MPa.

Exercise 8.2. The figure shows a pinned connection between two cylindrical bars. The bars are loaded by a tension force $P = 10$ kN, which is transferred through the connection by the cylindrical pin, indicated by the dashed line in the figure. The diameter of the circular cross section of the left and right bar is $a_l = 60$ mm and $a_r = 48$ mm, respectively, while it is $b = 30$ mm for the pin.

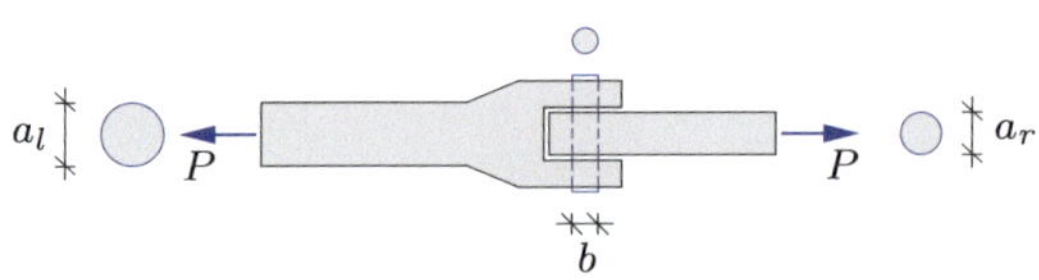

a) Determine the mean axial stress, σ_l and σ_r, in the bars.

b) Determine the mean shear stress τ in the pin.

Exercise 8.3. The figure shows a test specimen loaded in tension by the force P. The cross section is rectangular with area A. The figure shows a plane section inclined by θ relative to the loading direction. At this section the stress distribution is assumed to be homogeneous.

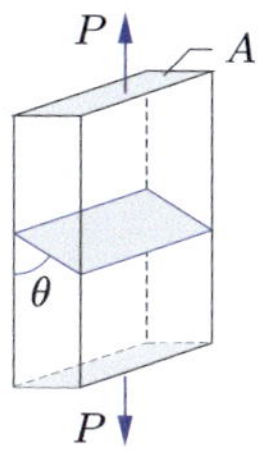

a) Determine the shear stress τ at the inclined section as function of θ.

b) Find the angle $\theta_{\max}$ that corresponds to the maximum shear stress.

c) Determine the maximum shear stress $\tau_{\max}$.

Exercise 8.4. The figure shows a test specimen loaded in compression by the force $P = 1\,\text{kN}$. The cross section is rectangular with width $b = 100\,\text{mm}$ and thickness $t = 50\,\text{mm}$. The figure shows a section inclined by $\theta = 30°$ relative to the cross section. At this section the stress distribution is assumed to be homogeneous.

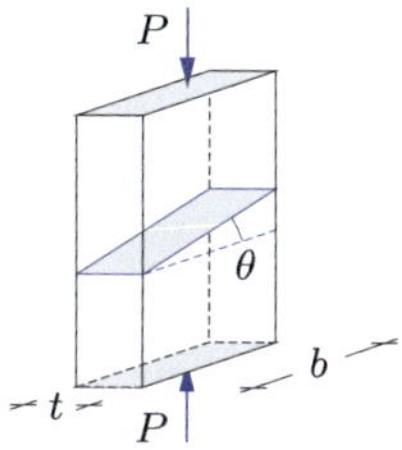

a) Determine the axial stress σ on the inclined section.

b) Determine the shear stress τ on the inclined section.

Exercise 8.5. The figure shows the stress state in terms of σ_* in plane stress conditions.

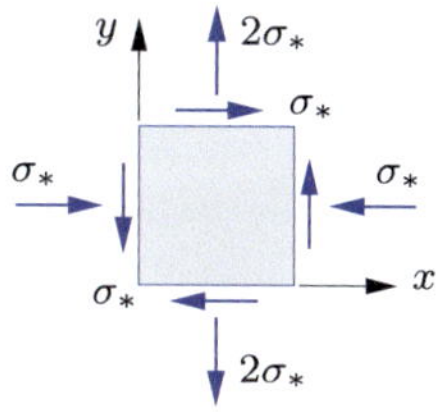

a) Identify the stress components σ_{xx}, σ_{yy} and σ_{xy}.

b) Determine the principal stresses σ_x and σ_y, and the orientation of the principal coordinate system.

c) Draw the Mohr's circle diagram and indicate the stress state of the figure in the diagram.

Exercise 8.6. The figure shows the plane stress state at a particular point in a structure.

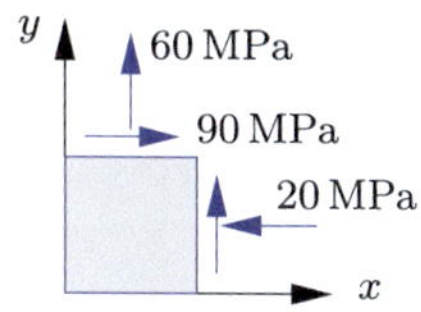

a) Identify the stress components σ_{xx}, σ_{yy} and σ_{xy}.

b) Determine the stress components in a coordinate system rotated by $\theta = 25°$.

c) Determine the principal stresses and the orientation of the principal coordinate system with respect to the $\{x, y\}$ coordinate system.

d) Draw the Mohr's circle diagram and indicate the stress state of the rotated components in b) in the diagram. Also determine the maximum shear stress $\tau_{\max}$.

Exercise 8.7. The figure shows an infinitely stiff beam AB with a simple support in A. The beam is supported by a wire CD at the center of the beam in D, and the application of a transverse tip load leads to an inclination of $\theta = 1°$ of the beam, as indicated in the figure.

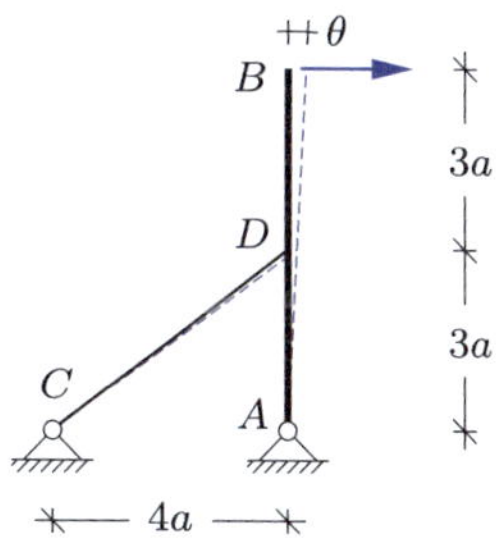

a) Determine the horizontal displacement of D and the elongation of the wire CD.

b) Determine the Green strain ε_G in (8.23).

c) Determine the linear strain ε in (8.30), and compare to ε_G.

Exercise 8.8. The figure shows a plate with dimensions $a_1 = 300\,\text{mm}$ and $a_2 = 250\,\text{mm}$. The deformation of the plate can be written in bi-linear form as

$$\begin{aligned} u(x,y) &= A_1 + B_1 x + C_1 y + D_1 xy\,, \\ v(x,y) &= A_2 + B_2 x + C_2 y + D_2 xy \end{aligned}$$

introducing eight constants A_j, B_j, C_j and D_j.

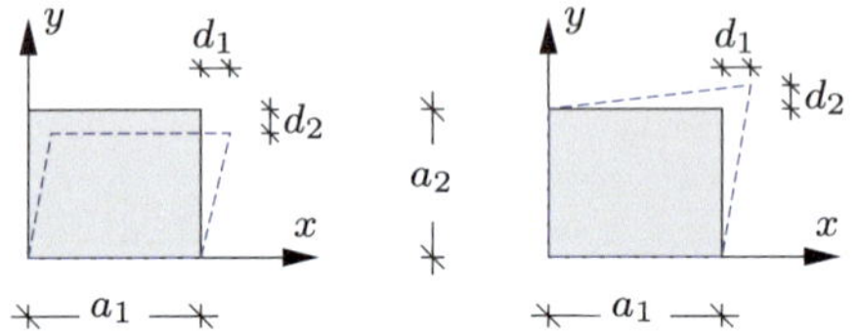

In the figure the deformation is indicated by the dashed line, where the corner displacement in both cases is determined by $d_1 = 3\,\text{mm}$ and $d_2 = 2\,\text{mm}$.

a) Find the expressions $u(x,y)$ and $v(x,y)$ for both deformation cases shown in the figure.

b) Determine an expression for the axial strains ε_{xx} and ε_{yy} and the shear strain ε_{xy}.

c) Find the value of the strains ε_{xx}, ε_{yy} and ε_{xy} at the center of the plate.

Exercise 8.9. The quadratic plate in the figure is fixed along AB and the dimensions are given by $a = 150\,\text{mm}$. The side CD is displaced $d = 2\,\text{mm}$ to the right, as indicated by the dashed line in the figure.

a) Determine the axial strain in the direction of AD.

b) Determine the axial strain in the direction of AC.

c) Determine the shear strain in the $\{xy\}$-coordinate system.

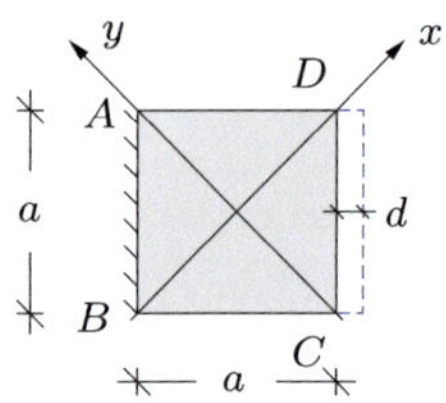

Exercise 8.10. The figure shows a cylindrical compression tank with diameter $D = 1.0\,\text{m}$ and wall thickness $t = 10\,\text{mm}$. The tank is composed of steel plates welded together at angle $\theta = 45°$ relative to longitudinal direction of the cylinder, as shown in the figure. The overpressure in the tank is $p = 10\,\text{MPa}$.

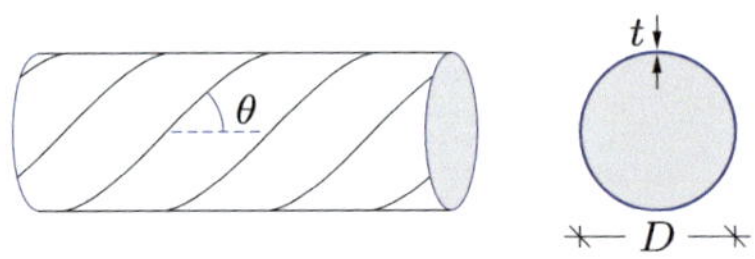

a) Determine the stresses in the longitudinal σ_l and circumferential σ_r direction, respectively.

b) Determine the axial and shear stress in the welding seam.

Exercise 8.11. At a point A in a compression tank the following strains have been measured:

$$\varepsilon_{xx} = 480 \cdot 10^{-6}\,, \qquad \varepsilon_{yy} = 720 \cdot 10^{-6}\,, \qquad \varepsilon_{xy} = 325 \cdot 10^{-6}\,.$$

It is assumed that $\varepsilon_{zz} = 0$.

a) Determine the principal strains ε_x and ε_y in A.

b) Find the maximum angular strain γ in the $\{x,y\}$ coordinate system.

c) Find the absolute maximum angular strain γ.

Exercise 8.12. The figure shows a strain gauge rosette with angle $\theta = 60°$ mounted on a plate. The strain gauge a is located along the x-axis and the individual strain readings of the rosette are:

$\varepsilon^a = 950 \cdot 10^{-6}\,, \quad \varepsilon^b = 380 \cdot 10^{-6}\,, \quad \varepsilon^c = -220 \cdot 10^{-6}\,.$

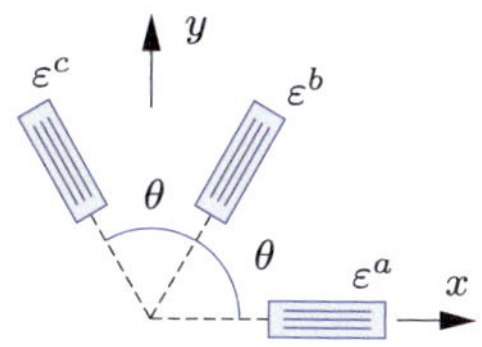

a) Determine the strain components in the x, y-plane.

b) Determine the magnitude and direction of the principal strains.

c) Draw Mohr's circle for the strains.

d) Find the magnitude of the maximum shear strain.

Exercise 8.13. The figure shows a strain gauge rosette with angle $\theta = 60°$ mounted on a plate. The strain gauge b is located along the x-axis.

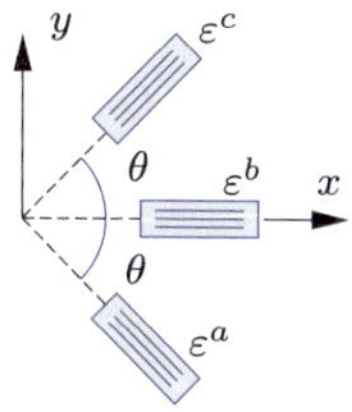

a) Express the strain components ε^a, ε^b and ε^c as function of ε_{xx}, ε_{yy} and ε_{xy}.

b) Invert the relations in a) to obtain the corresponding expressions for ε_{xx}, ε_{yy} and ε_{xy} as function of the measurements ε^a, ε^b and ε^c.

Exercise 8.14. Consider a thin-walled cylinder loaded by an axial force N and a torsion moment M as shown in Fig. 8.26 in Example 8.7. Let the cylinder have radius $r = 20t$, where t is the wall thickness. When the cylinder is thin-walled, higher order terms in t/r can be neglected, and the normal stress σ and the shear stress τ can be assumed uniformly distributed over the cross-section.

a) Express the normal stress σ in terms of N and t, and the shear stress τ in terms of M and t.

b) Consider the case $M = aN$, and determine the principal stresses σ_x and σ_y, and their angle θ with respect to the longitudinal axis of the cylinder.

c) Illustrate the stress state by a Mohr's circle diagram, and identify the various stresses and the angle 2θ in the diagram.

Exercise 8.15. The figure to the left shows a simple linear deformation field encountered in finite element analysis of beams and plates using only the axial displacement

$$u(x, y) = c\,xy\,.$$

In the figure to the right a quadratic vertical displacement has been added to give the displacement field

$$u(x, y) = c\,xy \quad , \quad v(x, y) = -\tfrac{1}{2}c\,x^2$$

representative of bending of a beam.

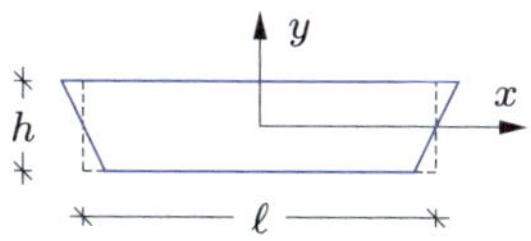

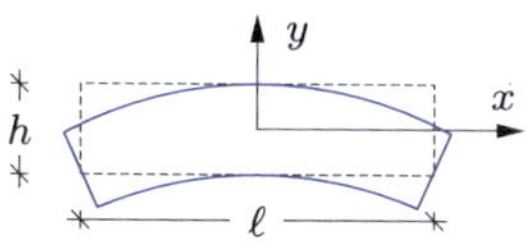

a) Determine the axial strain ε_{xx} and the shear strain $\varepsilon_{xy} = \varepsilon_{yx}$ for the linear deformation field in the left figure.

b) Determine the axial strain ε_{xx} and the shear strain $\varepsilon_{xy} = \varepsilon_{yx}$ for the quadratic deformation field in the right figure.

c) Make a sketch that illustrates the two strain fields.

Material Behavior | 9

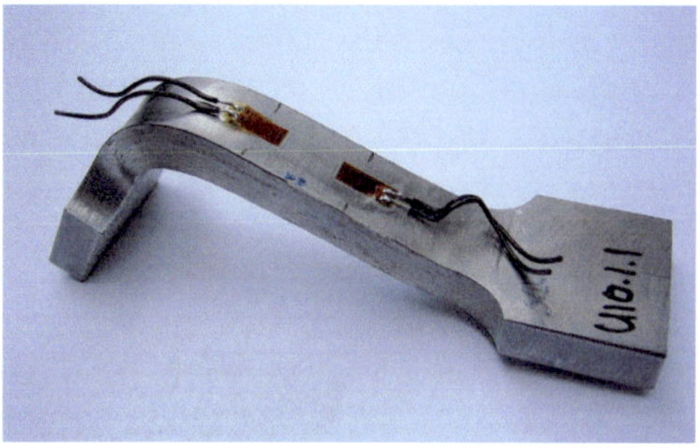

The mechanical behavior of materials depends on the relation between stresses and strains in the material. In the previous chapter stresses and strains have been discussed independently, but in order to represent material behavior, models that relate the stresses and strains must be established. The problem is illustrated in Fig. 9.1, showing a typical tension test on a metal bar. The test specimen is designed such that the load P leads to a uniformly distributed uniaxial stress σ_x in the central part of the cylindrical test specimen. This state of stress leads to an axial strain ε_x, but in most materials also to transverse contraction described by the strain components $\varepsilon_y = \varepsilon_z$. The typical behavior registered in the test is the development of strain(s), when the load and thereby the axial stress σ_x is increased, as illustrated in the right part of the figure. Typically, the first part of the stress-strain curve is reversible, meaning that it is traced backward if the specimen is unloaded. In many materials, like e.g. metals, additional irreversible strains develop, when the load exceeds a certain level. In metals this phenomenon is called yielding and is associated with movement of dislocations in the crystal structure of the material. However, also materials like concrete and soil exhibit development of nonlinear irreversible strain at sufficiently high load levels, but due to different mechanisms.

This chapter presents a brief introduction to two aspects of material behavior: elastic behavior typical of working conditions in structures, and criteria describing the states of stress limited by material failure. In both cases a central point is the systematic generalization of the theories beyond the simple uniaxial case illustrated in the figure. Section 9.1 gives a concise introduction

S. Krenk, J. Høgsberg, *Statics and Mechanics of Structures*,
DOI 10.1007/978-94-007-6113-1_9,

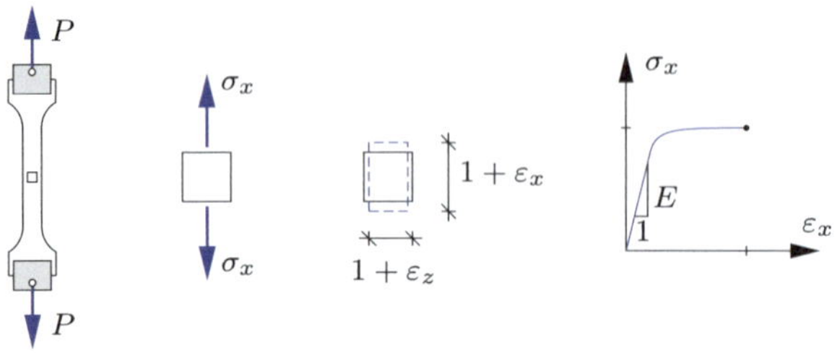

Fig. 9.1: Tension test on metal with elastic elongation, yield and failure.

to the theory of linear elasticity for a three-dimensional continuum. Simple special cases of this theory form an important tool in structural design, and the general theory constitutes the basis of most Finite Element programs, extensively used for analysis of structures, see e.g. Cook et al. (2002) and Zienkiewicz and Taylor (2000). After a brief general introduction the presentation is specialized to linear elasticity of isotropic materials, i.e. materials with properties that do not depend on the orientation. Anisotropy plays an important role in connection with composite materials. The basic theory has been presented by Lekhnitskii (1963), while applications to composites are dealt with e.g. by Jones (1999) and Reddy (2004).

Nonlinear models of material behavior play an increasing role in connection with analysis of structures by numerical methods, and a broad survey of the various models has been given e.g. by Lemaitre and Chaboche (1990) and Ottosen and Ristinmaa (2005). Many of these models provide a gradual nonlinear development of strains with increasing stress up to failure of the material. A general presentation of such models is outside the scope of the present book, but the more limited subject of typical conditions for yield and failure of metals is discussed in Section 9.3. These conditions are closely related to the theory of plasticity, which provides the transition from the initial development of additional irreversible strain to failure of the material, see e.g. Chakrabarty (2009) and Ottosen and Ristinmaa (2005). There is a marked difference between the yield and failure of metals and the similar characteristics in materials like concrete and soil. This is primarily due to the fact that while the mean stress has no or at most moderate influence on yield and failure of metals, an additional triaxial compression will lead to a marked increase in the strength of materials like concrete and soil. A simple explanation of this effect is given in Section 9.4, presenting the theory of friction materials.

9.1 Elastic materials

The simplest material model consists of the assumption that there exists a linear relation between the stress components and the strain components at any point of the body. As suggested by Fig. 9.1c this behavior is often found for limited stresses and strains, and as these stress and strain levels are often typical of the operating conditions of structures, the theory of linear elasticity plays an important role in the design of structures. The present section concentrates on the linear theory of elasticity for isotropic materials, but in order to put the properties of this theory into perspective it is derived from the internal energy of the material.

9.1.1 Internal elastic energy

An elastic material is characterized by the property that any deformation imposed by the application of stresses vanish, if the stresses are removed, whereby the elastic body returns to its original shape. This property applies to any stage of the deformation process, and it is therefore reversible. This in turn implies, that the work performed on any small volume of the material by application of stresses will be stored in the form of internal energy in the material. The work performed by the stresses $\boldsymbol{\sigma}$ on a small unit cube, when it changes its state of strain by the incremental strains $d\boldsymbol{\varepsilon}$ has already been treated in connection with the virtual work equation in Section 8.3.2. The internal work per unit volume is

$$\begin{aligned} dW &= \sigma_{xx}\, d\varepsilon_{xx} + \sigma_{yy}\, d\varepsilon_{yy} + \sigma_{zz}\, d\varepsilon_{zz} \\ &\quad + \sigma_{yz}\, d\gamma_{yz} + \sigma_{zx}\, d\gamma_{zx} + \sigma_{xy}\, d\gamma_{xy} \;=\; \boldsymbol{\sigma}^T d\boldsymbol{\varepsilon}\,. \end{aligned} \tag{9.1}$$

where the last expression appears as a scalar product, when using the six-component vector notation of Section 8.3.2 for stresses and strains. The full symmetric nine-component format can also be used, when respecting the symmetry conditions.

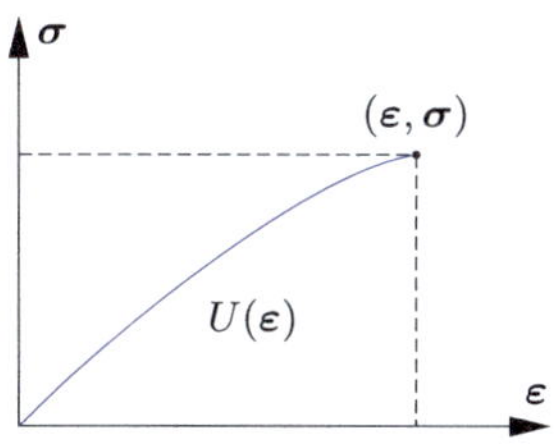

Fig. 9.2: Elastic stress–strain curve with internal energy density $U(\boldsymbol{\varepsilon})$.

In an elastic material there exists a specific internal energy density $U(\boldsymbol{\varepsilon})$, which represents the internal energy per unit volume, when the material is in a state of deformation described by the strain components $\boldsymbol{\varepsilon}$. The internal energy density as a function of the state of strain is illustrated in Fig. 9.2. When changing the state of strain in an elastic material the external work

dW is equal to the change in internal energy,

$$dW = dU(\varepsilon) = \frac{\partial U}{\partial \varepsilon}\, d\varepsilon\,. \tag{9.2}$$

The strain increment $d\varepsilon$ is arbitrary, and comparison of the expressions (9.1) and (9.2) for dW and dU then leads to the following relation between the stress components and the partial derivatives of the internal energy function,

$$\boldsymbol{\sigma}^T = \frac{\partial U}{\partial \varepsilon} = \Big[\frac{\partial U}{\partial \varepsilon_{xx}}, \frac{\partial U}{\partial \varepsilon_{yy}}, \cdots, \frac{\partial U}{\partial \gamma_{xy}}\Big]\,. \tag{9.3}$$

It is noted, that by this relation symmetry of the stress components follows from the symmetry of the strain components.

In an elastic material the stress and strain components are connected by the relation (9.3). The relation between a small change in the six stress components $d\boldsymbol{\sigma}$ and the corresponding change $d\varepsilon$ of the six independent strains can then be found by differentiation,

$$d\boldsymbol{\sigma} = d\Big(\frac{\partial U}{\partial \varepsilon}\Big) = \frac{\partial^2 U}{\partial \varepsilon^T \partial \varepsilon} d\varepsilon\,. \tag{9.4}$$

Note, that the components from the second differentiation are combined with the increment components, while the first differentiation generates a column array. This is a relation of the form

$$d\boldsymbol{\sigma} = \mathbf{D}\, d\varepsilon\,, \tag{9.5}$$

with a 6×6 component elastic stiffness matrix

$$\mathbf{D} = \frac{\partial^2 U}{\partial \varepsilon^T \partial \varepsilon} = \begin{bmatrix} D_{11} & D_{12} & D_{13} & D_{14} & D_{15} & D_{16} \\ D_{21} & D_{22} & D_{23} & D_{24} & D_{25} & D_{26} \\ D_{31} & D_{32} & D_{33} & D_{34} & D_{35} & D_{36} \\ D_{41} & D_{42} & D_{43} & D_{44} & D_{45} & D_{46} \\ D_{51} & D_{52} & D_{53} & D_{54} & D_{55} & D_{56} \\ D_{61} & D_{62} & D_{63} & D_{64} & D_{65} & D_{66} \end{bmatrix}. \tag{9.6}$$

The origin of the matrix $\mathbf{D}$ via a double derivative of the internal energy potential $U(\varepsilon)$ implies symmetry,

$$\mathbf{D}^T = \mathbf{D}\,. \tag{9.7}$$

Thus, the stiffness is specified by $\frac{1}{2}6(6+1) = 21$ independent stiffness components $D_{ij}(\varepsilon)$. However, in most cases considerable simplifications occur.

In a linear elastic material the stiffness coefficients D_{ij} are constants, and the stress-strain relation can therefore immediately be integrated from an incremental relation to a relation between current stress and current strain,

$$\boldsymbol{\sigma} = \mathbf{D}\,\boldsymbol{\varepsilon} \tag{9.8}$$

with the component form

$$\begin{bmatrix} \sigma_{xx} \\ \sigma_{yy} \\ \sigma_{zz} \\ \sigma_{yz} \\ \sigma_{zx} \\ \sigma_{xy} \end{bmatrix} = \begin{bmatrix} D_{11} & D_{12} & D_{13} & D_{14} & D_{15} & D_{16} \\ D_{21} & D_{22} & D_{23} & D_{24} & D_{25} & D_{26} \\ D_{31} & D_{32} & D_{33} & D_{34} & D_{35} & D_{36} \\ D_{41} & D_{42} & D_{43} & D_{44} & D_{45} & D_{46} \\ D_{51} & D_{52} & D_{53} & D_{54} & D_{55} & D_{56} \\ D_{61} & D_{62} & D_{63} & D_{64} & D_{65} & D_{66} \end{bmatrix} \begin{bmatrix} \varepsilon_{xx} \\ \varepsilon_{yy} \\ \varepsilon_{zz} \\ \gamma_{yz} \\ \gamma_{zx} \\ \gamma_{xy} \end{bmatrix}. \tag{9.9}$$

Note, that this relation is formulated using the 6-component format for stresses and strains in terms of the angle strains $\gamma_{xy} = 2\varepsilon_{xy}$ etc. This implies that transformation to a different coordinate system requires reformulation of the transformation relations derived in Chapter 8. This is an important aspect of elasticity theory for anisotropic materials, such as laminated composites, see e.g. Reddy (2004), but will not be treated in detail here. When formulated carefully, the boldface notation for stresses and strains can set up to cover both the component matrix and the single-array formats within the same relations. This is often used in modern continuum mechanics and finite element texts, see e.g. Zienkiewicz and Taylor (2000) and Krenk (2009).

9.1.2 Linear isotropic elasticity

For isotropic materials the material properties are independent of the particular orientation of the material relative to the load and deformation states. This implies for example that tension will produce the same effect, irrespective of the direction of the tension. Isotropy is the ultimate condition of material symmetry. A material that is not isotropic is called anisotropic. Anisotropic materials are of increasing importance in connection with the increased use of composites with properties specially designed for particular applications. An introduction to anisotropic elasticity has been given by Lekhnitskii (1963). In this section it is demonstrated how various kinds of material symmetry gradually reduce the number of independent elastic parameters, until reaching the isotropic elastic material, described by only two independent elastic parameters.

Figure 9.3 shows a small cube of a material with symmetry about the three coordinate planes. The cube is loaded by the normal stress components $\sigma_{xx}, \sigma_{yy}, \sigma_{zz}$, but without shear stress components. The normal stress components are therefore principal stresses, $\sigma_x, \sigma_y, \sigma_z$. When the material properties are symmetric about the coordinate planes the cube must deform without shear strain components, $\gamma_{yz} = \gamma_{zx} = \gamma_{xy} = 0$, because shear strains would break the symmetry. This is most easily seen by imagining a shear strain component such as $\gamma_{xy} \neq 0$. If the figure were seen from the back side of the paper the stresses and the normal strains $\varepsilon_{xx}, \varepsilon_{yy}, \varepsilon_{zz}$ would remain the same, while the shear strain γ_{xy} would have changed sign. This would break the assumed material symmetry, and thus $\gamma_{xy} = 0$.

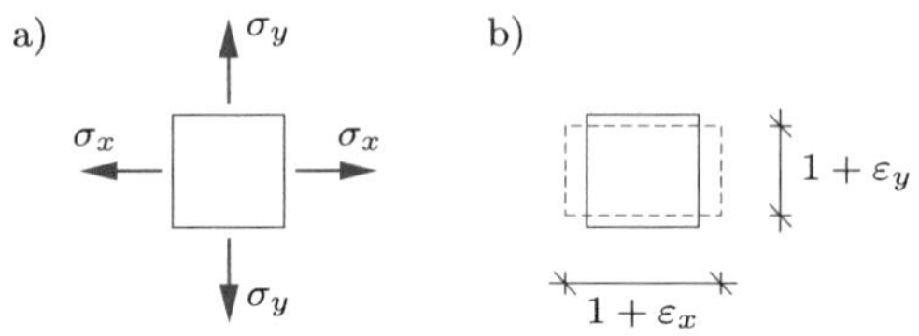

Fig. 9.3: a) Principal stresses, and b) principal strains.

In conclusion, if there is material symmetry about the coordinate planes, a state of normal stress on the coordinate planes leads to a state of normal strain on these coordinate planes. This requires that the lower left 3×3 block of the elastic stiffness matrix $\mathbf{D}$ matrix vanishes, as otherwise the normal strains would produce shear stresses. By symmetry of the $\mathbf{D}$ matrix this in turn implies that the upper right 3×3 block vanishes. This leaves the following elasticity relation for materials with symmetry about the coordinate planes

$$\begin{bmatrix} \sigma_{xx} \\ \sigma_{yy} \\ \sigma_{zz} \\ \sigma_{yz} \\ \sigma_{zx} \\ \sigma_{xy} \end{bmatrix} = \begin{bmatrix} D_{11} & D_{12} & D_{13} & 0 & 0 & 0 \\ D_{21} & D_{22} & D_{23} & 0 & 0 & 0 \\ D_{31} & D_{32} & D_{33} & 0 & 0 & 0 \\ 0 & 0 & 0 & D_{44} & D_{45} & D_{46} \\ 0 & 0 & 0 & D_{54} & D_{55} & D_{56} \\ 0 & 0 & 0 & D_{64} & D_{65} & D_{66} \end{bmatrix} \begin{bmatrix} \varepsilon_{xx} \\ \varepsilon_{yy} \\ \varepsilon_{zz} \\ \gamma_{yz} \\ \gamma_{zx} \\ \gamma_{xy} \end{bmatrix}. \tag{9.10}$$

It is seen that the block matrix format with vanishing of-diagonal blocks implies that normal stresses couple with normal strains, and shear stresses with shear strains, while there is no coupling between normal stresses and shear strains or shear stresses and normal strains. For an isotropic material any set of coordinate axes will have material symmetry, and thus the elasticity relation will always be in the form (9.10).

Before proceeding with the reduction it is convenient to introduce the inverse relation in terms of the elastic flexibility matrix $\mathbf{C} = \mathbf{D}^{-1}$. In terms of the elastic flexibility coefficients C_{ij} the relation (9.10) is

$$\begin{bmatrix} \varepsilon_{xx} \\ \varepsilon_{yy} \\ \varepsilon_{zz} \\ \gamma_{yz} \\ \gamma_{zx} \\ \gamma_{xy} \end{bmatrix} = \begin{bmatrix} C_{11} & C_{12} & C_{13} & 0 & 0 & 0 \\ C_{21} & C_{22} & C_{23} & 0 & 0 & 0 \\ C_{31} & C_{32} & C_{33} & 0 & 0 & 0 \\ 0 & 0 & 0 & C_{44} & C_{45} & C_{46} \\ 0 & 0 & 0 & C_{54} & C_{55} & C_{56} \\ 0 & 0 & 0 & C_{64} & C_{65} & C_{66} \end{bmatrix} \begin{bmatrix} \sigma_{xx} \\ \sigma_{yy} \\ \sigma_{zz} \\ \sigma_{yz} \\ \sigma_{zx} \\ \sigma_{xy} \end{bmatrix}. \tag{9.11}$$

The upper and lower blocks are now determined for isotropic materials by considering uniaxial tension and shear, respectively.

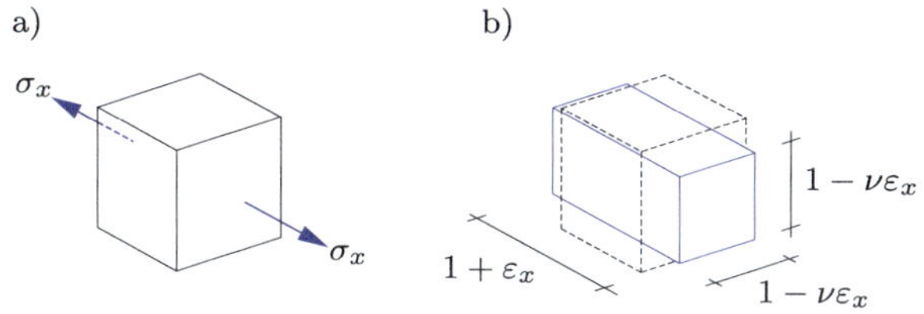

Fig. 9.4: Tension σ_x with elongation ε_x and transverse contraction $\nu\varepsilon_x$.

Figure 9.4 shows a state of uniaxial tension σ_x with $\sigma_y = \sigma_z = 0$. This state of stress leads to principal strains $\varepsilon_x, \varepsilon_y, \varepsilon_z$. The magnitude of the axial strain ε_x is described by the *modulus of elasticity* E, defined such that $\varepsilon_x = \sigma_x/E$. By symmetry the two transverse directions are identical, and thus the transverse strain may be described by a parameter ν as $\varepsilon_y = \varepsilon_z = -\nu\varepsilon_x$. The parameter ν is called *Poisson's ratio.* This state of stress and strain determines the first column in the elasticity relation (9.11),

$$\begin{bmatrix} \varepsilon_x \\ \varepsilon_y \\ \varepsilon_z \\ \vdots \end{bmatrix} = \left[\begin{array}{c|c} \begin{matrix} 1/E & & \\ -\nu/E & & \\ -\nu/E & & \end{matrix} & \\ \hline & \ddots \end{array}\right] \begin{bmatrix} \sigma_x \\ 0 \\ 0 \\ \vdots \end{bmatrix}. \tag{9.12}$$

Similarly, uniaxial stress states along the other two axes determine the next two columns. In an isotropic material the properties for the axes are identical, and thus

$$\begin{bmatrix} \varepsilon_x \\ \varepsilon_y \\ \varepsilon_z \\ \vdots \end{bmatrix} = \left[\begin{array}{c|c} \begin{matrix} 1/E & -\nu/E & -\nu/E \\ -\nu/E & 1/E & -\nu/E \\ -\nu/E & -\nu/E & 1/E \end{matrix} & \\ \hline & \ddots \end{array}\right] \begin{bmatrix} \sigma_x \\ \sigma_y \\ \sigma_z \\ \vdots \end{bmatrix}. \tag{9.13}$$

This completes the determination of the upper 3×3 block of the elasticity matrix in (9.11) in terms of the modulus of elasticity E and Poisson's ratio ν. These parameters have a direct physical interpretation in relation to a uniaxial state of stress, often used in testing of materials.

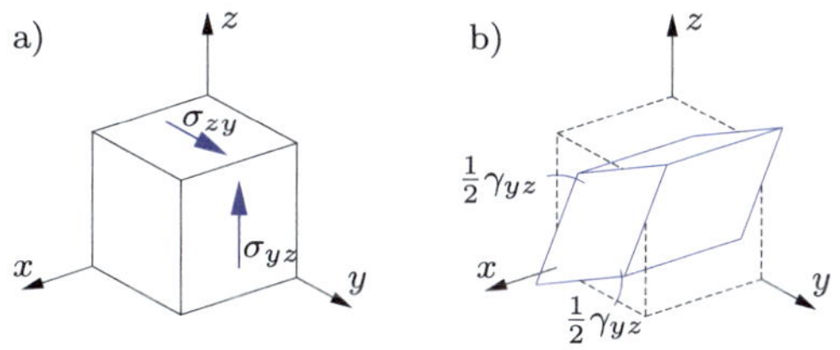

Fig. 9.5: Shear deformation in the yz-plane by angle strain γ_{yz}.

The lower 3×3 block of the elasticity matrix in (9.11) is determined by considering a state of shear, shown in Fig. 9.5. The shear stresses σ_{yz} and σ_{zy} are symmetric with respect to the yz-plane, and thus the deformation must also be symmetric with respect to this plane. This excludes the shear components γ_{xy} and γ_{zx}, leaving only a relation of the form $\gamma_{yz} = \sigma_{yz}/G$, where G is the *shear modulus*. By considering shear in each of the coordinate planes in turn, the lower block is determined as

$$\begin{bmatrix} \vdots \\ \gamma_{yz} \\ \gamma_{zx} \\ \gamma_{xy} \end{bmatrix} = \left[\begin{array}{c|ccc} \ddots & & & \\ \hline & 1/G & & \\ & & 1/G & \\ & & & 1/G \end{array}\right] \begin{bmatrix} \vdots \\ \sigma_{yz} \\ \sigma_{zx} \\ \sigma_{xy} \end{bmatrix}. \tag{9.14}$$

Thus, the lower block is diagonal, containing only a single parameter, the shear modulus G.

The complete set of stress-strain relations for an isotropic linear elastic material follows from combination of (9.13) and (9.14),

$$\begin{bmatrix} \varepsilon_{xx} \\ \varepsilon_{yy} \\ \varepsilon_{zz} \\ \gamma_{yz} \\ \gamma_{zx} \\ \gamma_{xy} \end{bmatrix} = \begin{bmatrix} 1/E & -\nu/E & -\nu/E & 0 & 0 & 0 \\ -\nu/E & 1/E & -\nu/E & 0 & 0 & 0 \\ -\nu/E & -\nu/E & 1/E & 0 & 0 & 0 \\ 0 & 0 & 0 & 1/G & 0 & 0 \\ 0 & 0 & 0 & 0 & 1/G & 0 \\ 0 & 0 & 0 & 0 & 0 & 1/G \end{bmatrix} \begin{bmatrix} \sigma_{xx} \\ \sigma_{yy} \\ \sigma_{zz} \\ \sigma_{yz} \\ \sigma_{zx} \\ \sigma_{xy} \end{bmatrix}. \tag{9.15}$$

This relation is sometimes called the generalized Hooke's law, in honor of ROBERT HOOKE (1635–1703), who considered one-dimensional extension and identified the law of proportionality.

The generalized Hooke's law (9.15) for isotropic linear elastic materials contains three elastic parameters: the modulus of elasticity E, the shear modulus G, and Poisson's ratio ν. The modulus of elasticity and the shear modulus have dimension of stress, i.e. [Force/Area] $=$ [N/m^2], while Poisson's ratio is non-dimensional. A closer analysis reveals that only two of the elasticity parameters of linear isotropic elasticity are independent. The connection between the three original parameters E, G and ν is established by considering a state of shear, generated by a combination of equal compression and tension along two of the coordinate directions. This state of stress has already been considered in Examples 8.17 and 8.19.

Figure 9.6 shows a state of shear established by combining equal compression and tension along two of the coordinate axes. The third principal stress component is assumed to vanish. The first figure shows the state of opposite principal stresses of equal magnitude, $-\sigma_x = \sigma_y = \sigma$. As illustrated in the upper right figure this corresponds to a state of shear $\sigma_{x'y'} = \sigma$ on a set of planes rotated by $\theta = 45°$ relative to the principal axes. The lower part of

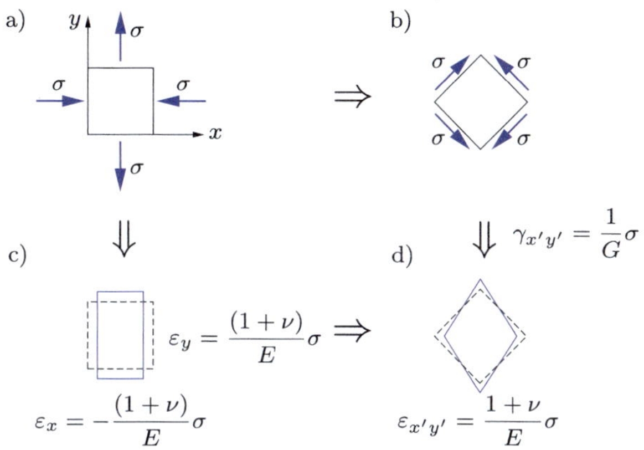

Fig. 9.6: State of shear established via equal compression and tension.

the figure shows the corresponding state of strain, calculated in the principal axis system to the left and in the rotated system to the right. In the principal axis system the strains follow from the first two rows of (9.15) as

$$-\varepsilon_x = \varepsilon_y = \frac{1+\nu}{E}\sigma\,. \tag{9.16}$$

These strains are of opposite sign and equal magnitude, and therefore transform in a similar way as the stresses to give a state of shear deformation on axes rotated by $\theta = 45^\circ$ with shear strain

$$\varepsilon_{x'y'} = \frac{1+\nu}{E}\sigma\,. \tag{9.17}$$

The shear strain in the rotated coordinate system could also have been determined directly from the shear stress $\sigma_{x'y'}$. In fact it follows directly from the last row of (9.15), when applied in the rotated coordinate system, that

$$\gamma_{x'y'} = \frac{1}{G}\sigma_{x'y'} = \frac{1}{G}\sigma\,. \tag{9.18}$$

When using that $\gamma_{x'y'} = 2\varepsilon_{x'y'}$ it follows from comparison of (9.17) and (9.18) that

$$G = \frac{E}{2(1+\nu)}\,. \tag{9.19}$$

A common value of Poisson's ratio for isotropic elastic materials is $\nu \simeq 0.25$, giving $G \simeq 0.4E$.

Volume change and the bulk modulus

It was demonstrated in Section 8.2.1 that within a linear theory the relative change of volume is given by the volume strain defined by (8.33),

$$e = \varepsilon_{xx} + \varepsilon_{yy} + \varepsilon_{zz} \,. \tag{9.20}$$

For an isotropic linear elastic material the volume strain is expressed in terms of the stresses by addition of the first three rows of the elasticity relation (9.15),

$$e = \frac{1-2\nu}{E}\,(\sigma_{xx} + \sigma_{yy} + \sigma_{zz})\,. \tag{9.21}$$

This is a relation between the volume strain e and the mean stress

$$\sigma_m = \tfrac{1}{3}(\sigma_{xx} + \sigma_{yy} + \sigma_{zz})\,. \tag{9.22}$$

The pressure is the negative of the mean stress, and the relation (9.21) therefore shows the stiffness with respect to volume changes. It is customary to express this relation in the form

$$e = \frac{\sigma_m}{K} \tag{9.23}$$

where the stiffness parameter K is the *bulk modulus*,

$$K = \frac{1}{3}\frac{E}{1-2\nu}\,. \tag{9.24}$$

The elastic modulus E, the shear modulus G and the bulk modulus K all represent the stiffness in a simple experiment. If the material presents resistance to deformation all these parameters must be positive. The relations (9.19) and (9.24) then impose restrictions on permissible values of Poisson's ratio. In fact, it follows from these relations that

$$-1 < \nu \le \tfrac{1}{2}\,. \tag{9.25}$$

In practice it is found that $0 < \nu < \frac{1}{2}$, where the upper limit corresponds to an incompressible material.

The bulk modulus K and the shear modulus G are closely related to the physical behavior of the material. The bulk modulus represents the resistance to change of volume, as expressed by the relation between the mean stress σ_m and the volume strain e. It was demonstrated in Section 8.4 that the volume strain e and the mean stress σ_m can be expressed in terms of the first invariant of the strain tensor and the stress tensor, respectively. Thus, they do not depend on the orientation of the coordinate system selected for the individual components. This confirms that the bulk modulus K is a scalar invariant, independent of any particular coordinate system. The shear modulus describes the stiffness in a shear deformation in an arbitrary plane in the material. Thus, the shear mechanism is oriented in space, but in an isotropic material the shear stiffness is independent of this orientation. Thus, the stiffness properties of an isotropic elastic material can be described by two scalar parameters K and G, each representing a well defined physical mechanism. The separation of deformation mechanisms into change of volume and change of shape is discussed further in Section 9.2.

Tension with constrained transverse contraction

It was illustrated in Fig. 9.4 that the deformation associated with uniaxial tension consists of an elongation, determined by the elastic modulus E, and a transverse contraction, determined by Poisson's ratio ν. In some cases the transverse contraction is prevented by surrounding material. This is the case e.g. for concrete used to fill the tubular members when strengthening off-shore structures, or when pulling on a bar that is cast into another material. In these cases the transverse contraction is reduced and a transverse stress appears. The magnitude of this effect depends on the stiffness of the surrounding material. The extreme case, in which transverse contraction is completely prevented, is illustrated in Fig. 9.7.

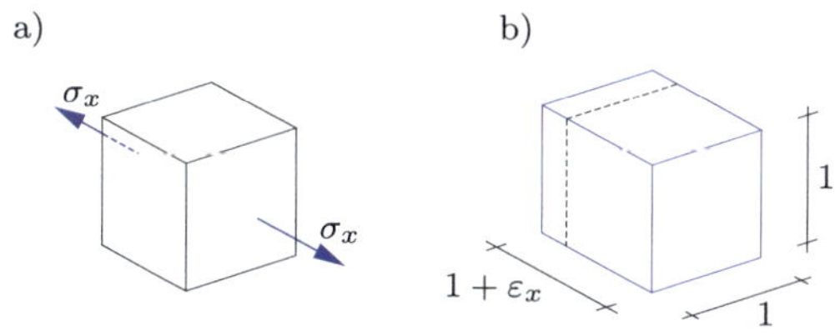

Fig. 9.7: Uni-axial tension with prevented transverse contraction.

For uniaxial tension σ_x complete prevention of the transverse contraction corresponds to $\varepsilon_y = \varepsilon_z = 0$. It follows from the symmetry of the problem that the transverse stress components are equal, $\sigma_y = \sigma_z$. Their magnitude is found by using the second equation in (9.15),

$$\varepsilon_y = -\frac{\nu}{E}\sigma_x + \frac{1-\nu}{E}\sigma_y = 0\,. \tag{9.26}$$

It follows from this equation that the transverse stress is

$$\sigma_y = \sigma_z = \frac{\nu}{1-\nu}\sigma_x\,. \tag{9.27}$$

Thus, axial tension σ_x leads to transverse tension with magnitude determined by Poisson's ratio. In the extreme case of an incompressible material $\nu = \frac{1}{2}$, whereby $\sigma_y = \sigma_z = \sigma_x$. It is seen that for a large value of Poisson's ratio

($\nu \simeq 0.45$–0.50) the transverse stress components are of the same order as the directly imposed axial stress. In most practical situations the effect is somewhat smaller. The transverse stresses act to reduce the elongation, and thus the material appears with a larger axial stiffness than in the unconstrained case. Substitution of the transverse stresses into the first equation of (9.15) gives

$$\varepsilon_x = \frac{1}{E}\sigma_x - \frac{2\nu}{E}\sigma_y = \frac{(1+\nu)(1-2\nu)}{1-\nu}\frac{1}{E}\sigma_x\,. \tag{9.28}$$

For $\nu = 0.25$ the relation becomes $\varepsilon_x = \frac{5}{6}\sigma_x/E$, corresponding to 20% increase of the axial stiffness. This is a fairly moderate effect. However, for ν approaching the incompressible value $\frac{1}{2}$ the stiffness increases to infinity. Thus, it is important that a large value of Poisson's value is estimated carefully before use, as it may have a large effect on the influence of constraints.

Example 9.1. Transverse beam deformation due to Poisson's ratio. An example of the effect of Poisson's ratio is the bending of a homogeneous isotropic beam with rectangular cross-section shown in Fig. 9.8a. For constant bending moment there are no shear stresses, and the normal stresses are

$$\sigma_x = -E\,\kappa\,y\,, \qquad \sigma_y = \sigma_z = 0\,,$$

where the elastic modulus E is included to give a convenient dimension, when κ is a representative curvature. The corresponding strains follow from the generalized Hooke's law (9.15) as

$$\varepsilon_x = -\kappa\,y\,, \qquad \varepsilon_y = \varepsilon_z = \kappa\,\nu\,y\,,$$

while all shear strain components vanish.

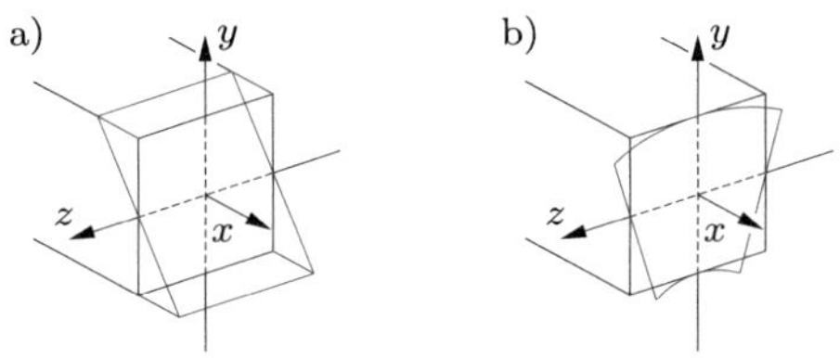

Fig. 9.8: Bending of a beam with rectangular cross-section.

The displacement component u can be integrated directly from the strain definition

$$\frac{\partial u}{\partial x} = \varepsilon_x = -\kappa\,y\,.$$

For symmetry about the yz-plane the solution is

$$u = -\kappa\,xy\,.$$

For the displacement component w the normal and shear strain conditions $\varepsilon_z = -\nu\varepsilon_x$ and $\gamma_{zx} = 0$ are

$$\frac{\partial w}{\partial z} = \varepsilon_z = \kappa\,\nu\,y\,, \qquad \frac{\partial w}{\partial x} = -\frac{\partial u}{\partial z} = 0\,.$$

The corresponding integral is

$$w = \nu \kappa yz + f(y)$$

where $f(y)$ is an arbitrary function of y. It follows from symmetry of the problem about the xy-plane that $w = 0$ for $z = 0$, and thus $f(y)$ must vanish identically. This establishes the displacement component

$$w = \nu \kappa yz .$$

The final displacement component v is found by combining the normal strain condition

$$\frac{\partial v}{\partial y} = \varepsilon_y = \nu \kappa y$$

with the two shear strain conditions $\gamma_{xy} = 0$ and $\gamma_{yz} = 0$,

$$\frac{\partial v}{\partial x} = -\frac{\partial u}{\partial y} = \kappa x , \qquad \frac{\partial v}{\partial z} = -\frac{\partial w}{\partial y} = -\nu \kappa z .$$

An integral to these partial differential equations is

$$v = \tfrac{1}{2}\kappa\big(x^2 + \nu y^2 - \nu z^2 \big)$$

where the arbitrary constant vanishes for $v(0,0,0) = 0$. It is seen that the parameter κ is equal to the upward curvature of the x-axis.

The presence of Poisson's ratio in the generalized Hooke's law produces transverse contraction in the parts of the cross section with tension and transverse expansion in the parts with compression. The transverse displacements of the section $x = 0$ are illustrated in Fig. 9.8b. It is seen that the top and bottom planes of the beam become curved. For a homogeneous beam the effect is purely kinematic and is usually neglected in design calculations. □

Plane stress

The concept of plane stress was introduced in Section 8.4.1. It is used as a generally good approximation in problems involving in-plane forces in a plane body with a transverse dimension that is small relative to the in-plane dimensions. The assumption of plane stress implies that only the stress components σ_{xx}, σ_{xy} and σ_{xy} are considered to be non-zero. For plane stress the in-plane components of the generalized Hooke's law follow directly from (9.15) in the form

$$\begin{aligned} \varepsilon_{xx} &= \frac{1}{E}\big(\sigma_{xx} - \nu \sigma_{yy} \big) , \\ \varepsilon_{yy} &= \frac{1}{E}\big(\sigma_{yy} - \nu \sigma_{xx} \big) , \\ \varepsilon_{xy} &= \frac{1+\nu}{E} \sigma_{xy} . \end{aligned} \tag{9.29}$$

In plane stress the transverse strain component ε_{zz} does not generally vanish, but is given by

$$\varepsilon_{zz} = -\frac{\nu}{E}\big(\sigma_{xx} + \sigma_{yy}\big) = -\frac{\nu}{1-\nu}\big(\varepsilon_{xx} + \varepsilon_{yy}\big) . \tag{9.30}$$

Thus, in plane stress a body that is stretched ($\varepsilon_{xx}+\varepsilon_{yy}>0$) becomes thinner ($\varepsilon_{zz}<0$). The magnitude of this effect depends on Poisson's ratio.

Plane strain

In the case of plane strain, introduced in Section 8.4.1 for bodies with constrained deformation in the direction transverse to a plane, the only non-zero strain components are ε_{xx}, ε_{yy} and ε_{xy}. The transverse displacement constraint leads to a transverse stress component σ_{zz} that must be accounted for, when using the generalized Hooke's law in the form (9.15). It follows directly from the condition $\varepsilon_{zz}=0$ that in a state of plane strain the transverse stress component is determined by

$$\sigma_{zz} = \nu\,(\sigma_{xx}+\sigma_{yy})\,. \tag{9.31}$$

With this relation the stress component σ_{zz} can be eliminated from the relevant part of the generalized Hooke's law (9.15). In plane strain the in-plane components then satisfy the relations

$$\begin{aligned}
\varepsilon_{xx} &= \frac{1-\nu^2}{E}\Big(\sigma_{xx}-\frac{\nu}{1-\nu}\,\sigma_{yy}\Big)\,,\\
\varepsilon_{yy} &= \frac{1-\nu^2}{E}\Big(\sigma_{yy}-\frac{\nu}{1-\nu}\,\sigma_{xx}\Big)\,,\\
\varepsilon_{xy} &= \frac{1+\nu}{E}\,\sigma_{xy}\,.
\end{aligned} \tag{9.32}$$

It is seen that the shear relation is the same in plane stress and plane strain. The relations for the normal stress and strain components are of the same form, but in the plane strain problem the elastic modulus and Poisson's ratio are replaced by the equivalent parameters

$$E_* = \frac{E}{1-\nu^2}\,, \qquad \nu_* = \frac{\nu}{1-\nu}\,. \tag{9.33}$$

The shear modulus can be expressed in the usual way also in terms of these parameters, $G = E_*/2(1+\nu_*)$. It follows from this analogy that if a plane stress problem has been solved, the corresponding plane strain solution can be obtained by an appropriate change of the elastic parameters.

9.2 Mean and deviator components

While it is fairly straightforward to illustrate vectors and the interaction of vectors graphically, it is much harder for e.g. stresses and strains with matrix component structure. This section introduces two concepts that are useful in the development and understanding of models relating stresses and strains. The first is to use principal components for development and description of general properties. The second is to extract the mean value of the general states of stress and strain at a point, and to work with these mean values plus the remaining part of the stresses and strains. These concepts provide an explanation of the simple properties of linear elasticity and give some necessary background for the yield and failure conditions developed in Section 9.3.

Decomposition of stress

It was seen in Section 9.1.2 that in the linear theory of elasticity the change of volume was a relation between the volume strain and the mean stress. The linear volume strain e and the mean stress σ_m can be expressed by the first invariant of the stress tensor and the strain tensor, respectively, and thus represent an invariant scalar property. This suggests a decomposition of the stress and strain representation into an invariant component, plus a remainder called the deviator part. This representation is first discussed for the stresses. The first step is to define the mean stress component

$$\sigma_m = \tfrac{1}{3}\big(\sigma_{xx} + \sigma_{yy} + \sigma_{zz}\big)\,. \tag{9.34}$$

The deviator stress components $\boldsymbol{\sigma}'$ are then defined by subtraction of the mean stress from the normal stress components,

$$\boldsymbol{\sigma}' = \begin{bmatrix} \sigma_{xx} - \sigma_m & \sigma_{xy} & \sigma_{xz} \\ \sigma_{yx} & \sigma_{yy} - \sigma_m & \sigma_{yz} \\ \sigma_{zx} & \sigma_{zy} & \sigma_{zz} - \sigma_m \end{bmatrix}. \tag{9.35}$$

This definition can also be written with the help of the unit matrix $\mathbf{I}$ as

$$\boldsymbol{\sigma}' = \boldsymbol{\sigma} - \sigma_m \mathbf{I}\,. \tag{9.36}$$

Conversely, any stress state can be written as the sum of its deviator part and its mean stress,

$$\boldsymbol{\sigma} = \boldsymbol{\sigma}' + \sigma_m \mathbf{I}\,. \tag{9.37}$$

This decomposition into deviator and mean stress plays an important role in many models of material behavior, because the mean stress is closely associated with change of volume, while the deviator stress is associated with shear deformations.

The decomposition (9.37) of a general stress state into deviator stresses and mean stress is illustrated in principal stress space in Fig. 9.9. The principal stress state $[\sigma_x, \sigma_y, \sigma_z]$ is shown as a vector with the principal stresses as components. The figure also shows a plane with normal $[1, 1, 1]$. The plane is constructed by intersecting the axes at equal distance from the origin. In the principal stress space the deviator stresses are

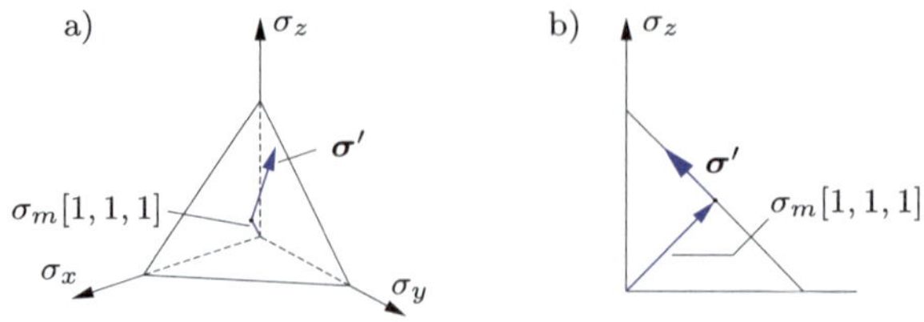

Fig. 9.9: Principal stress space with deviator plane.

$$\begin{bmatrix} \sigma'_x \\ \sigma'_y \\ \sigma'_z \end{bmatrix} = \frac{1}{3} \begin{bmatrix} 2\sigma_x - \sigma_y - \sigma_z \\ 2\sigma_y - \sigma_z - \sigma_x \\ 2\sigma_z - \sigma_x - \sigma_y \end{bmatrix} . \tag{9.38}$$

Multiplication of the deviator stress $[\sigma'_x, \sigma'_y, \sigma'_z]$ with the normal vector $[1,1,1]$ gives the projection on the normal as the sum of the principal deviator stresses,

$$\sigma'_x + \sigma'_y + \sigma'_z = 0 . \tag{9.39}$$

Thus, the deviator part of the stress state lies in the plane in principal stress space with normal $[1,1,1]$, and the plane is therefore called the deviator plane. Graphically the decomposition (9.37) of the stress state $[\sigma_x, \sigma_y, \sigma_z]$ appears as the sum of a component $\sigma_m[1,1,1]$ normal to the deviator plane, and a deviator part $[\sigma'_x, \sigma'_y, \sigma'_z]$ lying in the deviator plane.

In the graphic representation the length of the deviator stress $[\sigma'_x, \sigma'_y, \sigma'_z]$ is found by scalar multiplication of the vector with itself,

$$\begin{aligned} (\sigma'_x)^2 + (\sigma'_y)^2 + (\sigma'_z)^2 &= (\sigma_x - \sigma_m)^2 + (\sigma_y - \sigma_m)^2 + (\sigma_z - \sigma_m)^2 \\ &= \sigma_x^2 + \sigma_y^2 + \sigma_z^2 - 3\sigma_m^2 . \end{aligned} \tag{9.40}$$

The graphical interpretation of the decomposition into mean and deviator stress is much used in the development of theories for material behavior. However, also the theory of linear isotropic elasticity gains in clarity, when illustrated in terms of mean and deviator parts as shown next.

Decomposition of strain

A general state of strain can also be represented as the sum of two parts, a mean strain $\frac{1}{3}e = \frac{1}{3}(\varepsilon_{xx} + \varepsilon_{yy} + \varepsilon_{zz})$ and the deviator strain components $\boldsymbol{\varepsilon}'$. The deviator strain components are defined in the same way as the deviator stress components by subtraction of the mean strain. Thus, the deviator strain components are

$$\boldsymbol{\varepsilon}' = \begin{bmatrix} \varepsilon_{xx} - \frac{1}{3}e & \varepsilon_{xy} & \varepsilon_{xz} \\ \varepsilon_{yx} & \varepsilon_{yy} - \frac{1}{3}e & \varepsilon_{yz} \\ \varepsilon_{zx} & \varepsilon_{zy} & \varepsilon_{zz} - \frac{1}{3}e \end{bmatrix} . \tag{9.41}$$

This definition can also be written as

$$\boldsymbol{\varepsilon}' = \boldsymbol{\varepsilon} - \tfrac{1}{3}e\,\mathbf{I}\,. \tag{9.42}$$

The reason for the factor $\frac{1}{3}$ in the strain formulae is that this factor is not included in the volume strain $e = (\varepsilon_{xx} + \varepsilon_{yy} + \varepsilon_{zz})$. An arbitrary state of strain can then be represented as the sum of a volumetric strain and a deviator strain,

$$\boldsymbol{\varepsilon} = \boldsymbol{\varepsilon}' + \tfrac{1}{3}e\,\mathbf{I}\,. \tag{9.43}$$

The relations of linear isotropic elasticity take a particularly simple form, when both stresses and strains are represented as a sum of a mean value and a deviator part.

Linear isotropic elasticity

The linear elastic relation between the volume strain $e = \varepsilon_{xx} + \varepsilon_{yy} + \varepsilon_{zz}$ and the mean stress $\sigma_m = \frac{1}{3}(\sigma_{xx} + \sigma_{yy} + \sigma_{zz})$ was found in (9.23) as

$$e = \frac{\sigma_m}{K} = 3\frac{1-2\nu}{E}\,\sigma_m\,. \tag{9.44}$$

This is the volume relation between the two scalar components e and σ_m.

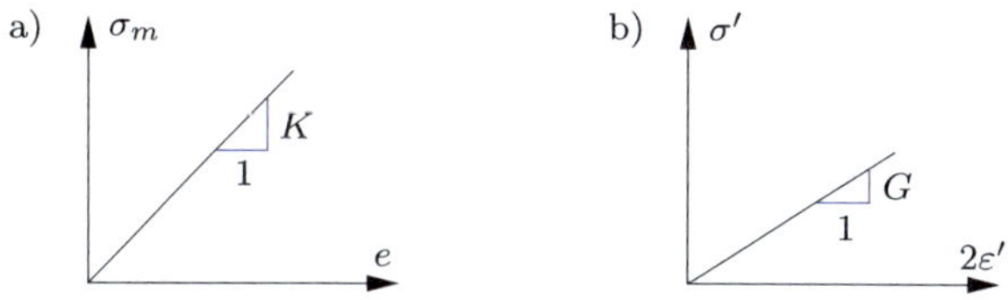

Fig. 9.10: Linear isotropic elasticity in (a) dilation and (b) generalized shear.

The elastic relation for the deviator normal strain components ε'_{xx} follows from the definition

$$\varepsilon'_{xx} = \varepsilon_{xx} - \tfrac{1}{3}(\varepsilon_{xx} + \varepsilon_{yy} + \varepsilon_{zz}) = \tfrac{2}{3}\varepsilon_{xx} - \tfrac{1}{3}\varepsilon_{yy} - \tfrac{1}{3}\varepsilon_{zz} \tag{9.45}$$

by substitution of the strains from the elasticity relation (9.15),

$$\varepsilon'_{xx} = \frac{2}{3}\frac{1+\nu}{E}\sigma_{xx} - \frac{1}{3}\frac{1+\nu}{E}\sigma_{yy} - \frac{1}{3}\frac{1+\nu}{E}\sigma_{zz}\,. \tag{9.46}$$

When collecting terms and using the definition of the deviator stress σ'_{xx}, this expression becomes

$$\varepsilon'_{xx} = \frac{1+\nu}{E}\big(\tfrac{2}{3}\sigma_{xx} - \tfrac{1}{3}\sigma_{yy} - \tfrac{1}{3}\sigma_{zz}\big) = \frac{1+\nu}{E}\,\sigma'_{xx}\,. \tag{9.47}$$

Similar relations apply for the deviator normal components ε_{yy} and ε_{zz}. However, a closer look reveals that the relation applies to *all* deviator components. This follows from the lower part of the elasticity relation (9.15) and the fact that for the shear components $\sigma'_{ij} = \sigma_{ij}$ and $\varepsilon'_{ij} = \varepsilon_{ij}$ when $i \neq j$. The isotropic elastic relation between the deviator strain and stress components can therefore be written as

$$\boldsymbol{\varepsilon}' = \frac{1+\nu}{E}\boldsymbol{\sigma}' = \frac{1}{2G}\boldsymbol{\sigma}' . \tag{9.48}$$

This relation implies that all linear elastic non-volumetric deformation processes are associated with the shear modulus G, while the volumetric deformation given by (9.44) is associated with the bulk modulus K. These two mechanisms are illustrated in Fig. 9.10. A similar split in volume and shape changing processes is found, when modeling yield and fracture of isotropic materials, as illustrated in the following sections.

9.3 Yield conditions for metals

Experimental evidence shows that metals yield, when they are loaded beyond a characteristic stress level. A simple example of this behavior is the uniaxial tension test shown in Fig. 9.1. When the uniaxial stress σ_x reaches a level denoted by σ_Y, the material develops additional deformation. This material behavior is called *yielding*, and the stress σ_Y is called the *yield stress*. Typically, yield would also occur at a uniaxial compressive stress of magnitude $-\sigma_Y$. This indicates that the condition of material yielding somehow bounds the stress states corresponding to elastic behavior. A general state of stress may also lead to material yielding, and it is important to have information of the general condition of yielding of a material, as this condition limits the stresses that can be sustained by the material. Two important yield conditions for metals, those of von Mises and of Tresca, are described below. These yield conditions form a part of the theory of plasticity, dealing with plastic deformation materials, see e.g. Chakrabarty (2009) and Ottosen and Ristinmaa (2005). While a presentation of this theory falls outside the present scope, the yield condition often serves as a limit for operating states of stress in structures, and thereby form a condition for the use of elastic analysis of structures.

9.3.1 Von Mises' yield condition

For metals it turns out that the occurrence of yielding is nearly independent of the mean stress σ_m. For these materials the condition of yielding can therefore be described solely in terms of the deviator stress components. For an isotropic material the yield condition must furthermore be independent of the particular coordinate system used to define the stress components. A simple yield condition that satisfies both of these conditions is described by a circular contour in the deviator plane in the principal stress space shown in Fig. 9.11. The defining equation of such a circle is

$$(\sigma'_x)^2 + (\sigma'_y)^2 + (\sigma'_z)^2 = \text{const}\,. \tag{9.49}$$

This relation can be expressed in terms of the principal stresses $[\sigma_x, \sigma_y, \sigma_z]$ and the yield stress σ_Y as

$$(\sigma_x - \sigma_y)^2 + (\sigma_x - \sigma_z)^2 + (\sigma_y - \sigma_z)^2 = 2\sigma_Y^2\,. \tag{9.50}$$

It is seen that a uniaxial state of stress $[\sigma_x, 0, 0]$ will lead to yield for $\sigma_x = \pm\sigma_Y$ in accordance with the definition of the yield stress σ_Y. This is the von Mises yield condition. It leads naturally to the definition of the so-called equivalent stress

$$\sigma_e = \frac{1}{\sqrt{2}}\sqrt{(\sigma_x - \sigma_y)^2 + (\sigma_x - \sigma_z)^2 + (\sigma_y - \sigma_z)^2}\,. \tag{9.51}$$

The von Mises yield condition can then be expressed in terms of the equivalent stress as

$$\sigma_e = \sigma_Y\,, \tag{9.52}$$

where the yield stress σ_Y is a material parameter, while the equivalent stress σ_e describes the stress state.

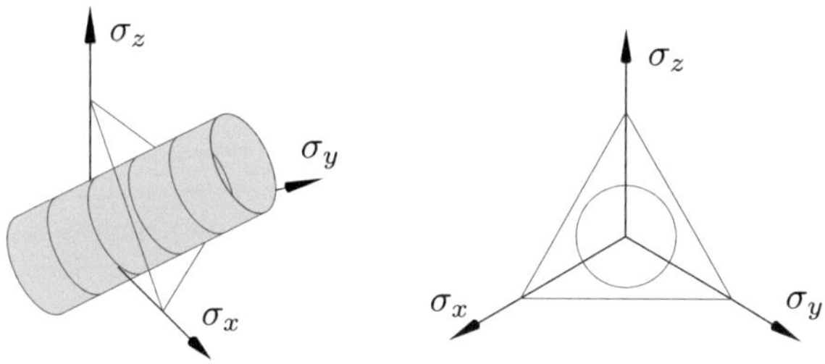

Fig. 9.11: Von Mises stress cylinder in principal stress space.

It follows from the definition (9.51) of the equivalent stress in terms of the principal stress components that it is an invariant. The yield criterion can therefore be generalized to components in an arbitrary coordinate system by expressing σ_e^2 in terms of the stress invariants I_1 and I_2. It follows from the principal component formulae (8.92) of the invariants that

$$\begin{aligned}\sigma_e^2 &= \tfrac{1}{2}\big((\sigma_x - \sigma_y)^2 + (\sigma_x - \sigma_z)^2 + (\sigma_y - \sigma_z)^2\big)\\ &= \sigma_x^2 + \sigma_y^2 + \sigma_z^2 - (\sigma_x\sigma_y + \sigma_z\sigma_x + \sigma_y\sigma_z) = I_1^2 - 3I_2\,.\end{aligned} \tag{9.53}$$

Substitution of the general component expressions (8.91) for the invariants then gives

$$\sigma_e^2 = \big(\sigma_{xx} + \sigma_{yy} + \sigma_{zz}\big)^2 - 3\big(\sigma_{yy}\sigma_{zz} + \\ + \sigma_{zz}\sigma_{xx} + \sigma_{xx}\sigma_{yy} - \sigma_{yz}^2 - \sigma_{xz}^2 - \sigma_{xy}^2\big)\,. \tag{9.54}$$

When this expression is rearranged in terms of the normal stress differences, the following general form of the equivalent stress is obtained,

$$\sigma_e = \frac{1}{\sqrt{2}}\sqrt{(\sigma_{xx}-\sigma_{yy})^2 + (\sigma_{xx}-\sigma_{zz})^2 + (\sigma_{yy}-\sigma_{zz})^2 + 6(\sigma_{yz}^2 + \sigma_{zx}^2 + \sigma_{xy}^2)}. \tag{9.55}$$

With this definition of the equivalent stress σ_e the von Mises yield criterion (9.50) has been extended to arbitrary stress components.

The equation (9.50) defines a circular cylinder in principal stress space. It has generators in the direction $[1,1,1]$ and intersects any deviator plane in a circle as illustrated in Fig. 9.11. In particular, it intersects the principal stress axes at $\pm\sigma_Y$. The cylinder is shown in three-dimensional principal stress space in Fig. 9.11a, and the circular trace in the deviator plane is shown in Fig. 9.11b. Stress states inside the cylinder correspond to elastic behavior, while stress states on the cylinder correspond to yield. If the material is perfectly plastic, yield will occur while the stress state remains on the yield surface, and thus stress states outside the cylinder can not occur. The von Mises yield criterion fits well with experimental data for metals, see e.g. Lemaitre and Chaboche (1990).

Example 9.2. Tension with constrained transverse contraction. In some situations secondary stresses, created by constraints from surrounding parts of a structure, may contribute to prevent yielding. In the uniaxial tension test with $\sigma_x > 0$ the transverse stress components σ_y and σ_z vanish, and yielding occurs at $\sigma_x = \sigma_Y$. If the transverse contraction of the test specimen is prevented, transverse tension stresses will occur. As shown in Section 9.1.2 the transverse stresses in an isotropic linear elastic material in the case of completely prevented transverse deformation are

$$\sigma_y = \sigma_z = \frac{\nu}{1-\nu}\,\sigma_x\,.$$

In this situation there is a positive mean stress of magnitude

$$\sigma_m = \tfrac{1}{3}(\sigma_x + \sigma_y + \sigma_z) = \frac{1+\nu}{1-\nu}\frac{\sigma_x}{3}\,.$$

This mean stress leads to an increase of the yield limit for completely constrained transverse deformation.

The equivalent stress σ_e follows from (9.51) as

$$\sigma_e = \frac{\sigma_x}{\sqrt{2}}\sqrt{\Big(1-\frac{\nu}{1-\nu}\Big)^2 + \Big(1-\frac{\nu}{1-\nu}\Big)^2} = \frac{1-2\nu}{1-\nu}\,\sigma_x\,.$$

The yield criterion $\sigma_e = \sigma_Y$ then gives the yield limit for the fully constrained specimen as

$$\sigma_x = \frac{1-\nu}{1-2\nu}\,\sigma_Y\,.$$

The constraint leads to an increase of the yield capacity. Theoretically, an incompressible material with $\nu = 0.5$ would have infinite yield limit for full transverse constraint. For steel with $\nu = 0.3$ the yield limit is $\sigma_x = 1.75\sigma_Y$, corresponding to an increase of 75% in the case of full transverse constraint. In practice the constraint may only be partial, leading to a somewhat smaller increase of the yield limit. □

The condition of plane stress occurs often in practice. If the stresses act in the xy-plane, plane stress implies $\sigma_{zz} = \sigma_{xz} = \sigma_{yz} = 0$. For this situation the equivalent stress is

$$\sigma_e = \sqrt{\sigma_{xx}^2 + \sigma_{yy}^2 - \sigma_{xx}\sigma_{yy} + 3\sigma_{xy}^2}\,. \tag{9.56}$$

This form of the equivalent stress finds wide application in the design of steel structures. In the principal stress space plane stress corresponds to stress states in the $\sigma_x\sigma_y$-plane. Thus the plane stress yield condition is the intersection of the circular cylinder of Fig. 9.11a with the $\sigma_x\sigma_y$-plane. The analytical expression follows from setting $\sigma_{xy} = 0$ in (9.56), whereby

$$\sigma_x^2 + \sigma_y^2 - \sigma_x\sigma_y = \sigma_Y^2\,. \tag{9.57}$$

This is the equation of an ellipse as shown in Fig. 9.12, with major axis on the line $\sigma_x = \sigma_y$.

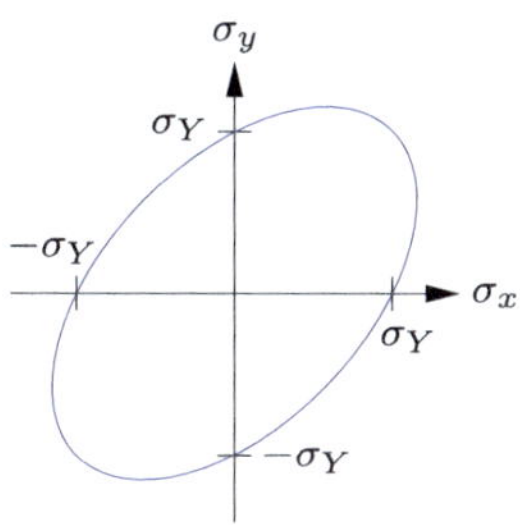

Fig. 9.12: Von Mises stress ellipse in plane stress.

Example 9.3. Von Mises biaxial stress. The von Mises ellipse for plane stress leads to the following three observations. In uniaxial stress, e.g. $\sigma_x \neq 0$ and $\sigma_y = 0$, yielding occurs for $\sigma_x = \pm\sigma_Y$. If the two in-plane stresses are equal yielding occurs at $\sigma_x = \sigma_y = \pm\sigma_Y$. If the two in-plane stress components are equal in magnitude but of opposite sign the stress state corresponds to pure shear with $\tau = -\sigma_x = \sigma_y$, see Example 8.4. Yielding then occurs at

$$\tau = \frac{\sigma_Y}{\sqrt{3}}\,.$$

It is seen that principal stresses of the same sign leads to increased yield limit, while principal stresses of opposite sign reduce the yield limit. □

9.3.2 Tresca's yield condition

An alternative yield criterion for metals was proposed by Tresca. It is based on the assumption that yielding is initiated, when the largest shear stress $\tau_{\max}$ exceeds a characteristic stress threshold. It was demonstrated in Section 8.4.4 that the maximum shear stress $\tau_{\max}$ is the radius of the Mohr's circle determined by the difference between largest and the smallest principal stress, see Fig. 8.31. Thus, the Tresca yield condition can be expressed as

$$2\tau_{\max} = \sigma_{\max} - \sigma_{\min} = \sigma_Y \,. \tag{9.58}$$

It follows from the definition in terms of the difference between the maximum and minimum principal stress, that according to the Tresca yield condition yielding is independent of the mean stress. Thus, the Tresca yield condition is also a cylindrical surface in principal stress space with generators in the direction $[1, 1, 1]$.

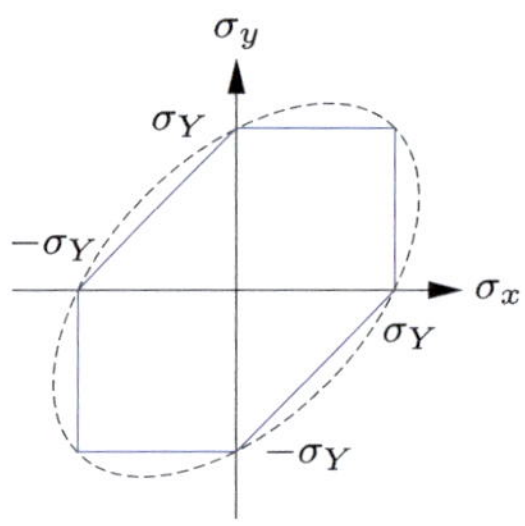

Fig. 9.13: Tresca stress polygon in plane stress.

Any of the three principal stresses σ_x, σ_y and σ_z may be the largest or the smallest, and the Tresca yield surface is therefore generated by the intersection of six planes in principal stress space. The easiest way of identifying these planes is probably by considering the trace of the yield surface in the $\sigma_x\sigma_y$-plane corresponding to plane stress. In this state of plane stress with $\sigma_z = 0$ there are three different situations to be considered. If σ_x and σ_y are of opposite sign they will be the maximum and minimum principal stress components, and the Tresca yield condition then is

$$2\tau_{\max} = |\,\sigma_x - \sigma_y\,| = \sigma_Y \,. \tag{9.59}$$

This gives the inclined lines in the second and fourth quadrant of the $\sigma_x\sigma_y$-plane shown in Fig. 9.13. If σ_x and σ_y are of the same sign, the numerically largest define the shear stress in combination with $\sigma_z = 0$ according to

$$2\tau_{\max} = |\,\sigma_x - 0\,| = \sigma_Y \qquad \text{or} \qquad 2\tau_{\max} = |\,\sigma_y - 0\,| = \sigma_Y \,. \tag{9.60}$$

These equations produce the vertical and horizontal lines in Fig. 9.13. In combination the conditions produce the Tresca hexagon for plane stress. The points of intersection with the axes are the same as the von Mises ellipse, shown by the dashed curve in the figure. In fact, the Tresca hexagon is inscribed in the von Mises ellipse. This implies that apart from stress states at the apexes of the hexagon the Tresca yield condition will predict a lower yield limit that the von Mises condition. However, in many cases the difference is small, and the choice between the two criteria is then a matter of convenience. The simple liner form of the Tresca condition, once the intermediate principal stress has been identified, enables explicit solution of a number of important problems with combined elastic-plastic material behavior, see e.g. Chakrabarty (2009).

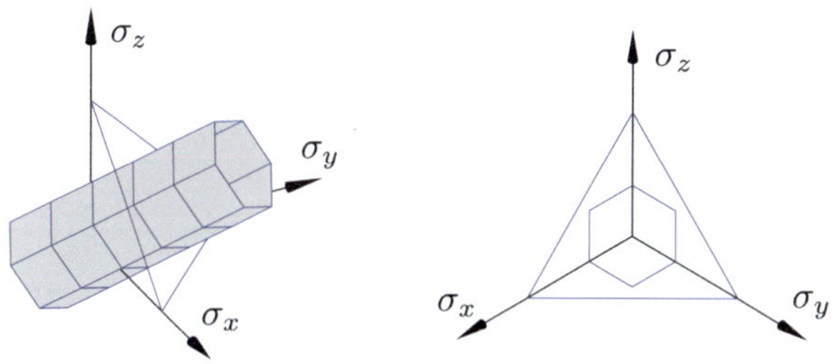

Fig. 9.14: Tresca stress surface in principal stress space.

The full Tresca yield surface in three-dimensional principal stress space follows by extending the plane stress hexagon of Fig. 9.13 along the generators [1, 1, 1]. The result is a regular hexagonal cylinder shown in Fig. 9.14a. The hexagon is inscribed in the von Mises circular cylinder, and the trace in the deviator stress plane is shown in Fig. 9.14b.

9.4 Coulomb's theory of friction materials

The Tresca yield condition (9.58) is based on the hypothesis that the maximum shear stress has a characteristic limit, independent of the mean stress. While this is a fair approximation to the behavior of metals, many building materials like concrete, brick and soil depend on the mechanism of friction. This implies that the shear capacity increases with compression mean stress. The Tresca yield condition may be seen as a special case of the more general theory of friction materials, originally developed by COULOMB (1736–1806), see e.g. Heyman (1997).

In Coulomb's theory of friction materials it is assumed that failure does not depend on the intermediate principal stress component, but only on the maximum and minimum component. Thus, the basis of the theory can be developed in a two-dimensional setting and then generalized to three-dimensional stress states subsequently. The principal stresses are compressive and are denoted σ_x and σ_z, with $|\sigma_z| > |\sigma_x|$. It is convenient to introduce the notation

$$\sigma = \tfrac{1}{2}(\sigma_z + \sigma_x) < 0\,, \qquad \tau = \tfrac{1}{2}(\sigma_z - \sigma_x) < 0 \tag{9.61}$$

for the (two-dimensional) mean stress, and the maximum shear stress, respectively.

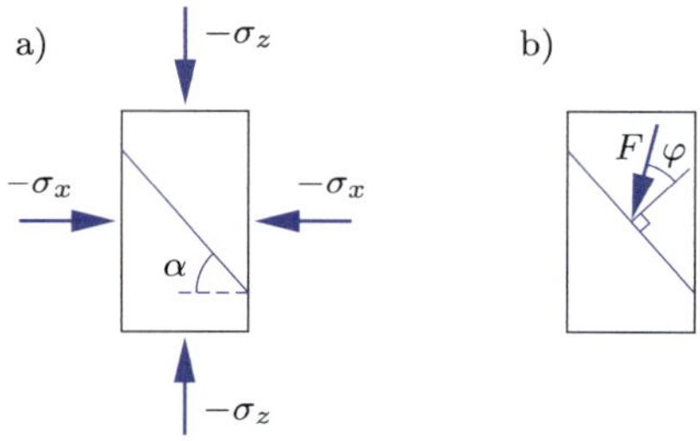

Fig. 9.15: a) Homogeneous stress state and b) friction condition.

Figure 9.15a shows a two-dimensional body in a homogeneous state of stress with principal stresses σ_x and σ_z. A plane section through the body is identified by the angle α shown in the figure. The stress vector acting on this plane is given by its normal and shear components σ_n and σ_t. These components are easily determined from equilibrium of a small triangle, whereby

$$\begin{aligned} \sigma_n &= \sigma + \tau\,\cos 2\alpha\,, \\ \sigma_t &= \qquad \tau\,\sin 2\alpha \end{aligned} \tag{9.62}$$

for any value of the angle α of the section.

9.4.1 Critical section and stress state

Friction failure on a critical section is assumed to occur, when two conditions are met: the ratio of tangential to normal stress on the critical section equals the coefficient of friction μ, and the ratio of tangential to normal stress is less than μ on all other sections. A simple cohesion contribution is included in the final theory later.

The coefficient of friction μ determines the inclination angle φ of the friction force **F** corresponding to the relation $\mu = \tan\varphi$, as shown in Fig. 9.15b. Thus, for the critical section the ratio between the shear stress σ_t and the normal stress σ_n on the cross section determines the friction angle, and the most critical section is determined by

$$\tan\varphi = \max_\alpha \frac{\sigma_t}{\sigma_n} = \max_\alpha \frac{\tau\,\sin 2\alpha}{\sigma + \tau\,\cos 2\alpha}\,. \tag{9.63}$$

Differentiation with respect to the angle α gives the condition

$$\cos 2\alpha = -\frac{\tau}{\sigma}. \tag{9.64}$$

When this expression is substituted into the friction equation (9.63) the following relation is obtained between the friction angle φ and the inclination angle α,

$$\tan\varphi = \frac{\tau \sin 2\alpha}{\sigma + \tau \cos 2\alpha} = -\cot 2\alpha. \tag{9.65}$$

By changing the argument the cot-function is transformed to a tan-function, and the following relation is obtained

$$\tan\varphi = -\tan(\tfrac{1}{2}\pi - 2\alpha) = \tan(2\alpha - \tfrac{1}{2}\pi). \tag{9.66}$$

This equation determines the critical angle as

$$\alpha = \tfrac{1}{4}\pi + \tfrac{1}{2}\varphi. \tag{9.67}$$

Thus, it is seen that for $|\sigma_z| > |\sigma_x|$ the angle of the critical section is always in the interval $\frac{1}{4}\pi < \alpha < \frac{1}{2}\pi$, corresponding to the section being steeper than 45°.

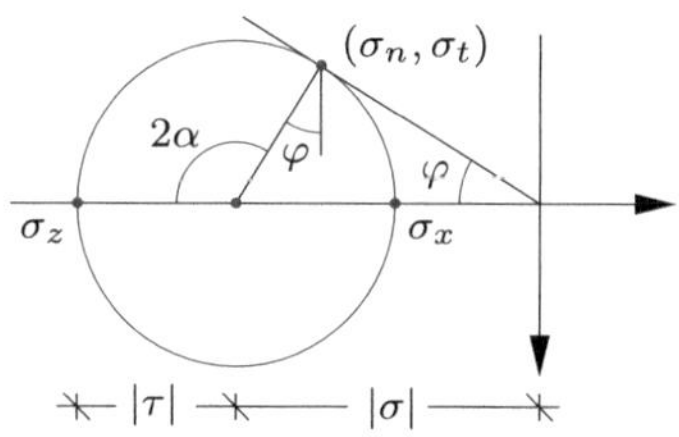

Fig. 9.16: Stress state at friction failure.

The stress state at friction failure is shown in a Mohr circle diagram in Fig. 9.16. The diagram is fully defined from equation (9.67), by which $2\alpha - \varphi = \frac{1}{2}\pi$, as shown in the figure. It is seen directly from the Mohr circle that the failure condition can be expressed in either of the to forms

$$\tan\varphi = \frac{\sigma_t}{\sigma_n} \qquad \text{or} \qquad \sin\varphi = \frac{\tau}{\sigma}, \tag{9.68}$$

where it is noted that all the stress components are negative, and their ratios therefore positive. The first equation is a direct expression of the friction condition in terms of the stresses σ_n and σ_t on the critical section, while the second is a relation between the prescribed stresses in terms of their sum σ and difference τ.

The friction failure stress state may also be represented directly in terms of the principal stresses σ_x and σ_z. Substitution of the definition (9.61) into the failure condition (9.68b) gives

$$(1+\sin\varphi)\sigma_x \;-\; (1-\sin\varphi)\sigma_z \;=\; 0\,, \tag{9.69}$$

or the stress ratio

$$\frac{\sigma_z}{\sigma_x} \;=\; \frac{1+\sin\varphi}{1-\sin\varphi} \;=\; \tan^2\alpha \;>\; 1\,, \tag{9.70}$$

where the section angle α has been substituted from (9.67).

Example 9.4. Inclination of failure planes. The inclination of the failure planes and the associated angle of the critical section can be illustrated with reference to an idealized retaining wall, supporting an ideal friction material. The problem is illustrated in Fig. 9.17, showing a vertical retaining wall in two situations.

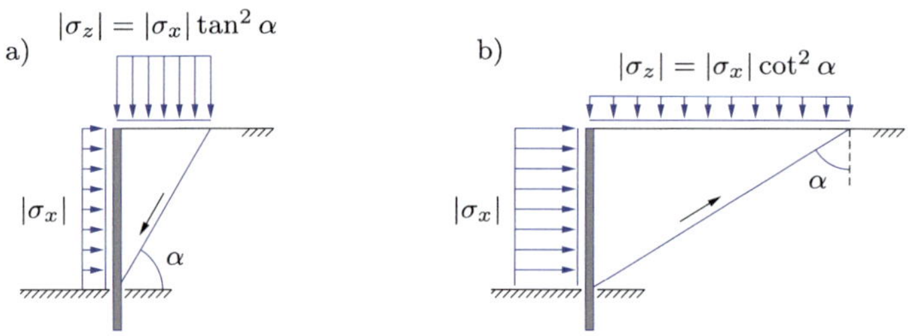

Fig. 9.17: Retaining wall pressure: a) passive wall, b) active wall.

In Fig. 9.17a the stress $|\sigma_x|$ represents the resistance capacity of the wall. The vertical pressure $|\sigma_z|$ on the upper horizontal soil surface is increased until the limit imposed by the friction failure criterion. In this problem the vertical stress $|\sigma_z|$ is imposed and increased until failure $|\sigma_z| > |\sigma_x|$, and $|\sigma_z|$ is called the *active stress*, while $|\sigma_x|$ is the *passive stress.* This corresponds to the roles of the stresses in the derivation of the theory and illustrated in Fig. 9.15. In Fig. 9.17b the roles of the stress components are interchanged. Here, the wall is pressed against the friction material until failure develops, and the material slides upwards towards the right. Here the $|\sigma_x|$ is the active stress component, and the resisting stress $|\sigma_z|$ is the passive component. □

9.4.2 Coulomb failure surface

It is important for a full understanding and general use of the Coulomb failure criterion to obtain a representation in the three-dimensional principal stress space. The Tresca stress surface shown in Fig. 9.14 is a regular hexagonal cylinder with the mean stress σ_m along the axis. Similarly, the Coulomb failure criterion has a hexagonal trace in the deviator stress plane, but with a size that increases linearly with compressive mean stress $-\sigma_m$. The Coulomb failure surface is characterized by two generators, each corresponding to a state of stress, in which two of the principal components are equal. As above, the stress σ_z is taken as the numerically largest, while σ_x is the numerically smallest. This defines the two characteristic stress states as: *triaxial compression* in which $\sigma_z < \sigma_y = \sigma_x$, and *triaxial tension* where $\sigma_z = \sigma_y < \sigma_x$. These two states can be considered as generated from an isostatic state of stress $\sigma_z = \sigma_y = \sigma_x$ – the first by adding *compression* in the z-direction, and the second by adding *tension* in the x-direction.

For the state of triaxial compression, with $\sigma_z < \sigma_y = \sigma_x$, the mean stress is

$$\sigma_m = \tfrac{1}{3}\sigma_z + \tfrac{2}{3}\sigma_x , \tag{9.71}$$

while the deviator stress components are

$$\sigma_z' = \tfrac{2}{3}(\sigma_z - \sigma_x), \qquad \sigma_x' = \sigma_y' = \tfrac{1}{3}(\sigma_x - \sigma_z). \tag{9.72}$$

Inversion of these relations determines the principal stress components as

$$\sigma_z = \sigma_m + \sigma_z' , \qquad \sigma_x = \sigma_y = \sigma_m - \tfrac{1}{2}\sigma_z' . \tag{9.73}$$

These expressions are now substituted into the friction failure criterion (9.69), that then takes the form

$$(1+\sin\varphi)(\sigma_m - \tfrac{1}{2}\sigma_z') - (1-\sin\varphi)(\sigma_m + \sigma_z') = 0 . \tag{9.74}$$

This equation gives the deviator stress σ_z' in terms of the mean stress as

$$\sigma_z' = \frac{4\sin\varphi}{3-\sin\varphi}\,\sigma_m . \tag{9.75}$$

This is the stress component, representing the projection of the triaxial compression stress state on the deviator plane. In this state of stress $\sigma_x = \sigma_y$, and thus the projection lies on the z-axis in the projected deviator plane in Fig. 9.18. The sign of σ_z' and σ_m are both negative, and thus the state of triaxial compression defines the top point of the stress contour in the deviator plane shown in Fig. 9.18. Similar points along the y- and x-axis follow from symmetry.

The state of triaxial tension, $\sigma_z = \sigma_y < \sigma_x$, can be analyzed in the same way. In this case the mean stress is

$$\sigma_m = \tfrac{2}{3}\sigma_z + \tfrac{1}{3}\sigma_x , \tag{9.76}$$

while the deviator stress components are

$$\sigma_z' = \sigma_y' = \tfrac{1}{3}(\sigma_z - \sigma_x), \qquad \sigma_x' = \tfrac{2}{3}(\sigma_x - \sigma_z). \tag{9.77}$$

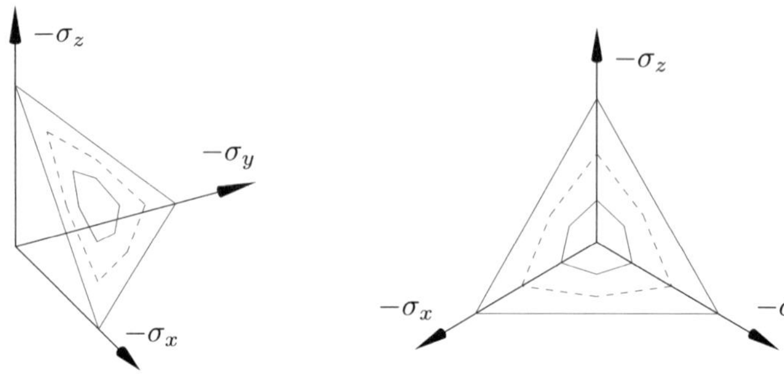

Fig. 9.18: Hexagonal trace of the friction criterion in the deviator plane.

Inversion of these relations gives the principal stress components for triaxial tension as

$$\sigma_z = \sigma_y = \sigma_m - \tfrac{1}{2}\sigma'_x, \qquad \sigma_x = \sigma_m + \sigma'_x. \tag{9.78}$$

These representations are now substituted into the friction failure criterion (9.69), whereby

$$(1 + \sin\varphi)(\sigma_m + \sigma'_x) - (1 - \sin\varphi)(\sigma_m - \tfrac{1}{2}\sigma'_x) = 0. \tag{9.79}$$

This equation gives the deviator stress σ'_x in terms of the mean stress as

$$\sigma'_x = -\frac{4\sin\varphi}{3 + \sin\varphi}\,\sigma_m. \tag{9.80}$$

This is the stress component, representing the projection of the triaxial tension stress state on the deviator plane. In this stress state σ'_x is positive, when σ_m is negative, and thus the state of triaxial tension represents the corner along the 'side' of the contour shown in Fig. 9.18. Symmetry identifies similar points along the other two 'sides'.

The shape of the deviator contour depends on the angle of friction as characterized by the ratio of the deviator components defining triaxial compression and triaxial tension, respectively. It follows from (9.75) and (9.80) that this ratio is

$$\left|\frac{\sigma'_{\rm com}}{\sigma'_{\rm ten}}\right| = \frac{3 + \sin\varphi}{3 - \sin\varphi} \to \begin{cases} 1 & \text{for } \varphi \to 0 \\ 2 & \text{for } \varphi \to \tfrac{1}{2}\pi \end{cases} \tag{9.81}$$

Thus, the deviator contour approaches a regular hexagon, as in the Tresca contour, for small values of the friction angle, while larger values of the friction angle makes the contour more and more triangular. In this way a family of Coulomb contours with increasing values of the friction angle can fill the triangle formed by the stress coordinate planes. This behavior corresponds closely to experimentally observed properties of soil, e.g. Lade and Duncan (1973), and concrete, see e.g. Ottosen and Ristinmaa (2005). However, most materials – like e.g. concrete – have an additional contribution to the strength, called *cohesion*. The effect of cohesion can be included in the theory of friction materials by increasing the shear resistance with an extra contribution, that does not depend on the mean stress. In essence, this amounts to a translation of the failure surface in Fig. 9.18 along the isostatic axis. This effect can be represented by replacing the mean stress by $\sigma_m - \sigma_c$, where σ_c represents the effect of cohesion.

The hexagonal pyramid shape of the Coulomb failure surface follows from the friction hypothesis, where any influence from the intermediate principal stress is omitted. Experimental evidence suggests more rounded shapes. For isotropic materials, satisfying full symmetry between the three principal stress components, yield and failure surfaces in the three-dimensional principal stress space can be represented in terms of the stress invariants introduced in Section 8.4.4. For a given mean stress σ_m the shape of the contour is represented by the stress invariants I_2 and I_3 or their deviator stress counterparts, see e.g. Ottosen and Ristinmaa (2005) and Krenk (2000). In the majority of material models the stress invariants are also used to represent the deformation mechanisms that generate non-linear behavior.

9.5 Exercises

Exercise 9.1. The figure shows an elastic bar of length $\ell = 200\,\text{mm}$. The cross section is circular with diameter $D = 15\,\text{mm}$. The material is a plastic compound with elastic modulus $E = 2.70\,\text{GPa}$ and Poisson's ratio $\nu = 0.4$. The bar is loaded by a tension force $P = 200\,\text{N}$.

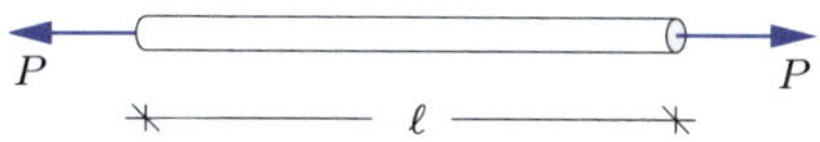

a) Calculate the elongation of the bar.

b) Determine the change in diameter of the cross section.

Exercise 9.2. The figure shows three (infinitely) stiff plates connected by two elastic rubber blocks of height h, thickness a and width b. The shear modulus of the elastic rubber material is G. The outer plates are supported and the central plate is loaded by a vertical force P.

a) Determine an expression for the vertical displacement u of the central plate.

b) Find u when $h = 60\,\text{mm}$, $a = 10\,\text{mm}$, $b = 20\,\text{mm}$, $G = 0.20\,\text{MPa}$ and $P = 50\,\text{N}$.

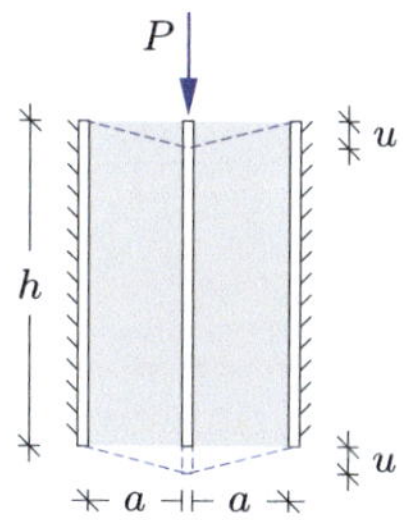

Exercise 9.3. A spring device is composed of a cylindrical bar attached to an outer tube by rubber ring of height h. The inner radius of the bar is r_i, while the outer radius is r_o. The rubber material is elastic and the shear modulus is G. The central cylindrical bar is loaded by a transverse force P.

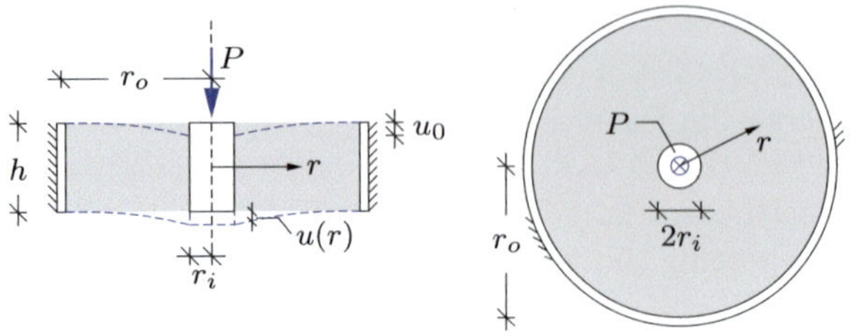

a) Determine the shear stress τ that acts in the rubber ring at a cylindrical section with radius r.

b) The transverse displacement $u(r)$ determines the angular strain γ by the relation:

$$\gamma = -\frac{du}{dr} = \frac{\tau}{G}.$$

Integrate this relation to obtain $u(r)$ and find the transverse displacement $u_0 = u(0)$ of the central cylindrical bar.

Exercise 9.4. The figure shows a vertical force P acting on a plate AB. This plate is attached to the supported plates CD and EF via two cubic blocks of rubber material with dimensions $a \times a \times a$. The dimension $a = 30\,\text{mm}$ and the force $P = 50\,\text{N}$. The shear modulus of the rubber material is $G = 0.20\,\text{MPa}$. Any deformations of the plates are neglected in the analysis.

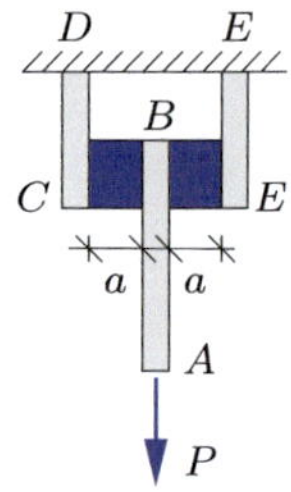

a) Determine the magnitude of the shear stress τ in the rubber blocks.

b) Determine the magnitude of the associated shear strains γ.

c) Determine the vertical displacement of the center plate AB.

Exercise 9.5. A cylindrical bar with diameter $d = 30\,\text{mm}$ is placed inside centrally inside a tube with inner diameter $D = 32\,\text{mm}$. The length of both inner bar and outer tube is $\ell = 50\,\text{mm}$. The material of the inner bar is elastic with elastic modulus $E = 5\,\text{MPa}$ and Poisson's ratio $\nu = 0.45$. The end section of the inner bar is loaded by a constant pressure p, as shown in the figure.

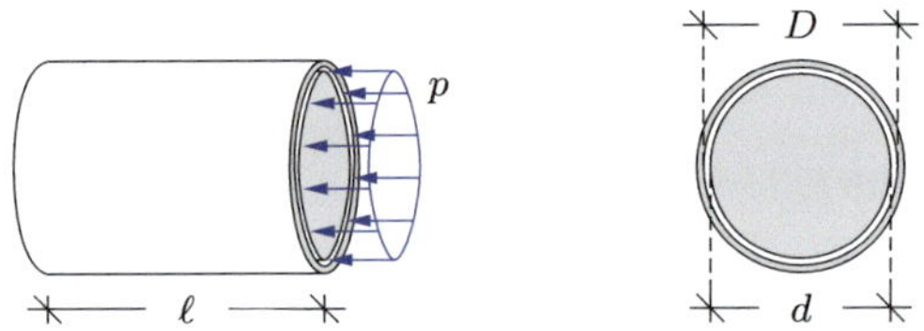

a) Determine the magnitude of the pressure p that exactly makes the inner bar touch the inside of the outer tube.

b) Determine how much the inner bar is shortened.

Exercise 9.6. The figure shows a 45° strain gate rosette with strain readings ε_a, ε_b and ε_c. The material is linear elastic and isotropic with elastic modulus $E = 100\,\text{MPa}$ and Poisson's ratio $\nu = 0.40$. A plane stress condition is assumed. The three strain gauge readings are:

$$\varepsilon_a = 600 \cdot 10^{-6} \quad , \quad \varepsilon_b = -320 \cdot 10^{-6} \quad , \quad \varepsilon_c = 450 \cdot 10^{-6} \, .$$

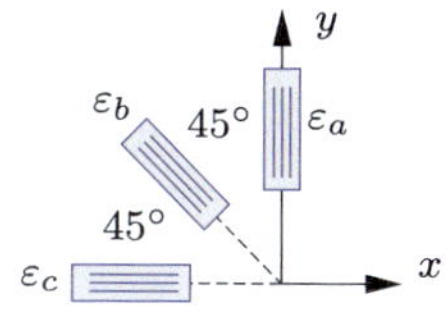

a) Determine the strain components ε_{xx}, ε_{yy} and ε_{xy}.

b) Determine the out-of-plane strain ε_{zz}.

c) Determine the plane stress components σ_{xx}, σ_{yy} and σ_{xy}, and find the mean stress.

Exercise 9.7. The figure shows a plate with uni-axial yield stress $\sigma_Y = 250\,\text{MPa}$. The plate is loaded in plane stress conditions with principal stresses σ_x and $\sigma_y = \frac{1}{2}\sigma_x$, as indicated in the figure.

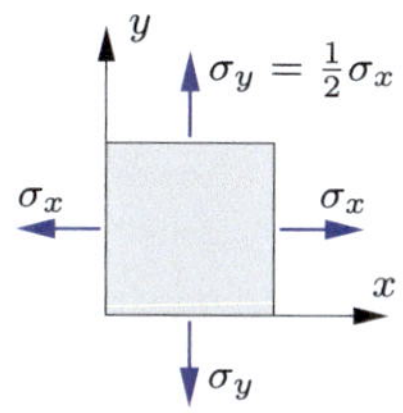

a) Determine the magnitude of σ_x that corresponds to yielding based on the von Mises criterion.

b) Determine the magnitude of σ_x that corresponds to yielding based on the Tresca criterion.

c) Illustrate the results in a graph that shows the yield surfaces in principal stresses.

Exercise 9.8. The figure shows a plate with uni-axial yield stress $\sigma_Y = 250\,\text{MPa}$. The plate is loaded in plane stress conditions with principal stresses σ_x and $\sigma_x = -\frac{1}{2}\sigma_y$, as indicated in the figure.

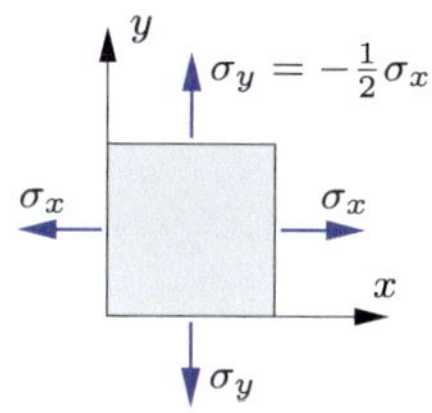

a) Determine the magnitude of σ_x that corresponds to yielding based on the von Mises criterion.

b) Determine the magnitude of σ_x that corresponds to yielding based on the Tresca criterion.

c) Illustrate the results in a graph that shows the yield surfaces in principal stresses.

Exercise 9.9. The figure shows the stress components in terms of σ_* in a plane stress condition. The material is linear elastic and isotropic with elastic modulus $E = 200\,\text{MPa}$ and Poisson's ratio $\nu = 0.37$. The yield stress of the material is $\sigma_Y = 4.0\,\text{MPa}$.

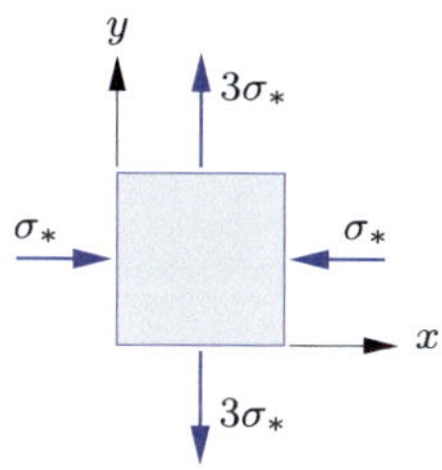

a) Determine the volume strain e for $\sigma_* = 0.5\,\text{MPa}$.

b) Determine the magnitude of σ_* that corresponds to yielding based on the Tresca criterion.

c) Determine the magnitude of σ_* that corresponds to yielding based on the von Mises criterium.

Exercise 9.10. The figure shows the stress components in terms of $\sigma_* = 10\,\text{MPa}$ in plane stress conditions. The material is linear elastic and isotropic with elastic modulus $E = 70\,\text{GPa}$ and Poisson's ratio $\nu = 0.37$.

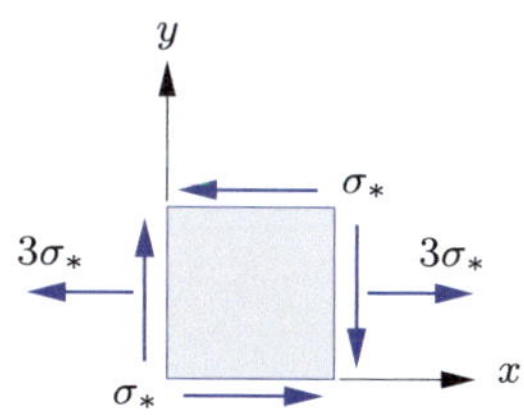

a) Determine the strains ε_{xx}, ε_{yy}, ε_{xy} and ε_{zz}.

b) Determine the principal strains ε_x and ε_y and the orientation θ of the associated principal axes.

c) Determine the magnitude of the von Mises stress σ_e.

Exercise 9.11. The figure shows the plane stress components. Note, that the identical stress components on the backside of the element are implied. The material is high-strength steel with yields stress $\sigma_Y = 500\,\text{MPa}$.

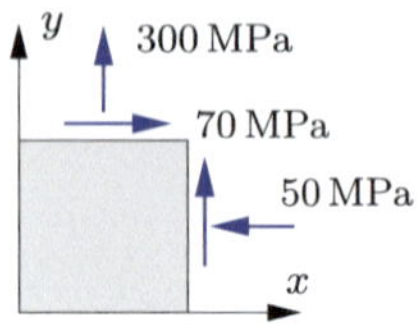

a) Determine the equivalent von Mises stress σ_e and find the stress ratio σ_e/σ_Y that represents the failure margin.

Exercise 9.12. Reconsider the plane stress state of the previous example. In this case the failure margin is evaluated based on Tresca's criterion.

a) Determine the principal stresses σ_x and σ_y.

b) Draw the Tresca yield surface in the principal stress plane with yield stress $\sigma_Y = 500\,\text{MPa}$. Plot also the current stress state and determine if the failure criterium is violated.

c) Discuss the possibility of a failure margin similar to the stress ratio σ_e/σ_Y of the von Mises criterion.

Exercise 9.13. The figure shows a thin-walled cylinder with radius r and wall thickness t. The cylinder is loaded by a tension force N and a torsion moment M, resulting in a tension stress σ and a shear stress τ, both assumed constant over the cross section. Failure of the material is assumed to agree with the von Mises criterion with yield stress σ_Y.

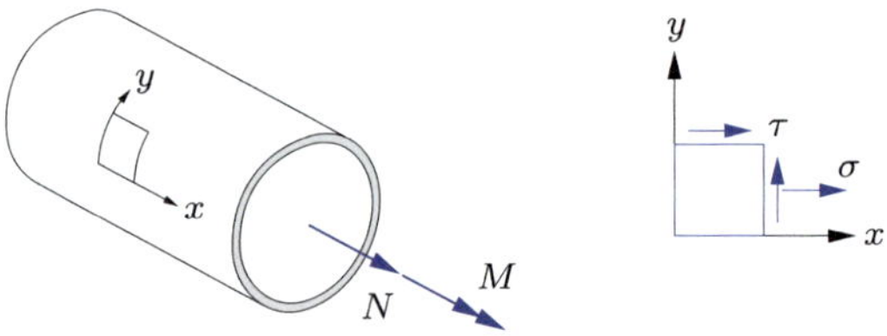

a) Determine an expression for the axial stress σ when $M = 0$, and find the tension force N_Y that corresponds to failure.

b) Determine an expression for the shear stress τ when $N = 0$, and find the torsion moment M_Y that corresponds to failure.

c) Demonstrate that tension force and torsion moment satisfy the combined failure criterion

$$\left(N/N_Y\right)^2 + \left(M/M_Y\right)^2 = 1\,,$$

and illustrate this relation graphically.

General Bending of Beams | 10

The theory of homogeneous bending of non-symmetric elastic beams with constant cross-section forms a central part of the mechanics of structures. The theory combines the possibility of general cross-section properties with the simultaneous bending about two axes, and thus constitutes a natural extension of the simple plane bending treated in Chapters 3–4 and developed into simple finite elements for analysis of plane frames in Chapter 7. The general theory of beam bending has wide application, e.g. to beams in buildings, bridge decks in concrete, steel or composites, or in a very general form to wind turbine blades with changing aerodynamic closed cross-section.

The basic problem is illustrated by the cantilever shown in Fig. 10.1. The load consists of a force $\mathbf{P} = [P_x, P_y, P_z]^T$ applied to the tip of the cantilever. The components are given with respect to a $\{x, y, z\}$ coordinate system with axial coordinate x and cross-section coordinates y and z as shown in the figure. The load introduces tension N and bending moments $M_y(x)$ and $M_z(x)$ in the beam, and this leads to extension and curvature. If the beam cross-section is symmetric with respect to the y and the z-axis and the force is applied to the intersection of the axes of symmetry, the problem is immediately resolved

S. Krenk, J. Høgsberg, *Statics and Mechanics of Structures*,
DOI 10.1007/978-94-007-6113-1_10,

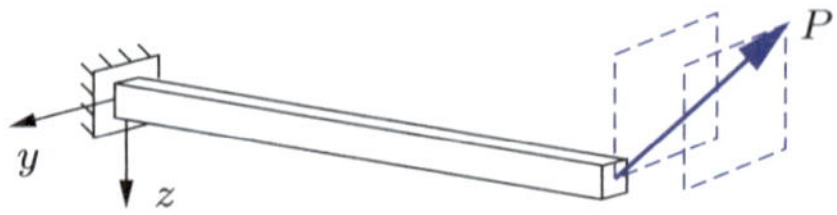

Fig. 10.1: Bending of cantilever.

into the three independent problems consisting of extension and bending in the xy and xz-planes as illustrated in Fig. 10.2. It will be demonstrated that this decomposition into extension and two plane bending problems can be obtained for a homogeneous elastic beam of general cross-section, when the axial force is applied to the *elastic center*, and the two planes of bending are determined as the *principal axes* of the cross-section. The theory is developed for beams with non-homogeneous distribution of elastic stiffness over the cross-section.

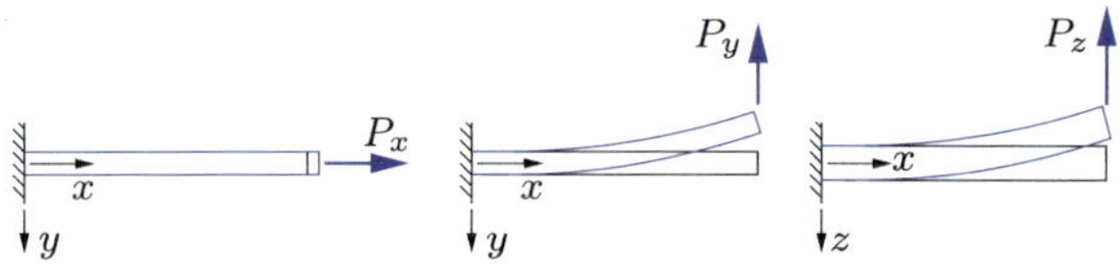

Fig. 10.2: Extension and bending of cantilever beam.

The bending of an elastic beam of constant but general cross-section is treated in Section 10.1. When bending a beam with symmetric cross-section in its plane of symmetry, the strain is proportional to the distance from the neutral axis. This concept is generalized by considering the strain distribution to be a linear function of both the cross-section coordinates. The result is a systematic theory of bending, originally developed by NAVIER (1785–1836). The theory relates the normal force N and the two bending moment components M_y and M_z at a cross-section to the corresponding set of kinematic quantities: the axial strain ε_0, and the two curvature components κ_y and κ_z. The constitutive relations that relate the section force components N, M_y, M_z to their kinematic counterparts $\varepsilon_0, \kappa_y, \kappa_z$ depend on the cross-section geometry and the distribution of elastic stiffness over the cross-section. These properties are contained in a set of cross-section parameters like area, moment of inertia etc. and Section 10.2 describes how the cross-section parameters are obtained from the geometry and stiffness distribution in the cross-section. This analysis identifies the *elastic center* where a normal force will only produce extension, and a set of *principal axes* that uncouples the two bending problems. Even if not using these axes directly the fact that bending can be uncoupled if choosing these axes constitutes an important characteristic of beam bending, that is essential for the understanding of the associated mechanisms. The bending parameters relating to different axes are connected by a transformation that is quite similar to that of the plane stress component transformation dealt with in Section 8.4. As a consequence the Mohr circle construction also illustrates the relation of bending stiffness around different axes, and concept of principal axes associated with minimum and maximum bending stiffness. As discussed in Section 10.3, the linear distribution of the axial strain over the cross-section in general bending leads to a simple explicit relation for the distribution of the axial stress component. The axial stress constitutes an important design parameter. The axial stress distribution also serves as a step in the determination of the shear stress distribution associated with non-homogeneous bending as discussed in Chapter 11.

10.1 Bending of non-symmetric beams

The theory of bending of beams with non-symmetric cross-section is based on a simple extension of the corresponding theory for plane bending. The basic assumption is that the beam is homogeneous and loaded by a combination of a normal force and two bending moment components, that are constant over the beam. This constitutes a logical generalization of the concept of homogeneous bending and permits the development of a fairly simple and rigorous theory of bending of homogeneous beams. Like in the case of plane bending, the idea is to develop the deformation characteristics from the ideal case of homogeneous bending, and then to assume its validity also in cases with moment variation along the beam. The procedure is to first define the kinematics and then to derive the associated stresses and section forces.

10.1.1 Kinematic formulation

When a beam is in a state of homogeneous bending and extension, symmetry suggests that plane cross-sections remain plane under the deformation. This implies that a cross-section that originally coincides with the yz-plane will have an axial displacement $u(y, z)$ in the form of a linear function of y and z as illustrated in Fig. 10.3. The mathematical form is

$$u(y, z) = u_0 + y\,\eta_y + z\,\eta_z\,, \tag{10.1}$$

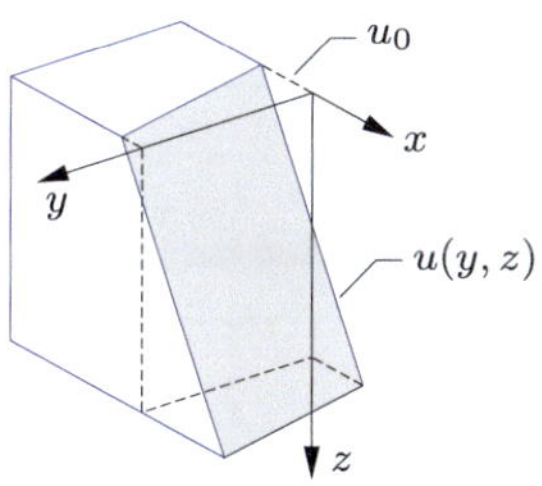

Fig. 10.3: Linear distribution of axial displacement $u(y, z)$.

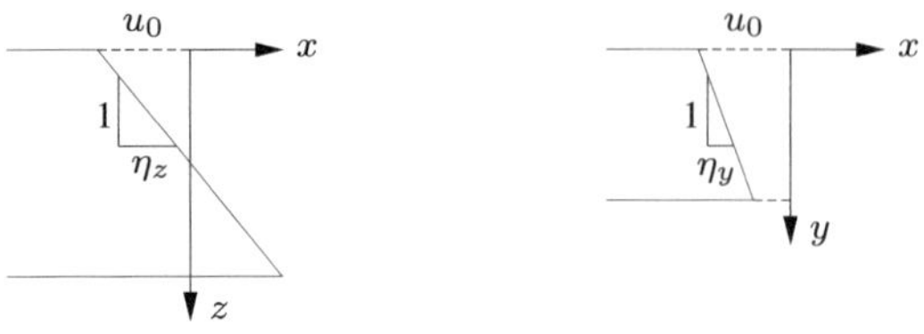

Fig. 10.4: Linear variation of axial displacement in coordinate planes.

where $u_0 = u(0,0)$ is the axial displacement of the origin of the coordinate system. Furthermore, it is found by differentiation that η_y and η_z are the inclinations of the cross-section in the y- and z-direction, respectively. The variation of the axial deformation along the cross-section axes y and z is illustrated in Fig. 10.4, where u_0 is shown as the axial displacement at the origin of the coordinate system, and η_y and η_z are the inclinations of the axes after deformation. Within a 'small deformation' theory a similar representation applies to the other cross-sections, and thus the axial displacement u_0 as well as the inclinations η_y and η_z are functions of the axial coordinate x,

$$u_0 = u_0(x)\,, \qquad \eta_y = \eta_y(x)\,, \qquad \eta_z = \eta_z(x)\,. \tag{10.2}$$

However, to keep the notation as compact as possible, the argument x is implied and not written explicitly in the following derivations.

The axial strain is the derivative of the axial displacement $u(x,y,z)$ with respect to the longitudinal coordinate x,

$$\varepsilon(y,z) = \frac{\partial u}{\partial x} = \varepsilon_0 + y\,\kappa_y + z\,\kappa_z\,, \tag{10.3}$$

where ε_0 is the axial strain at the origin, and κ_y and κ_z are defined by the inclinations,

$$\varepsilon_0 = \frac{du_0}{dx}\,, \qquad \kappa_y = \frac{d\eta_y}{dx}\,, \qquad \kappa_z = \frac{d\eta_z}{dx}\,. \tag{10.4}$$

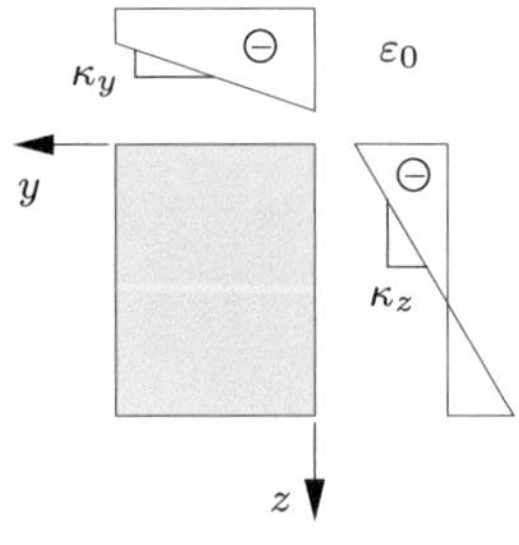

Fig. 10.5: Plot of linear strain variations along the y and z axes.

It follows from (10.3) that the distribution of the axial strain is linear over the cross-section, conveniently illustrated by the linear variation of the strain along the y- and z-axes, as shown in Fig. 10.5. The strain at the origin of the yz-coordinate system is ε_0, and the gradient of the normal strain is given by the components κ_y and κ_z.

The section inclinations η_y and η_z – and thereby the parameters κ_y and κ_z – are related to the transverse displacement of the beam. In the present case of pure bending there is no twist of the beam, and thus the transverse displacement is a translation, described by the motion of a single point, e.g. the origin of the yz-coordinate system,

$$v(x,y,z) = v_0(x)\,, \qquad w(x,y,z) = w_0(x)\,. \tag{10.5}$$

The two shear strain components associated with the x-component can now be calculated based on the displacement representations (10.1) and (10.5),

$$\gamma_{xy} = \frac{dv_0}{dx} + \eta_y\,, \qquad \gamma_{xz} = \frac{dw_0}{dx} + \eta_z\,. \tag{10.6}$$

These relations define the x-derivative of the transverse displacements as

$$\frac{dv_0}{dx} = -\eta_y + \gamma_{xy}\,, \qquad \frac{dw_0}{dx} = -\eta_z + \gamma_{xz}\,. \tag{10.7}$$

These relations are the general form of the plane bending relation (4.29) of Timoshenko beam theory.

The shear strains defined via (10.6) are constant over any cross-section, and thus represent only an average value. A more detailed analysis of the *shear stress* distribution over the cross-section is typically obtained directly from a static analysis as discussed in Chapter 11. As demonstrated in Section 4.3 the contribution from the shear strains to the beam displacements is often negligible, and this contribution is therefore often neglected in the kinematics of the beam. The result is the Bernoulli beam theory, in which the inclinations are given by

$$\eta_y = -\frac{dv_0}{dx}\,, \qquad \eta_z = -\frac{dw_0}{dx}\,. \tag{10.8}$$

The parameters κ_y and κ_z are then given as the curvatures of the beam axis by the relations

$$\kappa_y = -\frac{d^2v_0}{dx^2}\,, \qquad \kappa_z = -\frac{d^2w_0}{dx^2}\,. \tag{10.9}$$

These relations generalize the kinematic relations (4.17) and (4.18) of plane bending.

10.1.2 Stresses and section forces

In Bernoulli beam theory it is assumed that the deformation is not constrained in the transverse direction. The axial stress is then obtained directly from the axial strain (10.3) by multiplication with the elastic modulus E. This gives the expression

$$\sigma(y,z) \;=\; E\,\varepsilon(y,z) \;=\; E\,\big(\varepsilon_0 + y\,\kappa_y + z\,\kappa_z\big), \tag{10.10}$$

where the elastic modulus $E = E(y,z)$ may vary over the cross-section. While the strain distribution (10.3) is linear in the coordinates y and z a non-homogeneous elastic stifness may lead to a more complicated stress variation.

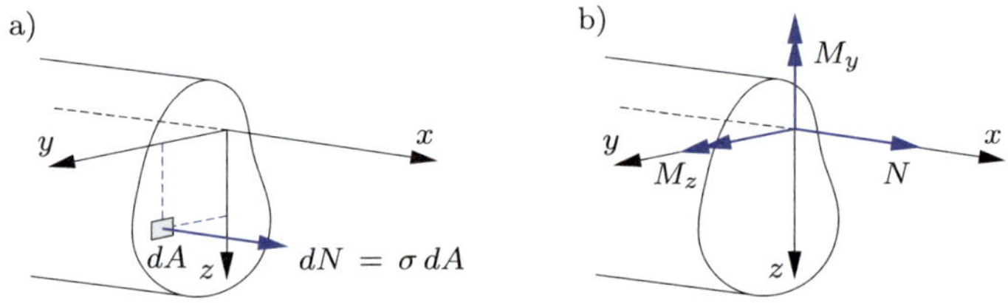

Fig. 10.6: Contribution of a) normal stress in dA to b) section forces.

The three section forces associated with the homogeneous beam bending problem are the normal force N and the two bending moments M_y and M_z with respect to axes of the cross-section coordinate system. Figure 10.6 shows an arbitrary cross-section with normal stress distribution σ given by the expression (10.10). Now, consider an infinitesimal part of the cross-section with area dA. As indicated in Fig. 10.6a the contribution to the normal force from this infinitesimal area is

$$dN \;=\; \sigma(y,z)\,dA\,. \tag{10.11}$$

The normal force component dN also introduces bending moments with respect to the reference coordinate system. In the present context it is convenient to introduce moment components such that M_y is formed via the moment arm y and M_z is formed via the moment arm z, and thus

$$dM_y \;=\; \sigma\,y\,dA\,, \qquad dM_z \;=\; \sigma\,z\,dA\,. \tag{10.12}$$

The corresponding moment vector components are shown in Fig. 10.6b. The resulting section forces are obtained by integration over the cross-section area,

$$N \;=\; \int_A \sigma\,dA\,, \quad M_y \;=\; \int_A \sigma\,y\,dA\,, \quad M_z \;=\; \int_A \sigma\,z\,dA\,. \tag{10.13}$$

By substitution of the expression for the normal stress (10.10) into (10.13a) the normal force is expressed as

$$N = \varepsilon_0 \int_A E\,dA + \kappa_y \int_A E\,y\,dA + \kappa_z \int_A E\,z\,dA \tag{10.14}$$

in terms of the reference strain ε_0 and the curvature components κ_y and κ_z. Similar expressions for the bending moments can be found by substitution of (10.10) into (10.13b,c), whereby

$$\begin{aligned} M_y &= \varepsilon_0 \int_A Ey\,dA + \kappa_y \int_A Ey^2\,dA + \kappa_z \int_A Ezy\,dA\,, \\ M_z &= \varepsilon_0 \int_A Ez\,dA + \kappa_y \int_A Eyz\,dA + \kappa_z \int_A Ez^2\,dA\,. \end{aligned} \tag{10.15}$$

The area integrals are cross-section parameters representative of axial, bending and coupling stiffness with respect to the yz-axes.

The integrals contain the elastic modulus E, that may depend on the location in the cross-section. To keep full generality E is therefore kept inside the integral. The beam stiffness parameters combine the elastic stiffness and the geometry of the cross-section and in order to keep this format a reference elastic modulus E_0 is introduced. The cross-section parameters can be expressed in matrix format as

$$E_0 \begin{bmatrix} F & S_y & S_z \\ S_y & I_{yy} & I_{yz} \\ S_z & I_{zy} & I_{zz} \end{bmatrix} = \int_A E \begin{bmatrix} 1 & y & z \\ y & y^2 & yz \\ z & zy & z^2 \end{bmatrix} dA\,. \tag{10.16}$$

Note, that the matrix is symmetric, thereby containing six cross-section parameters.

For homogeneous elastic stiffness with $E = E_0$ the notation implies that the cross-section parameters $F, S_y, \cdots$ represent geometric characteristics of the cross-section like area, static moment etc. In the general format F is defined by the weighted area integral

$$F = \frac{1}{E_0} \int_A E\,dA\,. \tag{10.17}$$

For a homogeneous cross-section with constant elastic modulus $E = E_0$ the weighted area F recovers the geometric area A, defined by

$$A = \int_A dA\,. \tag{10.18}$$

The parameters S_y and S_z are the weighted static moments,

$$S_y = \frac{1}{E_0} \int_A E\, y\, dA \,, \qquad S_z = \frac{1}{E_0} \int_A E\, z\, dA \,. \tag{10.19}$$

Finally, I_{yy}, I_{zz} and $I_{yz} = I_{zy}$ are the weighted moments of inertia,

$$I_{yy} = \frac{1}{E_0} \int_A E\, y^2\, dA \,, \quad I_{zz} = \frac{1}{E_0} \int_A E\, z^2\, dA \,, \quad I_{yz} = \frac{1}{E_0} \int_A E\, yz\, dA \,. \tag{10.20}$$

It is often convenient to choose E_0 as the elastic modulus of the dominant material in the cross-section. For instance, in the case of reinforced concrete beams E_0 is typically chosen as the elastic modulus of the concrete, whereby $E/E_0 = 1$ for the concrete part of the cross-section, while E/E_0 is approximately 15 for the small areas representing the steel reinforcement. This is illustrated in Example 10.3.

When introducing the cross-section parameters defined in (10.16) into the expressions (10.14) and (10.15) for the section forces, these can be combined into matrix form as

$$\begin{bmatrix} N \\ M_y \\ M_z \end{bmatrix} = E_0 \begin{bmatrix} F & S_y & S_z \\ S_y & I_{yy} & I_{yz} \\ S_z & I_{zy} & I_{zz} \end{bmatrix} \begin{bmatrix} \varepsilon_0 \\ \kappa_y \\ \kappa_z \end{bmatrix} . \tag{10.21}$$

This matrix relation represents the general constitutive relation between the section forces N, M_y and M_z and the associated generalized deformation measures ε_0, κ_y and κ_z. The normal force in (10.14) can be written directly as

$$N = E_0 \big(F\varepsilon_0 + S_y\kappa_y + S_z\kappa_z \big) \,, \tag{10.22}$$

while the bending moments are

$$\begin{aligned} M_y &= E_0 \big(S_y\varepsilon_0 + I_{yy}\kappa_y + I_{yz}\kappa_z \big) \,, \\ M_z &= E_0 \big(S_z\varepsilon_0 + I_{yz}\kappa_y + I_{zz}\kappa_z \big) \,. \end{aligned} \tag{10.23}$$

It is observed that for an arbitrarily located cross-section coordinate system all three deformation parameters ε_0, κ_y and κ_z contribute to each of the section forces N, M_y and M_z. However, a special coordinate system may be determined in which there are three independent deformation mechanisms – extension and two curvatures – each corresponding to a single section force – the normal force and bending moments about two suitably defined axes. This corresponds to the diagonal form of the matrix relation (10.21). The determination of this so-called *principal coordinate system* is an important part of the cross-section analysis, treated next.

10.2 Cross-section analysis

In general, diagonalization of the bending stiffness matrix in (10.21) consists of a translation of the cross-section coordinate system to the so-called *elastic center*, followed by a rotation. The translation leads to uncoupling of the normal force relation from the bending relations. Sometimes it is convenient to use the intermediate coordinate system through the elastic center, in which the normal force relation is uncoupled from the moment relations, while the coupled form of the two-component bending problem is retained.

10.2.1 Elastic center

Initially, the coupling between extension and bending is eliminated by a suitable translation of the coordinate system to a reference point $[c_y, c_z]$, as illustrated in Fig. 10.7. Hereby, new cross-section coordinates $\{\bar{y}, \bar{z}\}$ are introduced via

$$y = \bar{y} + c_y\,, \qquad z = \bar{z} + c_z\,. \tag{10.24}$$

Substitution of these relations into the kinematic expression for the axial displacement in (10.1) gives the axial displacement with respect to the new translated coordinate system,

$$u(\bar{y},\bar{z}) = u_0 + (\bar{y} + c_y)\eta_y + (\bar{z} + c_z)\,\eta_z = u_c + \bar{y}\,\eta_y + \bar{z}\,\eta_z\,. \tag{10.25}$$

In this expression u_c is the axial displacement at the origin of the translated coordinate system,

$$u_c = u_0 + \eta_y c_y + \eta_z c_z\,. \tag{10.26}$$

It is seen from (10.25) that the inclinations in the new coordinate system are still η_y and η_z. Thus, a translation of the coordinate system, as shown in Fig. 10.7, has no influence on the cross-section inclinations.

The corresponding axial strain is obtained by differentiation of the axial displacement in (10.25) with respect to the longitudinal coordinate x,

$$\varepsilon(\bar{y},\bar{z}) = \varepsilon_c + \bar{y}\,\kappa_y + \bar{z}\,\kappa_z\,, \tag{10.27}$$

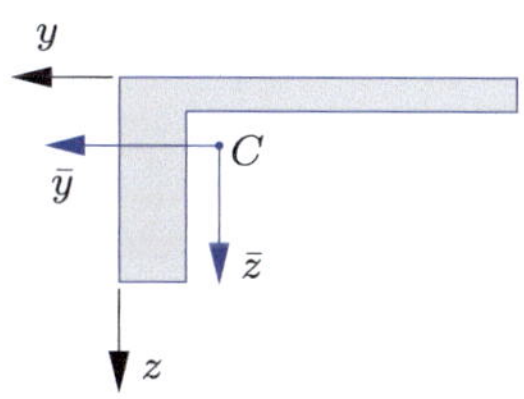

Fig. 10.7: Coordinate system centered at C.

where $\varepsilon_c = \partial u_c/\partial x$ is the axial strain at the reference point $[c_y, c_z]$, while $\kappa_y = \partial \eta_y/\partial x$ and $\kappa_z = \partial \eta_z/\partial x$ are the same curvatures as in the original coordinate system, illustrated in Fig. 10.5. The axial stress is obtained by multiplication of the strain ε by the elastic modulus E,

$$\sigma(\bar{y}, \bar{z}) = E\,\varepsilon(\bar{y}, \bar{z}) = E\,\varepsilon_c + E\,\bar{y}\,\kappa_y + E\,\bar{z}\,\kappa_z\,. \qquad (10.28)$$

The section forces in the new coordinate system are obtained by integration over the cross-section area as

$$N = \int_A \sigma\, dA\,, \quad M_{\bar{y}} = \int_A \sigma\,\bar{y}\, dA\,, \quad M_{\bar{z}} = \int_A \sigma\,\bar{z}\, dA\,. \qquad (10.29)$$

The bar on the moment subscripts indicate that the moments are with respect to the translated coordinate system $\{\bar{y}, \bar{z}\}$. Substitution of the axial stress σ in (10.28) into the expression for the section forces in (10.29) gives the constitutive relation with respect to the new translated coordinate system. It is of similar form as the original formulation (10.21), and can be written as

$$\begin{bmatrix} N \\ M_{\bar{y}} \\ M_{\bar{z}} \end{bmatrix} = E_0 \begin{bmatrix} F & S_{\bar{y}} & S_{\bar{y}} \\ S_{\bar{y}} & I_{\bar{y}\bar{y}} & I_{\bar{y}\bar{z}} \\ S_{\bar{z}} & I_{\bar{y}\bar{z}} & I_{\bar{z}\bar{z}} \end{bmatrix} \begin{bmatrix} \varepsilon_c \\ \kappa_y \\ \kappa_z \end{bmatrix}. \qquad (10.30)$$

The cross-section parameters in the above relations are defined and determined in the same way as in (10.16)–(10.20), but now with respect to the new coordinates $\{\bar{y}, \bar{z}\}$.

The location of the origin $[c_y, c_z]$ of the new coordinate system is now chosen such that axial extension creates no bending moments, or conversely that axial loading only produces axial extension and no curvature. Figure 10.8 illustrates the separation of the general linear deformation into a) pure extension and b) pure rotation about C. In the constitutive relation (10.30) the coupling parameters $S_{\bar{y}}$ and $S_{\bar{z}}$ between bending and extension are now eliminated by imposing the conditions

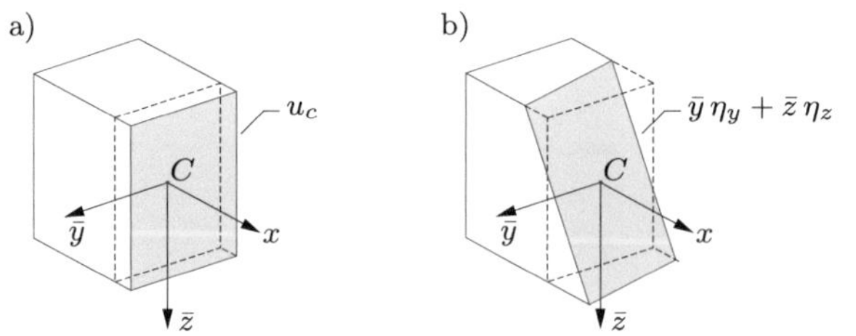

Fig. 10.8: Separation of axial displacement into a) pure extension, and b) pure bending.

$$S_{\bar{y}} = \frac{1}{E_0}\int_A E\bar{y}\,dA = \frac{1}{E_0}\int_A E(y-c_y)\,dA = S_y - c_yF = 0\,,$$
$$S_{\bar{z}} = \frac{1}{E_0}\int_A E\bar{z}\,dA = \frac{1}{E_0}\int_A E(z-c_z)\,dA = S_z - c_zF = 0\,. \tag{10.31}$$

The last equalities determine the origin $[c_y, c_z]$ of the new coordinate system as

$$c_y = \frac{S_y}{F}\,, \qquad c_z = \frac{S_z}{F}\,. \tag{10.32}$$

In the following, this point in the cross-section $[c_y, c_z]$ is referred to as the *elastic center*. For homogeneous cross-sections the elastic center coincides with the *geometric center*, while for inhomogeneous cross-sections these two centers are not equivalent. The elastic center is determined from the ratio between the weighted static moments S_y and S_z in the original reference coordinate system, and the weighted cross-section area F. Any cross-section analysis starts with the location of the elastic center, and if possible it is advantageous to choose the reference coordinate system $\{y, z\}$ to coincide with the elastic center or at least to facilitate evaluation of the integrals defining the cross-section parameters.

Static moment of a flange

Many thin-walled cross-sections are composed of rectangular parts, and the cross-section parameters are then conveniently determined by summation of the contribution from the individual rectangular parts. The basis of the procedure is be illustrated by a single rectangular flange, as shown in Fig. 10.9 together with the reference coordinate system $\{y, z\}$.

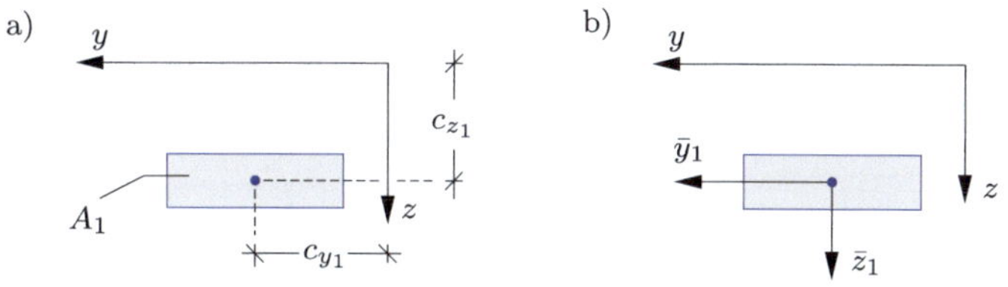

Fig. 10.9: Rectangular flange.

The weighted area of the flange is given by the area integral

$$F_1 = \frac{1}{E_0}\int_{A_1} Ey\,dA\,, \tag{10.33}$$

where the subscript 1 refers to this particular flange. For cross-sections with multiple flanges the subscripts indicate the results or parameters for the individual flanges. Assume that the position of the center of the flange is known, and given by c_{y_1} and c_{z_1} with respect to the reference coordinate system $\{y, z\}$. A change of coordinate system is introduced as in (10.24),

$$y = \bar{y}_1 + c_{y_1}\,, \qquad z = \bar{z}_1 + c_{z_1}\,, \tag{10.34}$$

where the local coordinate system $\{\bar{y}_1, \bar{z}_1\}$ for flange 1 is shown in Fig. 10.9b. The static moments S_y and S_z are defined in (10.19), and substitution of the coordinte representation (10.34) then gives

$$\begin{aligned} S_y &= \frac{1}{E_0}\int_{A_1} Ey\,dA = \frac{1}{E_0}\int_{A_1} E(\bar{y}_1 + c_{y_1})\,dA = S_{\bar{y}_1} + F_1 c_{y_1}\,, \\ S_z &= \frac{1}{E_0}\int_{A_1} Ez\,dA = \frac{1}{E_0}\int_{A_1} E(\bar{z}_1 + c_{z_1})\,dA = S_{\bar{z}_1} + F_1 c_{z_1}\,. \end{aligned} \tag{10.35}$$

When the origin of the local coordinate system $\{\bar{y}_1, \bar{z}_1\}$ is located at the center of the flange, the corresponding static moments vanish,

$$S_{\bar{y}_1} = 0\,, \qquad S_{\bar{z}_1} = 0\,. \tag{10.36}$$

Thus, the static moments of the flange with respect to the reference coordinate system $\{y, z\}$ are given by the weighted area of the flange times the distance to the center of the flange,

$$S_y = F_1\, c_{y_1}\,, \qquad S_z = F_1\, c_{z_1}\,. \tag{10.37}$$

It is important to use the correct sign of the coordinates c_{y_1} and c_{z_1} of the flange center. In Fig. 10.9 the center is located in the positive quadrant of the $\{y, z\}$ coordinate system, whereby both c_{y_1} and c_{z_1} are positive.

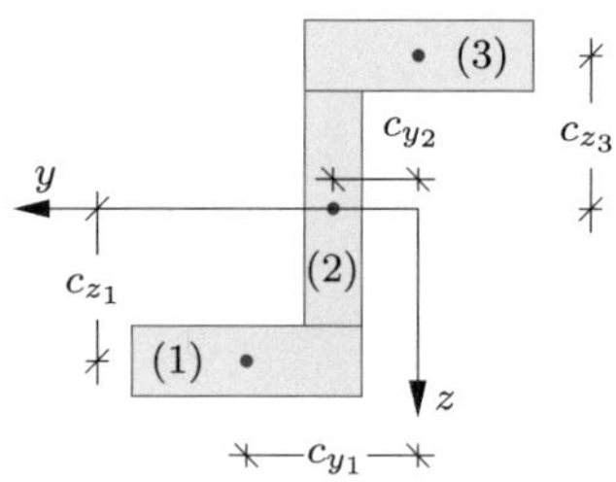

Fig. 10.10: Z-profile composed of flanges. Indication of non-zero distances to local centers.

Cross-sections are often composed of several flanges as illustrated in Fig. 10.10. The location of the elastic center of the full cross-section then depends on the resulting static moments, and thereby the sum of the contributions from the individual flanges. Based on the above result (10.37) for the single flange, the resulting static moments of a cross-section with n flanges are found by summation as

$$S_y = \sum_{j=1}^{n} F_j\, c_{y_j}\,, \quad S_z = \sum_{j=1}^{n} F_j\, c_{z_j}\,, \tag{10.38}$$

where F_j is the area weighted by the elastic modulus for flange j with center located in $[c_{y_j}, c_{z_j}]$. Thus, the determination of the static moments boils down to the determination of the weighted area and the location of the center for each individual flange.

If the elastic modulus is constant for each flange, $E(y,z) = E_j$, the expression for the static moments can be written as

$$S_y = \sum_{j=1}^{n} \frac{E_j}{E_0} A_j\, c_{y_j}\,, \quad S_z = \sum_{j=1}^{n} \frac{E_j}{E_0} A_j\, c_{z_j}\,, \tag{10.39}$$

where A_j is the geometric area and $[c_{y_j}, c_{z_j}]$ the geometric center of flange j.

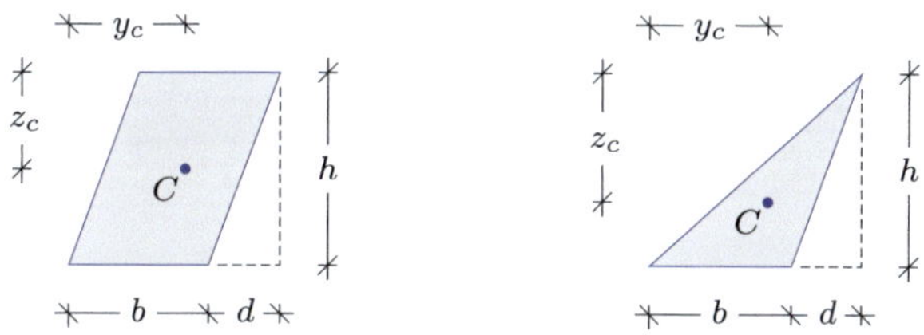

$$A = bh$$
$$y_c = \tfrac{1}{2}(b+d)$$
$$z_c = \tfrac{1}{2}h$$

$$A = \tfrac{1}{2}bh$$
$$y_c = \tfrac{1}{3}(2b+d)$$
$$z_c = \tfrac{2}{3}h$$

Fig. 10.11: Properties of different flange geometries.

Figure 10.11 shows the geometric area and the location of the center for an skew flange and a triangle. The calculation of static moments and the location of the elastic center is illustrated in the following by examples.

Example 10.1. Elastic center for angle profile. Figure 10.12 shows a cross-section with two flanges connected at a right angle. The origin of the coordinate system $\{y, z\}$ is placed where the flanges join, and such that the axes coincide with the respective centerlines of the flanges. Hereby, there is no contribution to the static moment S_z from the horizontal flange (1), while S_y contains no contribution from the vertical flange (2). The length, thickness and elastic modulus of the horizontal flange are a_1, t_1 and E_1, and similarly with subscript 2 for the vertical flange. The elastic modulus of the horizontal flange is chosen as the reference value, i.e. $E_0 = E_1$. The cross-section is assumed to be thin-walled with $t_1 \ll a_1$ and $t_2 \ll a_2$.

The effective area of the cross-section is defined in (10.17), and with $E_0 = E_1$

$$F = \sum_{j=1}^{2} \frac{E_j}{E_1} A_j = a_1 t_1 + a_2 t_2 \frac{E_2}{E_1}\,.$$

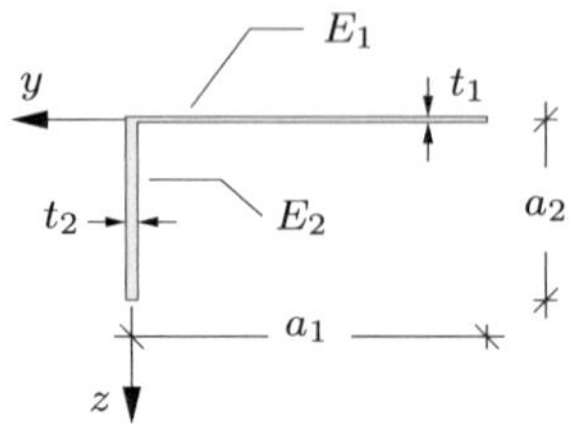

Fig. 10.12: Inhomogeneous thin-walled angle profile.

Note, that because the cross-section parameters are normalized with respect to the elastic modulus of the horizontal flange, the second term contains a scaling by the ratio of the elastic moduli. Following (10.39) the static moments are also determined by summation over the flanges. In the y-direction this gives

$$S_y = \sum_{j=1}^{2} \frac{E_j}{E_1} A_j c_{y_j} = a_1 t_1 (-\tfrac{1}{2} a_1) = -\tfrac{1}{2} a_1^2 t_1 ,$$

where the contribution from the vertical flange vanishes because the coordinate system is located such that $c_{y_2} = 0$. The center of the horizontal flange (1) is located in the negative y-direction, giving a negative value of c_{y_1} and S_y. The static moment in the z-direction is obtained in the same way as

$$S_z = \sum_{j=1}^{2} \frac{E_j}{E_1} A_j c_{z_j} = \frac{E_2}{E_1} a_2 t_2 \tfrac{1}{2} a_2 = \tfrac{1}{2} a_2^2 t_2 \frac{E_2}{E_1} ,$$

where the contribution from the horizontal flange vanishes because $c_{z_1} = 0$.

Following (10.32) the coordinates of the elastic center are found by the ratio between the weighted static moments and the weighted cross-section area. This gives the y-component

$$c_y = \frac{S_y}{F} = \frac{-\tfrac{1}{2} a_1^2 t_1}{a_1 t_1 + a_2 t_2 \dfrac{E_2}{E_1}} = -\tfrac{1}{2} a_1 \frac{1}{1 + \dfrac{a_2 t_2 E_2}{a_1 t_1 E_1}} = \frac{-\tfrac{1}{2} a_1}{1 + \dfrac{A_2 E_2}{A_1 E_1}} ,$$

and the z-component

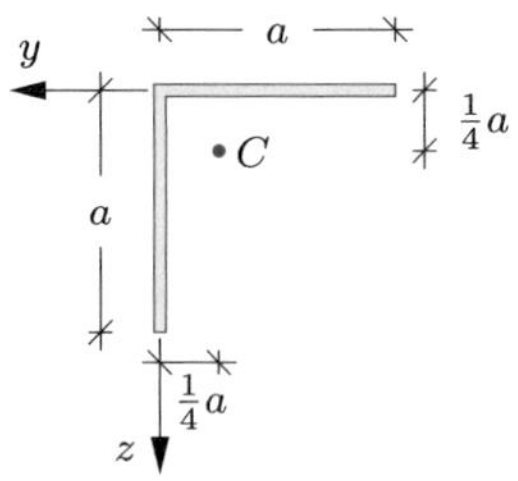

Fig. 10.13: Symmetric thin-walled angle profile.

$$c_z = \frac{S_z}{F} = \tfrac{1}{2}a_2 \frac{\dfrac{a_2 t_2 E_2}{a_1 t_1 E_1}}{1 + \dfrac{a_2 t_2 E_2}{a_1 t_1 E_1}} = \tfrac{1}{2}a_2 \frac{\dfrac{A_2 E_2}{A_1 E_1}}{1 + \dfrac{A_2 E_2}{A_1 E_1}} = \frac{\tfrac{1}{2}a_2}{\dfrac{A_1 E_1}{A_2 E_2} + 1} .$$

For the symmetric case with $a_1 = a_2 = a$, $t_1 = t_2$ and $E_1 = E_2$, shown in Fig. 10.13, the elastic center is located at the quarter point,

$$c_y = -\tfrac{1}{4}a , \qquad c_z = \tfrac{1}{4}a .$$

An important part of analytical cross-section analysis is the choice of reference coordinate system. Exercise 10.1 considers the location of the elastic center of the angle section in this example with respect to another reference system. □

Example 10.2. Inhomogeneous T-profile.

Figure 10.14 shows a thin-walled T-profile, where the length, thickness, elastic modulus of the flange and the web are a_f, t_f, E_f and a_w, t_w, E_w, respectively. The reference coordinate system is located such that the y- and z-axis coincide with the centerlines of the flange and web, respectively. The elastic modulus of the flange is chosen as the reference value, $E_0 = E_f$.

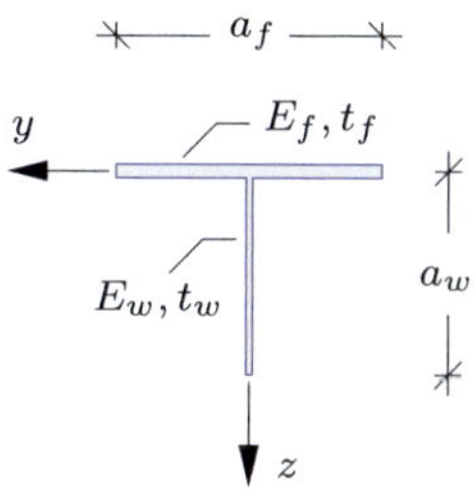

Fig. 10.14: Inhomogeneous T-profile.

The cross-section area is obtained by summation of the contributions from flange and web,

$$F = \frac{1}{E_f}(E_f\, a_f t_f + E_w\, a_w t_w) = a_f t_f + \frac{E_w}{E_f}\, a_w t_w = A_f + \frac{E_w}{E_f} A_w$$

where $A_f = a_f t_f$ and $A_w = a_w t_w$ are the geometric areas of flange and web, respectively. The static moments are determined as

$$S_y = 0 , \qquad S_z = \tfrac{1}{2}\, a_w\, A_w \frac{E_w}{E_f} = \tfrac{1}{2}\, a_w^2 t_w \frac{E_w}{E_f} .$$

Note, that $S_y = 0$ because of symmetry, and therefore it is usually advantageous to place one of the axes in the line of symmetry. The coordinates of the elastic center are

$$c_y = \frac{S_y}{F} = 0 , \quad c_z = \frac{S_z}{F} = \frac{\tfrac{1}{2}\, a_w\, A_w \dfrac{E_w}{E_f}}{A_f + \dfrac{E_w}{E_f} A_w} = \tfrac{1}{2}\, a_w \frac{\dfrac{E_w A_w}{E_f A_f}}{1 + \dfrac{E_w A_w}{E_f A_f}} = \frac{\tfrac{1}{2}\, a_w}{\dfrac{E_f A_f}{E_w A_w} + 1} .$$

A special case is $a_f = a_w = a$, $t_f = t_w$ and $E_f = E_w$, whereby $c_z = \tfrac{1}{4}\, a$.

The cross-section has a line of symmetry along the centerline of the web, and the elastic center is located on this line of symmetry. In general cross-sections with a common line of geometric and material symmetry will have the elastic center located on this line. □

Elastic center for symmetric cross-sections

For a cross-section with geometric and material double symmetry the location of the elastic center is at the intersection of the axes of symmetry, illustrated in Fig. 10.15a. However, also cross-sections formed by flanges or groups of flanges with point symmetry about a common center will have the elastic center at this point, as illustrated in Fig. 10.15b.

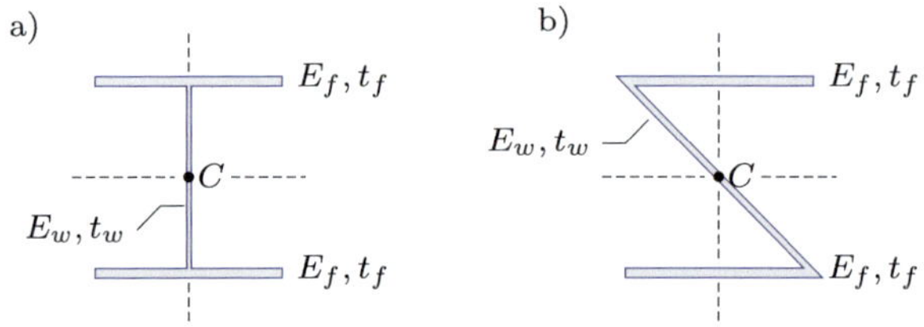

Fig. 10.15: Cross-sections with a) double symmetry, or b) point symmetry.

Example 10.3. Rectangular cross-section with reinforcement.

Figure 10.16 shows a rectangular cross-section in a concrete beam, with elastic modulus E_c and dimensions $a \times 2a$. Steel reinforcement is introduced at the bottom of the beam to improve the strength in tension. The steel reinforcement is located at vertical distance $\frac{1}{6}a$ from the bottom. The elastic modulus of steel is $E_s = 15E_c$ and the total area of the reinforcement is $A_s = \frac{1}{100}A_c = \frac{1}{50}a^2$. The cross-section is symmetric with respect to the vertical centerline. The reference coordinate system is therefore located at the top of the cross-section with the vertical z-axis coinciding with the line of symmetry. The elastic modulus of concrete is chosen as the reference value: $E_0 = E_c$.

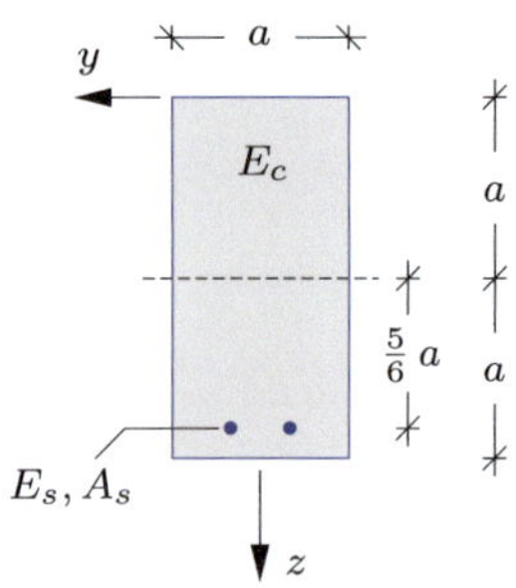

Fig. 10.16: Rectangular cross-section of concrete beam with steel reinforcement.

The weighted area is determined as the sum of the contributions from concrete and steel, and the concrete area is represented approximately by the full area of the cross-section,

$$F = \frac{1}{E_c} (E_c\, 2a^2 + E_s\, A_s) = a^2\big(2 + \tfrac{15}{50}\big) = \tfrac{23}{10}\, a^2 .$$

The center of the concrete part is located at vertical distance $z = a$, while the center of the steel reinforcement is located at $z = 2a - \frac{1}{6}a$. The static moment with respect to the z-axis can therefore be determined as

$$S_z = \frac{1}{E_c} \big(E_c\, 2a^2\, a + E_s\, A_s\, (2a - \tfrac{1}{6}a)\big) = \tfrac{51}{20}\, a^2 .$$

The elastic center is therefore located at

$$c_z = \frac{S_z}{F} = \tfrac{51}{46}\, a$$

which is slightly below the geometric center of the cross-section. □

10.2.2 Moments of inertia

By the translation of the coordinate system, whereby the origin of the coordinate system is located at the elastic center $[c_y, c_z]$, the extension and the bending problems uncouple. Hereby, the constitutive relation in (10.30) can be written as two separate relations,

$$N = E_0 F\, \varepsilon_c\, , \qquad \begin{bmatrix} M_{\bar{y}} \\ M_{\bar{z}} \end{bmatrix} = E_0 \begin{bmatrix} I_{\bar{y}\bar{y}} & I_{\bar{y}\bar{z}} \\ I_{\bar{y}\bar{z}} & I_{\bar{z}\bar{z}} \end{bmatrix} \begin{bmatrix} \kappa_y \\ \kappa_z \end{bmatrix} . \tag{10.40}$$

This leaves the determination of the moments of inertia $I_{\bar{y}\bar{y}}$, $I_{\bar{z}\bar{z}}$ and $I_{\bar{y}\bar{z}}$ with respect to the translated coordinate system. In the new coordinate system the moments of inertia are defined as

$$\begin{aligned} I_{\bar{y}\bar{y}} &= \frac{1}{E_0}\int_A E\bar{y}^2\, dA = \frac{1}{E_0}\int_A E(y-c_y)^2\, dA = I_{yy} + c_y^2 F - 2c_y S_y\, , \\ I_{\bar{z}\bar{z}} &= \frac{1}{E_0}\int_A E\bar{z}^2\, dA = \frac{1}{E_0}\int_A E(z-c_z)^2\, dA = I_{zz} + c_z^2 F - 2c_z S_z\, , \\ I_{\bar{y}\bar{z}} &= \frac{1}{E_0}\int_A E\bar{y}\bar{z}\, dA = \frac{1}{E_0}\int_A E(y-c_y)(z-c_z)\, dA \\ &\qquad\qquad\qquad\qquad\qquad = I_{yz} + c_y c_z F - c_y S_z - c_z S_y\, . \end{aligned} \tag{10.41}$$

The static moments S_y and S_z can be eliminated in terms of the coordinates of the elastic center $[c_y, c_z]$ by the relation (10.32). Hereby the expressions for the moments of inertia reduce to

$$I_{\bar{y}\bar{y}} = I_{yy} - c_y^2 F\, , \quad I_{\bar{z}\bar{z}} = I_{zz} - c_z^2 F\, , \quad I_{\bar{y}\bar{z}} = I_{yz} - c_y c_z F\, . \tag{10.42}$$

The separated format of the constitutive relations in (10.40) is quite suitable for analytical solutions, as the inverse of a 2×2 matrix has a fairly simple explicit form. This is illustrated in Section 10.3, where the axial strain and stress are determined.

Application of parallel axis theorem

The expression for the moments of inertia with respect to a translated coordinate system is often called the *parallel axis theorem*. The terms in (10.42) all involve integration over the cross-section area. As for the static moments the moments of inertia in (10.42) are determined by summation of the contribution to the moments of inertia from the individual flanges or parts of the cross-section.

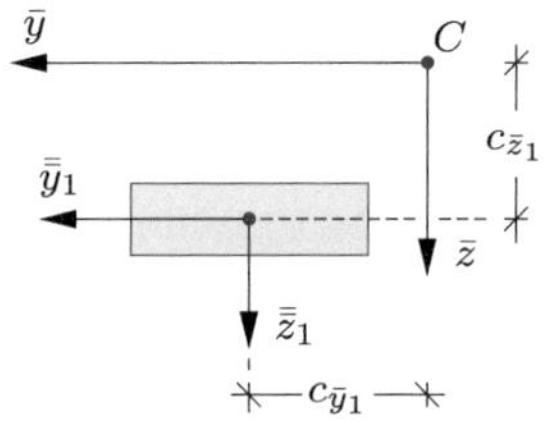

Fig. 10.17: Rectangular flange.

Figure 10.17 shows a simple rectangular flange with geometric area A_1. Assume that this flange is part of a larger cross-section, as indicated in Fig. 10.10, and that the coordinate system $\{\bar{y}, \bar{z}\}$ is located at the elastic center C of the full cross-section. A local coordinate system $\{\bar{\bar{y}}_1, \bar{\bar{z}}_1\}$ is placed at the center of the local flange in Fig. 10.17 with axes parallel to those of the $\{\bar{y}, \bar{z}\}$ coordinate system. The coordinates of the local center in the $\{\bar{y}, \bar{z}\}$ coordinate system are $c_{\bar{y}_1}$ and $c_{\bar{z}_1}$, as shown in Fig. 10.17. The relation between the two coordinate systems can be written as

$$\bar{y} = \bar{\bar{y}}_1 + c_{\bar{y}_1}\,, \qquad \bar{z} = \bar{\bar{z}}_1 + c_{\bar{z}_1}\,. \tag{10.43}$$

The contribution from the local flange to the moments of inertia with respect to the $\{\bar{y}, \bar{z}\}$ coordinate system can be determined by the integrals given in (10.41), where substitution of (10.43) gives

$$\begin{aligned}
I_{\bar{y}\bar{y}} &= \frac{1}{E_0}\int_{A_1} E(\bar{\bar{y}}_1 + c_{\bar{y}_1})^2\, dA &&= I_{\bar{\bar{y}}_1\bar{\bar{y}}_1} + c_{\bar{y}_1}^2 F_1\,,\\
I_{\bar{z}\bar{z}} &= \frac{1}{E_0}\int_{A_1} E(\bar{\bar{z}}_1 + c_{\bar{z}_1})^2\, dA &&= I_{\bar{\bar{z}}_1\bar{\bar{z}}_1} + c_{\bar{z}_1}^2 F_1\,,\\
I_{\bar{y}\bar{z}} &= \frac{1}{E_0}\int_{A_1} E(\bar{\bar{y}}_1 + c_{\bar{y}_1})(\bar{\bar{z}}_1 + c_{\bar{z}_1})\, dA &&= I_{\bar{\bar{y}}_1\bar{\bar{z}}_1} + c_{\bar{y}_1}c_{\bar{z}_1} F_1\,,
\end{aligned} \tag{10.44}$$

where it is used that in the local $\{\bar{\bar{y}}_1, \bar{\bar{z}}_1\}$ coordinate system $S_{\bar{\bar{y}}_1} = S_{\bar{\bar{z}}_1} = 0$.

For cross-sections composed of multiple flanges the summation of the contributions from each individual flange gives the resulting moments of inertia as

$$
\begin{aligned}
I_{\bar{y}\bar{y}} &= \sum_{j=1}^{n} \left(I_{\bar{\bar{y}}_j\bar{\bar{y}}_j} + c_{\bar{y}_j}^2 F_j \right) = \sum_{j=1}^{n} \left(I_{\bar{\bar{y}}_j\bar{\bar{y}}_j} + c_{\bar{y}_j}^2 A_j E_j / E_0 \right), \\
I_{\bar{z}\bar{z}} &= \sum_{j=1}^{n} \left(I_{\bar{\bar{z}}_j\bar{\bar{z}}_j} + c_{\bar{z}_j}^2 F_j \right) = \sum_{j=1}^{n} \left(I_{\bar{\bar{z}}_j\bar{\bar{z}}_j} + c_{\bar{z}_j}^2 A_j E_j / E_0 \right), \\
I_{\bar{y}\bar{z}} &= \sum_{j=1}^{n} \left(I_{\bar{\bar{y}}_j\bar{\bar{z}}_j} + c_{\bar{y}_j} c_{\bar{z}_j} F_j \right) = \sum_{j=1}^{n} \left(I_{\bar{\bar{y}}_j\bar{\bar{z}}_j} + c_{\bar{y}_j} c_{\bar{z}_j} A_j E_j / E_0 \right),
\end{aligned}
\qquad (10.45)
$$

where the last equalities apply to homogeneous flanges with area A_j. This is the general form of the parallel axis theorem.

Example 10.4. Moments of inertia for rectangular cross-section. Figure 10.18 shows a homogeneous rectangular cross-section with constant elastic modulus E, width b and height h. In this case the origin of the reference coordinate system, shown in the figure, coincides with the elastic center, whereby $[\bar{y}, \bar{z}] = [y, z]$. The elastic modulus E is normalized by a reference elastic modulus E_0 to present the results in a general form.

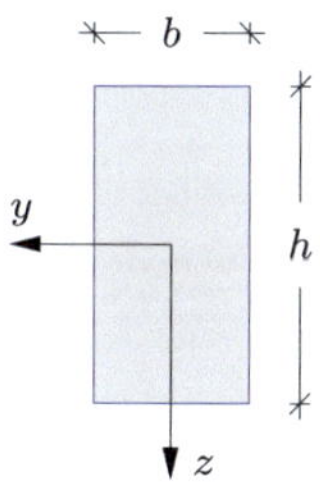

Fig. 10.18: Rectangular homogeneous cross-section.

The definition of the moments of inertia with respect to the reference axes are given by the integral in (10.16). This means that the moment of inertia in the z-direction is defined as

$$
I_{zz} = \frac{1}{E_0} \int_A E z^2 \, dA
$$

and because E is constant it can be taken outside the area integral, which can be represented by double integrals in the y- and z-coordinates,

$$
I_{zz} = \frac{E}{E_0} \int_{-\frac{b}{2}}^{\frac{b}{2}} \int_{-\frac{h}{2}}^{\frac{h}{2}} z^2 \, dz \, dy = \frac{Eb}{E_0} \int_{-\frac{h}{2}}^{\frac{h}{2}} z^2 \, dz = \frac{Eb}{E_0} \left[\tfrac{1}{3} z^3 \right]_{-\frac{h}{2}}^{\frac{h}{2}} = \tfrac{1}{12} b h^3 \, E/E_0 \, .
$$

In the case when $E = E_0$, the moment of inertia is $I_{zz} = \frac{1}{12} b h^3$. The moment of inertia with respect to the y-direction is determined similarly as

$$
I_{yy} = \tfrac{1}{12} h b^3 \, E/E_0 \, .
$$

For the coupling moment of inertia the integral separates into

$$I_{yz} = \frac{1}{E_0}\int_A Eyz\,dA = \frac{E}{E_0}\Big(\int_{-\frac{b}{2}}^{\frac{b}{2}} y\,dy\Big)\Big(\int_{-\frac{h}{2}}^{\frac{h}{2}} z\,dz\Big).$$

Because of symmetry both integrals vanish

$$\int_{-\frac{b}{2}}^{\frac{b}{2}} y\,dy = \Big[\tfrac{1}{2}y^2\Big]_{-\frac{b}{2}}^{\frac{b}{2}} = 0\,, \qquad \int_{-\frac{h}{2}}^{\frac{h}{2}} z\,dz = \Big[\tfrac{1}{2}z^2\Big]_{-\frac{h}{2}}^{\frac{h}{2}} = 0\,,$$

whereby

$$I_{yz} = 0\,.$$

Note, that only one of the integrals for I_{yz} have to vanish, and thus $I_{yz} = 0$ if at least one of the two coordinate axes is a line of symmetry. □

Example 10.5. Moments of inertia for skew cross-section. Figure 10.19 shows a skew homogeneous cross-section with elastic modulus $E = E_0$, width b and height h. The origin of the reference coordinate system in the figure coincides with the elastic center. The inclination of the cross-section is represented by the angle θ, as shown in the figure. The rectangular cross-section in Fig. 10.18 is recovered for $\theta = 0$.

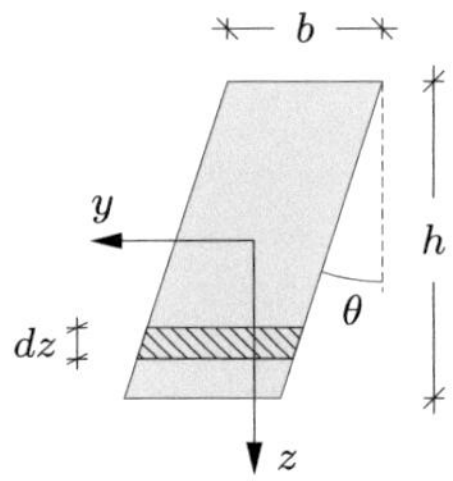

Fig. 10.19: Skew cross-section.

Consider a very thin strip of the cross-section, indicated by the hatched area in the figure. The width is b and the incremental height is dz. The coordinates of the center of the strip are $[y, z]$, with $y = z\tan\theta$. The moments of inertia for the strip are then calculated as

$$\begin{aligned} dI_{yy} &= \tfrac{1}{12}b^3\,dz + b\,dz\,y^2 = \tfrac{1}{12}b^3\,dz + bz^2\tan^2\theta\,dz\,,\\ dI_{zz} &= \tfrac{1}{12}(dz)^3 b + b\,dz\,z^2 = bz^2 dz\,,\\ dI_{yz} &= b\,dz\,zy = bz^2\tan\theta\,dz\,. \end{aligned}$$

As dz is infinitesimal, higher order terms in dz can be omitted. The moments of inertia are now obtained by integration over the height of the cross-section in the z-direction, whereby

$$I_{yy} = \tfrac{1}{12}b^3h + \tfrac{1}{12}h^3b\tan^2\theta\,, \quad I_{zz} = \tfrac{1}{12}h^3b\,, \quad I_{yz} = \tfrac{1}{12}h^3b\tan\theta\,.$$

It is seen that the results for the rectangular cross-section are recovered for $\theta = 0$. □

Example 10.6. Moments of inertia for angle profile. Figure 10.20a shows a cross-section with two thin-walled flanges connected at a right angle, similar to that in Example 10.1. In this case the cross-section is homogeneous with elastic modulus $E = E_0$. The dimensions of the horizontal and vertical flanges are $2a \times t$ and $a \times 2t$, respectively, as shown in the figure. The reference coordinate system $\{y, z\}$ is located as in Example 10.1, where the position of the elastic center was found as

$$c_y = -\tfrac{1}{2}a\,, \qquad c_z = \tfrac{1}{4}a\,.$$

The elastic center and the new coordinate system $\{\bar{y}, \bar{z}\}$ are both indicated in Fig. 10.20b.

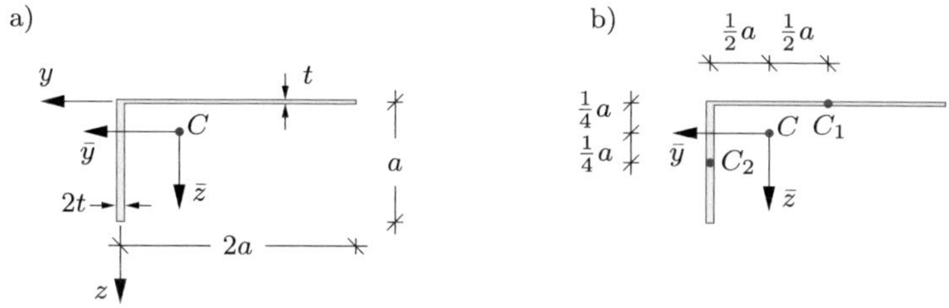

Fig. 10.20: Homogeneous thin-walled angle profile.

The centers of the two flanges are indicated in the figure as C_1 and C_2. The distance from the local center to the elastic center C is also shown in the figure. The local moments of inertia for the flanges are now determined. Both flanges are rectangular, whereby the results in Example 10.4 can be applied. For the horizontal flange 1 the local moments of inertia are:

$$I_{\bar{\bar{y}}_1\bar{\bar{y}}_1} = \tfrac{1}{12}(2a)^3 t = \tfrac{2}{3}a^3 t\,, \quad I_{\bar{\bar{z}}_1\bar{\bar{z}}_1} = \tfrac{1}{12}t^3 a \simeq 0\,, \quad I_{\bar{\bar{y}}_1\bar{\bar{z}}_1} = 0\,.$$

Since the flanges are thin-walled ($t \ll a$) any terms with higher powers of thickness are omitted because these contributions are insignificant. This is the reason for setting $I_{\bar{\bar{z}}_1\bar{\bar{z}}_1} = 0$, while the coupling moment of inertia $I_{\bar{\bar{y}}_1\bar{\bar{z}}_1}$ vanishes due to symmetry. The local moments of inertia for the vertical flange 2 are obtained similarly,

$$I_{\bar{\bar{y}}_2\bar{\bar{y}}_2} = \tfrac{1}{12}(2t)^3 a \simeq 0\,, \quad I_{\bar{\bar{z}}_2\bar{\bar{z}}_2} = \tfrac{1}{12}(a)^3 2t = \tfrac{1}{6}a^3 t\,, \quad I_{\bar{\bar{y}}_2\bar{\bar{z}}_2} = 0\,.$$

The resulting moments of inertia are found by a transformation via the parallel axis theorem to the actual coordinate system $\{\bar{y}, \bar{z}\}$, followed by summation of the contributions from the two flanges. By using (10.45) for homogeneous flanges,

$$\begin{aligned}
I_{\bar{y}\bar{y}} &= \big(\tfrac{2}{3}\,a^3 t + 2at(\tfrac{1}{2}a)^2\big) + \big(0 + 2at(\tfrac{1}{2}a)^2\big) = \tfrac{5}{3}a^3 t\,,\\
I_{\bar{z}\bar{z}} &= \big(0 + 2at(\tfrac{1}{4}a)^2\big) + \big(\tfrac{1}{6}a^3 t + 2at(\tfrac{1}{4}a)^2\big) = \tfrac{5}{12}a^3 t\,,\\
I_{\bar{y}\bar{z}} &= \big(0 + 2at\,(-\tfrac{1}{2}a)(-\tfrac{1}{4}a)\big) + \big(0 + 2at\,(\tfrac{1}{2}a)(\tfrac{1}{4}a)\big) = \tfrac{1}{2}a^3 t\,.
\end{aligned}$$

The first parenthesis represents the contribution from the horizontal flange, while the second represents the vertical flange. For the coupling moment of inertia $I_{\bar{y}\bar{z}}$ it is important to use the correct sign for the position of the local center with regard to the actual coordinate system. For the horizontal flange the local center C_1 is placed in the negative quadrant of the $\{\bar{y}, \bar{z}\}$ coordinate system, whereby both coordinates become negative. The contribution to $I_{\bar{y}\bar{z}}$ from the vertical flange 2 is also positive because the local center C_2 is located in the fully positive quadrant. □

Example 10.7. Moments of inertia for T-profile.

Figure 10.21 shows the thin-walled T-profile from Example 10.2. In this example the dimensions are chosen as follows:

$$a_w = a_f = a\,, \qquad E_w = E_f\,, \qquad t_w = \tfrac{1}{2}t_f = t\,.$$

By substitution of these relations into the expressions for the elastic center determined in Example 10.2 it is found that

$$c_y = 0\,, \qquad c_z = \tfrac{3}{16}a\,.$$

The coordinate system $\{\bar{y}, \bar{z}\}$ is placed with its origin in C, as shown in the figure.

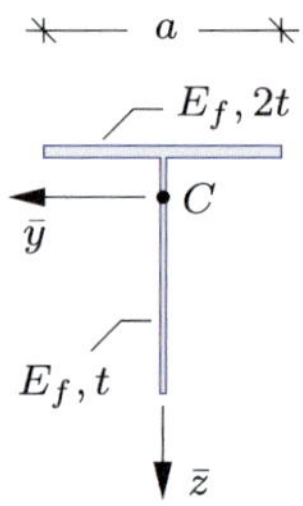

Fig. 10.21: Homogeneous T-profile.

The moments of inertia can now be determined by summation over the number of flanges following (10.45),

$$I_{\bar{y}\bar{y}} = \left(\tfrac{1}{12}a^3 2t + \tfrac{1}{12}t^3 a\right) = \tfrac{1}{6}\,a^3\,t\,,$$

$$I_{\bar{z}\bar{z}} = \left(\tfrac{1}{12}(2t)^3 a + 2at(\tfrac{3}{16}a)^2\right) + \left(\tfrac{1}{12}a^3 t + at\,(\tfrac{1}{2}a - \tfrac{3}{16}a)^2\right) = \tfrac{193}{768}\,a^3\,t \simeq 0.25\,a^3 t\,,$$

$$I_{\bar{y}\bar{z}} = 0 \quad \text{(symmetry)}\,.$$

The cross-section is assumed to be thin-walled, which implies that terms containing higher powers of thickness are omitted, whereby the only non-vanishing contribution to $I_{\bar{y}\bar{y}}$ given above is from the horizontal flange. Note also that the coupling moment of inertia is zero due to symmetry of the cross-section. □

Example 10.8. Moments of inertia for I-profile. Figure 10.22 shows an I-shaped cross-section. The flanges have dimensions $b \times t_f$ and elastic modulus E_f, while the web has dimensions $h \times t_w$ and elastic modulus E_w. Thus, the cross-section is inhomogeneous, and the reference elastic modulus is chosen as the flange value: $E_0 = E_f$. The particular geometry and material distribution implies that the cross-section is double symmetric and the elastic center is therefore located at the intersection of the lines of symmetry. In this case the origin of the reference coordinate system $\{y, z\}$ is therefore located directly at the elastic center.

Following (10.45) the moments of inertia are determined as

$$I_{yy} = \frac{1}{E_f}\left(E_w\,\tfrac{1}{12}t_w^3 h + 2E_f\,\tfrac{1}{12}b^3 t_f\right) = \tfrac{1}{6}b^3 t_f\,,$$

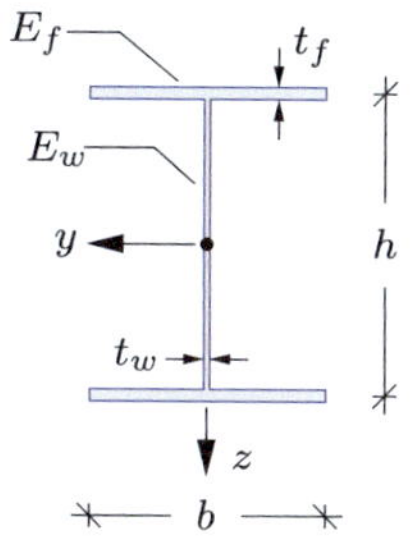

Fig. 10.22: Inhomogeneous thin-walled I-profile.

$$I_{zz} = \frac{1}{E_f}\Big(E_w \tfrac{1}{12}\, h^3\, t_w \,+\, 2E_f\big(\tfrac{1}{12}\, t_f^3\, b + b t_f (\tfrac{1}{2}\, h)^2\big)\Big) = \tfrac{1}{12} h^3 t_w \frac{E_w}{E_f} + \tfrac{1}{2} h^2 b t_f \,,$$

$$I_{yz} = 0 \quad \text{(symmetry)}\,.$$

Again, terms containing higher powers of thickness are omitted, and the coupling moment of inertia is zero due to symmetry.

The moment of inertia with respect to the z-coordinate contains contributions from both web and flanges. In the simple case where $E_w = E_f$, $h = b = a$ and $t_w = t_f = t$ the two contributions reduce to

$$\tfrac{1}{12} h^3 t_w \frac{E_w}{E_f} = \tfrac{1}{12} a^3 t \,, \qquad \tfrac{1}{2} h^2 b t_f = \tfrac{1}{2} a^3 t \,.$$

This indicates that the main contribution to the bending stiffness of an I-profile with respect to the beam deformation in the z-direction comes from the flanges. Note furthermore that the relative contribution from the flanges increase with increasing height h of the cross-section. □

10.2.3 Principal coordinate system

By using a coordinate system with origin at the elastic center the problems of extension and bending uncouple and the constitutive relations appear in the separated form (10.40). The constitutive relations can be completely uncoupled by rotating the coordinate system $\{\bar{y}, \bar{z}\}$ until the coupling moment of inertia vanishes. The coordinate system in which the inertia matrix is diagonal is called the *principal coordinate system*, analogous to the use of the term in the context principal stresses and strains discussed in Chapter 8.

In Fig. 10.23 the coordinate system $\{\bar{y}, \bar{z}\}$ is rotated by the angle θ. Hereby the coordinates $[y', z']$ in the rotated system are obtained from the corresponding coordinates $[\bar{y}, \bar{z}]$ in the original coordinate system by the relation

$$\begin{bmatrix} y' \\ z' \end{bmatrix} = \begin{bmatrix} \cos\theta & \sin\theta \\ -\sin\theta & \cos\theta \end{bmatrix} \begin{bmatrix} \bar{y} \\ \bar{z} \end{bmatrix} , \tag{10.46}$$

with the inverse relation

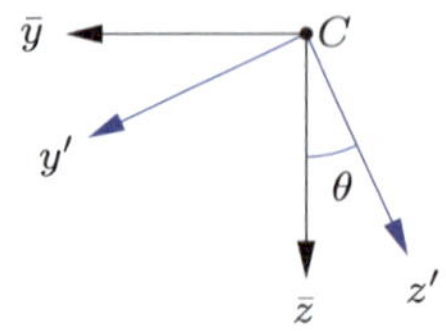

Fig. 10.23: Rotation of coordinate system.

$$\begin{bmatrix} \bar{y} \\ \bar{z} \end{bmatrix} = \begin{bmatrix} \cos\theta & -\sin\theta \\ \sin\theta & \cos\theta \end{bmatrix} \begin{bmatrix} y' \\ z' \end{bmatrix} . \tag{10.47}$$

Similar transform relations apply to other sets of vector components in the two coordinate systems, such as the inclinations $[\eta_y, \eta_z]$ with components $[\eta_{y'}, \eta_{z'}]$ in the rotated coordinate system.

The axial displacement field is given by (10.26) in the translated $\{\bar{y}, \bar{z}\}$ coordinate system, and when introducing the transforms (10.47) of both sets of components the following form is obtained

$$u = u_c + \bar{y}\,\eta_y + \bar{z}\,\eta_z = u_c + y'\,\eta_{y'} + z'\,\eta_{z'} . \tag{10.48}$$

The axial strain follows in a similar way from (10.27) as

$$\varepsilon = \varepsilon_c + \bar{y}\,\kappa_y + \bar{z}\,\kappa_z = \varepsilon_c + y'\,\kappa_{y'} + z'\,\kappa_{z'} , \tag{10.49}$$

with the axial strain at the elastic center C and the two curvatures defined as

$$\varepsilon_c = \frac{du_c}{dx} , \qquad \kappa_{y'} = \frac{d\eta_{y'}}{dx} , \qquad \kappa_{z'} = \frac{d\eta_{z'}}{dx} . \tag{10.50}$$

In fact, the relation (10.48) follows directly from the fact that the scalar product of two vectors is independent of the specific coordinate system used for the components.

The axial stress is obtained by multiplication of the axial strain with the elastic modulus, and the normal force is found by integration of the axial stress over the cross-section area,

$$N = \int_A E\varepsilon\, dA = E_0 F \varepsilon_c + E_0 S_{y'}\,\kappa_{y'} + E_0 S_{z'}\,\kappa_{z'} . \tag{10.51}$$

The static moments in the rotated coordinate system are obtained by substitution of (10.46) as

$$\begin{aligned} S_{y'} &= \frac{1}{E_0}\int_A E(\bar{y}\cos\theta + \bar{z}\sin\theta)\, dA &&= S_{\bar{y}}\cos\theta + S_{\bar{z}}\sin\theta = 0 , \\ S_{z'} &= \frac{1}{E_0}\int_A E(-\bar{y}\sin\theta + \bar{z}\cos\theta)\, dA &&= S_{\bar{y}}\cos\theta + S_{\bar{z}}\sin\theta = 0 , \end{aligned} \tag{10.52}$$

where the last equalities follow from the fact that static moments $S_{\bar{y}} = S_{\bar{z}} = 0$ in a coordinate system with origo at the elastic center. Thus, the constitutive relations in the rotated coordinate system have the form

$$N = E_0 F \varepsilon_c , \qquad \begin{bmatrix} M_{y'} \\ M_{z'} \end{bmatrix} = E_0 \begin{bmatrix} I_{y'y'} & I_{y'z'} \\ I_{y'z'} & I_{z'z'} \end{bmatrix} \begin{bmatrix} \kappa_{y'} \\ \kappa_{z'} \end{bmatrix} . \tag{10.53}$$

The moments of inertia in the rotated coordinate system are given as

$$\begin{aligned} I_{y'y'} &= \frac{1}{E_0} \int_A E y' y' \, dA = I_{\bar{y}\bar{y}} \cos^2\theta + I_{\bar{z}\bar{z}} \sin^2\theta + 2 I_{\bar{y}\bar{z}} \cos\theta \sin\theta , \\ I_{z'z'} &= \frac{1}{E_0} \int_A E z' z' \, dA = I_{\bar{y}\bar{y}} \sin^2\theta + I_{\bar{z}\bar{z}} \cos^2\theta - 2 I_{\bar{y}\bar{z}} \cos\theta \sin\theta , \\ I_{y'z'} &= \frac{1}{E_0} \int_A E y' z' \, dA = (I_{\bar{z}\bar{z}} - I_{yy}) \cos\theta \sin\theta + I_{\bar{y}\bar{z}} (\cos^2\theta - \sin^2\theta) , \end{aligned} \tag{10.54}$$

where the latter expressions are obtained by substitution of (10.46) followed by evaluation of the area integrals. The expressions in (10.54) are conveniently formulated in terms of the double angle, similar to the approach for stress components leading to the format in (8.68). Therefore, the following trigonometric relations are introduced,

$$2\cos\theta \sin\theta = \sin 2\theta , \qquad \cos^2\theta - \sin^2\theta = \cos 2\theta , \tag{10.55}$$

from which it follows that

$$\cos^2\theta = \tfrac{1}{2}\big(1 + \cos 2\theta\big) , \qquad \sin^2\theta = \tfrac{1}{2}\big(1 - \cos 2\theta\big) . \tag{10.56}$$

Hereby, the expressions (10.54) for the moments of inertia in the rotated coordinate system can be written as

$$\begin{aligned} I_{y'y'} &= \tfrac{1}{2}\big(I_{\bar{y}\bar{y}} + I_{\bar{z}\bar{z}}\big) + \tfrac{1}{2}\big(I_{\bar{y}\bar{y}} - I_{\bar{z}\bar{z}}\big) \cos 2\theta + I_{\bar{y}\bar{z}} \sin 2\theta , \\ I_{z'z'} &= \tfrac{1}{2}\big(I_{\bar{y}\bar{y}} + I_{\bar{z}\bar{z}}\big) - \tfrac{1}{2}\big(I_{\bar{y}\bar{y}} - I_{\bar{z}\bar{z}}\big) \cos 2\theta - I_{\bar{y}\bar{z}} \sin 2\theta , \\ I_{y'z'} &= \qquad\qquad\qquad - \tfrac{1}{2}\big(I_{\bar{y}\bar{y}} - I_{\bar{z}\bar{z}}\big) \sin 2\theta + I_{\bar{y}\bar{z}} \cos 2\theta . \end{aligned} \tag{10.57}$$

The polar moment of inertia I_p is defined with respect to the distance from the elastic center, $r^2 = \bar{y}^2 + \bar{z}^2 = {y'}^2 + {z'}^2$, and thus

$$I_p = I_{\bar{y}\bar{y}} + I_{\bar{z}\bar{z}} = I_{y'y'} + I_{z'z'} . \tag{10.58}$$

It is seen that I_p, and thereby the sum of the diagonal moments of inertia, are invariant with respect to rotation of the coordinate system.

The bending splits into two uncoupled problems when the matrix is diagonal, and the new coordinate system is therefore rotated such that the coupling moment of inertia is zero,

$$I_{y'z'} = -\tfrac{1}{2}\big(I_{\bar{y}\bar{y}} - I_{\bar{z}\bar{z}}\big)\sin 2\theta + I_{\bar{y}\bar{z}}\cos 2\theta = 0\,. \tag{10.59}$$

This particular set of axes is called the *principal axes* of the cross-section. The equation (10.59) corresponds to

$$\tan 2\theta = \frac{2I_{\bar{y}\bar{z}}}{I_{\bar{y}\bar{y}} - I_{\bar{z}\bar{z}}}\,. \tag{10.60}$$

The angle can therefore be found as

$$\theta = \tfrac{1}{2}\arctan\left(\frac{2I_{\bar{y}\bar{z}}}{I_{\bar{y}\bar{y}} - I_{\bar{z}\bar{z}}}\right) + \tfrac{1}{2}n\pi\,, \tag{10.61}$$

where the second term takes the periodicity of the tangent function into account. If the angle θ_0 is the principal solution ($n = 0$) to (10.61) in the interval $-\frac{1}{2}\pi \le \theta_0 \le \frac{1}{2}\pi$, a second solution ($n = 1$) is given as $\theta_1 = \theta_0 + \frac{1}{2}\pi$ or $\theta_1 = \theta_0 + 90^\circ$. Figure 10.24 illustrates the two orientations of the principal axes for an angle profile.

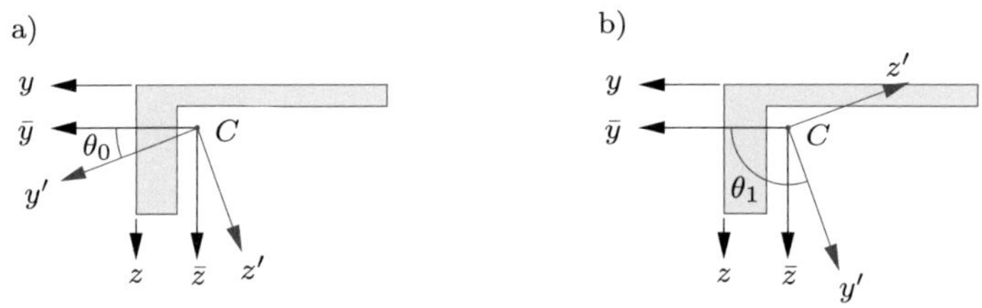

Fig. 10.24: Rotations of coordinate system.

In the coordinate system $\{y', z'\}$ the coupling moment of inertia vanishes, leaving only the two diagonal moments of inertia $I_{y'y'}$ and $I_{z'z'}$. The expression for these in terms of the original moments of inertia can be determined by the procedure used for stress components in Section 8.4.3 or by using the following relations for trigonometric functions of double angle,

$$\begin{aligned} \cos 2\theta &= \frac{\pm 1}{\sqrt{1+\tan^2 2\theta}} = \frac{\frac{1}{2}(I_{\bar{y}\bar{y}} - I_{\bar{z}\bar{z}})}{\sqrt{\frac{1}{4}(I_{\bar{y}\bar{y}} - I_{\bar{z}\bar{z}})^2 + I_{\bar{y}\bar{z}}^2}}\,,\\ \sin 2\theta &= \frac{\pm\tan 2\theta}{\sqrt{1+\tan^2 2\theta}} = \frac{I_{\bar{y}\bar{z}}}{\sqrt{\frac{1}{4}(I_{\bar{y}\bar{y}} - I_{\bar{z}\bar{z}})^2 + I_{\bar{y}\bar{z}}^2}}\,. \end{aligned} \tag{10.62}$$

By elimination of the trigonometric functions in (10.57) via the relations in (10.62) the moments of inertia with respect to the principal axes appear as

$$\left.\begin{matrix} I_{y'} \\ I_{z'} \end{matrix}\right\} = \frac{I_{\bar{y}\bar{y}} + I_{\bar{z}\bar{z}}}{2} \pm \sqrt{\left(\frac{I_{\bar{y}\bar{y}} - I_{\bar{z}\bar{z}}}{2}\right)^2 + I_{\bar{y}\bar{z}}^2}\,. \tag{10.63}$$

The plus sign represents the largest moment of inertia and the minus the smallest. The moments of inertia $I_{y'}$ and $I_{z'}$ with respect to the principal axes are denoted by a single subscript only, indicating that the coupling component is zero for this particular coordinate system orientation. Whether $I_{y'} > I_{z'}$ or $I_{z'} > I_{y'}$ depends on the choice of axes orientation. Figure 10.24a and b illustrate the rotation of the coordinate system by θ_0 and $\theta_1 = \theta_0 + \pi/2$, respectively. In the first case with θ_0 it is seen that $I_{y'} > I_{z'}$, whereby $I_{y'}$ is found by the plus in (10.63) and $I_{z'}$ by the minus. Conversely for θ_1, where it is seen that $I_{z'} > I_{y'}$, corresponding to $I_{z'}$ representing the plus in (10.63) and $I_{y'}$ the minus.

For cross-sections with symmetry with respect to one of the axes the coupling moment of inertia is already zero, and the axes therefore directly represent the principal coordinate system.

Principal values by eigenvalue problem

The principal moments of inertia of a cross-section can also be determined by an alternative strictly algebraic approach, similar to that used in Section 8.4.4 for three-dimensional states of stress. The principal directions are characterized by defining planes in which bending is uncoupled to bending out of the plane. Thus, if the components $[\kappa_{\bar{y}}, \kappa_{\bar{z}}]$ define curvatures in a principal plane, then the corresponding moment vector components $[M_{\bar{y}}, M_{\bar{z}}]$ will lie in the same plane. This property can be expressed by the relations

$$\begin{bmatrix} M_{\bar{y}} \\ M_{\bar{z}} \end{bmatrix} = E_0 \begin{bmatrix} I_{\bar{y}\bar{y}} & I_{\bar{y}\bar{z}} \\ I_{\bar{z}\bar{y}} & I_{\bar{z}\bar{z}} \end{bmatrix} \begin{bmatrix} \kappa_{\bar{y}} \\ \kappa_{\bar{z}} \end{bmatrix} = I\,E_0 \begin{bmatrix} \kappa_{\bar{y}} \\ \kappa_{\bar{z}} \end{bmatrix}, \tag{10.64}$$

where I is a proportionality factor representing the moment of inertia in the principal direction. The latter equality constitutes an eigenvalue problem of the form

$$\begin{bmatrix} I_{\bar{y}\bar{y}} - I & I_{\bar{y}\bar{z}} \\ I_{\bar{z}\bar{y}} & I_{\bar{z}\bar{z}} - I \end{bmatrix} \begin{bmatrix} \kappa_{\bar{y}} \\ \kappa_{\bar{z}} \end{bmatrix} = \begin{bmatrix} 0 \\ 0 \end{bmatrix}, \tag{10.65}$$

where the moment of inertia I appears as the eigenvalue, while the eigenvector $[\kappa_{\bar{y}}, \kappa_{\bar{z}}]$ defines the principal curvature direction. The equations (10.65) are homogeneous and the existence of non-trivial solutions requires the determinant of the matrix to vanish. This provides the characteristic equation

$$(I_{\bar{y}\bar{y}} - I)(I_{\bar{z}\bar{z}} - I) - I_{\bar{y}\bar{z}}^2 = 0\,. \tag{10.66}$$

This is a quadratic equation in the eigenvalue I,

$$I^2 - (I_{\bar{y}\bar{y}} + I_{\bar{z}\bar{z}})\, I + (I_{\bar{y}\bar{y}} I_{\bar{z}\bar{z}} - I_{\bar{y}\bar{z}}^2) = 0. \tag{10.67}$$

The solutions to this equation are similar to the expressions for the principal moments of inertia already derived by different means in (10.63),

$$\left.\begin{matrix} I_{y'} \\ I_{z'} \end{matrix}\right\} = \frac{I_{\bar{y}\bar{y}} + I_{\bar{z}\bar{z}}}{2} \pm \sqrt{\left(\frac{I_{\bar{y}\bar{y}} - I_{\bar{z}\bar{z}}}{2}\right)^2 + I_{\bar{y}\bar{z}}^2}\,. \tag{10.68}$$

The eigenvalue formulation is well suited to numerical solution, providing the principal moments of inertia as eigenvalues, and the principal directions via the eigenvectors.

Example 10.9. Principal properties of angle profile. Figure 10.25 shows the angle profile previously considered in Example 10.1, where the position of the elastic center was determined, and Example 10.6, where the corresponding moments of inertia were obtained as,

$$I_{\bar{y}\bar{y}} = \tfrac{5}{3}a^3 t\,, \qquad I_{\bar{z}\bar{z}} = \tfrac{5}{12}a^3 t\,, \qquad I_{\bar{y}\bar{z}} = \tfrac{1}{2}a^3 t\,.$$

Note that $I_{\bar{y}\bar{z}} \neq 0$, and thus the $\{\bar{y}, \bar{z}\}$ axes are not the principal axes.

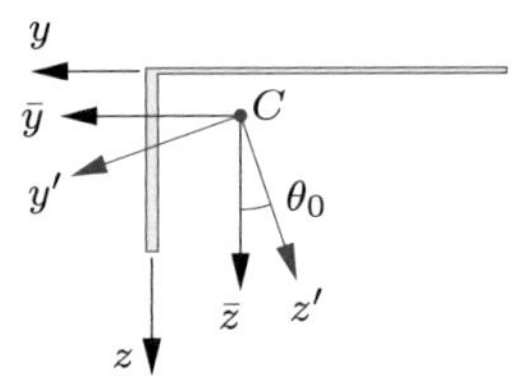

Fig. 10.25: Homogeneous thin-walled angle profile.

The counterclockwise angle of rotation of the principal coordinate system $\{y', z'\}$ relative to the $\{\bar{y}, \bar{z}\}$ coordinate system is given by (10.60),

$$\tan 2\theta = \frac{2 I_{\bar{y}\bar{z}}}{I_{\bar{y}\bar{y}} - I_{\bar{z}\bar{z}}} = \frac{2\,\frac{1}{2}\,a^3 t}{\frac{5}{3}\,a^3 t - \frac{5}{12}a^3 t} = \tfrac{4}{5}\,.$$

Thus the angle is

$$\theta_0 = 19.3^\circ$$

where subscript 0 indicates that the result corresponds to the principal value of the tangent function with $n = 0$. The orientation of the principal axes $\{y', z'\}$ is shown in Fig. 10.25.

The principal moments of inertia are given by the expressions in (10.63),

$$\left.\begin{matrix} I_{y'} \\ I_{z'} \end{matrix}\right\} = \left(\frac{25}{24} \pm \frac{\sqrt{41}}{8}\right) a^3 t\,.$$

With respect to the definition of the $\{y', z'\}$ axes it is seen that $I_{y'} > I_{z'}$, which means that $I_{y'}$ corresponds to the plus in the expression above, while $I_{z'}$ corresponds to the minus:

$$I_{y'} = 1.84\,a^3 t\,, \qquad I_{z'} = 0.24\,a^3 t\,.$$

In the case of $\theta = \theta + 90^\circ = 109.3^\circ$ these values would be interchanged. □

Example 10.10. Principal properties of Z-profile. Figure 10.26 shows a Z-profile with constant thickness t, height $2a$ and width $2a$. The specific dimensions are given in the figure. The cross-section has point symmetry about C, which is therefore the elastic center, as indicated in the figure. The coordinate axes $\{y, z\}$ are parallel to the flanges and web, respectively, and with origin at the elastic center. The cross-section is homogeneous, whereby the elastic modulus cancels in the expression for the moments of inertia. The cross-section is assumed to be thin-walled with $t \ll a$.

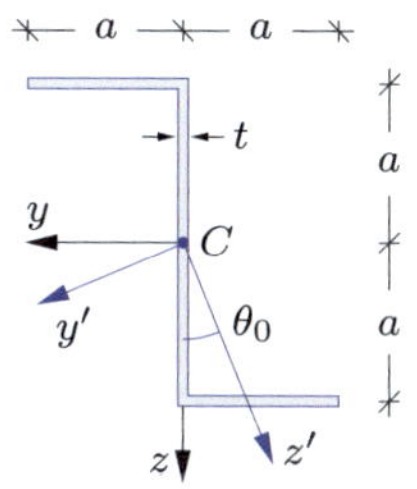

Fig. 10.26: Homogeneous Z-profile.

The moments of inertia with respect to the $\{y, z\}$ axes are determined as

$$I_{yy} = 2\left(\tfrac{1}{12}a^3t + at(\tfrac{1}{2}a)^2\right) = \tfrac{2}{3}a^3t\,,$$

$$I_{zz} = \tfrac{1}{12}(2a)^3t + 2ata^2 = \tfrac{8}{3}a^3t\,,$$

$$I_{yz} = at\,a\left(-\tfrac{1}{2}a\right) + at(-a)\,\tfrac{1}{2}a = -a^3t\,,$$

where terms containing higher powers in t are omitted. The coupling moment of inertia $I_{yz} \neq 0$ and thus $\{y, z\}$ are not principal axes. The orientation of the principal axes $\{y', z'\}$ is governed by the angle θ determined by (10.60) and (10.61),

$$\tan(2\theta) = 1 \qquad \Rightarrow \qquad \theta_n = 22.5^\circ + n\,90^\circ\,.$$

For $n = 0$ the angle is $\theta_0 = 22.5^\circ$, and the orientation of the principal axes for this angle is shown in Fig. 10.26.

The principal moments of inertia are determined by the expressions in (10.63),

$$\left.\begin{matrix} I_{y'} \\ I_{z'} \end{matrix}\right\} = \left(\tfrac{5}{3} \pm \sqrt{2}\right)a^3t\,.$$

From Fig. 10.26 it is seen that $I_{z'} > I_{y'}$, corresponding to

$$I_{y'} = \left(\tfrac{5}{3} - \sqrt{2}\right)a^3t = 0.25\,a^3t\,, \qquad I_{z'} = \left(\tfrac{5}{3} + \sqrt{2}\right)a^3t = 3.08\,a^3t\,.$$

The largest moment of inertia in the principal directions is larger than the moments of inertia in the original $\{y, z\}$ axes. Thus, for non-symmetric cross-sections the principal directions indicate the optimal orientation of the cross-section with respect to unidirectional transverse loading. This relates to the concept of Mohr's circle, as discussed next. □

Mohr's circle for moments of inertia

The transformation relations for moments of inertia in a rotated coordinate system (10.57) and the associated expression (10.63) for the principal moments of inertia are similar to the transformation relations (8.68) in plane stress and the expression (8.80) for the associated principal stresses. This indicates that the moments of inertia $I_{y'y'}$, $I_{z'z'}$ and $I_{y'z'}$ can be illustrated graphically by the Mohr's circle construction in terms of the principal moments of inertia.

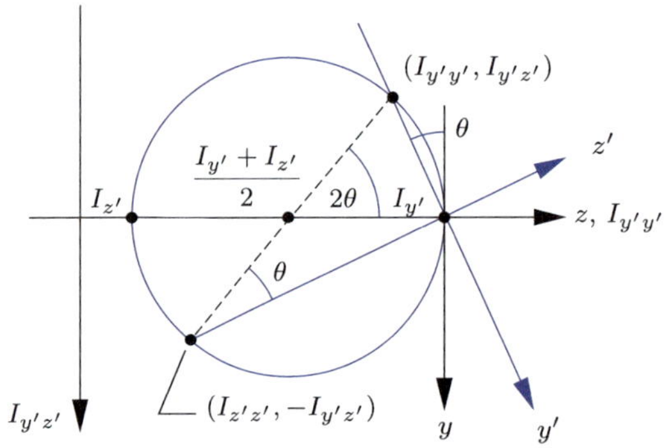

Fig. 10.27: Mohr's circle in the $(I_{y'y'}, I_{y'z'})$-plane with $I_{y'} > I_{z'}$.

Let $\{\bar{y}, \bar{z}\}$ take the role of the principal coordinate system with the principal moments of inertia $I_{y'}$ and $I_{z'}$. The moments of inertia in a coordinate system $\{y', z'\}$, obtained by a counter clockwise rotation θ, is then obtained from the transformation relations (10.57) as

$$\begin{aligned} I_{y'y'} &= \tfrac{1}{2}\big(I_{y'} + I_{z'}\big) + \tfrac{1}{2}\big(I_{y'} - I_{z'}\big)\cos 2\theta\,, \\ I_{z'z'} &= \tfrac{1}{2}\big(I_{y'} + I_{z'}\big) - \tfrac{1}{2}\big(I_{y'} - I_{z'}\big)\cos 2\theta\,, \\ I_{y'z'} &= \qquad\qquad\qquad\; - \tfrac{1}{2}\big(I_{y'} - I_{z'}\big)\sin 2\theta\,. \end{aligned} \tag{10.69}$$

The moments of inertia $I_{y'y'}$, $I_{z'z'}$ and $I_{y'z'}$ can now be illustrated graphically as a function of the angle θ as shown in Fig. 10.27. Let $I_{y'y'}$ serve as horizontal axis and $I_{y'z'}$ as vertical downward axis. The relations (10.69a) and (10.69c) constitute a parameter representation by which the point $[I_{y'y'}, I_{y'z'}]$ moves around in the counter clockwise direction as described by the center angle 2θ. The center of the circle is located at the horizontal axis,

$$C = [c, 0]\,, \quad c = \tfrac{1}{2}\big(I_{y'} + I_{z'}\big)\,, \tag{10.70}$$

and the radius is

$$R = \tfrac{1}{2}\big|I_{y'} - I_{z'}\big| . \tag{10.71}$$

It follows from the relations (10.69b) and (10.69c) that $[I_{z'z'}, -I_{y'z'}]$ is the diametrically opposite point on the circle. It is clear from this construction that the principal moments of inertia $I_{y'}$ and $I_{z'}$ constitute the maximum and minimum values of $I_{y'y'}$ and $I_{z'z'}$. These results correspond to those obtained for the plane stress components in (8.84).

Example 10.11. Mohr's circle for angle profile. Consider the angle profile in Fig. 10.25, where the original axes are now denoted $\{y', z'\}$ to indicate a non-principal coordinate system. The moments of inertia are

$$I_{y'y'} = \tfrac{5}{3}a^3t\,, \qquad I_{z'z'} = \tfrac{5}{12}a^3t\,, \qquad I_{y'z'} = \tfrac{1}{2}a^3t\,.$$

In Fig. 10.28 these values identify the two points $[I_{y'y'}, I_{y'z'}]$ and $[I_{z'z'}, -I_{y'z'}]$, shown in the figure normalized by $\frac{1}{12}a^3t$. The two points identify the center $c = \frac{1}{2}(\frac{5}{3} + \frac{5}{12})a^3t = \frac{25}{24}a^3t$ and thereby the radius. Alternatively, the circle may be constructed from the principal values

$$I_{y'} = 1.84\,a^3t\,, \qquad I_{z'} = 0.24\,a^3t\,,$$

obtained in Example 10.9. □

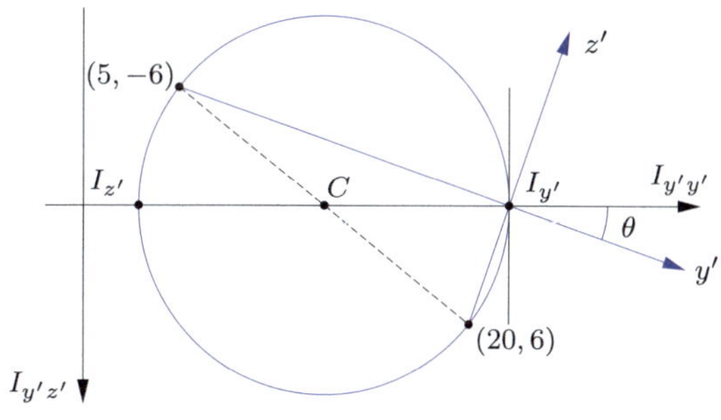

Fig. 10.28: Mohr's circle in $(I_{y'y'}, I_{y'z'})$-plane for angle profile.

10.3 Axial stresses and strains

In the general $\{y, z\}$ coordinate system the axial strain ε is given by (10.3) and the associated axial stress σ is found by multiplication with the elastic modulus E, as shown in (10.10). Both the strain and the stress distributions are functions of the deformation measures ε_0, κ_y and κ_z, as shown in Fig. 10.5. They are given in terms of the internal forces N, M_y and M_z with respect to the $\{y, z\}$ coordinate system by the general constitutive relation in (10.21). This requires inversion of the 3×3 stiffness matrix in (10.21), whereby the axial strain can be written as

$$\varepsilon = [1, y, z]\begin{bmatrix}\varepsilon_0\\ \kappa_y\\ \kappa_z\end{bmatrix} = \frac{1}{E_0}[1, y, z]\begin{bmatrix}F & S_y & S_z\\ S_y & I_{yy} & I_{yz}\\ S_z & I_{yz} & I_{zz}\end{bmatrix}^{-1}\begin{bmatrix}N\\ M_y\\ M_z\end{bmatrix}. \tag{10.72}$$

This procedure often requires numerical tools because of the inversion of the general cross-section stiffness matrix. The stress distribution is then obtained by multiplication with the elastic modulus,

$$\sigma = E\varepsilon\,. \tag{10.73}$$

The kinematic formulation implies a linear strain distribution, as shown in Fig. 10.3, and for homogeneous cross-sections the associated stress distribution is linear as well. However, for inhomogeneous cross-sections discontinuities may occur in the stress distribution due to sudden changes in the elastic modulus. Therefore, the basis of any stress analysis is the determination of the associated continuous strain distribution.

Separation of extension and bending

For analytical calculations it is convenient to separate the extension and bending problems by describing the axial strain and stress with respect to the (translated) coordinate system $\{\bar{y}, \bar{z}\}$ with origin at the elastic center. In that case the contributions to the axial deformation from pure extension and bending are separated, as illustrated in Fig. 10.8. The constitutive relations are given in (10.40) and the axial strain can be written as

$$\varepsilon(y, z) = \varepsilon_c + [\bar{y}, \bar{z}]\begin{bmatrix}\kappa_y\\ \kappa_z\end{bmatrix}, \tag{10.74}$$

where the extension and the bending problems are separated and appear in an additive format. Elimination of the deformation measures ε_c, κ_y and κ_z by the constitutive relations (10.40) gives the following expression for the axial strain in terms of the internal forces with respect to the $\{\bar{y}, \bar{z}\}$ coordinate system,

$$\varepsilon(y, z) = \frac{1}{E_0}\left(\frac{N}{F} + [\bar{y}, \bar{z}]\begin{bmatrix}I_{\bar{y}\bar{y}} & I_{\bar{y}\bar{z}}\\ I_{\bar{y}\bar{z}} & I_{\bar{z}\bar{z}}\end{bmatrix}^{-1}\begin{bmatrix}M_{\bar{y}}\\ M_{\bar{z}}\end{bmatrix}\right). \tag{10.75}$$

The inverse of the 2×2 bending stiffness matrix can be written in explicit form as

$$\begin{bmatrix}I_{\bar{y}\bar{y}} & I_{\bar{y}\bar{z}}\\ I_{\bar{y}\bar{z}} & I_{\bar{z}\bar{z}}\end{bmatrix}^{-1} = \frac{1}{I_{\bar{y}\bar{y}}I_{\bar{z}\bar{z}} - I_{\bar{y}\bar{z}}^2}\begin{bmatrix}I_{\bar{z}\bar{z}} & -I_{\bar{y}\bar{z}}\\ -I_{\bar{y}\bar{z}} & I_{\bar{y}\bar{y}}\end{bmatrix}, \tag{10.76}$$

where the denominator of the fraction is the determinant of the 2×2 bending moment of inertia matrix. The stress distribution is found by multiplication

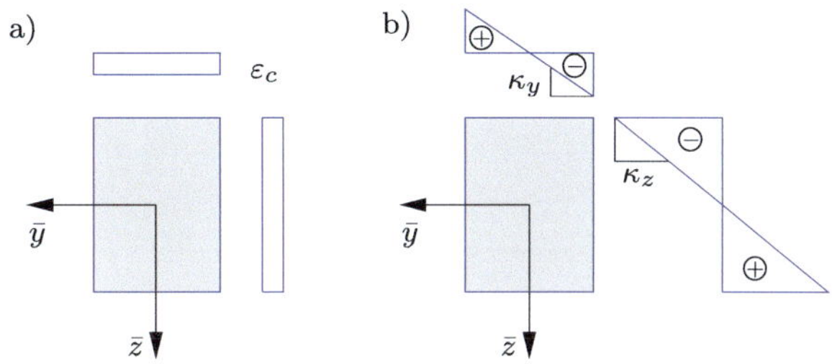

Fig. 10.29: Plot of linear strain variation: a) extension, and b) bending.

with the elastic modulus, as in (10.73). The separation of the strain distribution into contributions from extension and bending is illustrated in Fig. 10.29, where it is indicated that the strain contribution from bending, governed by the second term in (10.75), vanishes at the elastic center. The resulting strain distribution is the sum of the two contributions, as demonstrated by the expression in (10.75).

Example 10.12. Axial stress in cantilever with angle profile. Figure 10.30 shows a cantilever of length ℓ with tip loads P in the y direction and $\frac{1}{2}P$ in the z direction. The present example determines the axial stress at the fixed support, where the moments are:

$$M_y = -Pl, \qquad M_z = -\tfrac{1}{2}\,Pl.$$

Note, that the moments are negative following the sign convention introduced in Chapter 3. The cross-section of the cantilever is the angle profile shown in Fig. 10.31. The location of the elastic center has been determined in Example 10.1, and in the present case the coordinate system $\{y, z\}$ is located with origin at the elastic center. The precise location of the transverse force in the cross-section plane that does not introduce torsion depends on the shear stress distribution and is dealt with in Chapter 11.

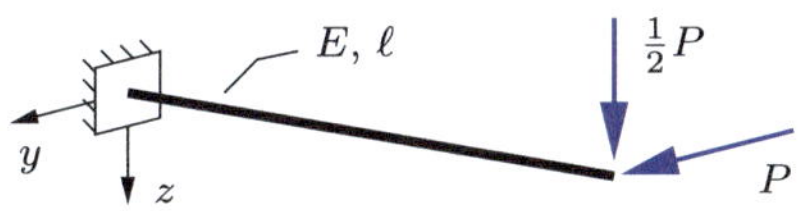

Fig. 10.30: Cantilever with tip loads P and $\frac{1}{2}P$.

The axial strain is given by the relation in (10.75). This expression involves the inverse of the bending stiffness matrix given in (10.76). For the angle profile the moments of inertia have been found in Example 10.6,

$$I_{yy} = \tfrac{5}{3}a^3t, \qquad I_{zz} = \tfrac{5}{12}a^3t, \qquad I_{yz} = \tfrac{1}{2}a^3t,$$

and the inverse matrix is

$$\begin{bmatrix} I_{yy} & I_{yz} \\ I_{yz} & I_{zz} \end{bmatrix}^{-1} = \frac{3}{16}\begin{bmatrix} 5 & -6 \\ -6 & 20 \end{bmatrix}\frac{1}{a^3t}.$$

The expression for the axial strain is given in (10.75). In the present example it leads to

$$\varepsilon(y,z) = -\frac{1}{E_0}\frac{3}{16}[y\,,\,z]\begin{bmatrix} 5 & -6 \\ -6 & 20 \end{bmatrix}\begin{bmatrix} 1 \\ \frac{1}{2} \end{bmatrix}\frac{P\ell}{a^3t} = -\frac{3}{8}\frac{P\ell}{E_0a^3t}(y+2z)\,,$$

where the minus follows from the negative bending moments M_y and M_z.

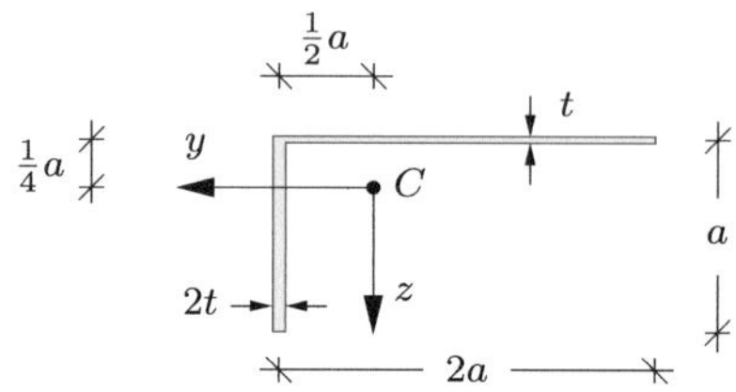

Fig. 10.31: Homogeneous thin-walled angle profile.

The linear axial strain distribution is plotted along the coordinate axes in Fig. 10.32. Along the y-axis, with $z = 0$, the strain variation is

$$\varepsilon(y,0) = -\frac{3}{8}\,y\,\frac{P\ell}{E_0a^2t}\,,$$

and along the z-axis, with $y = 0$, it is

$$\varepsilon(0,z) = -\frac{3}{4}\,z\,\frac{P\ell}{E_0a^2t}\,.$$

These linear variations of the axial strain along the two coordinate axes is shown in Fig. 10.32. The values given in the figure correspond to the normalized strain $\varepsilon\,E_0\,a^2t/(P\ell)$. The corresponding stress variations are found by multiplication with $E = E_0$.

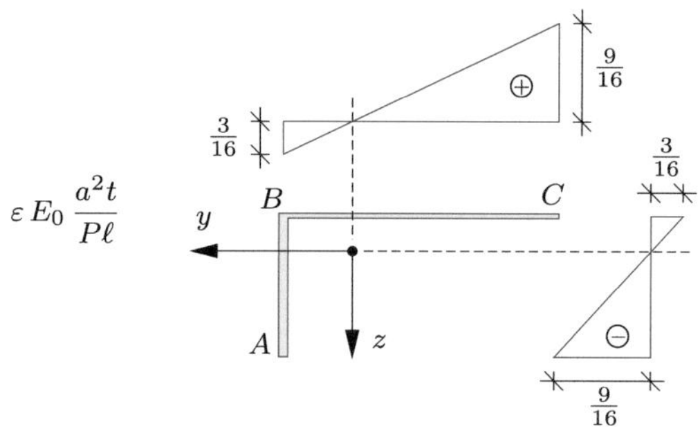

Fig. 10.32: Variation of normalized strain along coordinate axes.

The variations of the strain along the coordinate axes can be used to determine the strain and stress at end and corner points of the cross-section, which are denoted as points A, B and C in Fig. 10.32. The stress in A is found as

$$\sigma_A = E_0\,\varepsilon_A = -\Big(\frac{3}{16} + \frac{9}{16}\Big)\frac{P\ell}{a^2t} = -\frac{3}{4}\frac{P\ell}{a^2t}\,.$$

This result can be verified by substitution of $y = \frac{1}{2}a$ and $z = \frac{3}{4}a$ into the expression for $\varepsilon(y,z)$ given previously,

$$\sigma_A = E_0\,\varepsilon\big(\tfrac{1}{2}\,a,\,\tfrac{3}{4}\,a\big) = -\frac{3}{8}\,\frac{P\ell}{E_0 a^3 t}\,\big(\tfrac{1}{2}a + \tfrac{3}{2}a\big) = -\frac{3}{4}\,\frac{P\ell}{a^2 t}\,.$$

Similarly, the stress in B and C can be determined by combination of the axis values given in Fig. 10.32, or by substitution of the coordinates into the general expression for the strain. It is often convenient to set up a table containing the strains and stresses at end and corner points of the cross-section. In the present case the stress distribution is fully determined by the values in A, B and C, which are given in Table 10.1.

Table 10.1: Axial stress in angle profile.

point	y	z	$\sigma\,\dfrac{a^2 t}{P\ell}$
A	$\frac{1}{2}\,a$	$\frac{3}{4}\,a$	$-\frac{3}{4}$
B	$\frac{1}{2}\,a$	$-\frac{1}{4}\,a$	0
C	$-\frac{3}{2}\,a$	$-\frac{1}{4}\,a$	$\frac{3}{4}$

The axial stress can also be plotted along the contour of the cross-section, as shown in Fig. 10.33, using the linear variation along straight parts of the contour.

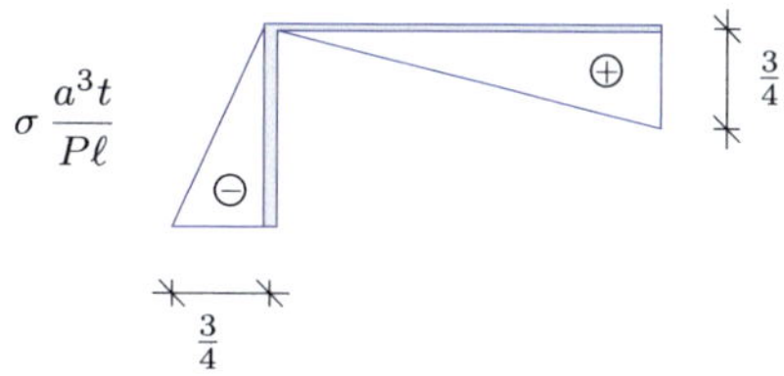

Fig. 10.33: Axial stress distribution along cross-section contour.

The particular loading of the cantilever yields zero stress at the corner B, while the maximum stresses are present at the end points A and B. Assume that σ_Y is the yield stress in the von Mises yield condition. The maximum force P associated with this yield stress is then found as

$$P_{\max} = \pm\,\frac{4}{3}\,\frac{a^3 t}{\ell}\,\sigma_Y\,.$$

Note, that because the shear stress is omitted here the Tresca failure condition gives the same yield limit. □

Example 10.13. Stresses in simply supported beam with Z-profile. Figure 10.34a shows a simply supported beam with distributed load in the z direction with constant intensity p. The present example determines the axial strain and stress from bending at the center of the span in C, where the moments are

$$M_y^C = 0\,, \qquad M_z^C = \tfrac{1}{8}p\ell^2\,.$$

The beam is homogeneous with elastic modulus $E = E_0$, and the cross-section is the Z-profile shown in Fig. 10.34b. The moments of inertia of this cross-section have been

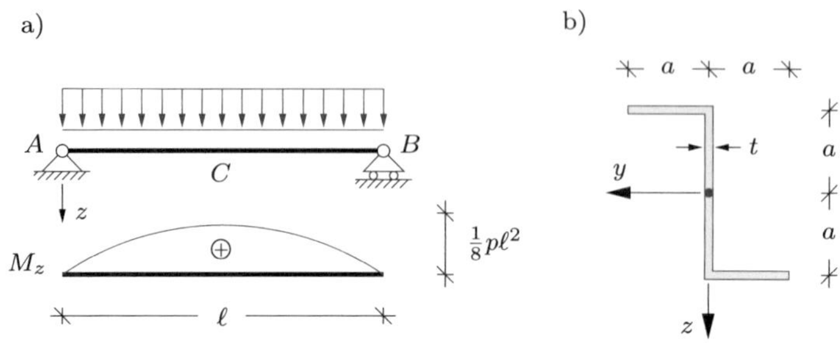

Fig. 10.34: Simply supported beam with distributed load.

determined in Example 10.10,

$$I_{yy} = \tfrac{2}{3}a^3t\,, \qquad I_{zz} = \tfrac{8}{3}a^3t\,, \qquad I_{yz} = -a^3t\,.$$

The inverse of the bending stiffness matrix is

$$\begin{bmatrix} I_{yy} & I_{yz} \\ I_{yz} & I_{zz} \end{bmatrix}^{-1} = \frac{3}{7}\begin{bmatrix} 8 & 3 \\ 3 & 2 \end{bmatrix}\frac{1}{a^3t}\,,$$

and the strain distribution is then obtained by (10.75),

$$\varepsilon(y,z) = \frac{1}{E_0}\,\frac{3}{56}\,[\,y\,,\,z\,]\begin{bmatrix} 8 & 3 \\ 3 & 2 \end{bmatrix}\begin{bmatrix} 0 \\ 1 \end{bmatrix}\frac{p\ell^2}{a^3t} = \frac{3}{56}\,\frac{p\ell^2}{E_0a^3t}\,(3y+2z)\,.$$

The strain distribution is shown in Fig. 10.35a with respect to the coordinate axes.

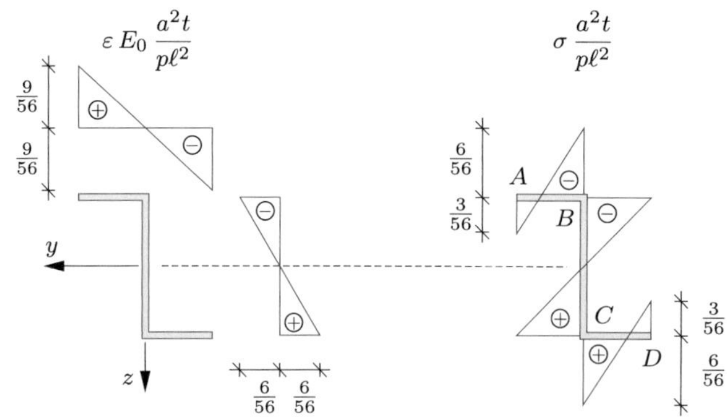

Fig. 10.35: a) Strain and b) stress distribution.

The stress distribution is obtained by multiplication with the elastic modulus, which is constant $E = E_0$ for the present homogeneous beam. The value of the axial stress is found at the ends and corners A to D of the profile as shown in Fig. 10.35b. In A the axial stress σ_A is found by superposition of the strain distribution in Fig. 10.35a,

$$\sigma_A = E_0\,\varepsilon_A = \left(\frac{9}{56} - \frac{6}{56}\right)\frac{p\ell^2}{a^2t} = \frac{3}{56}\,\frac{p\ell^2}{a^2t}\,.$$

Table 10.2: Axial stress in angle profile.

point	y	z	$\sigma\,\dfrac{a^2t}{p\ell^2}$
A	a	$-a$	$\frac{3}{56}$
B	0	$-a$	$-\frac{6}{56}$
C	0	a	$\frac{6}{56}$
D	$-a$	a	$-\frac{3}{56}$

Table 10.2 gives the axial stress at the four points on the cross-section. The full stress distribution is obtained by linear interpolation between the four values, as shown in Fig. 10.35b. The maximum axial stress occurs in points B and C, with $\sigma_{\max} = \pm\frac{6}{56}p\ell^2/(a^2t)$. □

Principal axes

In the principal coordinate system the coupling moment of inertia $I_{y'z'} = 0$, whereby the bending stiffness matrix becomes diagonal. The constitutive relations are hereby given as

$$N = E_0F\,\varepsilon_c\,, \qquad M_{y'} = E_0I_{y'}\kappa_{y'}\,, \qquad M_{z'} = E_0I_{z'}\kappa_{z'}\,. \tag{10.77}$$

The axial strain in the principal coordinate system is defined in (10.48). Elimination of axial strain and curvatures in terms of the section forces yields

$$\varepsilon = \frac{1}{E_0}\left(\frac{N}{F} + \frac{M_{y'}}{I_{y'}}\,y' + \frac{M_{z'}}{I_{z'}}\,z'\right). \tag{10.78}$$

The associated stress distribution is obtained by multiplication with the elastic modulus

$$\sigma = \frac{E}{E_0}\left(\frac{N}{F} + \frac{M_{y'}}{I_{y'}}\,y' + \frac{M_{z'}}{I_{z'}}\,z'\right). \tag{10.79}$$

The stress distribution expressed via the linear kinematic formulation and in terms of section forces is often referred to as the Navier stress distribution.

Example 10.14. Axial stresses in cantilever with I-profile. Figure 10.36a shows a cantilever loaded by a tip force P acting in the z direction. The beam is homogeneous with elastic modulus $E_0 = E$ and length ℓ. The distribution of strain and stress is determined at the fixed support, where the bending moment attains its maximum,

$$M_z = -P\ell\,.$$

The cross-section is an I-profile with height and width a and thickness t, as shown in Fig. 10.36b.

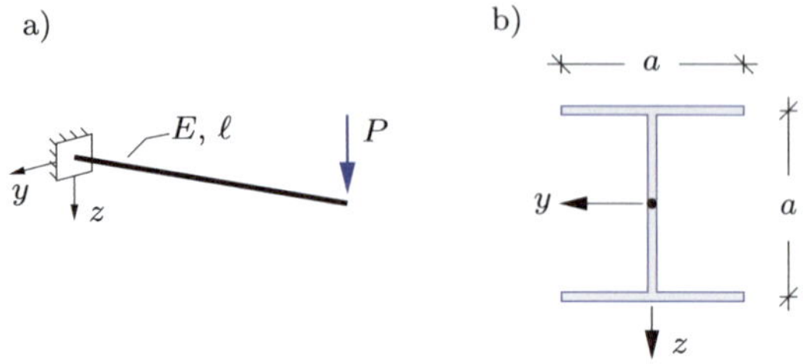

Fig. 10.36: a) Cantilever with tip load P, and b) I-profile cross-section.

The cross-section has double symmetry with respect to the coordinate system shown in the figure, which hereby is a principal coordinate system. Thus, the strain distribution is determined by the simplified expression (10.78). Since $N = 0$ and $M_y = 0$ this expression reduces to

$$\varepsilon = \frac{1}{E}\frac{M_z}{I_z}\,z\,.$$

The moment of inertia with respect to the z-direction is

$$I_z = \tfrac{1}{12}a^3t + 2at\big(\tfrac{1}{2}\,a\big)^2 = \big(\tfrac{1}{12}+\tfrac{1}{2}\big)a^3t = \tfrac{7}{12}\,a^3\,t\,.$$

This agrees with the result obtained in Example 10.8 for the inhomogeneous I-profile. The expression for the axial strain is now determined as

$$\varepsilon = -\frac{1}{E}\,\frac{12}{7}\,\frac{P\ell}{a^3t}\,z\,.$$

The corresponding axial stress is found by multiplication of the strain with the constant elastic modulus E, and thus the strain and stress distribution are similar in shape. Figure 10.37a shows the variation of the axial strain along the z-axis, whereas in Fig. 10.37b the axial stress is plotted along the center lines of the thin-walled cross-section. The negative strain and stress indicates that the bottom flange is in compression, in agreement with the negative sign of the bending moment M_z. □

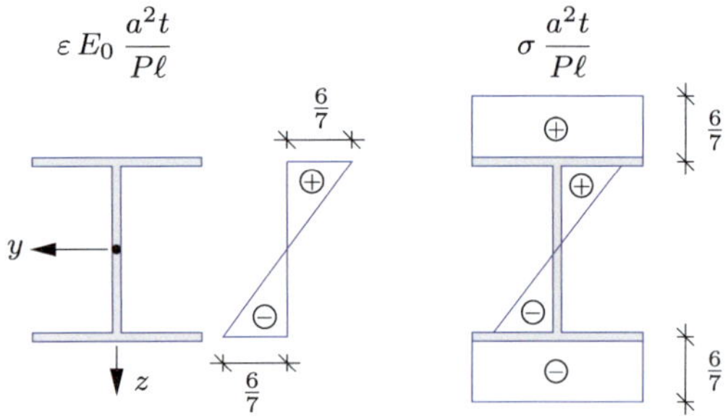

Fig. 10.37: Normalized distributions of axial strain and stress.

Example 10.15. Inhomogeneous I-profile. This example considers the inhomogeneous I-profile cross-section shown in Fig. 10.38. As in the previous example the beam is loaded by a transverse force P in the z-direction, producing the moment

$$M_z = -P\ell\,.$$

The height and width of the cross-section is a, the thickness of the flanges is t, while the thickness of the web is only $\frac{1}{5}t$. The elastic modulus of the flanges is E_f, while the web has elastic modulus $E_w = 5E_f$. In the following the reference elastic modulus is chosen as the flange value, $E_0 = E_f$.

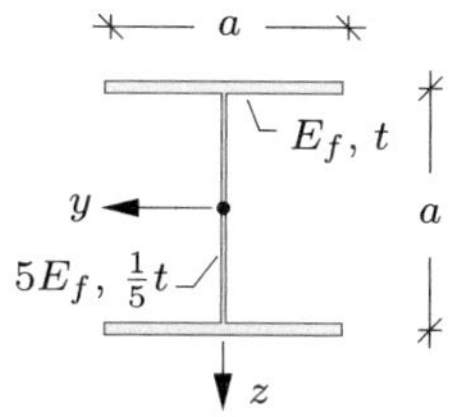

Fig. 10.38: Inhomogeneous I-profile cross-section.

The continuous strain distribution is determined by (10.78),

$$\varepsilon = \frac{1}{E_0} \frac{M_z}{I_z} z .$$

The moment of inertia in the z-direction is determined as

$$I_z = \frac{E_w}{E_f} \tfrac{1}{12} a^3 \tfrac{1}{5}t + 2at\left(\tfrac{1}{2}a\right)^2 = \tfrac{7}{12} a^3 t ,$$

which by design is equal to the moment of inertia for the homogeneous cross-section in the previous example. Substitution into the strain expression gives

$$\varepsilon = -\frac{1}{E_f} \frac{12}{7} \frac{P\ell}{a^3 t} z .$$

The linear variation of the strain ε along the z-direction is shown in Fig. 10.39a. The stress distribution is obtained by multiplication of the strain by the elastic modulus. Since the cross-section is inhomogeneous with different elastic moduli, discontinuities in the stress variation occur. Table 10.3 gives the strain and the associated stress values at the points A and D located in the flanges, and B and C located at the top and bottom of the web, as shown in Fig. 10.39. It is seen that the stresses in the web are significantly larger due to the large elastic modulus. The stress distribution is also shown in Fig. 10.39b.

Assume that the yield stress of the flange material is $\sigma_Y = \sigma_f$, while it is $\sigma_Y = 10\sigma_f$ for the web material. In the present case the only stress component is the axial stress σ,

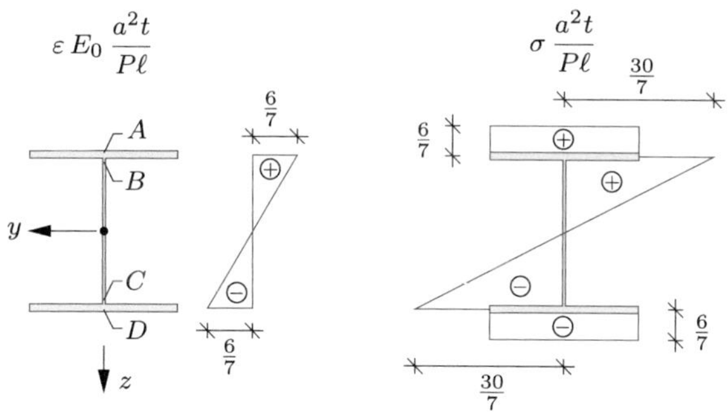

Fig. 10.39: Normalized distributions of axial strain and stress for inhomogeneous I-profile.

Table 10.3: Axial stress in angle profile.

point	y	z	$\varepsilon E_f \dfrac{a^2 t}{P\ell}$	$\sigma \dfrac{a^2 t}{P\ell}$
A	$-\frac{1}{2}a$	0	$\frac{6}{7}$	$\frac{6}{7}$
B	$-\frac{1}{2}a$	0	$\frac{6}{7}$	$\frac{30}{7}$
C	$\frac{1}{2}a$	0	$-\frac{6}{7}$	$-\frac{30}{7}$
D	$\frac{1}{2}a$	0	$-\frac{6}{7}$	$-\frac{6}{7}$

whereby the von Mises yield condition (9.57) gives $\sigma = \sigma_Y$. For the flanges the maximum stress is $\pm\frac{6}{7}P\ell/(a^2t)$, which for $\sigma_Y = \sigma_f$ gives the yield load

$$P^f_{\rm max} = \frac{7}{6}\frac{a^2t}{\ell}\sigma_f .$$

Thus, failure occurs in the flanges when the load P reaches $P^f_{\rm max}$. The maximum stress in the web is $\pm\frac{30}{7}P\ell/(a^2t)$, and failure occurs when this value reaches $10\sigma_f$. This gives

$$P^w_{\rm max} = \frac{7}{3}\frac{a^2t}{\ell}\sigma_f > P^f_{\rm max} .$$

This shows that although the stresses in the web are larger than those in the flanges, the significantly smaller yield stress of the flange material implies that it is in fact the flanges that limit the maximum loading of the beam by $P_{\rm max} = P^f_{\rm max}$. □

10.3.1 Neutral axis and line of curvature

A number of materials, such as concrete, have significantly larger strength in compression than in tension. For beams composed of this type of material it is therefore of interest to determine the regions of the cross-section that are in compression and tension, respectively. The linear form of the strain distribution implies that the transition between the compression and tension regions is described by a straight line, typically denoted as the *neutral axis.* In the coordinate system with the elastic center as origin the axial strain is given by (10.27). Thus, the neutral axis is given by the relation

$$\varepsilon_c + \bar{y}\,\kappa_y + \bar{z}\,\kappa_z = 0 , \tag{10.80}$$

where the curvature is given in terms of the bending moments by the constitutive relation (10.40b). The straight line representing the neutral axis is fixed by two points. These are conveniently chosen as the intersections with the y and z axes, respectively. It follows from (10.80) that

$$\begin{aligned} \bar{y}_n &= -\frac{\varepsilon_c}{\kappa_y} \qquad \text{for} \qquad z = 0 , \\ \bar{z}_n &= -\frac{\varepsilon_c}{\kappa_z} \qquad \text{for} \qquad y = 0 , \end{aligned} \tag{10.81}$$

where the subscript n indicates that these points define the intersections of the coordinate axes with the neutral axis. For pure bending $\varepsilon_c = 0$, and the neutral axis passes through the elastic center C.

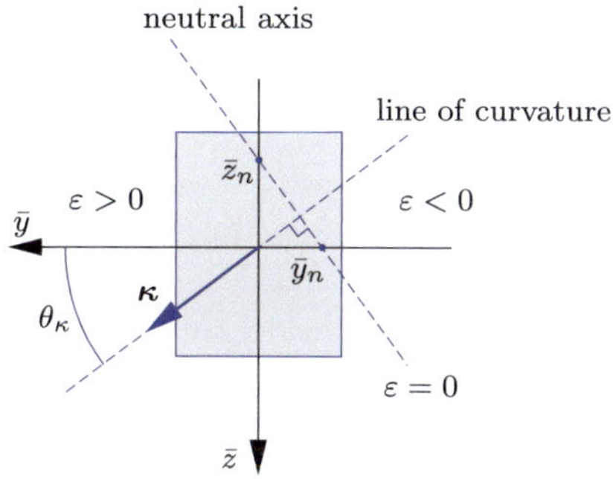

Fig. 10.40: Neutral axis perpendicular to line of curvature.

The direction of the neutral axis in the cross-section plane is determined by the two curvature terms in (10.80). When considering the coordinate increments $[d\bar{y}, d\bar{z}]$ along the neutral axis, it follows from differentiation of (10.80) that these satisfy the equation

$$[d\bar{y}, d\bar{z}] \begin{bmatrix} \kappa_y \\ \kappa_z \end{bmatrix} = 0\,. \tag{10.82}$$

It follows from this relation that the direction of the neutral axis $[d\bar{y}, dz]$ is perpendicular to the curvature vector $[\kappa_y, \kappa_z]$. In a coordinate system with the elastic center as origin the curvature vector components are given in terms of the moment vector components as

$$\begin{bmatrix} \kappa_y \\ \kappa_z \end{bmatrix} = \frac{1}{E_0} \begin{bmatrix} I_{\bar{y}\bar{y}} & I_{\bar{y}\bar{z}} \\ I_{\bar{y}\bar{z}} & I_{\bar{z}\bar{z}} \end{bmatrix}^{-1} \begin{bmatrix} M_{\bar{y}} \\ M_{\bar{z}} \end{bmatrix}. \tag{10.83}$$

The axial strain at the elastic center ε_c is simply an offset of the neutral axis along the normal.

In Fig. 10.40 the curvature vector $\boldsymbol{\kappa} = [\,\kappa_y\,,\,\kappa_z\,]$ determines the direction of the *line of curvature*. Hence, the neutral axis is perpendicular to the line of curvature and intersects the y and z axis at $\bar{y}_n$ and $\bar{z}_n$, respectively. Along the neutral axis $\varepsilon = 0$, and the direction of the curvature vector $\boldsymbol{\kappa}$ defines the region of the cross-section with positive strain, as shown in Fig. 10.40. The direction of the curvature vector is determined by for instance the angle relative to the y-axis, which is given by the tangent relation

$$\tan\theta_\kappa = \frac{\kappa_z}{\kappa_y}\,. \tag{10.84}$$

The neutral axis is not only used to identify the regions of tension and compression. In fact the neutral axis and the line of curvature constitute a local basis, in which the strain distribution is described particularly simple. This is demonstrated in the following.

Strain distribution

The neutral axis represents the line with vanishing axial strain, i.e. $\varepsilon = 0$. Because of the linear strain distribution the strain increases linearly in the direction perpendicular to the neutral axis along the line of maximum curvature. Thus, along lines perpendicular to the neutral axis the strain varies linearly and along lines parallel to the neutral axis the strain is constant. This implies that the strain distribution is fully determined by a single linear variation along the line of curvature, as illustrated in Fig. 10.41. In the expression for the axial strain (10.27) the slope of the linear strain distribution is given by the curvatures κ_y and κ_z, which are the components of the curvature vector $\boldsymbol{\kappa}$. The slope of the resulting strain distribution along the line of curvature corresponds to the length of the curvature vector, which is

$$\kappa = |\boldsymbol{\kappa}| = \sqrt{\kappa_y^2 + \kappa_z^2}\,. \tag{10.85}$$

The curvature vector is fully determined by the direction given in (10.84) and the length given above in (10.85).

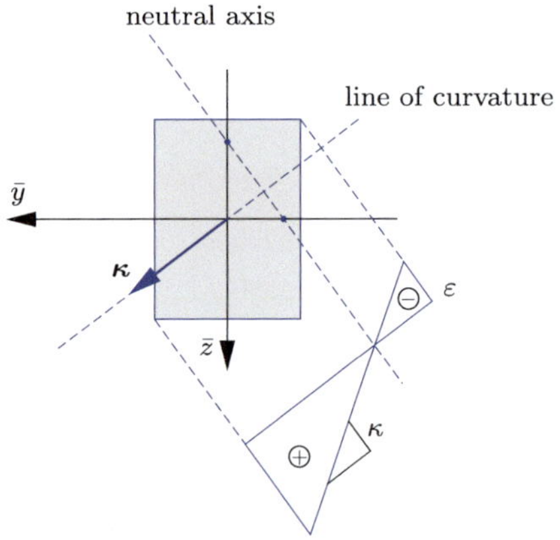

Fig. 10.41: Strain distribution with respect to neutral axis and line of curvature.

Example 10.16. Strain distribution for angle profile. Figure 10.42 shows the cantilever from Example 10.12, with tip loads P and $\frac{1}{2}P$ in the y and z direction, respectively. The moments at the fixed support are

$$M_y = -P\ell\,, \qquad M_z = -\tfrac{1}{2}P\ell\,,$$

and the inverse of the bending stiffness matrix has been found in Example 10.12:

$$\begin{bmatrix} I_{yy} & I_{yz} \\ I_{yz} & I_{zz} \end{bmatrix}^{-1} = \frac{3}{16} \begin{bmatrix} 5 & -6 \\ -6 & 20 \end{bmatrix} \frac{1}{a^3 t}.$$

The origin of the coordinate system $\{y, z\}$ is located at the elastic center C, determined in Example 10.1.

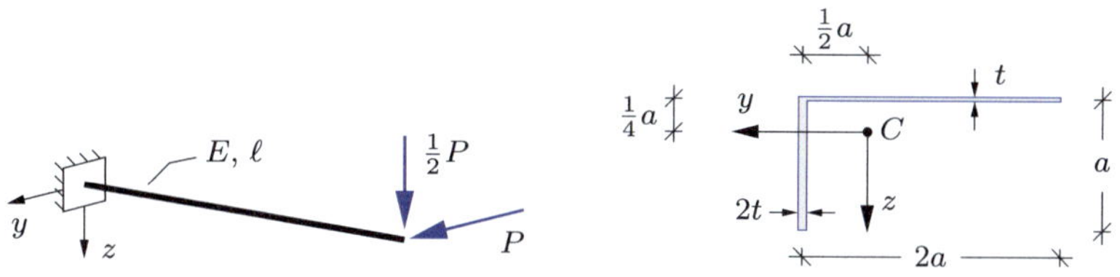

Fig. 10.42: Cantilever with angle profile and tip loads P and $\frac{1}{2}P$.

There is no axial loading of the beam, and thus

$$\varepsilon_C = \frac{1}{E_0} \frac{N}{F} = 0.$$

The neutral axis therefore passes through the elastic center and thereby the origin of the coordinate system. The direction of the neutral axis is determined via the curvature vector,

$$\begin{bmatrix} \kappa_y \\ \kappa_z \end{bmatrix} = \begin{bmatrix} I_{yy} & I_{yz} \\ I_{yz} & I_{zz} \end{bmatrix}^{-1} \begin{bmatrix} M_y \\ M_z \end{bmatrix} = -\frac{3}{16} \begin{bmatrix} 5 & -6 \\ -6 & 20 \end{bmatrix} \begin{bmatrix} 1 \\ \frac{1}{2} \end{bmatrix} \frac{P\ell}{E_0 a^3 t} = -\frac{3}{8} \begin{bmatrix} 1 \\ 2 \end{bmatrix} \frac{P\ell}{E_0 a^3 t}.$$

The angle of the curvature vector relative to the y-axis is found by (10.84),

$$\tan\theta_\kappa = \frac{\kappa_z}{\kappa_y} = 2 \qquad \Rightarrow \qquad \theta_\kappa = 63.4^\circ.$$

The curvature vector and the line of maximum curvature are shown in Fig. 10.43, and the neutral axis is shown as the line perpendicular to the line of maximum curvature. Both κ_y and κ_z are negative, and thus the curvature vector points into the fully negative quadrant of the coordinate system. This again implies that the region with tension (positive axial strain) is above the neutral axis, while the compression zone is below. The slope of the strain variation is obtained as the length of the curvature vector (10.85),

$$\kappa = \frac{3}{8}\sqrt{1^2 + 2^2}\, \frac{P\ell}{E_0 a^3 t} = \frac{3\sqrt{5}}{8} \frac{P\ell}{E_0 a^3 t} = 0.84 \frac{P\ell}{E_0 a^3 t}.$$

It is seen in Fig. 10.43 that the neutral axis intersects the corner of the cross-section in B, indicating that the axial strain is zero at this point. This agrees with the results obtained in Example 10.12, and shown in Table 10.1. The strain increases linearly in the direction of the line of curvature, and it is seen directly that the largest positive strain (tension) occurs in point C, while the largest negative strain (compression) occurs at the other free end in point A. The distance from the neutral axis to the point C is denoted as d in Fig. 10.43 and determined via the angle of the line of curvature:

$$\cos\theta_\kappa = \frac{d}{2a} \qquad \Rightarrow \qquad d = 2a\cos\theta_\kappa = \frac{2}{\sqrt{5}}\, a.$$

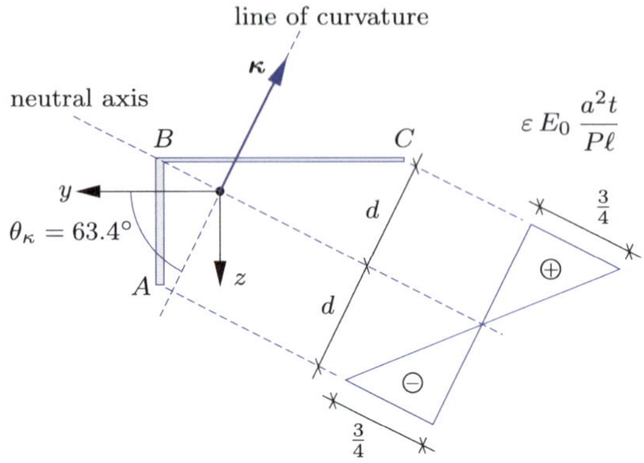

Fig. 10.43: Normalized strain distribution along line of curvature.

The strain at C is found as the distance d times the slope of the linear strain distribution κ:

$$\varepsilon_C = \kappa\, d = \frac{3\sqrt{5}}{8}\,\frac{P\ell}{E_0 a^3 t}\,\frac{2}{\sqrt{5}}\,a = \frac{3}{4}\,\frac{P\ell}{E_0 a^2 t}\,,$$

corresponding to the result obtained in Example 10.12. It turns out that the distance from the neutral axis to point A is also d, whereby $\varepsilon_A = -\varepsilon_C$. □

Example 10.17. Strain distribution for Z-profile. Consider the simply supported beam in Fig. 10.44 with a Z-profile as cross-section. The strain and stress distribution has been determined in Example 10.13, and the present example determines the strain distribution with respect to the neutral axis and the line of curvature. The loading of the beam implies that the bending moments at the center section are

$$M_y^C = 0\,, \qquad M_z^C = \tfrac{1}{8}p\ell^2\,.$$

The normal force is zero. The coordinate system $\{y, z\}$ has its origin at the elastic center.

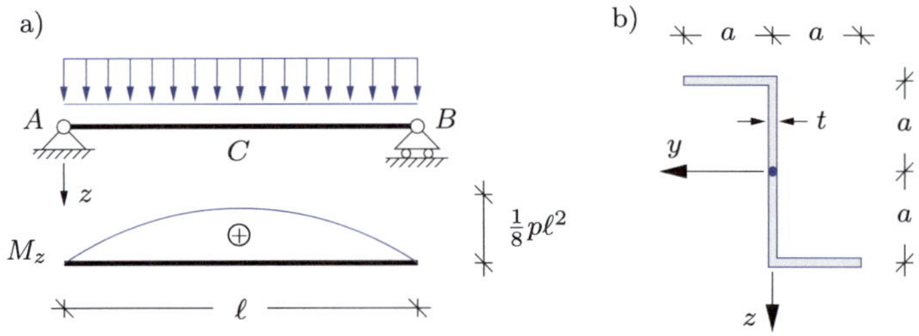

Fig. 10.44: Simply supported with with distributed load.

The moments of inertia with respect to the coordinate system, shown in Fig. 10.44b, have been determined in Example 10.13, and the curvature vector can therefore been found as

$$\begin{bmatrix} \kappa_y \\ \kappa_z \end{bmatrix} = \begin{bmatrix} I_{yy} & I_{yz} \\ I_{yz} & I_{zz} \end{bmatrix}^{-1} \begin{bmatrix} M_y^C \\ M_z^C \end{bmatrix} = \frac{3}{56}\begin{bmatrix} 8 & 3 \\ 3 & 2 \end{bmatrix}\begin{bmatrix} 0 \\ 1 \end{bmatrix}\frac{p\ell^2}{E_0 a^3 t} = \frac{3}{56}\begin{bmatrix} 3 \\ 2 \end{bmatrix}\frac{p\ell^2}{E_0 a^3 t}\,.$$

The direction of the curvature vector is defined by the tangent relation

$$\tan\theta_\kappa = \frac{\kappa_z}{\kappa_y} = \tfrac{2}{3} \qquad \Rightarrow \qquad \theta_\kappa = 33.7^\circ\,.$$

The curvature vector $\boldsymbol{\kappa}$ and the associated line of curvature are shown in Fig. 10.45. Because the components of the curvature vector are positive it is located in the positive $\{y,z\}$ quadrant. The associated neutral axis is perpendicular to the line of curvature, and because $N = 0$ it contains the elastic center.

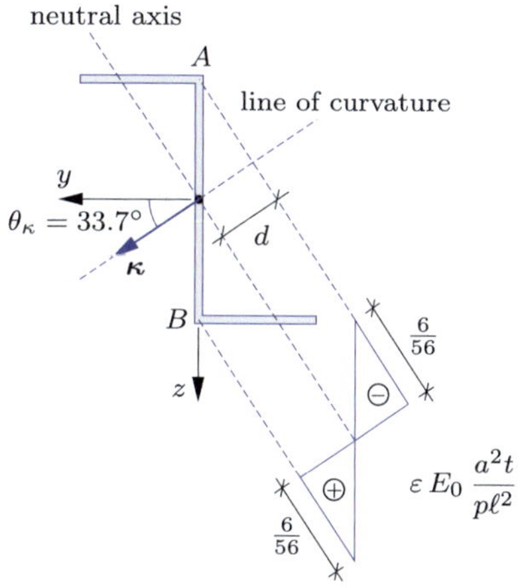

Fig. 10.45: Normalized strain distribution along line of curvature.

The linear strain variation along the line of maximum curvature is now determined. The slope of the linear variation is determined as the length of the curvature vector:

$$\kappa = \frac{3}{56}\sqrt{3^2+2^2}\,\frac{p\ell^2}{E_0 a^3 t} = \frac{3\sqrt{13}}{56}\,\frac{p\ell^2}{E_0 a^3 t} = 0.193\,\frac{p\ell^2}{E_0 a^3 t}\,.$$

The axial strain variation along the line of curvature is hereby determined as the slope κ times the distance from the neutral axis. Consequently, the largest strains occur at the points on the cross-section with greatest distance d to the neutral axis, which are the corners A and B, as shown in Fig. 10.45. The distance d is determined by the geometric relation:

$$\sin\theta_\kappa = \frac{d}{a} \qquad \Rightarrow \qquad d = a\,\sin\theta_\kappa = \tfrac{2}{\sqrt{13}}\,a\,,$$

whereby the strains at corners A and B are:

$$\varepsilon_A = -\,\kappa\, d = -\,\frac{6}{56}\,\frac{p\ell^2}{E_0 a^2 t}\,, \qquad \varepsilon_B = \kappa\, d = \frac{6}{56}\,\frac{p\ell^2}{E_0 a^2 t}\,.$$

The sign of the axial strain follows from the direction of the curvature vector, see Fig. 10.45. The above values for the strain in A and B correspond to the results obtained in Example 10.13, see Fig. 10.35b. □

10.4 Exercises

Exercise 10.1. The figure shows the angle profile considered in Example 10.1. In this exercise the reference coordinate system is shown in the figure with origin at the right end of the horizontal flange. Assume that the cross-section is homogeneous with $E_1 = E_2 = E_0$, and $t_1 = t_2 = t$, $a_1 = 2a$ and $a_2 = a$. The cross-section can be considered as thin-walled with $t \ll a$.

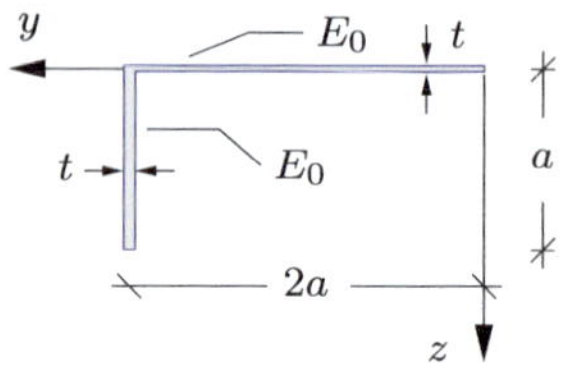

a) Determine the static moments S_y and S_z with respect to the reference axes.

b) Determine the location of the elastic center and compare with the results of Example 10.1.

Exercise 10.2. The figure shows the T-profile considered in Example 10.2. In this exercise the reference coordinate system is shown in the figure with origin at the right end of the horizontal flange. Assume that the cross-section is homogeneous with $E_f = E_w = E_0$, and $a_f = a_w = a$, $t_w = t$ and $t_f = 3t$. The cross-section can be considered as thin-walled with $t \ll a$.

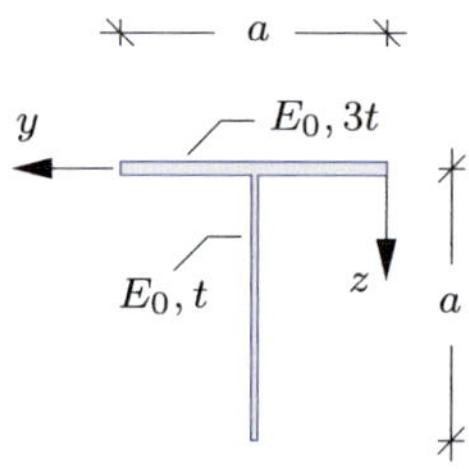

a) Determine the static moments S_y and S_z with respect to the reference axes.

b) Determine the location of the elastic center and compare with the results of Example 10.2.

c) Locate the coordinate system $\{\bar{y}, \bar{z}\}$ with origin in the elastic center, and determine the moments of inertia $I_{\bar{y}\bar{y}}$, $I_{\bar{z}\bar{z}}$ and $I_{\bar{y}\bar{z}}$.

Exercise 10.3. The figure shows a homogeneous C-profile with elastic modulus $E = E_0$. The height of the cross-section is h and the width is b. The thickness of the web is t, while the flange thickness is $2t$. The cross-section can be considered as thin-walled with $t \ll h, b$.

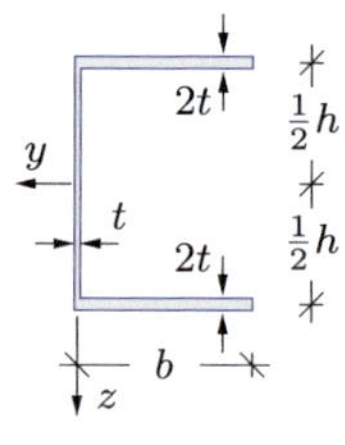

a) Choose the coordinate system $\{y, z\}$ shown in the figure and determine the location of the elastic center $[c_y, c_z]$.

b) Locate the coordinate system $\{\bar{y}, \bar{z}\}$ with origin in the elastic center, and determine the moments of inertia $I_{\bar{y}\bar{y}}$, $I_{\bar{z}\bar{z}}$ and $I_{\bar{y}\bar{z}}$ when $h = b = a$.

Exercise 10.4. The figure shows a homogeneous C-profile with elastic modulus $E = E_0$. The height of the cross-section is $3a$, the width is $2a$ and the thickness is t, as shown in the figure. The cross-section can be considered as thin-walled with $t \ll a$.

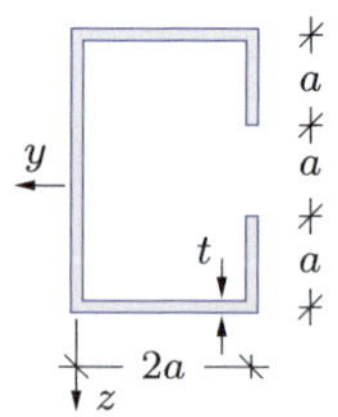

a) Choose a suitable coordinate system $\{y, z\}$ and determine the location of the elastic center $[c_y, c_z]$.

b) Locate the coordinate system $\{\bar{y}, \bar{z}\}$ with origin in the elastic center, and determine the moments of inertia $I_{\bar{y}\bar{y}}$, $I_{\bar{z}\bar{z}}$ and $I_{\bar{y}\bar{z}}$.

c) Consider the section forces: $N = 0$ and $M_{\bar{y}} = M_{\bar{z}} = \bar{M}$, and determine the axial strain distribution ε.

d) Determine the maximum axial stress $\sigma_{\max}$.

Exercise 10.5. The figure shows an inhomogeneous I-profile cross-section. The top flange has dimensions $2a \times t$ and elastic modulus E_f. The web has dimensions $2a \times \frac{1}{5}t$ and elastic modulus $10E_f$. And finally, the bottom flange has dimensions $a \times \frac{1}{5}t$ and elastic modulus $10E_f$. The cross-section can be considered as thin-walled with $t \ll a$.

a) Choose a suitable coordinate system $\{y, z\}$ and determine the location of the elastic center $[c_y, c_z]$.

b) Locate the coordinate system $(\bar{y}, \bar{z})$ with origin in the elastic center, and determine the moments of inertia $I_{\bar{y}\bar{y}}$, $I_{\bar{z}\bar{z}}$ and $I_{\bar{y}\bar{z}}$.

c) Assume the section forces: $N = 0$, $M_{\bar{y}} = 0$ and $M_{\bar{z}} = \bar{M}$, and determine the axial strain distribution ε.

d) Find the axial stress in the top flange and in the bottom flange, respectively.

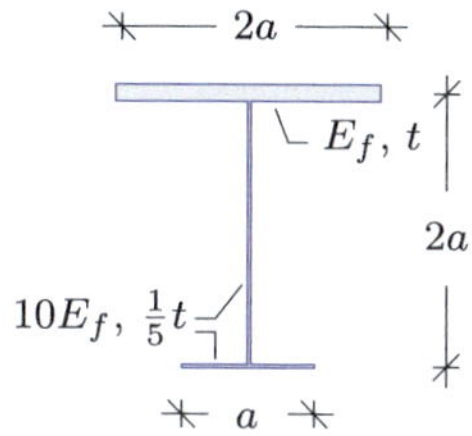

Exercise 10.6. The figure shows the rectangular cross-section made of concrete with steel reinforcement. The elastic modulus of concrete is E_c, and the elastic modulus of the steel reinforcement is $E_s = 15E_c$. The total area of reinforcement is $A_s = a^2/50$. The location of the elastic center has been determine in Example 10.3.

a) Locate the coordinate system $\{\bar{y}, \bar{z}\}$ with origin in the elastic center, and determine the moments of inertia $I_{\bar{y}\bar{y}}$, $I_{\bar{z}\bar{z}}$ and $I_{\bar{y}\bar{z}}$.

b) Consider the section forces: $N = 0$, $M_{\bar{y}} = 0$ and $M_{\bar{z}} = \bar{M}$, and determine the axial strain distribution ε.

c) Determine both the maximum tension and compression stress in the concrete, and the maximum stress in the steel reinforcement.

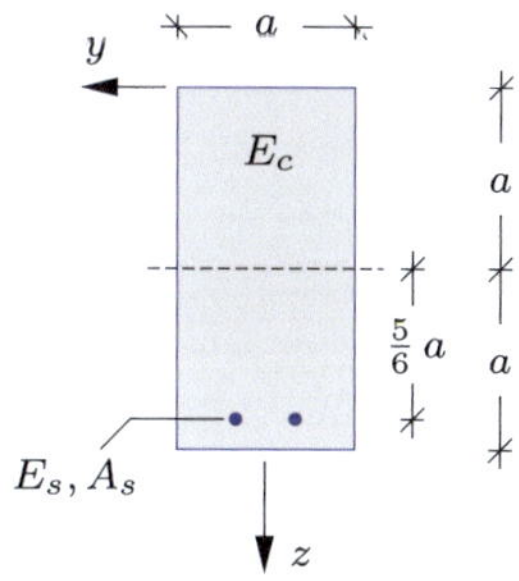

Exercise 10.7. Consider the T-profile in Example 10.7, that is shown in the figure below. The moments of inertia with respect to the axes with origin at the elastic center is given in Example 10.7. In the following the normal force $N = 0$.

a) Determine the distribution of axial strain ε and the maximum axial stress $\sigma_{\max}$ for the loading $M_y = \bar{M}$ and $M_z = 0$.

b) Repeat a) for $M_y = 0$ and $M_z = -2\bar{M}$.

c) Use the results in a) and b) to determine the distribution of axial strain ε and the maximum axial stress $\sigma_{\max}$ for the combined load case $M_y = \bar{M}$ and $M_z = -2\bar{M}$.

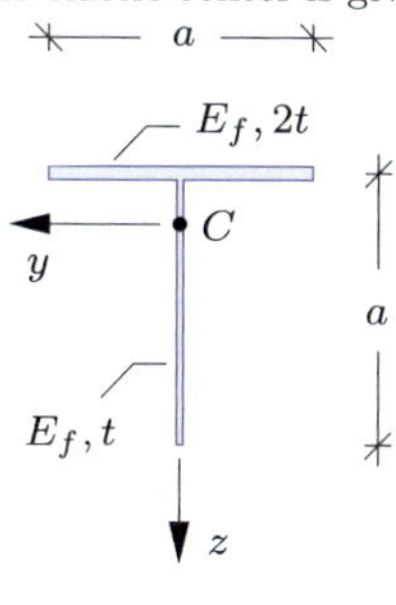

Exercise 10.8. Consider the Z-profile in Example 10.10. Draw Mohr's circle and determine the principal moments of inertia $I_{y'}$ and $I_{z'}$ graphically. Compare with the result obtained in the example.

Exercise 10.9. The figure shows a Z-profile with height $2a$ and width a, and thickness t. The cross-section is homogeneous with elastic modulus $E = E_0$. It is loaded by bending moments $M_y = \bar{M}$ and $M_z = 2\bar{M}$. The cross-section can be considered as thin-walled with $t \ll a$.

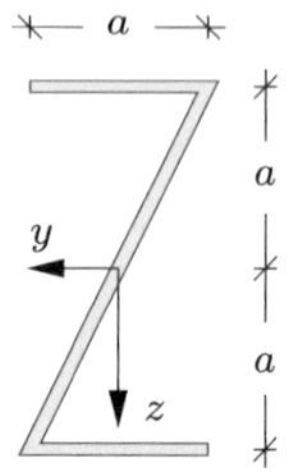

a) Determine the moments of inertia I_{yy}, I_{zz} and I_{yz}.

b) Determine the principal moments of inertia $I_{y'}$ and $I_{z'}$, and the orientation of the principal axes.

c) Draw Mohr's circle.

d) Determine the curvatures κ_y and κ_z, and draw the line of curvature and the neutral axis.

e) Determine the distribution of the axial strain ε and find the maximum axial stress $\sigma_{\max}$.

Exercise 10.10. The figure shows a Z-profile similar to that in the previous exercise, but with additional vertical flanges. The geometry is shown in the figure below and the thickness is t. The cross-section is homogeneous with elastic modulus $E = E_0$. It is loaded by bending moments $M_y = \bar{M}$ and $M_z = 2\bar{M}$. The cross-section can be considered as thin-walled with $t_f \ll a$.

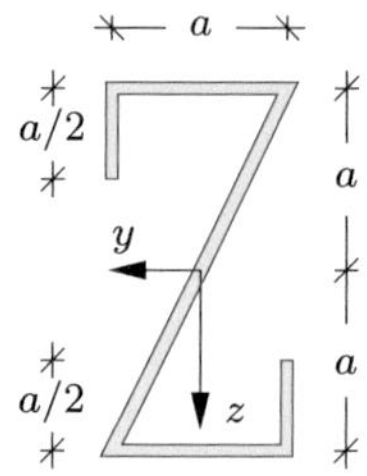

a) Determine the moments of inertia I_{yy}, I_{zz} and I_{yz}.

b) Determine the principal moments of inertia $I_{y'}$ and $I_{z'}$, and the orientation of the principal axes.

c) Draw Mohr's circle.

d) Determine the curvatures κ_y and κ_z, and draw the line of curvature and the neutral axis.

e) Determine the distribution of the axial strain ε and find the maximum axial stress $\sigma_{\max}$.

11 Flexure and Torsion of Beams

In homogeneous bending a beam is exposed to a constant bending moment along the beam. The constant bending moment does not generate any shear force, and shear stresses play only a minor role in connection with non-homogeneous cross-section properties. However, the bending theory is often used in the somewhat more general context in which the bending moment varies along the beam. In that case it follows from the equilibrium conditions discussed in Chapter 3 that shear forces will occur, and these shear forces in turn introduce shear stresses. The shear force is the total effect of the corresponding shear stresses over the cross-section, and thus the shear stress distribution determines the line of action of the shear force. The theory of nonhomogeneous bending – or flexure – of a beam thereby corresponds to a particular location of the transverse force with respect to the beam cross-section. If the load is offset from this line, it also produces a torsion moment, and the beam sections will rotate in twist. Modern structures often make use of beams with non-symmetric cross-sections, and it is important to identify possible contributions from a transverse load to torsion of the beam. The present chapter deals with the properties of flexure and torsion of beams and the proper separation of the two problems. A key point is the location of the so-called shear center – the point of action of the shear forces associated with beam flexure.

S. Krenk, J. Høgsberg, *Statics and Mechanics of Structures*,
DOI 10.1007/978-94-007-6113-1_11,

A central aspect of beam flexure and torsion is the distribution of the associated shear stresses over the beam cross-section. The calculation of shear stresses in flexure is discussed in Section 11.1. They can be determined approximately in terms of their resultant shear flow by the so-called Grashof formula. For thin-walled cross-sections the relation between shear flow and local stresses is rather accurate, and details of the shear stresses and the shear center in thin-walled cross-sections are given in Section 11.2. The remainder of the chapter deals with the complementary problem of beam torsion. First, the torsion problem is introduced in a simple setting in the form of homogeneous torsion of solid and hollow circular cylinders in Section 11.3. This setting is particularly simple, because the initially plane cross-sections remain plane during twist. For non-circular cross-sections the initially plane sections develop warping in connection with torsion, and the general theory of homogeneous torsion is developed in Section 11.4. As for the flexure problem, particularly simple and rather robust results can also be developed for homogeneous torsion of thin-walled beams as demonstrated in Section 11.5 for open as well as closed thin-walled cross-sections.

11.1 Shear stresses in beam flexure

Beam flexure is the static problem associated with beam bending under transverse load. A beam in flexure develops shear forces and associated shear stresses. The origin of these shear stresses is illustrated in Fig. 11.1 showing flexure of a homogeneous cantilever beam loaded by a transverse tip force P, acting in a plane of symmetry. The actual beam is shown in Fig. 11.1a together with the normal stress distribution over the cross-section. Now, assume that a longitudinal section is made in the beam, whereby the upper and lower parts of the beam act as individual beams as illustrated in Fig.11.1b. The transverse load is distributed between these two new beams according to their bending stiffness, and thus they will exhibit the same transverse displacement. The bending stiffness is represented by the moment of inertia about a line through the elastic center, and thus the total bending stiffness is

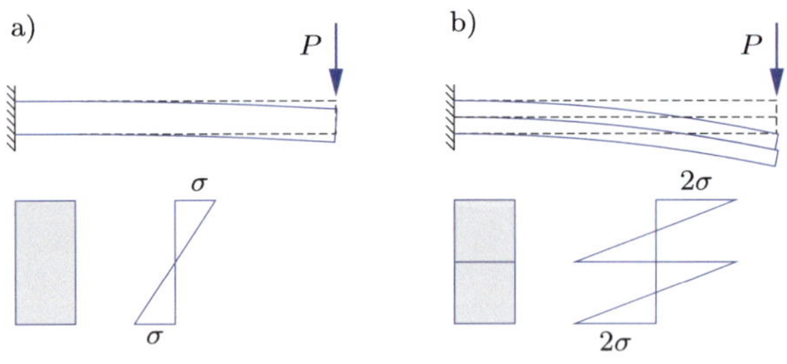

Fig. 11.1: Deformation of (a) single and (b) double layer cantilever.

reduced when splitting the original section into two separate parts, and the tip displacement as well as the normal stresses in the beam increase.

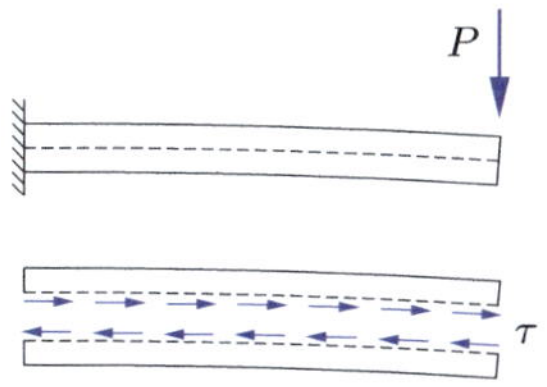

Fig. 11.2: Longitudinal shear stress τ in beam.

The axial displacement is discontinuous across the horizontal section, with extension in the lower beam and contraction in the upper beam. This relative sliding across the horizontal section can be eliminated by applying a suitable shear stress distribution $\pm\tau(x, y)$, acting in the positive x-direction on the upper part and in the negative direction on the lower part, as illustrated in Fig. 11.2. An approximate theory for the shear stress distribution in a beam in flexure can be obtained by using the normal stress distribution from the homogeneous bending theory developed in Chapter 10, supplemented with conditions of longitudinal equilibrium. This approach, due to JOURAWSKI (1821-1891) and often associated with GRASHOF (1826–1893), is developed in the following section.

11.1.1 Shear flow – Grashof's formula

Figure 11.3a shows a cantilever beam with cross-section shown in Fig. 11.3b. The origin of the coordinate system $\{\bar{y}, \bar{z}\}$ is placed at the elastic center C, and the coordinate axes are initially assumed be principal axes of the cross-section. The beam is loaded in the z-direction, and thus $M_y = 0$. The axial stress σ is given by the Navier's stress distribution formula (10.79), which for plane bending reduces to

$$\sigma = \frac{E}{E_0}\Big(\frac{N}{F} + \frac{M_{\bar{z}}}{I_{\bar{z}}}\,\bar{z}\Big). \tag{11.1}$$

In this formula the vertical coordinate is denoted $\bar{z}$ to indicate that the origin of the coordinate system is at the elastic center. A thin segment of the beam of width dx is shown in Fig. 11.3. Figure 11.3c shows this segment in detail, with axial stress distribution σ on the left cross section and $\sigma + d\sigma$ on the right, with $d\sigma$ representing the change in axial stress over the incremental length dx.

The small segment is now divided into two parts by a introducing longitudinal section, as shown by the dashed line in Fig. 11.3b. The area of the upper part

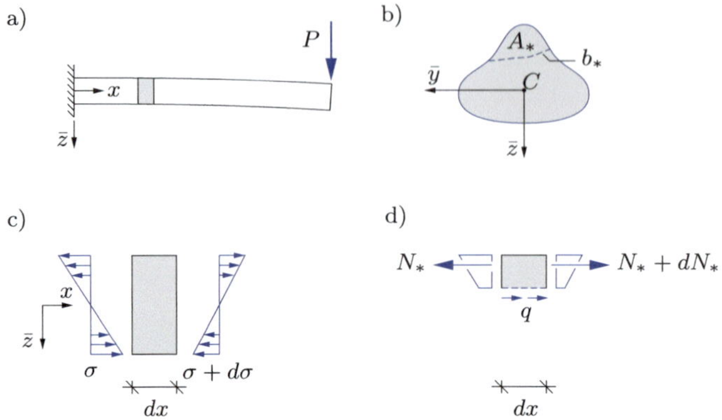

Fig. 11.3: Force equilibrium between normal stress σ and shear flow q.

is A_*, and the length of the section is b_*. The forces acting on this small part of the beam are shown in Fig. 11.3d. They consist of a normal force N_* at the left cross-section, the force $N_* + dN_*$ at the right cross-section, and a distributed force of intensity q acting along the longitudinal section. The distributed force with intensity q is called the *shear flow*. The normal force N_* is found by integrating the normal stresses over the area A_*,

$$N_* = \int_{A_*} \sigma \, dA \,. \tag{11.2}$$

Equilibrium in the longitudinal direction of the beam requires that

$$\rightarrow \qquad -N_* + N_* + dN_* + q\, dx = 0 \,.$$

The first two terms cancel, and after division by dx the shear flow q is found as

$$q = -\frac{dN_*}{dx} = -\int_{A_*} \frac{d\sigma}{dx} \, dA \,, \tag{11.3}$$

where the last expression follows from differentiation of (11.2). In the present case of symmetric bending the longitudinal derivative of the axial stress is found by differentiation of (11.1),

$$\frac{d\sigma}{dx} = \frac{E}{E_0} \frac{dM_{\bar{z}}}{dx} \frac{\bar{z}}{I_{\bar{z}}} = \frac{E}{E_0} \frac{\bar{z}}{I_{\bar{z}}} Q_{\bar{z}} \,, \tag{11.4}$$

where the shear force is the derivative of the corresponding moment, $Q_{\bar{z}} = dM_{\bar{z}}/dx$, as given in (3.10). Note, that a constant normal force N in the beam does not contribute to the shear flow. When substituting the gradient (11.4) back into (11.3), the shear flow is expressed as

$$q = -\frac{1}{E_0}\Big(\int_{A_*} E\,\bar{z}\,dA\Big)\,\frac{1}{I_{\bar{z}}}\,Q_{\bar{z}}\,. \tag{11.5}$$

The integral represents the static moment of the part of the beam cross-section with area A_*, and it is therefore convenient to introduce the notation

$$S_{\bar{z}}^* = \frac{1}{E_0}\int_{A_*} E\,\bar{z}\,dA\,. \tag{11.6}$$

Hereby the final form of the shear flow in symmetric bending is obtained as

$$q = -\frac{S_{\bar{z}}^*}{I_{\bar{z}}}\,Q_{\bar{z}}\,. \tag{11.7}$$

This result is often termed Grashof's formula. When the shear flow is represented in terms of the mean value of the shear stress as $q = b_*\tau$, the mean shear stress over the section is determined as

$$\tau_m = \frac{q}{b_*} = -\frac{S_{\bar{z}}^*}{b_* I_{\bar{z}}}\,Q_{\bar{z}}\,. \tag{11.8}$$

The subscript on m on τ_m refers to the stress as a mean value over the section b_*. The approximation by the average value is adequate as long as the shear stresses are approximately constant along the section.

For simplicity the derivation of Grashof's formula in the form (11.8) was based on the case of symmetric bending. However, the argument is based on the general relation (11.3) between shear flow q and the longitudinal derivative of the normal stress σ. Thus, the general form of the Grashof formula is obtained when the axial stress is obtained from the general strain distribution (10.75) by multiplication with the local elastic modulus E,

$$\sigma = \frac{E}{E_0}\left(\frac{N}{F} + [\,\bar{y}\,,\,\bar{z}\,]\begin{bmatrix} I_{\bar{y}\bar{y}} & I_{\bar{y}\bar{z}} \\ I_{\bar{y}\bar{z}} & I_{\bar{z}\bar{z}} \end{bmatrix}^{-1}\begin{bmatrix} M_{\bar{y}} \\ M_{\bar{z}} \end{bmatrix}\right). \tag{11.9}$$

Substitution of this expression for the axial stress into (11.3) gives

$$q = [\,S_{\bar{y}}^*\,,\,S_{\bar{z}}^*\,]\begin{bmatrix} I_{\bar{y}\bar{y}} & I_{\bar{y}\bar{z}} \\ I_{\bar{y}\bar{z}} & I_{\bar{z}\bar{z}} \end{bmatrix}^{-1}\begin{bmatrix} Q_{\bar{y}} \\ Q_{\bar{z}} \end{bmatrix}, \tag{11.10}$$

extending Grashof's formula to general bending about non-principal axes.

In the case of principal axes the inertia matrix – and thereby its inverse – assume diagonal form, and the shear flow formula takes the somewhat simpler form

$$q = -\frac{S_{\bar{y}}^*}{I_{\bar{y}}}\,Q_{\bar{y}} - \frac{S_{\bar{z}}^*}{I_{\bar{z}}}\,Q_{\bar{z}}\,, \tag{11.11}$$

from which the mean stress follows by division with the length of the section,

$$\tau_m = \frac{q}{b_*} = -\frac{S_{\bar{y}}^*}{b_* I_{\bar{y}}} Q_{\bar{y}} - \frac{S_{\bar{z}}^*}{b_* I_{\bar{z}}} Q_{\bar{z}} . \tag{11.12}$$

This formula is used to determine the interface stress in a composite beam in the following example, before turning to its use in determining the distribution of shear stresses over the beam cross-section.

Example 11.1. Shear stress in interface of composite beam. Figure 11.4a shows a cantilever beam with a tip load P acting in the z-direction. The cross-section of the beam is composed of a soft core material with elastic modulus E_c, and two stiff flanges with elastic modulus $E_f = 50E_c$ and thickness $t = \frac{1}{25}h$. The length of the beam is ℓ, and the reference elastic modulus is $E_0 = E_c$. The tip load generates a shear force in the z-direction

$$Q_z = P ,$$

that is positive and constant with respect to the longitudinal axis x.

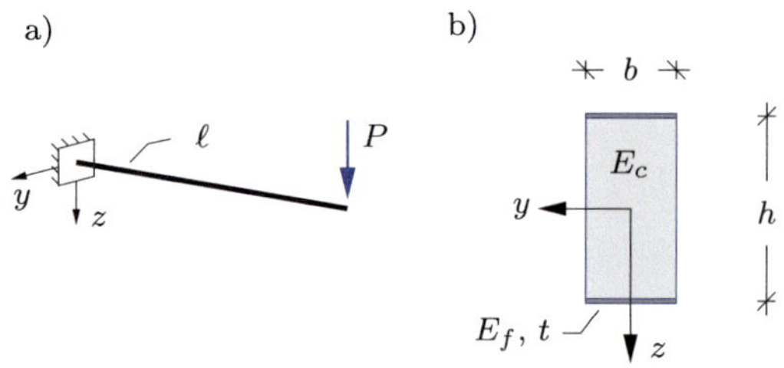

Fig. 11.4: a) Cantilever with tip load P, and b) composite cross-section.

The flanges are glued to the core material, and it is of interest to determine the shear stress between the core and flanges. The glue should be strong enough to withstand the shear stresses due to the loading of the cantilever. In Fig. 11.5 a longitudinal section is placed in the cantilever between the upper flange and the core material. Longitudinal shear stresses τ act at the section, and the intensity is given by (11.8).

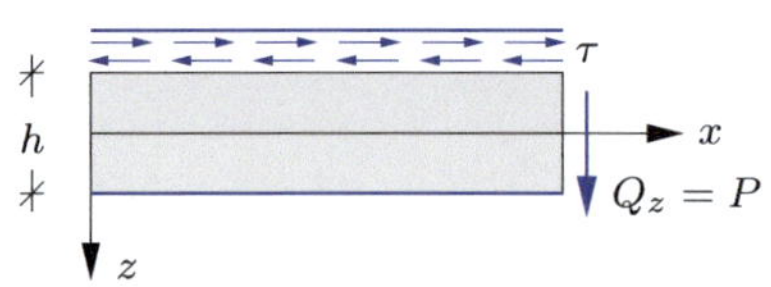

Fig. 11.5: Longitudinal shear stress τ at interface between top flange and core.

The cross-section is symmetric and the moment of inertia in the z-direction is a principal moment of inertia,

$$I_z = \tfrac{1}{12}\, h^3\, b + \frac{E_f}{E_c}\, 2bt \left(\tfrac{1}{2}h\right)^2 = \tfrac{13}{12}\, h^3 b .$$

Note, that because of the fairly large elastic modulus of the flanges they constitute the main part of the bending stiffness. The static moment of the top flange is

$$S_z^* = \frac{E_f}{E_c}\, bt\,(-\tfrac{1}{2}h) = -\,bh^2,$$

which is negative because the top flange is located with negative z-value. The shear flow follows from Grashof's formula (11.7) as

$$q = -\frac{S_z^*}{I_z}\, Q_z = \frac{bh^2}{\frac{13}{12}h^3 b}\, P = \frac{12}{13}\frac{P}{h}\,.$$

Note, that the shear flow is positive because the analysis is conducted for the top flange. As indicated in Fig. 11.5 the longitudinal shear stresses act in the negative x-direction on the core. It is left as an exercise to show that Grashof's formula actually leads to the same result as above, but with opposite sign, for the part of the cross-section below the section. The average shear stress is obtained by division of the shear flow with the length $b_* = b$ of the section,

$$\tau_m = \frac{12}{13}\frac{P}{bh}\,.$$

Assume that the maximum allowable shear stress in the glue is τ_g. This leads to an estimate of the corresponding maximum tip force as $P_{\max} = \frac{13}{12}\tau_g hb$. □

11.1.2 Shear stress on cross-section

It was demonstrated in Section 8.1.2 that the shear stress components are symmetric, as illustrated in Fig. 8.7. This implies that the shear stresses on two intersecting perpendicular planes are equal and either both point towards the line of intersection or away from this line, as illustrated in Fig. 11.6. In the present context this implies that the shear stress, initially determined on a longitudinal section through the beam, also determines the shear stress in the cross-section plane along the line of intersection.

Fig. 11.6: Shear stress components at line of intersecting orthogonal planes.

Figure 11.7 shows a segment of a beam of thickness dx. The longitudinal shear flow q is determined from Grashof's formula. The shear flow is represented in terms of the mean shear stress τ_m as $q = b_*\tau_m$, where b_* is the length of the section in the cross-section plane. It then follows from the equality of the shear stress in the longitudinal plane and the intersecting cross-section plane that τ_m, and thereby the shear flow q, have a component in the cross-section plane that is equal to the longitudinal component. The shear stress component in the cross-section plane is normal to the section used to define the partial area A_* in Grashof's formula. The direction is determined by the associated direction of the longitudinal shear flow at the section – both

Fig. 11.7: Relation between longitudinal and cross-sectional shear flow.

pointing towards or away from the line of intersection. In an actual analysis, the determination of the cross-section shear stress component is quite straightforward, as demonstrated in the following examples.

Example 11.2. Shear stress in rectangular cross-section. Figure 11.8a shows a cantilever of length ℓ and with elastic modulus E_0. A tip force P is acting in the z-direction, whereby the shear forces are

$$Q_y = 0, \qquad Q_z = P.$$

The cross-section of the beam is rectangular as shown in Fig. 11.8b, with height h and width b.

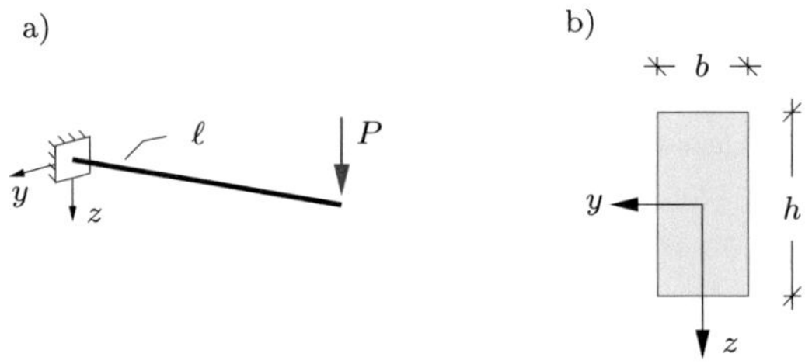

Fig. 11.8: a) Cantilever with tip load P, and b) rectangular cross-section.

A longitudinal section is introduced in the beam as shown in Fig. 11.9 at the distance d from the top of the cross-section. The shear flow q at the section is found by Grashof's formula, where the simple form (11.7) can be used because $Q_y = 0$,

$$q = -\frac{S_z^*}{I_z} Q_z .$$

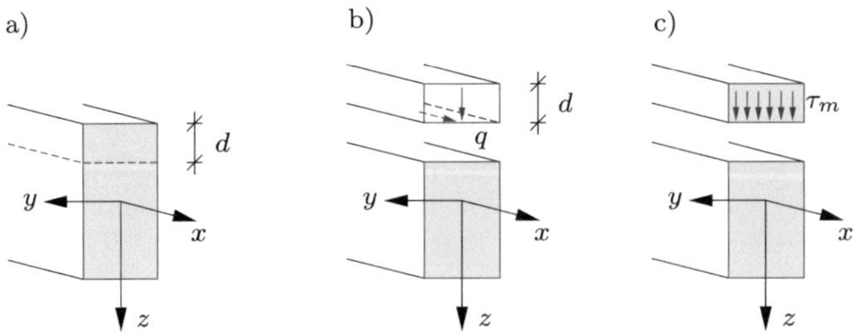

Fig. 11.9: Section with shear flow q and average shear stress τ_m.

In the present case Grashof's formula is applied using the upper segment of the cross-section. The longitudinal shear flow is positive in the x-direction. Thus, the shear flow on the upper segment of the cross-section is positive in the downward z-direction, as shown in Fig. 11.9b.

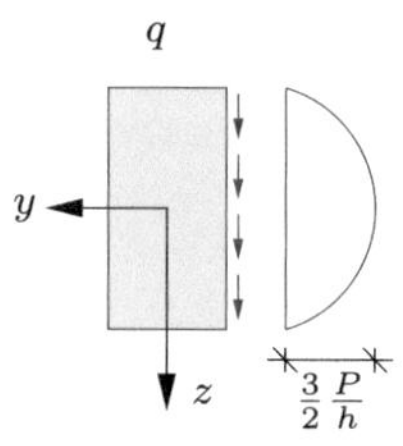

Fig. 11.10: Distribution of shear flow q_*.

The moment of inertia with respect to the z-direction of the full cross-section is

$$I_z = \tfrac{1}{12}h^3 b\,.$$

The static moment of the upper segment with respect to z is the area of the segment times the distance to the center of the segment,

$$S_z^* = bd(-\tfrac{1}{2}h + \tfrac{1}{2}d) = -\tfrac{1}{2}bd(h-d)\,.$$

Substitution into the expression for the shear flow gives

$$q = \frac{\tfrac{1}{2}bd(h-d)}{\tfrac{1}{12}h^3 b}\,P = 6\,\frac{d}{h}\Big(1-\frac{d}{h}\Big)\frac{P}{h}\,.$$

This represents a parabolic variation of the shear flow q with the distance d from the top of the cross-section, as illustrated in Fig. 11.10, where the parabolic curve gives the magnitude, while small arrows indicate the direction of the shear flow. The average shear stress over the width is then obtained by division with the length of the section b,

$$\tau_m = 6\,\frac{d}{h}\Big(1-\frac{d}{h}\Big)\frac{P}{hb}\,.$$

The maximum shear stress occurs at the center of the cross-section where $d = \frac{1}{2}$,

$$\tau_m^{\max} = \frac{3}{2}\,\frac{P}{hb}\,.$$

The shear flow in the y-direction due to the shear force $Q_z = P$ can be investigated by placing a vertical section in the rectangular cross-section at distance d from the right side,

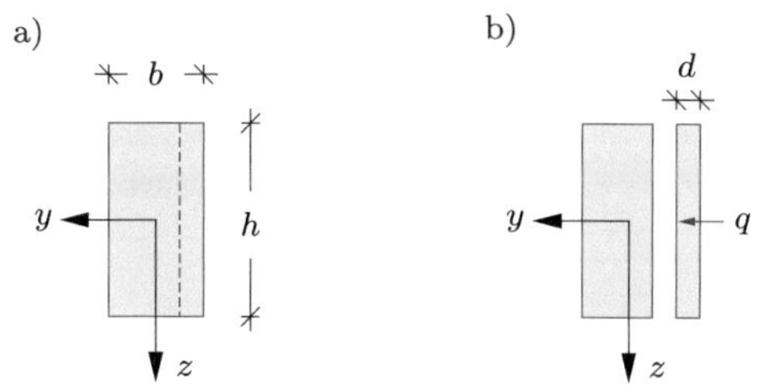

Fig. 11.11: Determination of shear flow in the horizontal direction.

see Fig. 11.11. The shear flow q is determined by the expression (11.7), where the static moment is

$$S_z^* = 0\,.$$

This means that for the rectangular cross-section loaded in the z-direction, the shear flow in the y-direction vanishes. Thus, the y-component of the average shear stress also vanishes. However, the present method, based on Grashof's formula, only determines average shear stress, and this does not exclude variations in the local shear stress along the section line. For simple cross-sections with homogeneous material properties analytical methods can be used, while for more general cross-sections recourse is taken to analysis by the finite element method for a detailed stress analysis. □

Properties of the shear flow

As shown in Fig. 11.10 the shear flow vanishes at the upper and the lower boundary of the cross-section, because the static moment of the isolated segment vanishes with vanishing area. Thus, the shear flow component normal to the cross-section boundary vanishes, in accordance with the implicit assumption that no shear stresses act on the boundary of the beam. The approximate shear stress distributions resulting from Grashof's formula have some simple properties, summarized in Table 11.1.

Table 11.1: Properties the approximate τ_m shear stress distribution.

1) The shear stress at the boundary of the cross-section is parallel to the boundary.
2) For parts of a cross-section of constant width the shear stress τ varies linearly over cross-section areas with constant axial stress σ.
3) For parts of a cross-section of constant width the stress τ has a parabolic variation over cross section areas with linear variation of the axial stress σ.
4) The shear flow attains its maximum value at a section through the elastic center.

Example 11.3. Shear stress in solid T-profile. Figure 11.8 shows a solid T-profile with constant elastic modulus E_0. The height of the profile is $5a$ and the width is $6a$. The T-profile represents the cross-section of a cantilever beam loaded by a tip force P in the z-direction, see e.g. Fig. 11.8a. This implies that the shear forces are:

$$Q_y = 0\,, \qquad Q_z = P\,.$$

The origin of the coordinate system $\{y, z\}$ is placed in the elastic center, which is located on the vertical line of symmetry and $\frac{1}{2}a$ below the bottom side of the flange, as indicated in Fig. 11.12.

The distribution of the shear flow over the cross-section is found by placing suitable sections in the flange and web of the cross-section. First a vertical section is placed in the right part of the flange, with distance d_1 from the right end of the flange, as shown in Fig. 11.13.

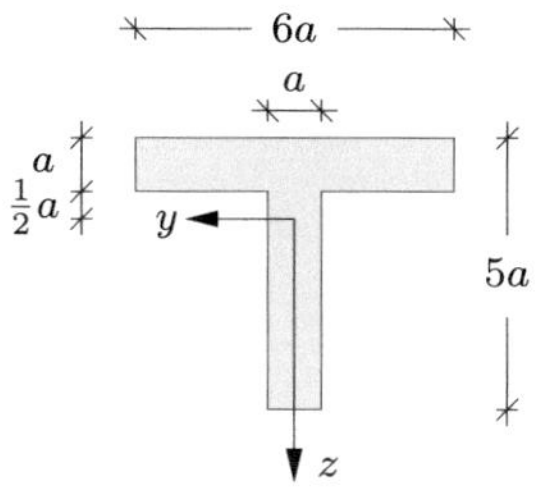

Fig. 11.12: Geometry and dimensions of massive T-profile.

The shear flow in the flange is given by Grashof's formula (11.7),

$$q = -\frac{S_z^*}{I_z} Q_z .$$

The moment of inertia with respect to the z-direction is

$$I_z = \tfrac{1}{12}a^3 6a + 6a^2 a^2 + \tfrac{1}{12}(4a)^3 a + 4a^2\left(\tfrac{3}{2}a\right)^2 = \left(\tfrac{1}{2} + 6 + \tfrac{16}{3} + 9\right)a^4 = \tfrac{125}{6}a^4 .$$

The static moment with respect to z for the segment in Fig. 11.13b is

$$S_z^* = a d_1(-a) = -a^2 d_1 .$$

Substitution into the expression for the shear flow gives

$$q_1 = \frac{a^2 d_1}{\frac{125}{6}a^4} P = \frac{6}{125}\frac{d_1}{a}\frac{P}{a},$$

where the subscript 1 indicates that this expression is for the right part of the flange with local coordinate d_1. The shear flow variation is linear with respect to the local coordinate d_1 in agreement with point 2 in Table 11.1, as the axial stress σ is constant in the horizontal direction of the shear flow. Moreover, it is found that the shear flow vanishes at the right end of the flange, i.e. for $d_1 = 0$. This corresponds to point 1 in Table 11.1. At the joint of the cross-section, corresponding to $d_1 = \frac{5}{2}$, the shear flow in the right part of the flange is

$$q_1\left(\tfrac{5}{2}a\right) = \frac{3}{25}\frac{P}{a} .$$

The same results can be obtained for the left part of the flange, where positive direction is simply in the opposite direction, as shown in Fig. 11.14b.

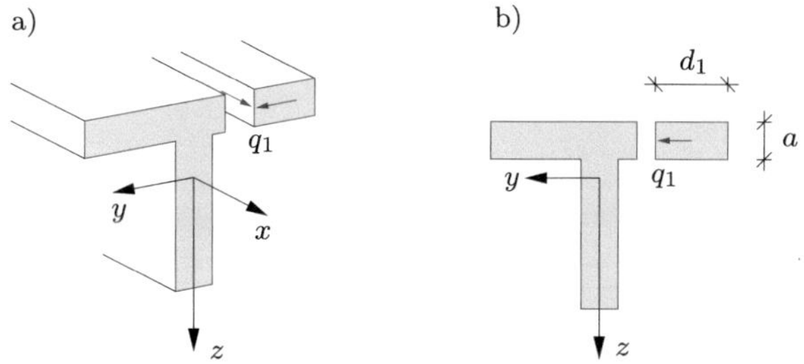

Fig. 11.13: Shear flow in right part of horizontal flange.

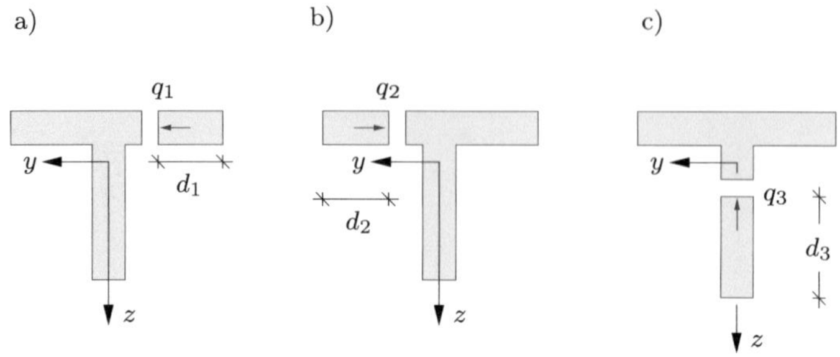

Fig. 11.14: Shear flow in flange and web.

In the vertical web the shear flow is determined by placing a horizontal section at distance d_3 from the bottom, see Fig. 11.14c. The static moment for this segment is

$$S_z^* = ad_3\left(\tfrac{7}{2}a - \tfrac{1}{2}d_3\right) = \tfrac{1}{2}ad_3\,(7a - d_3)\,.$$

The shear flow in the web is then given as

$$q_3 = -\frac{\tfrac{1}{2}ad_3(7a-d_3)}{\tfrac{125}{6}a^4}\,P = \frac{3}{125}\frac{d_3}{a}\left(\frac{d_3}{a} - 7\right)\frac{P}{a}\,.$$

At the bottom of the web, where $d_3 = 0$, the shear flow is zero. At the joint of the cross-section, which corresponds to $d_3 = 4a$, the shear flow is

$$q_3(4a) = -\frac{36}{125}\frac{P}{a}\,.$$

The vertical shear flow continues into the horizontal flange. The direction and magnitude of the shear flow is determined by placing a horizontal section in the flanges at distance d_4 from the top of the flange, as shown in Fig. 11.15. The shear flow is

$$q_4 = \frac{18}{125}\frac{d_4}{a}\left(3 - \frac{d_4}{a}\right)\frac{P}{a}\,.$$

At the top of the flange the shear flow vanishes, while at the bottom of the flange it is

$$q_4(a) = \frac{36}{125}\frac{P}{a} = -q_3(4a)\,.$$

Thus, there is continuity in the shear flow.

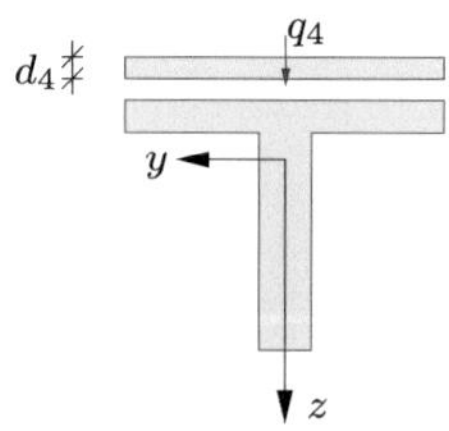

Fig. 11.15: Vertical shear flow in flange.

The distribution along the flange and web is illustrated in Fig. 11.16, where it is seen that the maximum shear flow occurs at $z = 0$, i.e. at the level of the elastic center. This can be

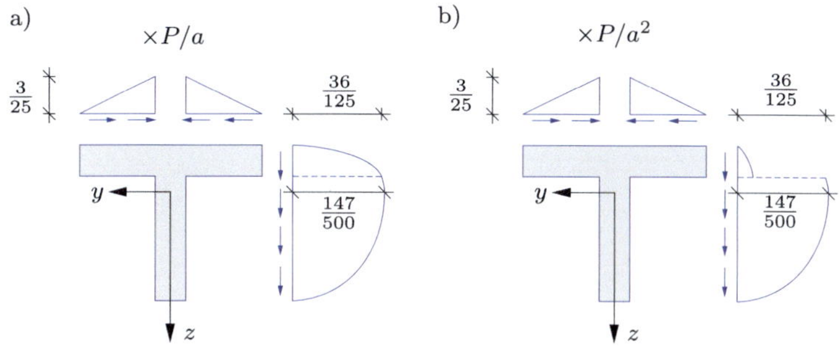

Fig. 11.16: Normalized a) shear flow q, and b) shear stress τ_m in flange and web.

verified in terms of the gradient of q_3,

$$\frac{d(q_3)}{d(d_3)} = \frac{3}{125}\Big(2\frac{d_3}{a} - 7\Big)\frac{P}{a^2} = 0 \qquad \Rightarrow \qquad d_3^{\max} = \tfrac{7}{2}a\,.$$

This value of d_3 corresponds to the level of the elastic center at $z = 0$. The corresponding maximum value of the shear flow is

$$q_3^{\max} = q_3\big(\tfrac{7}{2}a\big) = -\frac{147}{500}\frac{P}{a}\,.$$

The mean shear stresses τ_m are obtained from (11.12) by division with $b_* = a$. The distribution of the mean shear stress is shown in Fig. 11.16b. It is seen that the vertical component of the mean shear stress is very small in the horizontal flange because the shear flow in this part is divided by $b_* = 6a$. The shear on the lower side of the flange vanishes, and thus the mean vertical shear stress over the flanges is not very representative of the local stress distribution. However, the shear stresses are typically dominated by the component in the direction of the individual flanges, and often only this component is determined. This is the case in particular for thin-walled cross-sections, where the thickness of the individual flanges is small compared to the overall dimensions of the cross-section. □

11.2 Thin-walled cross-sections in shear

Thin-walled cross-sections are composed of individual flanges with thickness much smaller than the overall dimensions of the cross-section. The distribution of shear flow and shear stresses is simplified for this type of cross-section. As demonstrated in the previous example the shear flow of interest is in the direction of the individual flange. Furthermore, the shear flow fulfills a continuity condition at joints in the cross-section, as demonstrated next.

Continuity of shear flow at joints

Consider a typical joint in a thin-walled cross-section. Figure 11.17a shows a beam section of length dx, with a cross-section that contains a joint connecting three flanges. The shear flow in the three flanges at the joint are denoted q_1, q_2 and q_3, respectively. The positive direction of the shear flow on the

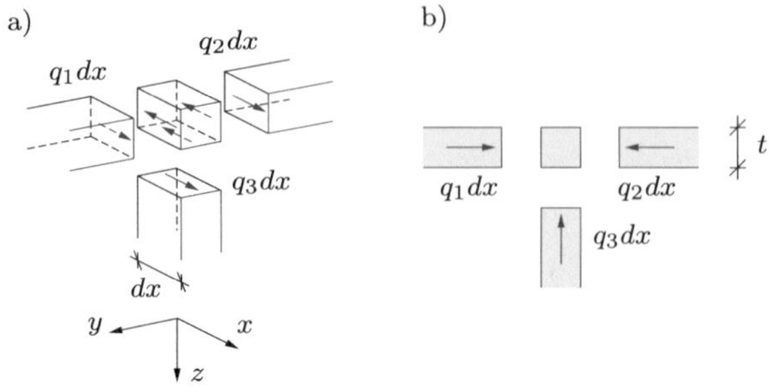

Fig. 11.17: Shear flow at joint of thin-walled flanges.

flanges is in the direction of the longitudinal coordinate x. The shear flow is the force per unit length on the sections, and the resulting force in the longitudinal x-direction for each of the three sections is obtained as the shear flow times the incremental length dx, giving $q_1 dx$, $q_2 dx$ and $q_3 dx$. Each section must be in equilibrium, which means that the three resulting forces are transferred to the joint when changing the direction, as shown in Fig. 11.17a. Longitudinal equilibrium of the joint then requires that

$$q_1 dx \,+\, q_2 dx \,+\, q_3 dx \,-\, d\sigma\, t^2 \,=\, 0\,,$$

where $d\sigma$ is the increment of the axial strain over the length dx. Division by the increment dx gives

$$q_1 \,+\, q_2 \,+\, q_3 \,=\, t^2\, \frac{d\sigma}{dx}\,.$$

The gradient of the axial stress has been determined in (11.4). For thin-walled cross-sections the area t^2 is of higher order and is therefore omitted. If the shear flows were determined via inclined sections, meeting at the intersection of of the flange center-lines, the term would vanish exactly. Thus, longitudinal equilibrium of the joint requires that the sum of the shear flows must vanish. This can be generalized for a joint connecting n flanges,

$$\sum_{j=1}^{n} q_j \,=\, 0\,. \tag{11.13}$$

This result can be given a continuity argument, similar to a flow in a channel with varying cross-section: What comes in balances what goes out. Let $q_{\rm in}$ be the total shear flow that runs into the joint, and let $q_{\rm out}$ be the total shear flow that runs out of the joint. Equilibrium then states that what runs in, must run out:

$$q_{\rm in} \,=\, q_{\rm out}\,. \tag{11.14}$$

This simple condition can be used to determine the shear flow at corners and joints, and to verify results.

Example 11.4. Shear stress in I-profile. Figure 11.18a shows a cantilever of length ℓ with tip force P in the z-direction. The shear forces are

$$Q_y = 0\,, \qquad Q_z = P\,.$$

The cross-section of the beam is the I-profile shown in Fig. 11.18b. The height is a, the width is a, the thickness of the flanges is $2t$ and the thickness of the web is t. The beam is homogeneous with constant elastic modulus $E_0 = E$. The cross-section is assumed to be thin-walled, $t \ll a$.

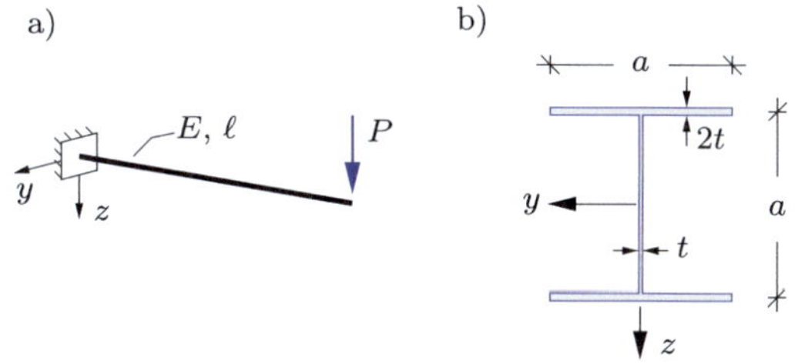

Fig. 11.18: Cantilever with I-profile cross-section.

The moment of inertia of the entire cross-section is:

$$I_z = 2\big(2at(\tfrac{1}{2}a)^2\big) + \tfrac{1}{12}a^3t = \tfrac{13}{12}\,a^3t\,.$$

The shear flow in the flanges and the web are found by introducing sections in the individual parts. First, a vertical section is introduced in the left part of the top flange at distance d_1 from the left end of the flange, as shown in Fig. 11.19a. The static moment of the segment with respect to the z-coordinate is

$$S_z^* = 2td_1(-\tfrac{1}{2}a) = -\,at\,d_1\,,$$

and the shear flow follows from (11.7) as

$$q_1 = -\frac{S_z^*}{I_z}\,Q_z = \frac{at\,d_1}{\frac{13}{12}\,a^3t}\,P = \frac{12}{13}\,\frac{d_1}{a}\,\frac{P}{a}\,.$$

At the left end of the flange the shear flow is zero, while at the center it is

$$q_1(\tfrac{1}{2}a) = \frac{6}{13}\,\frac{P}{a}\,.$$

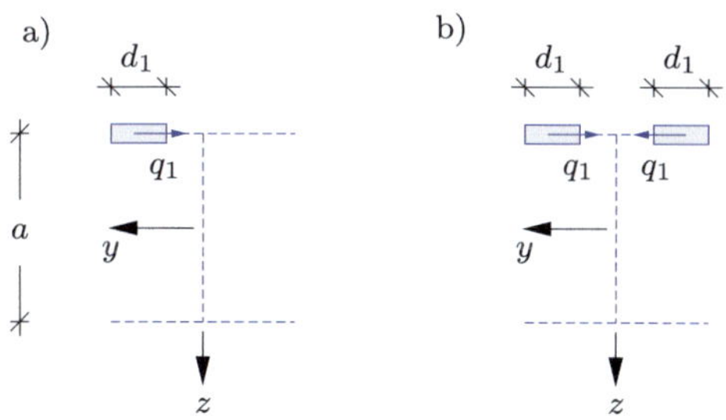

Fig. 11.19: Shear flow in top flange: a) section in left part, and b) double section.

Because of symmetry the shear flow in the right part of the flange is similar to that in the left part of the flange. In fact, because of symmetry, a double section could be placed at distance d_1 from both the left and the right end of the flange, as shown in Fig. 11.19b. Hereby the magnitude of the static moment S_z^* doubles. However, as the resulting shear flow is $2q_1$, the expression for q_1 recovers the expression for the single section given above. The introduction of a double section is particularly useful for closed cross-section, as demonstrated in Example 11.5 for a single cell box section.

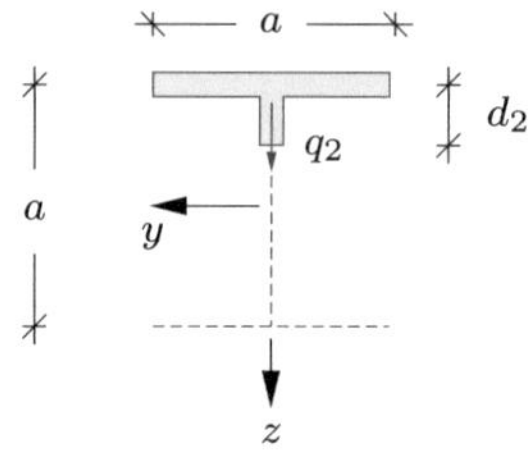

Fig. 11.20: Shear flow in web.

The shear flow in the web is determined by introducing a horizontal section in the web, at distance d_2 below the centerline of the top flange, as shown in Fig. 11.20. The shear flow is positive in the downward direction, as indicated in the figure. The static moment of the segment contains contributions from the top flange and part of the web,

$$S_z^* = 2at(-\tfrac{1}{2}a) + td_2(-\tfrac{1}{2}a + \tfrac{1}{2}d_2) = \tfrac{1}{2}td_2(d_2 - a) - a^2 t .$$

The terms are grouped to illustrate the contribution $-a^2t$ from the upper flange, and the contribution $\frac{1}{2}td_2(d_2 - a)$ from the web. The contribution from the web vanishes for $d_2 = 0$ because no web area is included, and for $d_2 = a$ because the full web area is centered at the elastic center. The expression for the shear flow in the web now is

$$q_2 = -\frac{S_z^*}{I_z} Q_z = \frac{12}{13}\Big(1 - \frac{d_2(a - d_2)}{2a^2}\Big)\frac{P}{a} .$$

At the two joints the last term vanishes, leaving the contribution from the flange

$$q_2(0) = q_2(a) = \frac{12}{13}\frac{P}{a} ,$$

while the maximum occurs at the level of the elastic center:

$$q_2^{\max} = q_2\big(\tfrac{1}{2}a\big) = \frac{27}{26}\frac{P}{a} .$$

Note, that $q_2^{\max} \simeq 1.1 q_2(0)$, indicating that the shear flow in the web is almost constant with a modest parabolic variation.

Figure 11.21a shows the variation of the shear flow over the cross-section, normalized by the factor $\frac{1}{26}P/a$. In the bottom flange the variation of the shear flow is similar to that in the top flange, but because the static moments have opposite sign, the direction of the shear flow in the bottom flange is opposite that in the top flange. The shear stress is determined as the shear flow divided with the thickness of the individual flanges. Thus, the shear flow is divided by $2t$ in the top and bottom flanges and by t in the web. The distribution of shear stresses, normalized by $\frac{1}{26}P/(at)$, is shown in Fig. 11.21b, illustrating that the maximum shear stress occurs in the web, and that the double thickness of the flanges reduces the stress in the flanges.

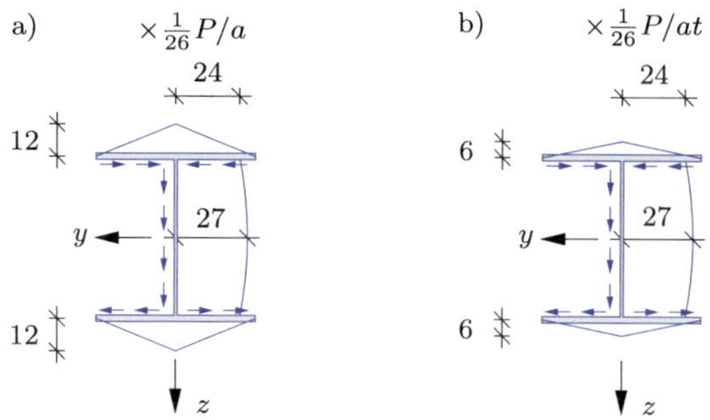

Fig. 11.21: Distribution of a) shear flow q, and b) shear stress τ_m.

It is seen from Fig. 11.21 that the total shear flow running into the top joint comes from the two parts of the flange:

$$q_{\text{in}} = 2\,\frac{6}{13}\frac{P}{a}\,,$$

while the resulting shear flow running out of the joint in the web is

$$q_{\text{out}} = \frac{12}{26}\frac{P}{a}\,.$$

Thus, the balance condition (11.14) is satisfied.

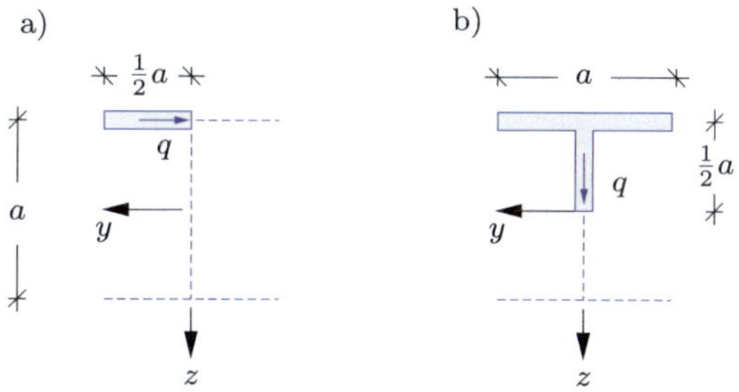

Fig. 11.22: Shear flow by sections at a) corner and b) elastic center.

With the knowledge of the general characteristics of the shear flow variation in thin-walled cross-sections, the variation of the shear flow can be determined by using only the two sections shown in Fig. 11.22. By the first section in the left flange at the joint, see Fig. 11.22a, the shear flow is determined as

$$q = \frac{6}{13}\frac{P}{a}\,.$$

Because of symmetry with respect to the z-axis the shear flow at the joint in the right part of the top flange is the same. The only unknown shear flow at the joint now is the shear flow in the web. The continuity condition (11.13) gives the shear flow in the web at the upper joint as

$$q = -\,2\,\frac{6}{13}\frac{P}{a} = -\,\frac{12}{13}\frac{P}{a}\,.$$

From the section in Fig. 11.22b the shear flow at the elastic center is

$$q = \frac{27}{26} \frac{P}{a} .$$

Because of symmetry with respect to the y-axis, the shear flow in the lower half of the cross-section can be mirrored from the upper half. When the shear flow has been determined at the critical points of the cross-section, the complete shear stress distribution is determined by linear curves in the horizontal direction, where the axial stress is constant, and by a parabolic curve with maximum at the elastic center in the vertical direction, where the axial stress varies linearly. □

Example 11.5. Shear stress in box section. Figure 11.23a shows a cantilever of length ℓ with tip force P in the z-direction, whereby

$$Q_y = 0 , \qquad Q_z = P .$$

The cross-section of the beam is a box-profile, as shown in Fig. 11.23b. The height is $4a$, the width is $3a$ and the thickness is t in all flanges. The beam is homogeneous with constant elastic modulus $E_0 = E$. The cross-section is assumed to be thin-walled.

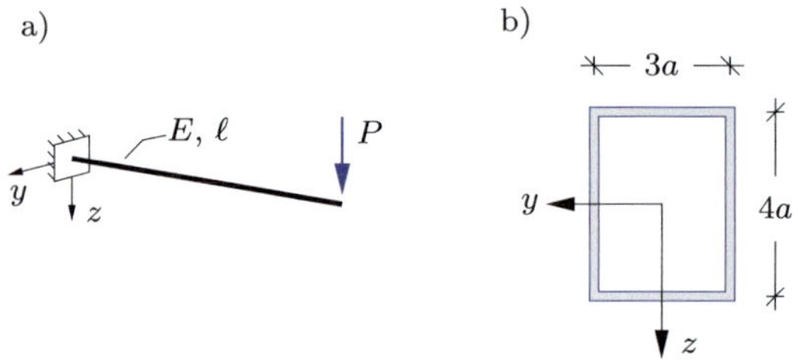

Fig. 11.23: Shear stresses in I-profile cross-section.

The beam is only loaded in the vertical z-direction and the variation of the axial stress is therefore constant in the top and bottom flanges and of linear variation in the vertical webs. The variation of the shear flow in the cross-section is fully determined by the two sections shown in Fig. 11.24a,c. Because the box profile is a closed cross-section, a single segment can not be isolated by a single section. Instead a double section is placed symmetrically in the top flange, as shown in Fig. 11.24a.

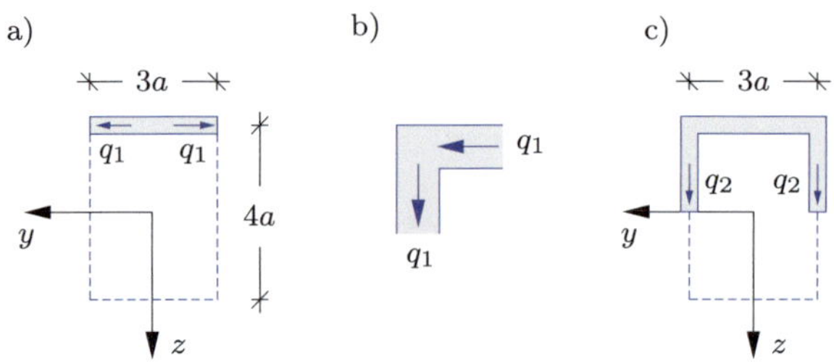

Fig. 11.24: Shear flow in box-profile.

The variation of the shear flow in the top flange is linear, and thus the distribution in the flange is fully determined when the shear flow is known at the corners. A double section can be applied because the shear flow q_1 at the two sections is identical. The total shear flow is then given as

$$2q_1 = -\frac{S_z^*}{I_z} Q_z \,,$$

where the moment of inertia for the full cross-section is

$$I_z = \tfrac{104}{3}\, a^3 \, t \,.$$

The static moment of the segment in Fig. 11.24a is

$$S_z^* = 3at(-2a) = -\,6a^2 t \,.$$

Hereby, the shear flow in the top flange at the corners, shown in Fig. 11.24a, is given as

$$q_1 = \frac{1}{2} \frac{6a^2 t}{\frac{104}{3}\, a^3 \, t} P = \frac{9}{104} \frac{P}{a} \,.$$

By the continuity condition (11.14) the shear flow in the left web at the top corner is also q_1 as indicated in Fig. 11.24b. Finally, the maximum shear flow is found at $z = 0$ by the double section along the y-axis shown in Fig. 11.24c. The static moment of this segment is

$$S_z^* = 3at(-2a) + 2(2at)(-a) = -10a^2 t \,,$$

whereby the shear flow is determined as

$$q_2 = \frac{1}{2} \frac{10a^2 t}{\frac{104}{3}\, a^3 \, t} P = \frac{15}{104} \frac{P}{a} \,.$$

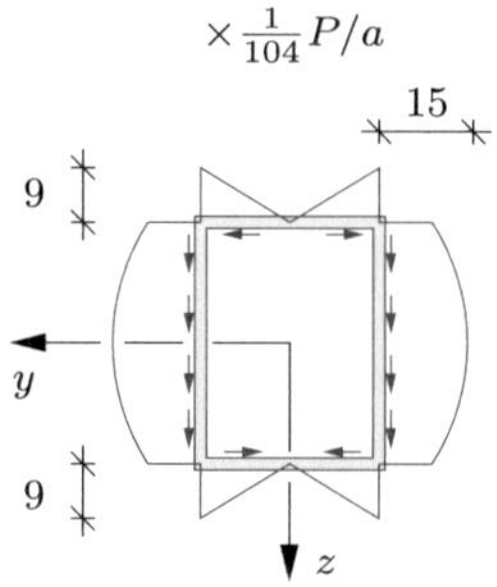

Fig. 11.25: Distribution of shear flow q in box section.

Because of symmetry with respect to the y-axis the solution for the upper half of the box-profile can be mirrored to the bottom half. The variation of the shear flow in the box-section is shown in Fig. 11.25. Note, that the shear flow in the vertical webs is running in the same direction as the shear force $Q_z = P$, whereby continuity at the corners implies that the shear flow in the top flange runs inside-out, while it runs outside-in in the bottom flange. □

11.2.1 Shear center

The shear force components Q_y and Q_z represent the accumulated effect of the shear stresses distributed over the corresponding cross-section. As forces in the cross-section plane they are represented by a magnitude and a line of action. The magnitude of the shear forces follow from integrating the projection of the shear stresses on the y- and z-axis, respectively. Conversely, in the Grashof procedure the shear flow, and thereby the corresponding shear stresses, are determined from the magnitude and direction of the shear force. However, it is remarkable that the Grashof procedure for the shear flow and mean shear stresses does not depend on the line of action of the shear force, but only its direction and magnitude. The Grashof procedure is based on equilibrium and the resulting shear flow therefore represent the corresponding shear force exactly. However, in order to determine the line of action the shear stresses on the various parts of the section must be considered, and the full shear stress of shear flow distribution must therefore be evaluated in order to calculate the line of action. Before discussing this in detail the following example briefly illustrates integration of the shear force from the shear flow distribution.

Example 11.6. Shear force in I-profile. Figure 11.26a shows the shear flow distribution over the I-profile considered in Example 11.4. The variation is linear in the flanges and parabolic in the web. The resulting forces on the individual flanges are obtained by integrating the individual distributions. Some simple area integrals are provided in Fig. 11.27. The linear variations are all identical, and the horizontal force in one half flange denoted Q_1 and shown in Fig. 11.26b. The integral of the shear flow corresponds to the area under the curve, and for the linear variation this gives:

$$Q_1 = \frac{1}{2}\frac{a}{2}\Big(\frac{6}{13}\frac{P}{a}\Big) = \frac{3}{26}P\,.$$

The positive direction of the resulting force Q_1 corresponds to that of the shear flow. Force equilibrium in the y-direction follows from the fact that the resulting forces in the two halves of each flange are in opposite direction and of equal magnitude.

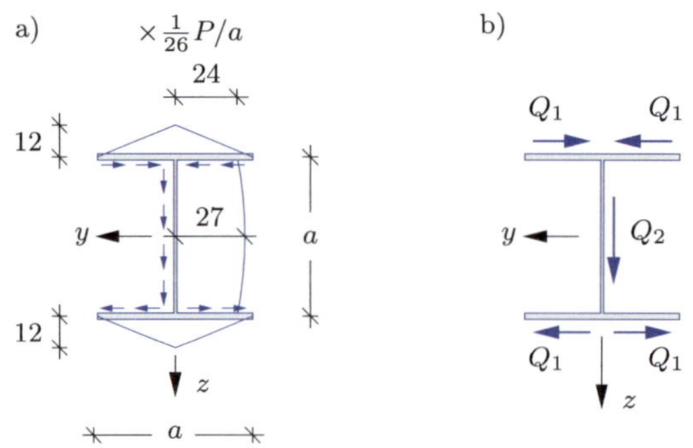

Fig. 11.26: Resulting shear forces on flanges of I-profile.

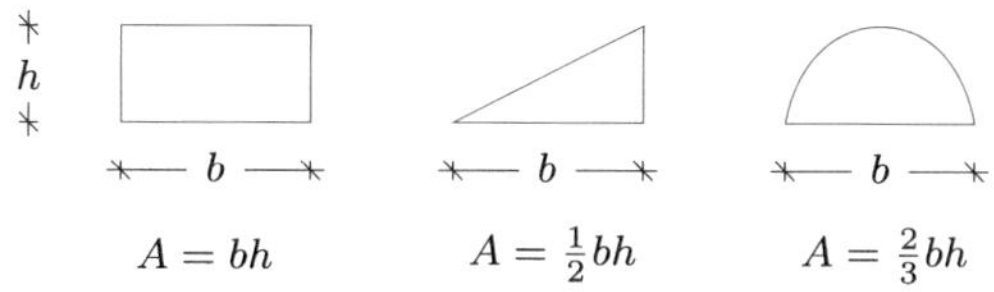

Fig. 11.27: Simple area integrals.

Only the resulting force from the web contributes to the force in the z-direction. This resulting force Q_2 is obtained by the area under the parabolic distribution, which is separated into the area of the constant part and the area of the pure parabola:

$$Q_2 = a\left(\frac{12}{13}\frac{P}{a}\right) + \frac{2}{3}a\left(\frac{3}{26}\frac{P}{a}\right) = P.$$

This corresponds to the magnitude and direction of the applied shear force $Q_z = P$. □

In the beam flexure problem the shear stress distribution is determined from a shear force, defined by its magnitude and direction. Once the shear stress distribution has been determined, the line of action can be determined from moment equilibrium. Assume that an imposed shear force Q_y has a line of action passing through the point $[0, a_z]$, and similarly that the shear force Q_z has a line of action passing through the point $[a_y, 0]$. It then follows from the argument presented in Section 1.2.1 and illustrated in Fig. 1.7 that the combined shear force $[Q_y, Q_z]$ passes through the point $[a_y, a_z]$. This is a characteristic point of the cross-section, called the *shear center*. If a transverse force is applied at a point not coinciding with the shear center, it can be decomposed into a force of the same direction and magnitude through the shear center plus a torsion moment about the x-direction, as demonstrated in Section 1.2.3. Clearly, it is important to know, to what extent a transversely loaded beam is exposed to torsion. In particular thin-walled beams with open cross-section have very low torsion stiffness, and the determination of the shear center therefore constitutes an important part of the analysis of thin-walled beams.

The location of the shear center $A = [a_y, a_z]$ can be determined from the two shear stress distributions associated with shear forces Q_y and Q_z along the y- and the z-axis, respectively. The corresponding torsion moment is

$$M_x = a_y Q_z - a_z Q_y. \tag{11.15}$$

Let the shear stress distribution associated with a shear force Q_z give a torsion moment M_x about the longitudinal x-axis. The coordinate a_y then follows directly from (11.15) with $Q_y = 0$. Similarly, the moment associated with shear stresses generated by the shear force Q_y determines the coordinate a_z. The procedure is illustrated in the following example and in some of the exercises.

Example 11.7. Shear center of I-profile. Consider the I-profile of Example 11.4, shown in Fig. 11.18. The distribution of the shear flow and the resulting forces on the flanges for the load case $Q_y = 0$ and $Q_z = P$ were determined in Example 11.4, and are shown in Fig. 11.28a,b for $Q_z = 1$. The results for the other load case $Q_y = 1$ and $Q_z = 0$ is also presented in the figure, while the analysis is left as an exercise. Figure 11.28c shows the shear flow distribution, while Fig. 11.28d gives the resulting flange forces.

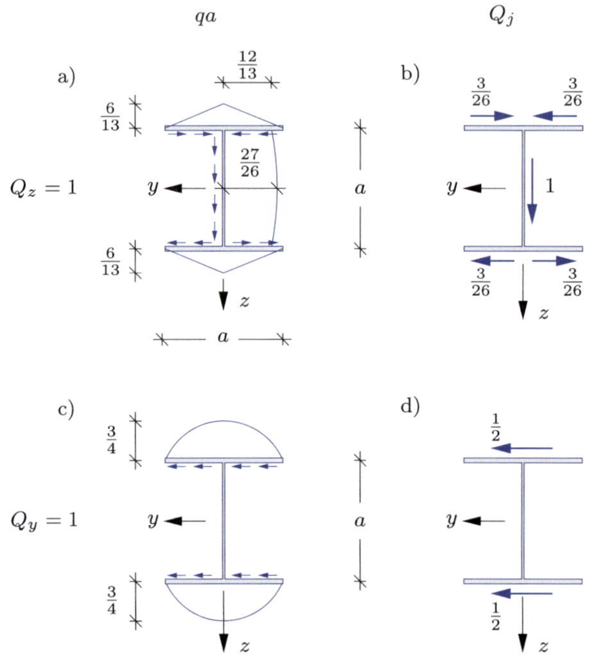

Fig. 11.28: Shear flow distribution for normalized load cases: $Q_z = 1$ and $Q_y = 1$.

In the case $Q_z = 1$ it is seen directly the resulting force in each of the flanges vanish, leaving only the vertical force in the web. As this force passes through the origin of the coordinate system $M_x = 0$. Thus, the coordinate follows from (11.15) as

$$M_x = 0 = a_y \, 1 \quad \Rightarrow \quad a_y = 0 \, .$$

For the load case $Q_y = 1$ the moment equation is

$$M_x = \tfrac{1}{2} a \tfrac{1}{2} + \tfrac{1}{2} a \left(-\tfrac{1}{2}\right) = 0 = -a_z \, 1 \quad \Rightarrow \quad a_z = 0 \, .$$

The results follow directly from the double symmetry of the cross-section, whereby $M_x = 0$ for both shear force components. □

The position of the shear center is given directly for some particular types of cross-sections. As illustrated in Example 11.7 the shear center is located on the line of symmetry. This implies that if the cross-section has a single line of symmetry, e.g. the C-profile shown in Fig. 11.29a, the shear center A is located on that line. Furthermore, for double symmetric cross-sections, such

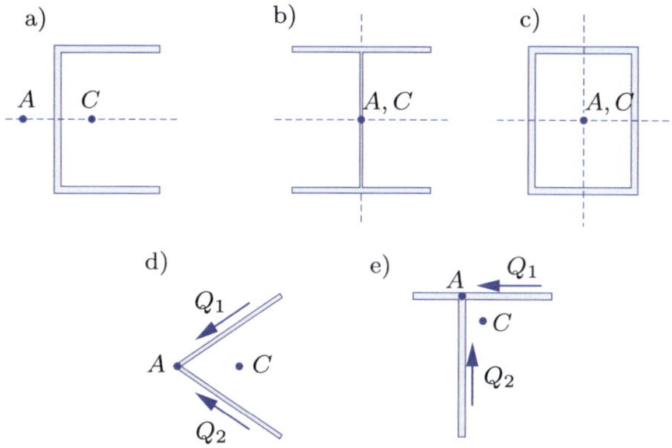

Fig. 11.29: Shear center A and elastic center A for some simple cross-sections.

as the I- and box-profiles considered in Examples 11.6 and 11.7, the shear center is located at the intersection of the two lines of symmetry, whereby it coincides with the elastic center, see Fig. 11.29b,c. In cross-sections composed of straight flanges with only a single point of intersection, as shown in Fig. 11.29d,e, the forces in the flanges intersect at the common intersection point, which is thereby the shear center.

11.2.2 Shear flexibility

The possibility of including shear flexibility in a beam theory was discussed in Section 4.3. The shear flexibility for deformation in the z-direction is expressed by the relation (4.32),

$$Q_z = G_0 A_z \bar{\gamma}_z \,, \tag{11.16}$$

where Q_z is the shear force and $\bar{\gamma}_z$ is the corresponding beam shear strain. In this relation the shear stiffness is represented as the product of a reference shear modulus G_0 and the representative shear area A_z. For a homogeneous cross-section G_0 is simply the shear modulus, and the shear stiffness is thereby represented via the shear area A_z. The calculation of a representative shear area A_z requires knowledge of the shear stress distribution over the cross-section, and was therefore not included in the formulation of the theory in Section 4.3. The present theory can fairly easily be extended to include both shear components and non-symmetric cross-sections. However, this is outside the present scope.

As seen in Section 4.4 the principle of virtual work constitutes a basis for the formulation of consistent beam theories, in which all or only some of the deformation mechanisms are included. Therefore a consistent way of defining and calculating the shear flexibility, and thereby the shear area A_z, is to use the elastic energy of the corresponding deformation mechanism. The elastic energy in a unit length of a beam associated with the shear deformation mechanism can be expressed in terms of the shear force and the representative beam shear strain as

$$\tfrac{1}{2}\bar{\gamma}_z Q_z = \tfrac{1}{2} G_0 A_z \bar{\gamma}_z^2 = \frac{1}{2} \frac{Q_z^2}{G_0 A_z}. \tag{11.17}$$

An approximate shear stress distribution corresponding to the shear force Q_z has been obtained in the previous sections by use of Grashof's formula. Thus, the elastic energy of a unit length of the beam can be expressed as an integral of the elastic energy per unit volume, integrated over the cross-section. By equating this integral with the last expression in (11.17) in terms of the shear force, the following equation is obtained

$$\frac{Q_z^2}{G_0 A_z} = \int_A \frac{\tau^2}{G} \, dA. \tag{11.18}$$

In this equation G is the shear modulus at a particular point in the cross-section, while G_0 is a reference shear modulus. The particular value chosen for G_0 is not important, as it only appears in combination with the shear area as the shear stiffness $G_0 A_z$. The equation (11.18) defines the shear area A_z via the integral

$$\frac{1}{A_z} = \frac{1}{Q_z^2} \int_A \frac{G_0}{G} \tau^2 \, dA. \tag{11.19}$$

For constant shear modulus it is natural to take $G_0 = G$. The integral accounts for the non-uniformity of the shear stress distribution. In the hypothetical case of a completely uniform distribution of the shear stress, the shear area A_z would equal the cross-section area A. For any variation of the shear stress over the cross-section $A_z < A$. Some fairly general conclusions can be drawn from the following simple examples for symmetric cross-sections.

Example 11.8. Shear area of rectangular cross-section. The shear stress distribution over a rectangular cross-section with with dimensions b and h in flexure was determined in Example 11.2 and illustrated in Fig. 11.10. The shear stress $\gamma = \gamma_z$ has a parabolic distribution given by

$$\tau(z) = \frac{3}{2} \frac{Q_z}{A} \Big[1 - \Big(\frac{2z}{h} \Big)^2 \Big]$$

with maximum value $\tau_{\max} = \frac{3}{2} Q_z / A$ for $z = 0$. The shear area A_z is defined via the integral (11.19),

$$\frac{1}{A_z} = \frac{1}{A^2} \Big(\frac{3}{2} \Big)^2 \int_A \Big[1 - \Big(\frac{2z}{h} \Big)^2 \Big]^2 dz.$$

Introduce the area $A = bh$ and the non-dimensional coordinate $\zeta = 2z/h$ to get

$$\frac{1}{A_z} = \frac{1}{A} \frac{9}{8} \int_{-1}^{1} (1 - \zeta^2)^2 \, d\zeta = \frac{6}{5} \frac{1}{A}.$$

Hereby the shear area of a rectangle is determined as

$$A_z = \tfrac{5}{6}A\,.$$

This is slightly smaller than the geometric area, due to the non-uniform shear stress distribution. □

Example 11.9. Shear area of I-profile. Beams with I-profile cross-sections are commonly used, and typically the flanges are thicker than the web. The shear stress distribution of an I-profile of height and width a, web thickness t and flange thickness $2t$ was determined in Example 11.4 and shown in Fig. 11.21b. The shear stress is given in the form

$$\tau(s) = \frac{1}{26}\frac{Q_z}{at}\,f(s)\,,$$

where the non-dimensional distribution $f(s)$ is shown in Fig. 11.21b. It is noted that the maximum value in the flanges is 6, which jumps to 24 in the web, because two parts of the flange meet, and the web has only half the thickness of the flanges. Thus, the shear stress is considerably larger in the web. The shear area A_z is defined via the integral (11.19),

$$\frac{1}{A_z} = \frac{1}{26^2}\frac{1}{a^2t^2}\int_A f(s)^2\,dz = \frac{1}{26^2}\frac{1}{a^2t^2}\Big(4\big(\tfrac{1}{3}6^2\big)\big(\tfrac{1}{2}a\big)(2t) + t\int_{-h/2}^{h/2} f(z)^2\,dz\Big).$$

The first term is the contribution from the flanges, consisting of four pieces of length $\frac{1}{2}a$ with a squared triangular distribution, while the last integral is the contribution from the web. Along the web the parabolic variation is

$$f_{\rm web}(z) = 27 - 3(2z/a)^2 = 27 - 3\zeta^2$$

in terms of the non-dimensional coordinate $\zeta = 2z/a$. The integral over the web then is

$$\frac{1}{a}\int_{-h/2}^{h/2} f(z)^2\,dz = \frac{1}{2}\int_{-1}^{1}\big(27 \quad 3\zeta^2\big)^2\,d\zeta = \frac{9}{5}\,376\,.$$

Upon substitution of this result into the area expression above,

$$\frac{1}{A_z} = \frac{1}{at}\,\frac{6}{5}\,\frac{151}{13^2} = \frac{1.072}{at}\,.$$

This determines the shear area as

$$A_z = 0.933\,at\,.$$

As at is the area of the web, the result suggests that a fair approximation for the shear area is the area of the web – an approximation often used in practice. □

11.3 Torsion of circular cylinders

Torsion of cylinders constitutes a particularly simple case that is treated first as an illustration. The problem is illustrated for a full cylinder in Fig. 11.30. The cylinder is loaded by opposing torsion moments $\pm M_x$ at the ends, generating a constant section moment M_x in the beam – the so-called homogeneous torsion problem. The constant section moment corresponds to a particular distribution of the shear stresses over the cross-section, and the problem is formulated such that the distribution of the shear stresses is identical for all cross-sections. In the present case the axial symmetry of the problem implies that the shear stress distribution depends only on the radial distance r from the axis of the cylinder, but not on the location in the circumferential direction. Furthermore, the shear stress distribution only has a circumferential component, here denoted $\tau(r)$. The identification of the stress distribution $\tau(r)$ is a central part of the problem.

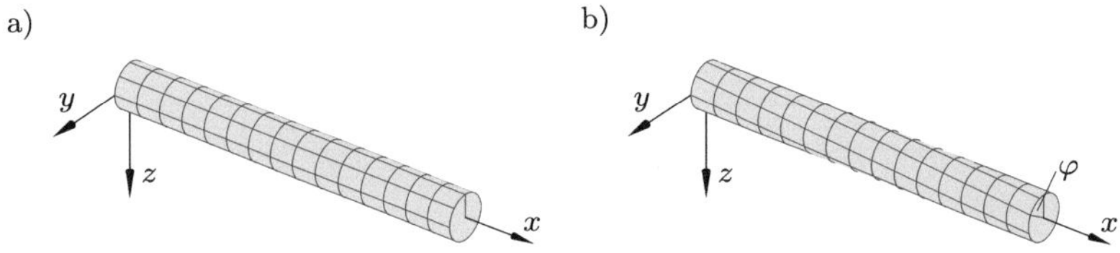

Fig. 11.30: Homogeneous torsion of circular cylinder.

Figure 11.30a shows the circular cylinder in its undeformed state before application of the moments $\pm M_x$. To illustrate the deformation imposed via the torsion moment a grid consisting of lengthwise and circumferential lines is drawn on the surface of the cylinder. Due to symmetry each point on a circumferential line will only move in the circumferential direction, and thus the circle remains a circle in the cross-section plane. It might be imagined that the cross-section develops an axisymmetric warping, but due to symmetry there can be no preference for warping in the positive or the negative axial direction, and thus the cross-sections must remain plane. This leads to the deformation illustrated in Figure 11.30b, where each cross-section remains plane but rotates about the longitudinal x-axis, thereby changing the lines originally parallel to the axis into helices.

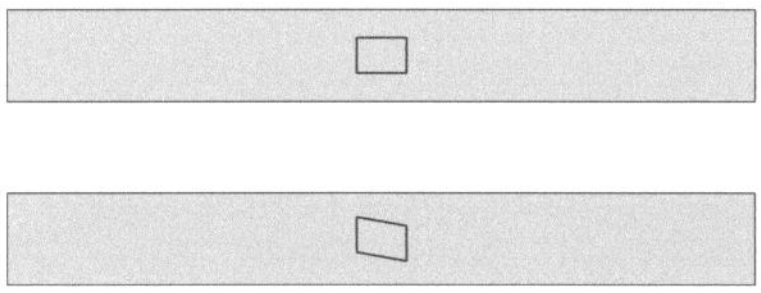

Fig. 11.31: Angle between circumferential trace and generator changes in torsion.

Torsion of the cylinder leads to a deformation field called twist, illustrated in Fig. 11.31. The circumferential displacement component v_θ at the radial distance r from the cylinder axis is given in terms of the local angle of rotation $\varphi(x)$ as

$$v_\theta = r\,\varphi(x)\,. \tag{11.20}$$

This is the only non-vanishing displacement component, and the shear strain in the circumferential direction then follows by differentiation as

$$\gamma_{x\theta} = \frac{dv_\theta}{dx} = r\,\frac{d\varphi}{dx}\,. \tag{11.21}$$

Thus, according to the assumed displacement field (11.20) the only strain is the circumferential shear strain component (11.21), increasing linearly with the distance from the axis of the cylinder.

For an isotropic elastic cylinder the shear stress follows from Hooke's law (9.15) as

$$\tau_{x\theta} = G\gamma_{x\theta}\,, \tag{11.22}$$

where G is the shear modulus, here assumed constant. The linear distribution of the shear strain then implies a linear distribution of the shear stress, and the relations (11.21) and (11.22) can be expressed in compact form in terms of their maximum value, attained at the surface,

$$\frac{\tau_{x\theta}}{\tau_{x\theta}^{\max}} = \frac{\gamma_{x\theta}}{\gamma_{x\theta}^{\max}} = \frac{r}{r_o}\,, \tag{11.23}$$

with $r_o = r_{\max}$ as the radius of the cylinder. The shear stress distribution is illustrated in Fig. 11.32. It is noted that $\tau_{x\theta}$ is the only non-vanishing stress component. This stress distribution leaves the cylinder surface stress free, and thereby confirms that the assumed displacement field (11.20) is indeed correct.

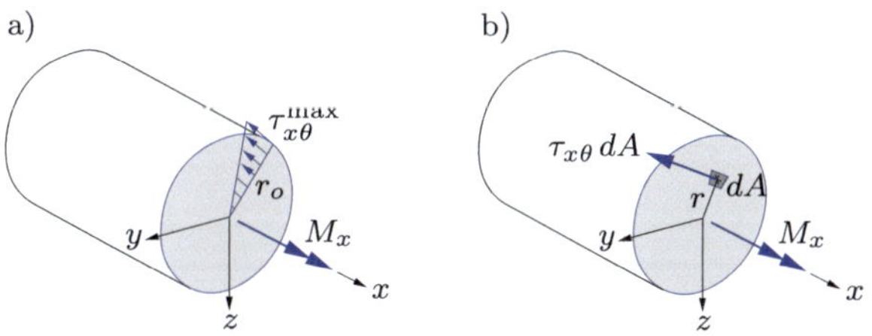

Fig. 11.32: Linear radial dependence of shear stress in cylinder.

The torsion moment M_x is determined by integration of the shear stress distribution multiplied by the moment arm r. When using the linear variation relations (11.23) the moment integral takes the form

$$M_x = \int_A \tau_{x\theta}\, r\; dA = \frac{\tau_{x\theta}^{\max}}{r_o}\int_A r^2\; dA = \frac{\tau_{x\theta}^{\max}}{r_o}\, I_p\,, \tag{11.24}$$

with I_p denoting the polar moment of inertia,

$$I_p = \int_A r^2\; dA\,. \tag{11.25}$$

It follows from (11.24) that the maximum shear stress is given by

$$\tau_{x\theta}^{\max} = \frac{r_o\, M_x}{I_p}\,, \tag{11.26}$$

and it furthermore follows from the linear variation of the shear stress that the stress at radius r is

$$\tau_{x\theta}(r) = \frac{r\, M_x}{I_p}\,. \tag{11.27}$$

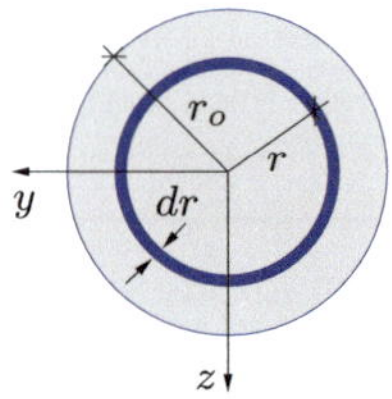

Fig. 11.33: Polar representation and integration.

The integral for the polar moment of inertia (11.25) is calculated by dividing the circle into rings of width dr with area $dA = 2\pi r\, dr$ as illustrated in Fig. 11.33, whereby

$$I_p = \int_0^{r_o} r^2 (2\pi r\, dr) = 2\pi \int_0^{r_o} r^3\, dr = \frac{\pi}{2}\, r_o^4\,. \tag{11.28}$$

This completes the determination of the stress distribution in a cylinder subjected to homogeneous torsion by the moment M_x.

Torsion stiffness

The torsion stiffness of a beam relates the rate of twist $\varphi' = d\varphi/dx$ to the torsion moment M_x. It is evaluated by substituting the strain distribution (11.21) into the stress integral (11.24) for the torsion moment M_x,

$$M_x = \int_A \tau_{x\theta}\, r\, dA = \int_A (G\gamma_{x\theta})\, r\, dA = G\Big(\int_A r^2\, dA\Big)\frac{d\varphi}{dx}\,. \tag{11.29}$$

The integral is recognized as the polar moment of inertia I_p, whereby the relation takes the form

$$M_x = G\,K\,\frac{d\varphi}{dx} = G\,I_p\,\frac{d\varphi}{dx}\,. \tag{11.30}$$

K is the torsion stiffness with the value $K = I_p = \frac{1}{2}\pi r_o^4$ for a homogeneous isotropic elastic cylinder. It should be noted that for general cross-sections the torsion stiffness K can not be expressed in simple terms via the polar moment of inertia, but requires an independent evaluation as discussed in the following sections.

Hollow cylinder

The stress distribution in a cylinder in homogeneous torsion does not generate stresses on the exterior surface of the cylinder. Similarly it does not generate stresses on any smaller cylindrical surface. As a consequence, the stress state in a hollow cylinder may be deduced directly from that of a similar full cylinder with the same outer radius as illustrated in Fig. 11.34.

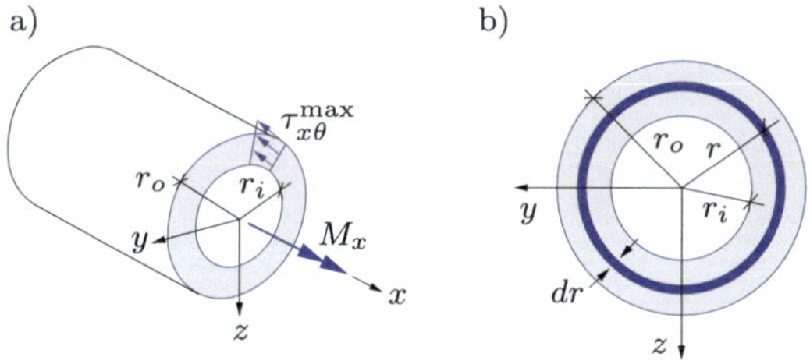

Fig. 11.34: Linear variation of stress $\tau_{x\theta}$ and strain $\gamma_{x\theta}$ in a hollow cylinder.

The moment M_x can be expressed in terms of the stress integral (11.24) taken over the area of the cross-section,

$$M_x = \int_A (G\gamma_{x\theta})\, r\, dA = G\Big(\int_A r^2\, dA\Big)\frac{d\varphi}{dx} = G\, I_p\, \frac{d\varphi}{dx}. \tag{11.31}$$

Here I_p is now the polar moment of the hollow cylinder, given by

$$I_p = \int_{r_i}^{r_o} r^2(2\pi\, r)\, dr = 2\pi \int_{r_i}^{r_o} r^3\, dr = \frac{\pi}{2}\big(r_o^4 - r_i^4\big). \tag{11.32}$$

Thus $K = I_p$, also for a hollow circular section.

An impression of the polar moment of inertia may be gained by writing it in the form

$$I_p = \frac{\pi}{2}\big(r_o^4 - r_i^4\big) = \frac{\pi}{2}\big(r_o^2 + r_i^2\big)\big(r_o^2 - r_i^2\big) = \tfrac{1}{2}\big(r_o^2 + r_i^2\big)\, A\,, \tag{11.33}$$

where A is the cross-section area. In this formula it is seen that the polar moment of inertia is expressed as the product of area and the square of a characteristic radius – the radius of gyration. The square of the radius of gyration is obtained by the mean value of the squares of the outer and the inner radius.

11.4 General homogeneous torsion of beams

In the homogeneous torsion problem for a cylinder the cross-sections remain plane during deformation. This is a special case, and for beams with non-cylindrical cross-section the cross-sections exhibit deformation in the axial direction, the so-called warping. This effect determines the distribution of stresses and strains over the cross-section, and therefore constitutes a central part of the torsion problem.

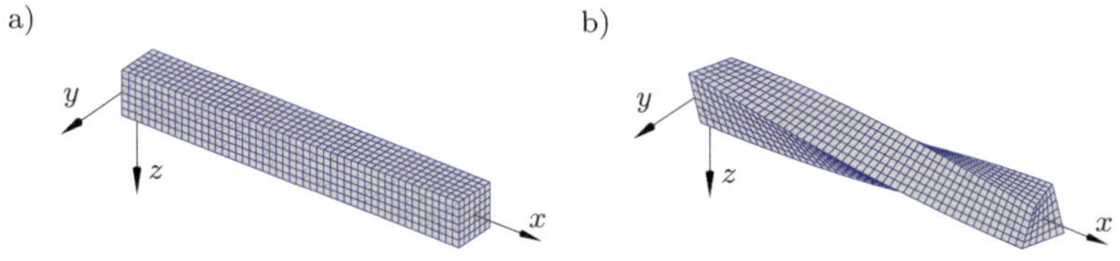

Fig. 11.35: Torsion of beam with square cross-section.

The warping effect is illustrated in Fig. 11.35, showing homogeneous torsion of a beam with square cross-section. The circumferential curves, originally designating plane cross-sections, deform axially in warping, and the angle between the axial and the circumferential lines in the deformed state depends on the location on the cross-section contour. At the corners the shear stress components vanish, and thus the two sets of curves remain orthogonal here, while they develop an angle representing the shear strain with maximum at the center of the sides. The shear stress distribution is illustrated in Fig. 11.36a, with maximum at the center of the sides and vanishing at the corners. The corresponding warping of the cross-section is illustrated in Fig. 11.36b, showing a characteristic S-shape of the originally straight sides of the cross-section.

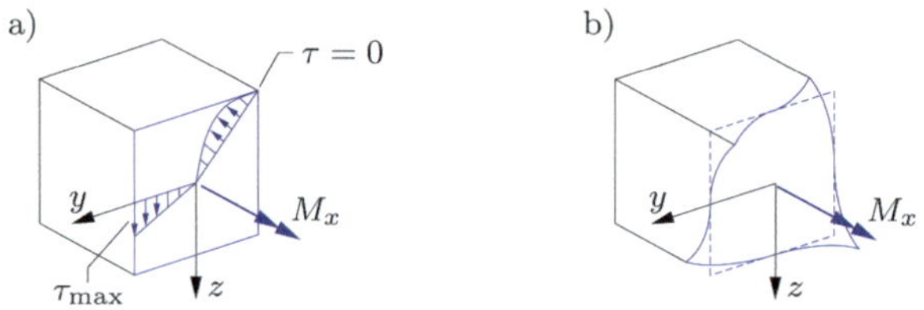

Fig. 11.36: a) Shear stress distribution and b) warping in a square cross-section in torsion.

The analysis of the shear stress and shear strain distribution over the cross-section can take two different routes: either a kinematic analysis in which the warping of the cross-section is determined first, or a static analysis in which the shear stress distribution is determined via a potential function. The stress potential formulation is often most convenient for hand calculation type analysis, while the kinematic analysis is more easily adapted to numerical analysis of general cross-sections. Here, the basic kinematic relations are derived, and the stress potential method is then treated in detail in Section 11.4.1.

Kinematics of homogeneous torsion

The theory of the homogeneous torsion problem of homogeneous elastic beams with arbitrary cross-section was initially formulated by BARRÉ DE ST. VENANT in 1855, see e.g Timoshenko (1983). The basic idea is that there exists a state of homogeneous torsion in which the distribution of stresses and strains are identical for all cross-sections. It can furthermore be shown that this homogeneous torsion problem does not involve deformation of the cross-sections in their own plane, but only an in-plane rotation as a rigid body. In addition each cross-section exhibits identical axial deformation, the so-called warping. A displacement field with these properties can be described by an in-plane displacement corresponding to a rotation φ of the cross-sections about a point A with coordinates $[a_y, a_z]$, as illustrated in Fig. 11.37. This in-plane displacement has the form

$$v = -(z - a_z)\,\varphi\,, \qquad w = (y - a_y)\,\varphi\,, \tag{11.34}$$

where $\varphi(x)$ is the angle of rotation of the cross-section at the axial coordinate x. A constant value of the rotation φ simply represents a rigid body rotation about an axis through A. Twist is generated when the rotation angle φ depends on x, and the axial displacement is proportional to the rate of twist $\varphi' = d\varphi/dx$, as expressed by the relation

$$u = \omega(y, z)\,\varphi', \tag{11.35}$$

where the function $\omega(y, z)$ describes the warping of the cross-section. In homogeneous torsion the rate of twist φ' is constant and each cross-section exhibits identical warping.

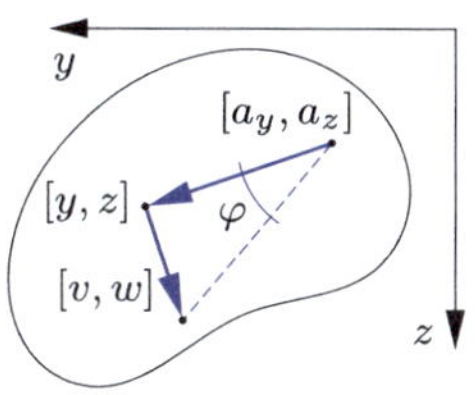

Fig. 11.37: Infinitesimal cross-section rotation about the point A.

The in-plane displacement field (11.34) corresponds to an infinitesimally small rigid-body rotation, and thus the three in-plane strain components vanish, $\varepsilon_{yy} = \varepsilon_{zz} = \varepsilon_{yz} = 0$. In homogeneous torsion the rate of twist φ' is constant, and thus also the axial strain vanishes, $\varepsilon_{xx} = 0$. The only non-vanishing strain components γ_{xy} and γ_{xz} can then be obtained from (8.34) in the form

$$\begin{aligned} \gamma_{xy} &= \frac{\partial u}{\partial y} + \frac{\partial v}{\partial x} = \Big[\frac{\partial \omega}{\partial y} - (z - a_z)\Big]\varphi' , \\ \gamma_{xz} &= \frac{\partial u}{\partial z} + \frac{\partial w}{\partial x} = \Big[\frac{\partial \omega}{\partial z} + (y - a_y)\Big]\varphi' . \end{aligned} \tag{11.36}$$

For constant rate of twist these strains are identical for all cross-sections of the beam. The shear strain distribution is determined by the gradient of the warping function $\omega(y,z)$ and the position $[a_y, a_z]$ of the axis of twist. The warping function in turn is determined by local equilibrium.

Equilibrium in homogeneous torsion

It follows from the kinematics of homogeneous torsion presented above that the only non-vanishing strain components are γ_{xy} and γ_{xz}. For an isotropic elastic beam the corresponding shear stress components are

$$\tau_{xy} = G\,\gamma_{xy} , \qquad \tau_{xz} = G\,\gamma_{xz}, \tag{11.37}$$

where G is the shear modulus. These stresses must satisfy the axial equilibrium equation (8.19a). In the present case the axial stress σ_{xx} vanishes, and the equilibrium equation then takes the simple form

$$\frac{\partial \tau_{xy}}{\partial y} + \frac{\partial \tau_{xz}}{\partial z} = 0 . \tag{11.38}$$

This equilibrium equation is illustrated in Fig. 11.38, showing a prismatic cut-out with side lengths dy and dz. The equilibrium equation (11.38) is obtained by summation of the forces on the four sides and normalizing with the area $dA = dy\,dz$.

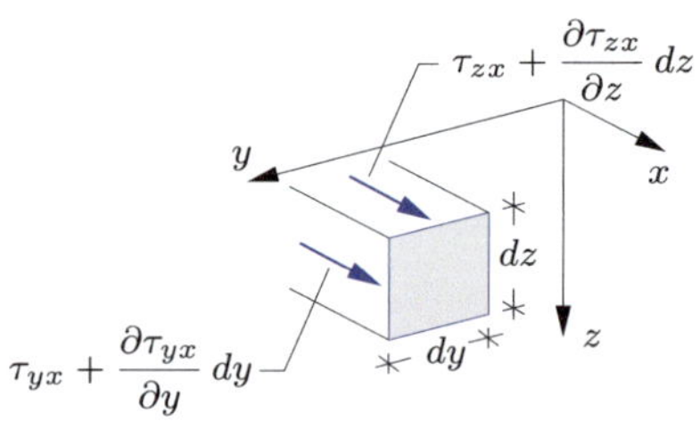

Fig. 11.38: Equilibrium in the axial direction.

In addition to the equilibrium equation the stress distribution must also satisfy the boundary condition of vanishing stress on the free surfaces. The surface of the beam has the normal vector $[0, n_y, n_z]$, and thus the axial component of the surface stress vector follows from (8.11a) as

$$\tau_n = \big[\,\tau_{xy}, \tau_{xz}\big]\begin{bmatrix} n_y \\ n_z \end{bmatrix} = 0\,. \tag{11.39}$$

The combination of the internal axial equilibrium equation (11.38) and the boundary condition (11.39) constitutes the static equations of the homogeneous torsion problem.

The equilibrium conditions involve the two stress components τ_{xy} and τ_{xz}. When the stresses are represented in terms of the shear strains by (11.37), and the strains are represented in terms of the warping function by (11.36), the following partial differential equation is obtained for the warping function of a beam with constant shear modulus G,

$$\frac{\partial^2\omega}{\partial y^2} + \frac{\partial^2\omega}{\partial z^2} = 0\,. \tag{11.40}$$

This is the two-dimensional Laplace equation for which several solution methods exist. However, the boundary conditions obtained by introducing the strain representation (11.36) into the boundary condition (11.39) depend on the particular properties of the boundary contour in a non-standard way, and direct analytical solution of torsion problems in terms of the warping function $\omega(y,z)$ can be rather complicated. Analytical solutions to the torsion problem are typically easier to obtain by introducing a stress potential – the Prandtl stress function – as described in the following sections. However, the kinematic formulation in terms of the warping function is eminently suitable in the context of numerical solution by the Finite Element Method with appropriate boundary conditions.

11.4.1 The Prandtl stress function

While the problem of homogeneous torsion of elastic beams was originally formulated together with selected analytical solutions by Barré de St. Venant in 1855, the alternative formulation in terms of a stress potential was developed later by L. Prandtl in 1903 – see e.g. Timoshenko (1983). In the kinematic approach of St. Venant the key is the representation (11.36) of the strain components in terms of the gradient of the warping function plus some extra explicit terms. In Prandtl's approach this is replaced by a representation of the stress components as derivatives of a potential $\Phi(y,z)$ in such a way that the equilibrium equation (11.38) is satisfied identically. This leads to the potential representation

$$\tau_{xy} = \varphi'\,\frac{\partial\Phi}{\partial z}\,, \qquad \tau_{xz} = -\,\varphi'\,\frac{\partial\Phi}{\partial y} \tag{11.41}$$

in terms of the so-called Prandtl stress function $\Phi(y,z)$. The rate of twist φ' appears as a scaling factor.

While the equilibrium equation has now been satisfied identically, it is necessary to ensure that the corresponding stresses, and thereby strains, can be derived from a suitably chosen displacement field. This is the compatibility condition. This condition follows from the kinematic relations (11.36) by elimination of the warping function $\omega(y,z)$,

$$\frac{\partial \gamma_{xy}}{\partial z} - \frac{\partial \gamma_{xz}}{\partial y} = -\,2\varphi'. \tag{11.42}$$

The strains are now expressed in terms of the corresponding stress components via Hooke's law in the form (11.37), and the stresses are expressed in terms of the Prandtl stress function by (11.41). In the case of constant shear modulus G the resulting equation for the Prandtl stress function is

$$\frac{\partial^2 \Phi}{\partial y^2} + \frac{\partial^2 \Phi}{\partial z^2} = -\,2G. \tag{11.43}$$

This is the non-homogeneous equivalent of the Laplace equation, called the Poisson equation. Like the Laplace equation this is one of the classic equations of mathematical physics, and several analytical and numerical solution methods have been developed.

An important aspect of the stress function approach is the associated boundary condition. The shear stress at the bounding surface was expressed by (11.39). In this relation the stresses are expressed in terms of the stress function, and the normal vector components are introduced in the form $[n_y, n_z] = [dz/ds, -dy/ds]$. Hereby the boundary condition takes the form

$$\tau_n = \tau_{xy} n_y + \tau_{xz} n_z = \varphi' \Big(\frac{\partial \Phi}{\partial z}\frac{dz}{ds} + \frac{\partial \Phi}{\partial y}\frac{dy}{ds} \Big) = \varphi' \, \frac{d\Phi}{ds} = 0. \tag{11.44}$$

It follows from this equation that $\Phi = const$ on any connected part of the bounding surface. For a single connected cross-section – i.e. a cross-section without holes – the boundary condition can be introduced in the form $\Phi = 0$. For a multiply connected cross-section – i.e. a cross-section with one or more holes – the stress function Φ takes a constant value on each of the closed bounding contours, but the value may be different for the different contours, and the determination of the individual values is part of the problem to be solved, a problem discussed in more detail in Section 11.5.3 on multi-cell cross-sections.

Torsion stiffness

The torsion stiffness is determined by integrating the contributions from the stresses to the moment M_x over the cross-section as illustrated in Fig. 11.39. The stress vector $[\tau_y, \tau_z]$ acts at the infinitesimal area $dA = dy\, dz$ located at $[y, z]$. The moment about the point $[a_y, a_z]$ then follows from integration over the cross-section area as

$$\begin{aligned} M_x &= \int_A \Big(\tau_{xz}(y - a_y) \;-\; \tau_{xy}(z - a_z)\Big)\, dA \\ &= -\,\varphi' \int_A \Big((y - a_y)\frac{\partial \Phi}{\partial y} \;+\; (z - a_z)\frac{\partial \Phi}{\partial z}\Big)\, dA \end{aligned} \tag{11.45}$$

where the stress components have been represented via the Prandtl stress function (11.41).

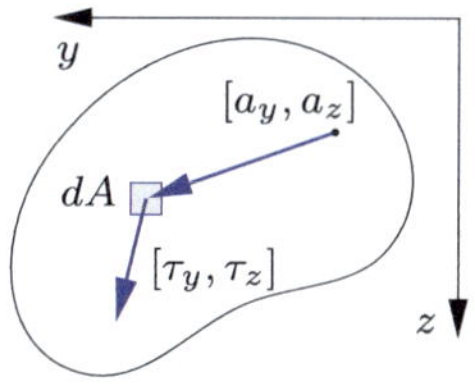

Fig. 11.39: Integration of stresses for torsion moment.

The subscript notation with $\alpha = 2, 3$ and $[y, z] = [x_2, x_3]$ is now introduced in order to simplify the following use of the divergence theorem. First the integrand is reformulated by a simple identity,

$$\begin{aligned} M_x &= -\,\varphi' \int_A \Big((x_\alpha - a_\alpha)\frac{\partial \Phi}{\partial x_\alpha}\Big)\, dA \\ &= \varphi' \int_A \Big(2\Phi \;-\; \frac{\partial}{\partial x_\alpha}\big[(x_\alpha - a_\alpha)\Phi\big]\Big)\, dA\,, \end{aligned} \tag{11.46}$$

where terms containing repeated subscripts α are added with $\alpha = 2$ and $\alpha = 3$. The second term in the integrand is identified as the divergence of the product in the square brackets, and the divergence theorem is used to transform this term into a flux through the boundary,

$$M_x \;=\; \varphi'\Big(2\int_A \Phi\, dA \;-\; \int_C \Phi\,(x_\alpha - a_\alpha) n_\alpha\, ds\Big)\,. \tag{11.47}$$

The boundary conditions on on the potential function Φ are now introduced by choosing $\Phi = 0$ on the exterior boundary C_0, and $\Phi = \Phi_i$ on any internal boundaries C_i. The contour integral can then be evaluated by use of the divergence theorem on the areas A_i within the internal contours,

$$\begin{aligned}\int_C \Phi\,(x_\alpha - a_\alpha)n_\alpha\, ds &= \sum_i \Phi_i \int_{C_i} (x_\alpha - a_\alpha)n_\alpha\, ds \\ &= -\sum_i \Phi_i \int_{A_i} \frac{\partial(x_\alpha - a_\alpha)}{\partial x_\alpha}\, dA = \sum_i 2\Phi_i A_i .\end{aligned} \tag{11.48}$$

The last equality follows from the differentiation rule $\partial x_\alpha/\partial x_\alpha = \partial x_2/\partial x_2 + \partial x_3/\partial x_3 = 2$, where the contributions from $\alpha = 2$ and $\alpha = 3$ are added, because the subscript is repeated. The potential function Φ has the boundary value Φ_i on the contour C_i surrounding the cut-out area A_i, and it may therefore be extended to have this value within the area A_i. When substituting the last term in the torsion moment expression (11.47) from (11.48) the result takes the simple form

$$M_x = 2\varphi' \int_{A_0} \Phi\, dA \tag{11.49}$$

where the integral is over the total area A_0 contained within the exterior contour C_0 of the cross-section. The corresponding torsion stiffness is

$$GK = 2\int_{A_0} \Phi\, dA\,. \tag{11.50}$$

The form of this formula suggests that the torsion stiffness of a cross-section can be increased by giving it a cellular shape with maximum enclosed area. This effect is clearly illustrated by the torsion stiffness of a hollow cylinder given by I_p in (11.33).

Example 11.10. Elliptic cross-section. Consider an elliptic cross-section with semi-axes a and b. The elliptic contour is represented by the equation

$$\frac{y^2}{a^2} + \frac{z^2}{b^2} = 1$$

and is illustrated in Fig. 11.40.

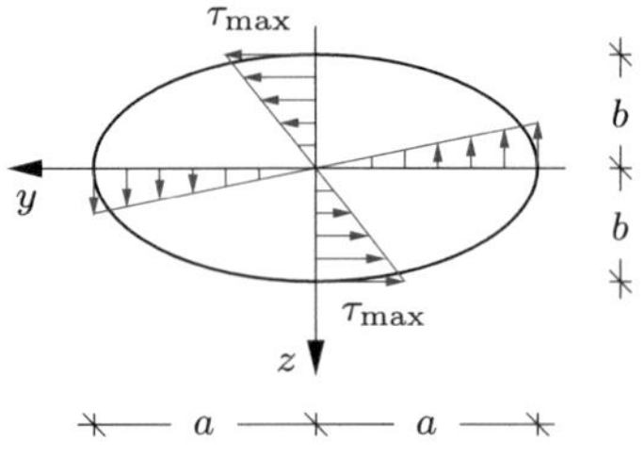

Fig. 11.40: Torsion of elliptic cross-section.

The Prandtl stress function satisfies the boundary condition $\Phi = 0$ on the contour C_0 and according to the Poisson equation (11.43) the sum of the second derivatives must be constant. These conditions are satisfied by the function

$$\Phi = C\left[1 - \frac{y^2}{a^2} - \frac{z^2}{b^2}\right].$$

The boundary condition is seen to be satisfied and the differential equation gives

$$\frac{\partial^2\Phi}{\partial y^2} + \frac{\partial^2\Phi}{\partial z^2} = -2C\left[\frac{1}{a^2} + \frac{1}{b^2}\right] = -2G,$$

from which the constant C is determined as

$$C = \frac{a^2b^2}{a^2+b^2}\,G\,.$$

The torsion stiffness then follows from evaluation of the area integral (11.50),

$$K = \frac{2}{G}\int_A \Phi\, dA = \frac{2a^2b^2}{a^2+b^2}\int_A\left[1 - \frac{y^2}{a^2} - \frac{z^2}{b^2}\right]dA = \frac{2a^2b^2}{a^2+b^2}\,A\left[1 - \frac{1}{4} - \frac{1}{4}\right].$$

With the area of the ellipse $A = \pi ab$ this gives the torsion stiffness parameter

$$K = \frac{\pi a^3 b^3}{a^2+b^2}\,.$$

The special case of the circle $a = b$ has already treated in Section 11.3 with $K = \frac{1}{2}\pi a^4$.

The stress distribution follows from differentiation of the stress potential as

$$[\tau_{xy}, \tau_{xz}] = \varphi'\left[\frac{\partial\Phi}{\partial z}, -\frac{\partial\Phi}{\partial y}\right] = \frac{a^2b^2}{a^2+b^2}\left[\frac{-2z}{b^2}, \frac{2y}{a^2}\right]G\varphi'\,.$$

The stress components increase linearly with distance from the center of the cross-section, as illustrated in Fig. 11.40. The maximum stress is located at the ends of the minor axis, and with $b < a$ the maximum shear stress is

$$\tau_{xy}^{\max} = \frac{2a^2b}{a^2+b^2}\,G\varphi' = \frac{2}{\pi}\frac{M_x}{ab^2} = \frac{2}{b}\frac{M_x}{A}\,.$$

It follows from the last expression that the maximum stress increases with decreasing width of the ellipse. □

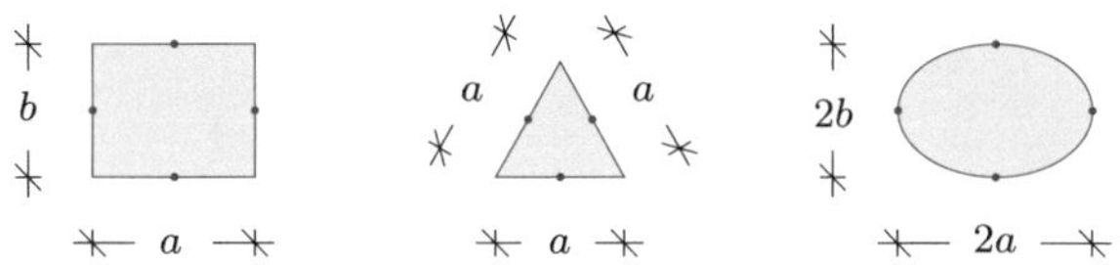

$$\tau_{\max} = 4.8\,\frac{M_x}{a^3} \qquad \tau_{\max} = 20\,\frac{M_x}{a^3} \qquad \tau_{\max} = 2\,\frac{M_x}{\pi ab^2}$$

$$\varphi = 7.10\,\frac{M_x\ell}{a^4G} \qquad \varphi = 46\,\frac{M_x\ell}{a^4G} \qquad \varphi = (a^2+b^2)\frac{M_x\ell}{\pi a^3b^3G}$$

Fig. 11.41: Maximum shear stress and torsion stiffness for three cross-sections.

There are rather few cross-sections whose torsion stiffness and stress distribution can be determined by analytic means. Three examples are summarized in Fig. 11.41: a rectangle, an equilateral triangle, and an ellipse. Details of the analysis may be found in e.g. Timoshenko and Goodier (1970) or Boresi et al. (2011).

11.5 Torsion of thin-walled beams

Beams with thin-walled cross-section find wide application in engineering structures as they often exhibit efficient use of the material by placing the material in its most useful location in the cross-section. In bending this typically implies the use of sections with well developed flanges connected by one or more webs of more moderate dimension. Although torsion is often a secondary load it is important to ensure the necessary strength and stiffness also for this load. The torsion problem for beams with thin-walled cross-section is presented in three steps. First the case of an open cross-section such as I-beams and typical cold-formed sections of e.g. C-shape is dealt with and results for the stress distribution and stiffness are derived. If the walls of the cross-section are joined to form closed cells torsion generates a stress flow around these cells that contributes to a substantial increase of the torsion stiffness and a corresponding reduction of the maximum stress. The simple case of a single-cell cross-section is treated first, and then extended to the general case of multi-cell cross-sections.

11.5.1 Open sections

The basic case of a thin-walled cross-section is shown in Fig. 11.42 in the form of a rectangle of length ℓ along the z-axis and width t along the y-axis with $t \ll \ell$. The Prandtl stress function Φ is determined by the differential equation

$$\frac{\partial^2 \Phi}{\partial y^2} + \frac{\partial^2 \Phi}{\partial z^2} = -2G. \tag{11.51}$$

As illustrated in the figure, the curvature of the surface formed by the function $\Phi(y,z)$ is much larger in the transverse direction, whereby

$$\frac{\partial^2 \Phi}{\partial z^2} \ll \frac{\partial^2 \Phi}{\partial y^2}. \tag{11.52}$$

When neglecting the contribution from the lengthwise curvature this gives the approximate differential equation

$$\frac{\partial^2 \Phi}{\partial y^2} \simeq -2G, \tag{11.53}$$

with the simple solution

$$\Phi \simeq G\left(\tfrac{1}{4}t^2 - y^2\right). \tag{11.54}$$

Clearly, this approximation loses its validity near the ends of the cross-section, while it is fairly representative over the central part, forming most of the section.

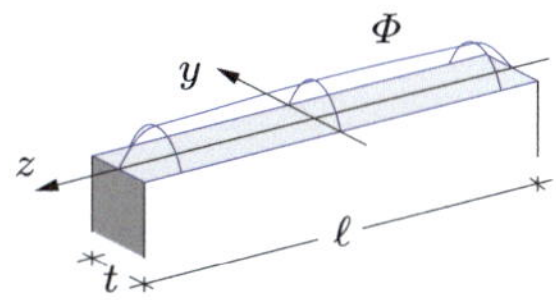

Fig. 11.42: Prandtl's stress function for thin-walled rectangular cross-section.

The torsion stiffness follows from (11.50) as the integral of the stress function, and substitution of the approximate expression (11.54) gives

$$GK = 2\int_A \Phi\, dA \simeq 2\ell \int_{-t/2}^{t/2} G\left(\tfrac{1}{4}t^2 - y^2\right) dy\,, \tag{11.55}$$

whereby an approximate expression for the stiffness parameter is

$$K = \tfrac{1}{3}\ell\, t^3. \tag{11.56}$$

This formula also finds application to open thin-walled cross-sections of more general shape such as e.g. I-, T, and C-shaped sections, where the different parts may have different wall thickness as well as different shear modulus. The general formula is

$$GK = \sum \tfrac{1}{3} G_j \ell_j t_j^3\,, \tag{11.57}$$

with the subscript j indicating properties associated with the part j of the cross-section. The shear modulus G is a selected reference value, only used in the context GK. It follows from the formula (11.56) – and its generalized form (11.57) – that the torsion stiffness parameter K for an open thin-walled section is of the order $\ell\, t^3$. In contrast, the polar moment of inertia I_p is of order $\ell^3 t$, and thus $K \ll I_p$ for thin-walled open cross-sections.

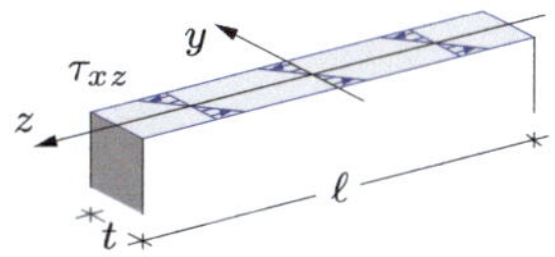

Fig. 11.43: Stress distribution in thin-walled rectangular cross-section.

An approximate representation of the stress distribution follows from the Prandtl stress function (11.54) by differentiation,

$$\tau_{xz} = -\varphi' \frac{\partial \Phi}{\partial y} \simeq 2y\, G\varphi' , \qquad \tau_{xy} = \varphi' \frac{\partial \Phi}{\partial z} \ll \tau_{xz}, \tag{11.58}$$

where the last result follows from the smaller change of Φ in the lengthwise direction. The maximum shear stress will occur at the center of the sides of the section. At these points the approximation is rather accurate, giving the maximum shear stress as

$$\tau_{xz}^{\max} \simeq t\, G\varphi' . \tag{11.59}$$

This corresponds to the maximum stress in an elliptical cross-section in the limiting case of a very slender section.

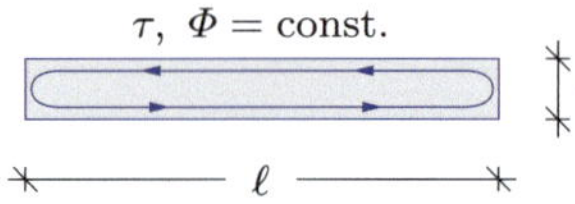

Fig. 11.44: Shear flow in thin-walled rectangular cross-section.

Additional qualitative information about the shear stress distribution can be gained from a sketch of the contours $\Phi = const$ shown in Fig. 11.44. It follows from the definition (11.41) of the stress function, that the shear stress at any point is in the direction of the corresponding contour at that point. The contours $\Phi = const$ are closed, and thus it is seen from Fig. 11.44 that there will be transverse shear stresses near the ends of the section. These stress components are neglected, when deriving the stress distribution from the approximate stress function (11.54).

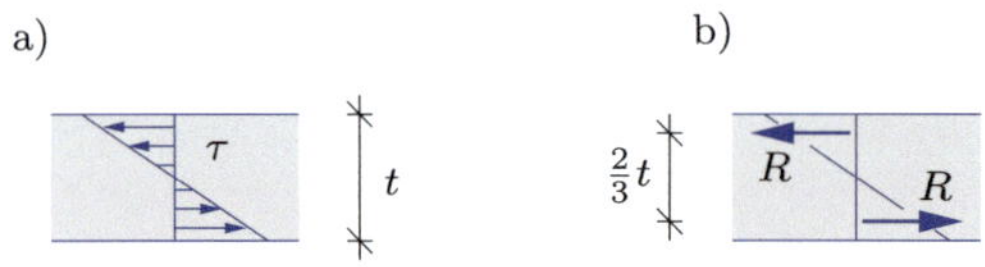

Fig. 11.45: Moment of primary shear stresses.

It is important to realize that the torsion stiffness is best evaluated directly as the integral of the stress function Φ over the cross-section area, and not by integration of the approximate stress field without the transverse stress components. The approximate stress field (11.58) corresponds to a linear variation of the shear stress over the thickness as illustrated in Fig. 11.45. This stress distribution corresponds to resultants of the two triangular parts of magnitude $R = \frac{1}{4} t\tau_{\max}$, acting at a distance of $\frac{2}{3}t$. The moment from this approximate stress distribution is

$$M_x^* = \tfrac{1}{6}\ell t^2 \tau_{\max} = \tfrac{1}{6}\ell\, t^3\, G\,\varphi' = \tfrac{1}{2} GK\,\varphi' \tag{11.60}$$

where $\tau_{\max}$ has been substituted from (11.59) and K from (11.56). Thus, the lengthwise stress components determined by (11.58) only account for half of the torsion stiffness. The other half is contributed by the transverse components that, although relatively smaller, act at a greater distance.

Example 11.11. Thin-walled T-profile. Figure 11.46a shows a T-profile with height h and width b. The thickness of the vertical web is t_w, while the thickness of the flange is t_f. The cross-section is assumed thin-walled, so that $t_w, t_f \ll h, b$. The cross-section is loaded in torsion by the moment M_x.

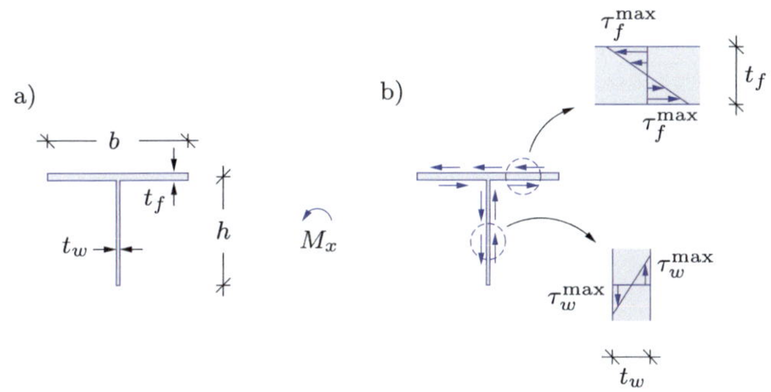

Fig. 11.46: Thin-walled T-section.

The torsion stiffness for cross-sections composed of thin-walled flanges with different thickness is obtained by summation of the contribution from each individual flange,

$$GK = \tfrac{1}{3}\sum G\,\ell_j\,t_j^3 = \tfrac{1}{3}G\big(h\,t_w^3 + b\,t_f^3\big).$$

In each part of the cross-section the shear stress has a linear variation between $\pm\tau^{\max}$. The maximum shear stress is determined by (11.58), using the local wall thickness. For the flange this gives

$$\tau_f^{\max} = t_f\,G\,\varphi' = \frac{t_f}{K}\,M_x\,,$$

while for the web

$$\tau_w^{\max} = t_w\,G\,\varphi' = \frac{t_w}{K}\,M_x\,.$$

The magnitude and direction of the shear stress is shown in Fig. 11.46b. Note, that in an open thin-walled cross-section the largest stress occurs at the greatest wall thickness. □

Example 11.12. Thin-walled C-profile. Figure 11.47a shows a C-profile with height h and width b. The thickness of the vertical web is $t_w = t$, while the thickness of the horizontal flanges is $t_f = 2t$. The cross-section is assumed thin-walled, so that $t_w, t_f \ll h, b$.

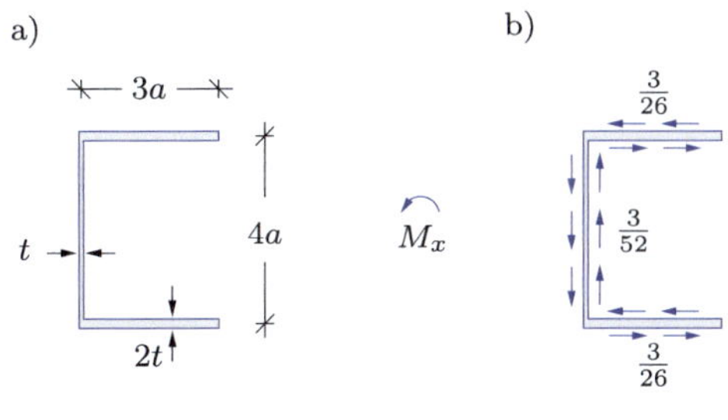

Fig. 11.47: a) Thin-walled C-profile, b) normalized stress distribution: $\tau a t^2/M_x$.

As in the previous example the torsion stiffness GK is found by summation of the contribution from the two flanges and the web,

$$GK = \tfrac{1}{3}\sum G\,\ell_j\,t_j^3 = \tfrac{1}{3}G\big(4a\,t^3 + 2(3a)\,(2t)^3\big) = \tfrac{52}{3}\,G\,at^3\,,$$

where the main contribution comes from the two flanges with thickness $2t$.

The maximum stress in the flange follows from (11.59) as

$$\tau_f^{\max} = t_f\,G\varphi' = \frac{t_f}{K}\,M_x = \frac{3}{26}\,\frac{M_x}{at^2}\,,$$

while for the web it is

$$\tau_w^{\max} = t_w\,G\varphi' = \frac{t_w}{K}\,M_x = \frac{3}{52}\,\frac{M_x}{at^2}\,.$$

The maximum shear stress ratio is identical to the ratio of the web thickness, and thus the maximum shear stress of the present C-profile occurs in the flanges. □

11.5.2 Single-cell sections

Figure 11.48a shows a thin-walled beam with single-cell cross-section, loaded by a torsion moment M_x. The shear stresses are dominated by the component τ parallel to the wall. It turns out to be convenient to introduce the shear flow q as the integral of this shear stress τ over the wall thickness,

$$q = \int_t \tau\,dn = \tau_m\,t \tag{11.61}$$

where n is a coordinate normal to the wall, and τ_m the mean value of the shear stress at the location.

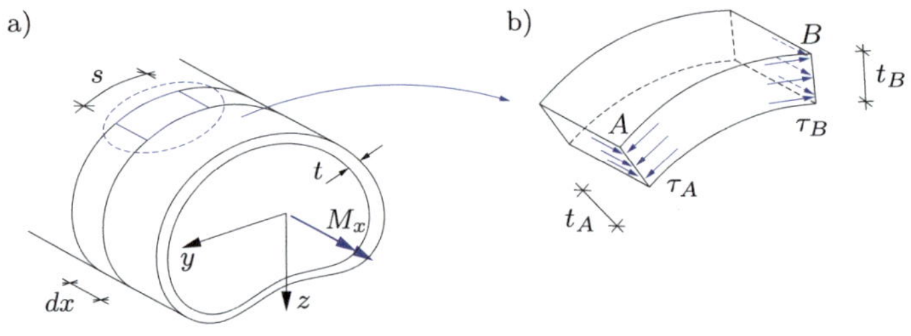

Fig. 11.48: Thin-walled single-cell cross-section.

The shear flow has an important property, identified by considering equilibrium of a small part of the wall illustrated in Fig. 11.48b. This part is limited by cross-sections at x and $x + dx$ long the beam axis and contains the part of the wall between the points A and B. It follows from symmetry of the shear stress components that the component τ_{xs} acting on the cross-section is equal to the component τ_{sx} acting on the axial section, e.g. at B. As there is no axial stress σ_{xx} the integral of the stresses over the wall thickness at A and B must be of equal magnitude in order to uphold axial equilibrium of the small part of the wall shown in Fig. 11.48b. Thus, the shear flow can be evaluated either at A or at B as

$$q = \int_A \tau \, dn = \int_B \tau \, dn \,. \tag{11.62}$$

As A and B represent arbitrary points along the cell wall, it follows that the shear flow q is a constant.

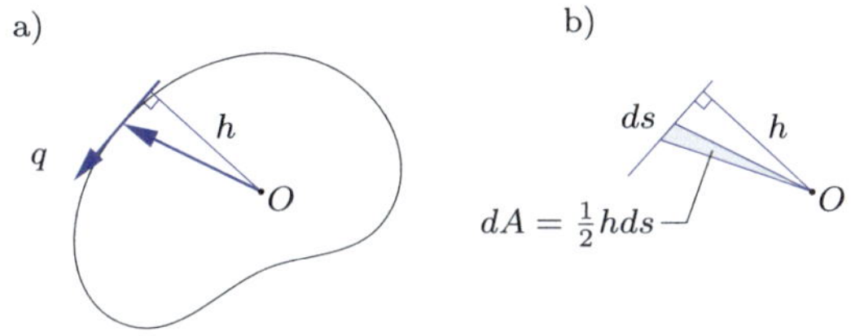

Fig. 11.49: Shear flow integral as cell area.

The relation between the shear flow q and the torsion moment M_x is found by integrating the contributions along the contour of the cell as illustrated in Fig. 11.49. The force on an infinitesimal part of length ds along the contour is $dF = \tau_m \, t \, ds = q \, ds$. This infinitesimal force contributes to the moment about a point O in the cross-section via the distance h of this point from the line of action of the force as illustrated in Fig. 11.49a. The total moment is found by integration along the cell contour,

$$M_x = \oint_C h \, dF = q \oint_C h \, ds = 2q \, A_m \tag{11.63}$$

where the last equality follows from the observation that $\frac{1}{2}h \, ds = dA$ is the infinitesimal area spanned by the part ds along the cell contour as illustrated in Fig. 11.49b. A_m denotes a mean area of the cell, typically corresponding to a center line of the cell wall.

When the torsion moment M_x is given, the shear flow q follows from (11.63) as

$$q = \frac{M_x}{2A_m} \,. \tag{11.64}$$

The mean stress τ_m then follows from the definition of the shear flow as

$$\tau_m = \frac{M_x}{2t \, A_m} \,. \tag{11.65}$$

It is seen that the mean shear stress τ_m is inversely proportional to the local wall thickness t.

Stress distribution over the wall thickness

The stress distribution over a cross-section from homogeneous torsion can be expressed in terms of the Prandtl stress function Φ as discussed in Section 11.4.1. It was demonstrated in (11.44) that the stress function Φ takes a constant value along each individual contour of the cross-section. For single-cell cross-sections this implies that the boundary values can be characterized by $\Phi = 0$ at the exterior boundary and $\Phi = \Phi_1$ at the boundary within the cell. Between these two boundaries the stress function Φ is determined by the Poisson differential equation (11.43) from which it follows that the variation of Φ over the cell wall is as illustrated in Fig. 11.50.

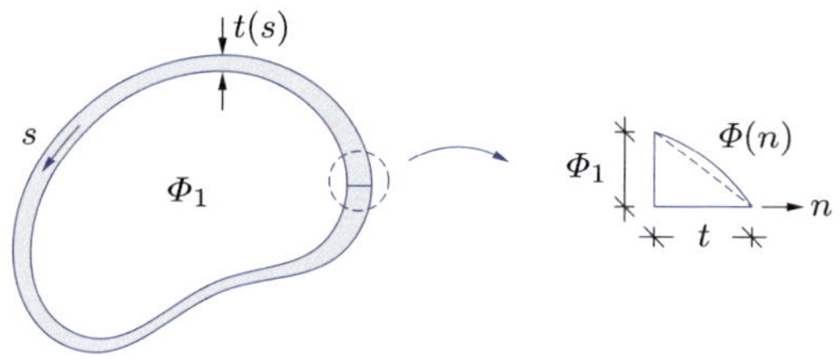

Fig. 11.50: Single-cell section with Prandtl stress function Φ.

The value Φ_1 of the stress function within the cell can be expressed in terms of the shear flow q via an integral over the wall thickness as

$$\Phi_1 = -\int_{-t/2}^{t/2} \frac{\partial \Phi}{\partial n}\, dn = \int_{-t/2}^{t/2} \frac{\tau_{xs}}{\varphi'}\, dn = \frac{q}{\varphi'}\,. \tag{11.66}$$

When using the definition of the shear flow in terms of the mean stress this corresponds to

$$\tau_m = \frac{\Phi_1}{t}\,\varphi'. \tag{11.67}$$

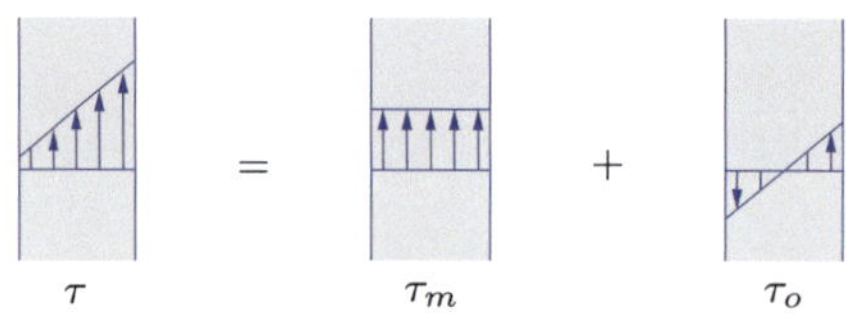

Fig. 11.51: Stress distribution over wall thickness of cell-section.

A constant stress of magnitude of τ_m corresponds to a linear variation of the stress function Φ over the cell wall as illustrated in Fig. 11.50. In addition to this stress there is a contribution corresponding the curvature of Φ. This corresponds to the stress distribution in a similar open cross-section as discussed in Section 11.5.1, and given by

$$\tau_o = 2n\, G\, \varphi', \tag{11.68}$$

where n is a coordinate along the normal to the cell wall, with origin at the center of the wall. Thus, the total stress distribution in a single cell cross-section consists of a uniformly distributed part of magnitude τ_m and a part τ_o with linear variation over the wall thickness and zero mean value, corresponding to torsion of a similar open section. These contributions are illustrated in Fig. 11.51. For a typical thin-walled cell section the component τ_o with linear variation is considerably smaller than the mean stress τ_m.

Torsion stiffness

The basic mechanism generating the shear flow is illustrated in Fig. 11.52 where the discontinuity that would occur in the section with a slit is counteracted by imposing a shear flow q of suitable magnitude along the slit. This condition amounts to the condition that the axial displacement u must vary continuously around the contour.

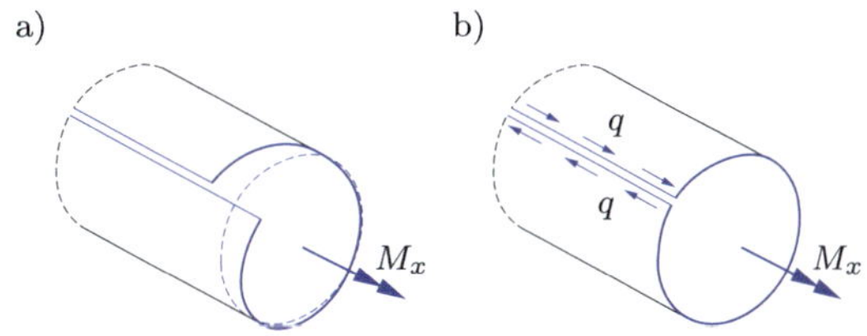

Fig. 11.52: a) Open single-cell beam, b) Closed by shear flow q.

The condition of continuity of the axial displacement is expressed via the shear strain γ_{xs} along the center line of the cell wall. The displacement components in the homogeneous torsion problem consist of an axial displacement u and an in-plane displacement along the tangent of magnitude $h\varphi$ generated by rotation about a point O as illustrated in Fig. 11.49a. The corresponding shear strain along the center line of the wall is

$$\gamma_{xs} = \frac{\partial u}{\partial s} + \frac{\partial (h\varphi)}{\partial x} = \frac{\partial u}{\partial s} + h\varphi'. \tag{11.69}$$

The shear strain γ_{xs} at the center line of the wall is determined by the mean shear stress shear τ_m, and γ_{xs} can therefore be expressed in terms of Φ_1 by (11.67), whereby

$$\frac{\partial u}{\partial s} = \left(\frac{\Phi_1}{G\,t} - h\right)\varphi'. \tag{11.70}$$

For a cellular cross-section the axial displacement varies continuously around each cell, and thus the integral of its derivative around a cell must vanish,

$$\oint \frac{\partial u}{\partial s}\, ds = \varphi' \oint \Big(\frac{\Phi_1}{G\,t} - h \Big) ds = 0. \tag{11.71}$$

In the last integral Φ_1 is a constant that can be moved outside the integral sign, whereby the condition takes the form

$$\Phi_1 \oint_C \frac{ds}{G\,t} - \oint_C h\, ds = 0. \tag{11.72}$$

The second integral is an expression of the double area enclosed within the integration contour as illustrated in Fig. 11.49b, and thus the value Φ_1 of the stress function inside the cell is determined as

$$\Phi_1 = \frac{2A_m}{\displaystyle\oint_C \frac{ds}{G\,t}}\,. \tag{11.73}$$

In this formula A_m is the area within the center line contour.

By (11.49) the torsion moment is expressed by the integral of the stress function Φ over the area enclosed by the exterior contour of the cross-section, whereby

$$M_x = 2\varphi' \int_{A_m} \Phi\, dA \simeq 2\varphi'\, \Phi_1 A_m, \tag{11.74}$$

when the contribution from the local stress variation, represented by the curvature of Φ over the wall thickness, is neglected. When substituting Φ_1 from (11.73) the torsion moment is expressed in terms of the rate of twist φ' as

$$M_x = \frac{4A_m^2}{\displaystyle\oint_C \frac{ds}{G\,t}}\, \varphi'\,. \tag{11.75}$$

This determines the torsion stiffness of a thin-walled cross-section by the so-called Bredt's formula

$$GK = \frac{4A_m^2}{\displaystyle\oint_C \frac{ds}{G\,t}}\,. \tag{11.76}$$

For cross-sections with varying shear modulus, G on the left hand side represents a selected reference value, while G^{-1} appearing in the integral is the local value.

If one imagines a family of cross-sections with the same cell contour C, but different values of the thickness t, it follows from Bredt's formula that the torsion stiffness scales like $K \propto \ell^3 t$, where ℓ is a characteristic length e.g. the dimension across the cess-section. This is in marked contrast to an open thin-walled section, where $K \propto \ell t^3$. The increased torsion stiffness is one of the reasons, that modern bridge cross-sections often contain closed cells.

Example 11.13. Thin-walled circular section. Figure 11.53 shows a thin-walled cylindrical cross-section with radius r and wall thickness $t \ll r$. This example presents an analysis of this important problem according to the thin-walled assumptions as presented above, as well as a comparison with the results from the full analysis from Section 11.3.

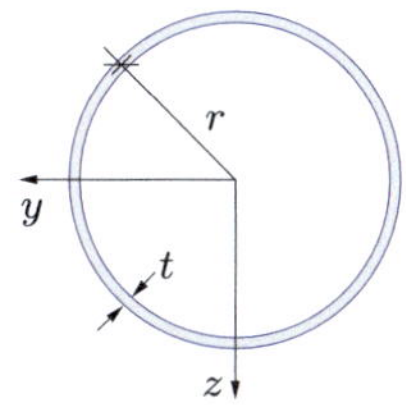

Fig. 11.53: Thin-walled cylinder with radius r.

When r represents the radius of the center line of the circular cylinder the area enclosed by the cross-section is

$$A_m = \pi r^2,$$

and the mean stress τ_m is determined from (11.65) as

$$\tau_m = \frac{M_x}{2tA_m} = \frac{M_x}{2\pi r^2 t}.$$

The full stress distribution also includes the linear variation over the wall thickness from the open section solution. This stress distribution is given in terms of the rate of twist φ', and thus the torsion stiffness is needed to obtain a common expression for the detailed stress distribution.

The torsion stiffness of the section follows from Bredt's formula (11.76), by which

$$K = \frac{4A_m^2}{\oint_C ds/t} = 4(\pi r^2)^2 \frac{t}{2\pi r} = 2\pi r^3 t.$$

The stress distribution over the wall thickness contains two contributions as illustrated in Fig. 11.51: the mean value and a linear variation over the wall thickness. Thus the stress distribution over the thickness is represented by a linear variation between the values

$$\tau_{\pm} = \frac{M_x}{2\pi r^2 t} \pm t\,G\,\varphi' = \Big(\frac{1}{2\pi r^2 t} \pm \frac{1}{2\pi r^3}\Big) M_x = \big(1 \pm (t/r)\big)\frac{M_x}{2\pi r^2 t}.$$

The contribution from the local stress distribution over the wall thickness is linear in t/r.

According to the full theory the torsion parameter is given as the polar moment of inertia $K = I_p$. The expression given above is in fact the usual approximation of the polar moment of inertia for thin walled cylinders. The ratio between the two results is

$$\frac{K_{\text{thin}}}{K_{\text{full}}} = \frac{2\pi r^3 t}{(r_o^2 + r_i^2)\pi r t} = \frac{2r^2}{r_o^2 + r_i^2} = \frac{r^2}{r^2 + (t/2)^2}.$$

Thus, in respect to the torsion stiffness the relative error involved in the thin-wall approximation is of the order $(t/2r)^2$. □

Example 11.14. Thin-walled rectangular section. Figure 11.54 shows a thin-walled rectangular cross-section with height h and width b, measured to the center lines of the section walls. The corresponding flange thicknesses are t_h and t_b, respectively as shown in the figure.

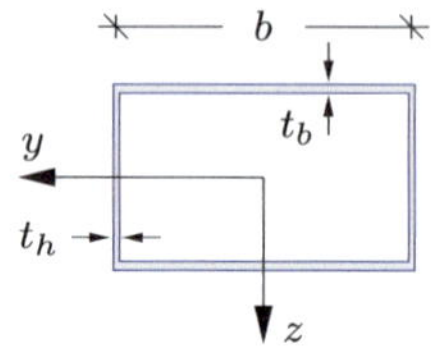

Fig. 11.54: Thin-walled rectangular section.

The area of the rectangle is

$$A_m = hb.$$

The mean stress τ_m is determined from (11.65) as

$$\tau_m = \frac{M_x}{2tA_m} = \frac{M_x}{2thb},$$

with $t = t_h$ in the vertical webs, and $t = t_b$ in the horizontal webs.

The torsion stiffness of the section follows from Bredt's formula (11.76) with constant G,

$$K = \frac{4A_m^2}{\oint_C ds/t} = 4(hb)^2 \frac{1}{2\big(h/t_h + b/t_b\big)} = \frac{2h^2b^2}{h/t_h + b/t_b}.$$

For identical thickness t of all cell walls, this corresponds to

$$K = \frac{2h^2b^2t}{h+b}.$$

As for other single-cell profiles the torsion stiffness is of the order $K \propto \ell^3 t$, where ℓ represents h or b, e.g. in the form $\ell^2 = hb = A_m$. □

11.5.3 Multi-cell sections

In torsion of multi-cell cross-sections a shear flow is generated in each cell in order to generate a continuous axial displacement. Within each cell the stress function Φ takes a constant value Φ_j. In the exterior wall this corresponds to the shear flow

$$q_j = \Phi_j \varphi', \quad j = 1, \cdots \tag{11.77}$$

The shear flows are positive in the counter clockwise direction as shown in Fig. 11.55. In the walls between cells there are two contributing shear flows. This is illustrated in the figure, where the wall AD supports the shear flow q_1 from A to D, and the shear flow q_2 in the opposite direction. The resulting shear flow is

$$q_{AD} = q_1 - q_2 = (\Phi_1 - \Phi_2)\varphi', \tag{11.78}$$

and thus the shear flow is simply given by the increase of the stress function Φ over the wall thickness.

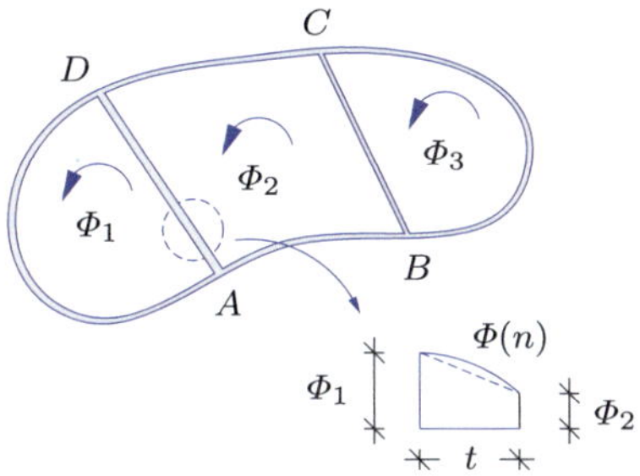

Fig. 11.55: Thin-walled three-cell section.

The shear strain along the center line of the walls is given by (11.69), and continuity of the axial displacement around the center line contour C_j is therefore expressed by the integral

$$\oint_{C_j} \frac{\partial u}{\partial s}\, ds = \varphi' \oint_{C_j} \Big(\frac{\Phi_j - \Phi_k}{G\,t} - h \Big) ds = 0, \tag{11.79}$$

where Φ_k denotes the value of the stress function in the neighboring cell, and $\Phi_0 = 0$ for the exterior contour. The integral of h is equal to $2A_j$ as illustrated in Fig. 11.49, and thus the uniqueness condition of cell j takes the form

$$\Phi_j \oint_{C_j} \frac{ds}{G\,t} - \sum_{k \neq j} \Phi_k \int_{C_{jk}} \frac{ds}{G\,t} = 2A_j\,. \tag{11.80}$$

In this relation C_j denotes the full center line contour around cell j, while C_{jk} is the part of this contour that borders the neighbor cell k.

There is an equation of the form (11.80) for each cell, and each cell is associated with the stress function value Φ_j within the cell. Thus, continuity of the axial displacement around each cell provides the necessary number of equations to determine these values of the stress function. It is convenient to introduce the coefficients

$$B_{jj} = \oint_{C_j} \frac{ds}{G\,t}\,, \qquad B_{jk} = -\int_{C_{jk}} \frac{ds}{G\,t}\,. \tag{11.81}$$

The values Φ_j of the stress function within the cells are then determined from the equations

$$\sum_k B_{jk}\,\Phi_k = 2A_j\,, \qquad j = 1, \cdots . \tag{11.82}$$

When the values Φ_j have been determined, the torsion stiffness follows from the integral of the stress function as

$$GK = 2\int_{C_0} \Phi \, dA \simeq 2\sum_j \Phi_j A_j \,. \tag{11.83}$$

If the shear modulus G changes over the cross-section, the local values are used in the integrals (11.81) defining the equation coefficients. The torsion stiffness GK is given by (11.83) and can be split into a representative value of the shear modulus G and a corresponding torsion constant K. Finally, the mean stress in the walls follows from the shear flows as

$$\tau_{jk} = \frac{1}{t}(\Phi_j - \Phi_k)\,\varphi', \tag{11.84}$$

where t is the local wall thickness.

Example 11.15. Three-cell box profile. The procedure is illustrated by application to the three-cell box profile shown in Fig. 11.56. The cell size is defined by the length a, while the web thickness is t and the flange thickness is $2t$. The shear modulus is G in the full section.

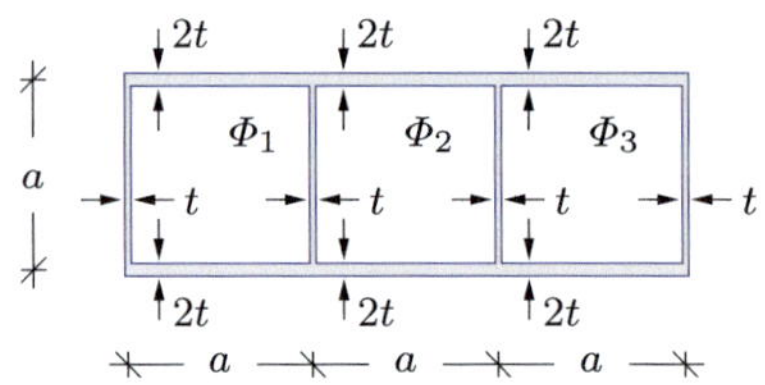

Fig. 11.56: Rectangular cross-section with three cells.

The coefficients for the individual cells follow from (11.81a) as

$$B_{11} = B_{22} = B_{33} = \frac{2}{G}\Big(\frac{a}{2t} + \frac{a}{t}\Big) = \frac{3}{G}\frac{a}{t}\,,$$

while the coefficients for neighboring cells follow from (11.81b),

$$B_{12} = B_{21} = B_{23} = B_{32} = -\frac{1}{G}\frac{a}{t}\,.$$

The cell areas are

$$A_1 = A_2 = A_3 = a^2,$$

whereby the equations (11.82) take the form

$$\begin{bmatrix} 3 & -1 & \\ -1 & 3 & -1 \\ & -1 & 3 \end{bmatrix} \begin{bmatrix} \Phi_1 \\ \Phi_2 \\ \Phi_3 \end{bmatrix} = 2Gat \begin{bmatrix} 1 \\ 1 \\ 1 \end{bmatrix}.$$

It follows from the symmetry of the cross-section that $\Phi_3 = \Phi_1$, and the solution then follows immediately as

$$\Phi_1 = \Phi_3 = \tfrac{8}{7} Gat\,, \qquad \Phi_2 = \tfrac{10}{7} Gat\,.$$

This gives the torsion stiffness

$$GK = 2a^2(\Phi_1 + \Phi_2 + \Phi_3) = \tfrac{52}{7} Ga^3 t\,.$$

The mean stress in the exterior and interior webs follow from (11.84) as

$$\tau_{\rm ex} = \frac{\Phi_1}{t}\,\varphi' = \tfrac{8}{7} Ga\,\varphi'\,, \qquad \tau_{\rm in} = \frac{\Phi_1 - \Phi_2}{t}\,\varphi' = -\tfrac{2}{7} Ga\,\varphi'\,.$$

The mean shear stress in the flanges follow in the same way, using the wall thickness $2t$. □

11.6 Exercises

Exercise 11.1. Consider the cantilever beam with the composite cross-section in Example 11.1.

a) Determine magnitude and direction of the longitudinal shear stress in the core material at the upper interface.

b) Determine the distribution of the shear flow and the mean shear stress τ_m in the cross-section.

Exercise 11.2. The figure shows a cantilever beam with an I-profile loaded by a tip force in the y-direction. The beam is homogeneous with elastic modulus E_0, and the dimensions of the cross-section are given in terms of a and t, as shown in the figure. The cross-section is thin-walled with $t \ll a$.

a) Determine the distribution of the shear flow q, and compare with Fig. 11.28c.

b) Find the magnitude and location of the maximum shear stress.

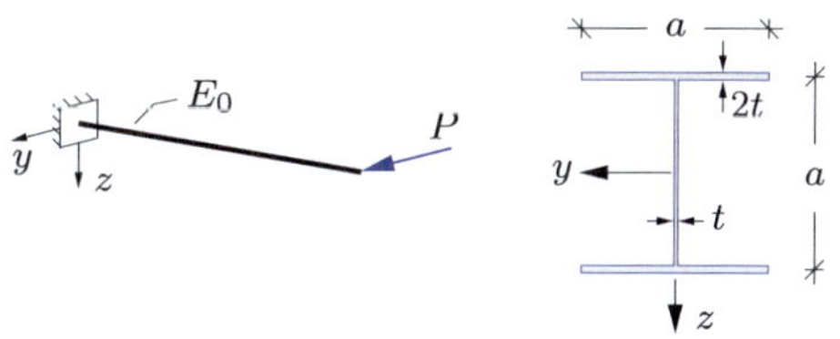

c) Assume the combined load case: $Q_y = P$ and $Q_z = P$. Use the results in Example 11.4 to find the maximum shear stress corresponding to this load combination.

Exercise 11.3. The figure shows a simply supported beam of length ℓ. A transverse force P is acting on the mid section of the beam in the z-direction. The beam is homogeneous with elastic modulus E_0. The cross-section of the beam is rectangular with height h and width b.

a) Determine the distribution of the moment M_z and the shear force Q_z.

b) Identify the cross-section with maximum moment and shear force.

c) Find the maximum normal stress $\sigma_{\rm max}$ in the cross-section.

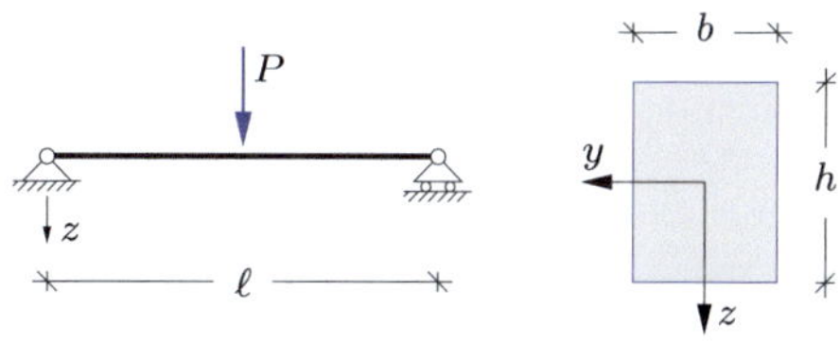

d) Determine the distribution of the shear flow and find the maximum shear stress $\tau_{\rm max}$.

e) Find the ratio $\sigma_{\rm max}/\tau_{\rm max}$ for $\ell/h = 15$.

Exercise 11.4. The figure shows a simply supported beam of length $3a$. A transverse force $3P$ is acting in the z-direction at distance a from the left support. The beam is homogeneous with elastic modulus E_0. The cross-section of the beam is a U-profile with height h and width h. The thickness of the horizontal upper flange is $2t$, while the thickness of the two vertical flanges is t, as shown in the figure. The cross-section is thin-walled with $t \ll h$.

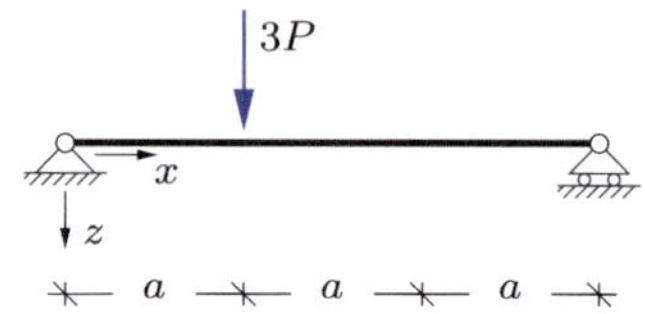

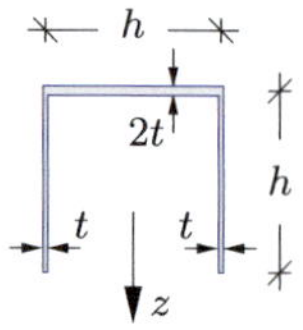

a) Determine the location of the elastic center.
b) Determine the relevant cross-section moment(s) of inertia.
c) Determine the distribution of the shear flow and the shear stress in the cross-section.
d) Determine the shear flow in the x direction at the right corner of the cross-section.

Exercise 11.5. The figure shows an I-profile cross-section with wall thickness t and overall dimensions given in terms of a. The cross-section is loaded by a shear force $Q_z = P$.

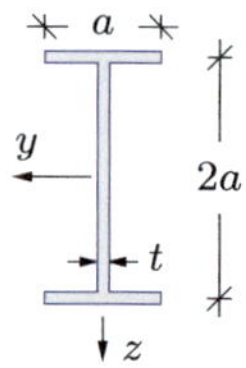

a) Determine the relevant cross-section moment(s) of inertia.
b) Determine the distribution of the shear flow q and the shear stress τ in the cross-section.
c) Determine the ratio between the maximum shear stress in the flange $\tau_f^{\max}$ and in the web $\tau_w^{\max}$.

Exercise 11.6. The figure shows a C-profile with height $2a$, width a, thickness t and constant elastic modulus E_0. The cross-section is initially loaded by the shear force component $Q_z = P$.

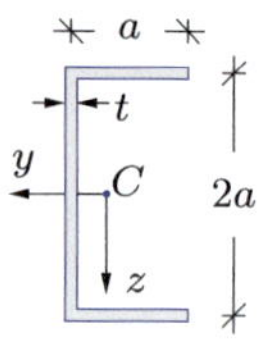

a) Find the location of the elastic center C and determine the relevant moment(s) of inertia.
b) Determine the distribution of the shear flow q and find the magnitude and location of the maximum shear stress $\tau_{\max}$ in the cross-section.
c) Determine the distribution of the shear flow for the load case $Q_y = P$ and $Q_z = 0$.
d) Find the maximum shear stress for the load case in c).
e) Find the location of the shear center A.

Exercise 11.7. The figure to the left shows a cantilever beam with a tip load P. The length of the beam is $40a$ and the dimensions of the cross-section are given in terms of a, as shown in the figure to the right. The thickness of the horizontal upper flange is $5t$, while the thickness of the remaining flanges is t. The cross-section is thin-walled with $t \ll a$.

a) Determine the distribution of the moment M_z and the shear force Q_z.

b) Determine the location of the elastic center C in the cross-section.

c) Determine the moment of inertia $I_{\bar{z}}$.

d) Determine the distribution of the normal stress σ in the cross-section.

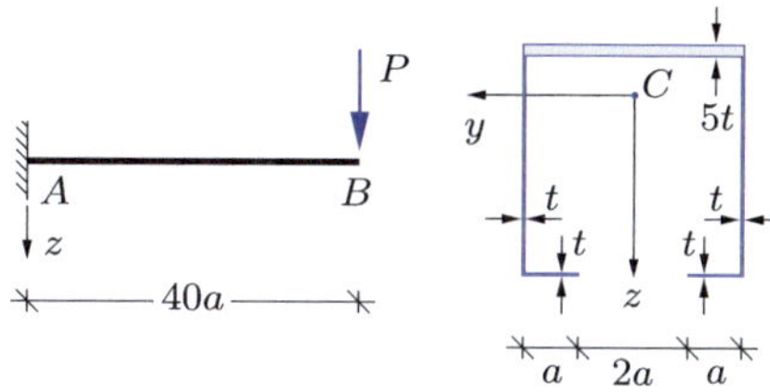

e) Determine the distribution of the shear stress τ in the cross-section.

Exercise 11.8. The figure shows a simply supported beam of length $10a$, which is loaded by a transverse force P at the center of the beam. The cross-section is a double symmetric box profile with overall dimensions given in terms of a and wall thickness t. The profile is homogeneous and thin-walled with $t \ll a$.

a) Find the distribution of the internal forces and the moment M_z and shear force Q_z at the section indicated by the dashed line immediately to the left of the force P.

b) Find the distribution of the normal stress σ and shear stress τ in the cross-section.

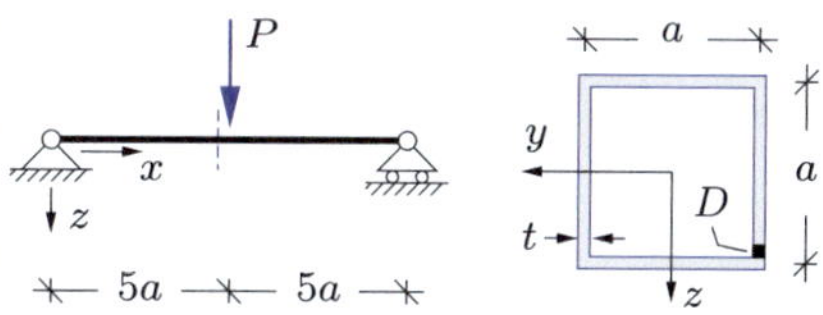

c) Determine the largest principal stress component σ_1 at point D in the cross-section, and find the orientation of σ_1 with respect to the longitudinal x axis.

d) Determine the von Mises stress at point D in the cross-section.

Exercise 11.9. The figure shows a cantilever beam of length $20a$ loaded by a tip force P in B. The beam is homogeneous and the cross-section is a symmetric thin-walled profile with overall dimensions given in terms of a, as shown in the figure to the right. The horizontal flange thickness is $2t$, while the thickness of the vertical flanges is $5t$.

a) Determine the moment M_z and shear force Q_z at the support A.

b) Find the location of the elastic center C and determine the cross-section moment of inertia $I_{\bar{z}}$.

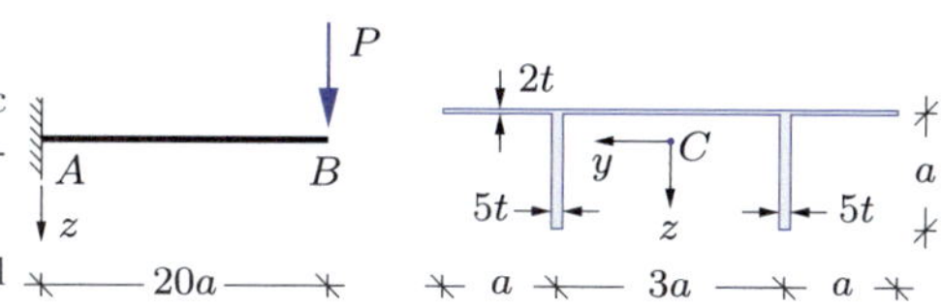

c) Find the distribution of the normal stress σ and shear stress τ in the cross-section.

d) Determine the maximum von Mises stress in the top flange, and compare with the von Mises stress at the bottom of the vertical flanges.

Exercise 11.10. The figure shows a cantilever beam of length ℓ with torsion stiffness GK and a torsion moment $\bar{M}$ applied at $x = \ell$.

a) Integrate the stiffness relation $GK\varphi'(x) = M_x(x)$ to obtain an expression for $\varphi(x)$.

b) Determine the arbitrary integration constant in a) via the boundary condition in $x = 0$.

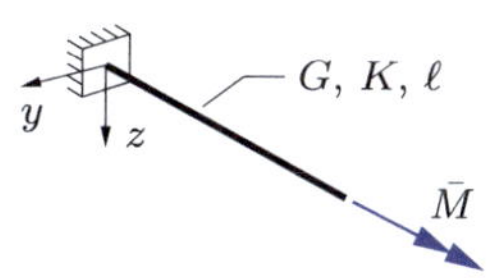

c) Find the expression for the angle of twist $\varphi(\ell)$ at the tip of the beam.

Exercise 11.11. The figure shows two thin-walled circular cross-sections with radius a and wall thickness t.

a) Determine the torsion stiffness parameter K_{closed} of the closed cross-section in Fig. a).

b) Determine the torsion stiffness parameter K_{open} of the open cross-section in Fig. b).

c) Determine and comment on the torsion stiffness ratio $K_{\text{open}}/K_{\text{closed}}$.

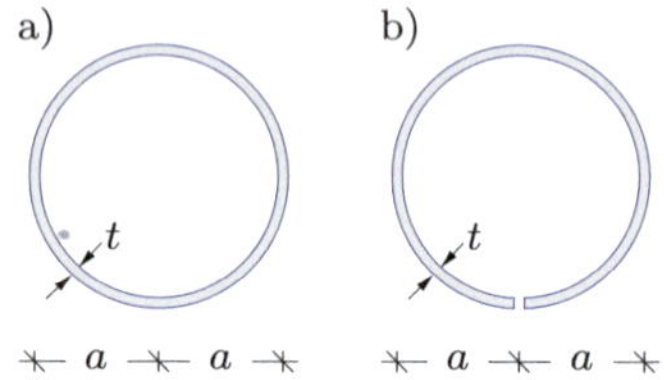

Exercise 11.12. The figure shows two thin-walled single cell cross-sections with characteristic dimension a and wall thickness t.

a) Determine and compare the shear flow q in the two cross-sections when exposed to a torsion moment M_x.

b) Determine and compare the torsion stiffness parameter K of the cross-sections.

c) Determine and compare the stiffness to material ratio K/A of the cross-sections.

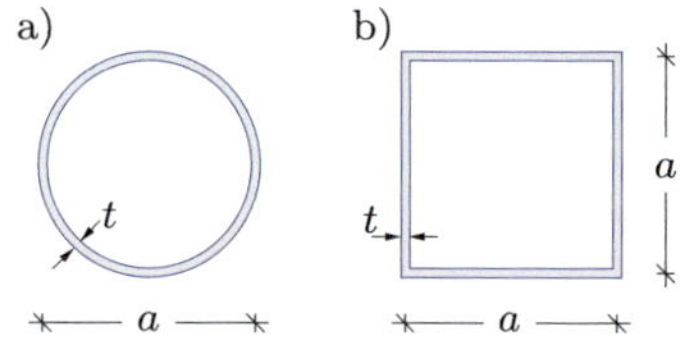

Exercise 11.13. The figure shows two thin-walled box sections with overall dimensions given in terms of a. The thickness of the vertical flanges is t, while the thickness of the top and bottom flanges is $2t$, and $a = 12t$. Both sections are loaded by a torsion moment M_x.

a) Determine the torsion stiffness parameter K of the closed cross-section in Fig. a).

b) Determine the maximum shear stress $\tau_{\max}$ in each of the flanges of the closed cross-section.

c) Repeat a)–b) for the open cross-section in Fig. b) and compare the results.

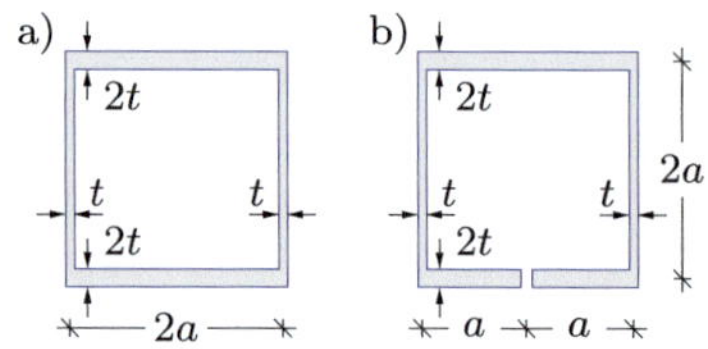

Exercise 11.14. The figure shows the cross-section of a box girder with overall dimensions given in terms of a in the figure. The thickness of the upper horizontal flange is $2t$, while the thickness of the remaining three flanges is t. The cross-section is loaded by a torsion moment M_x. The material is linear elastic with shear modulus G. The cross-section is thin-walled with $t \ll a$.

a) Determine the shear flow q in the cross-section.

b) Determine the mean shear stress τ_m in each of the flanges of the cross-section.

c) Determine the torsion stiffness GK.

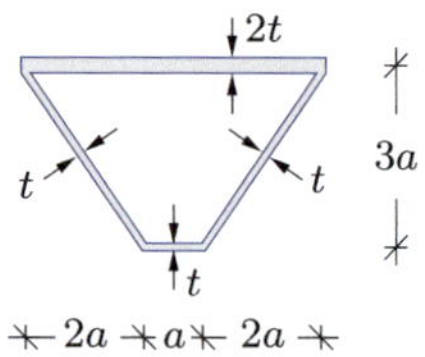

Exercise 11.15. The figure shows a cantilever beam with a thin-walled box cross-section. The beam is loaded by a downward tip force P with vertical line of action in $y = -a$. The

dimensions of the cross-section are given in terms of a as shown in the figure and the wall thickness is t for all flanges.

a) Find the distribution of the shear force Q_z and the torsion moment M_x.

b) Determine the shear stress distribution from the shear force Q_z.

c) Determine the shear stress distribution from the torsion moment M_x.

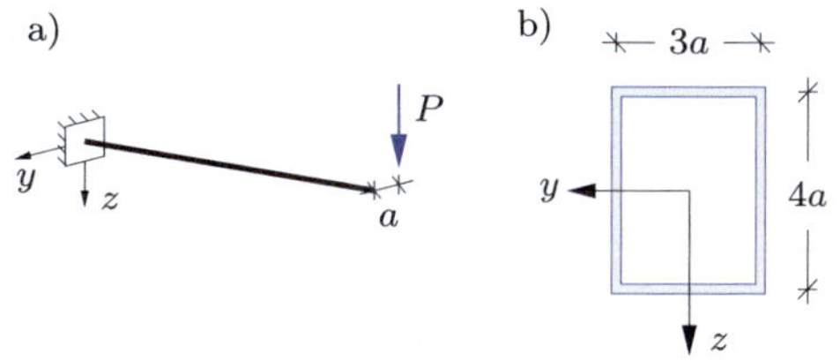

d) Determine the resulting shear stress distribution by combining the results in b) and c), and find the maximum shear stress $\tau_{\max}$.

Exercise 11.16. The figure shows a thin-walled C-shaped channel section loaded by a downward force P with line of action coinciding with the web of the channel section. The overall dimensions of the cross-section are given in terms of a as shown in the figure and the wall thickness is $t = a/10$ for the web and $2t$ for the flanges. The shear center A of the cross-section is located on the line of symmetry at distance $e = \frac{3}{14}a$ behind the web.

a) Determine magnitude and direction of the shear force and the torsion moment acting on the cross-section.

b) Determine the shear stress distribution from the shear force.

c) Determine the shear stress distribution from the torsion moment.

d) Determine the resulting shear stress distribution by combining the results in b) and c), and find the maximum shear stress $\tau_{\max}$.

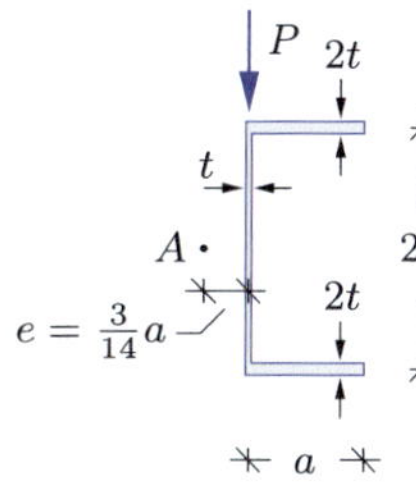

Exercise 11.17. The figure shows a cantilever beam loaded by a normal force $N = P$ and a torsion moment $M_x = 0.4Pa$. The thin-walled cross-section is a double symmetric box profile with overall dimensions given in terms of a as shown in the figure. The thickness of the top and bottom flanges is $2t$, while the thickness of the side flanges is t. The material is linear elastic with elastic modulus E and shear modulus $G = E/2.6$.

a) Determine the distribution of the normal stress from N.

b) Determine the shear stress distribution from M_x.

b) Find the largest principal stress σ_1 in each of the four flanges.

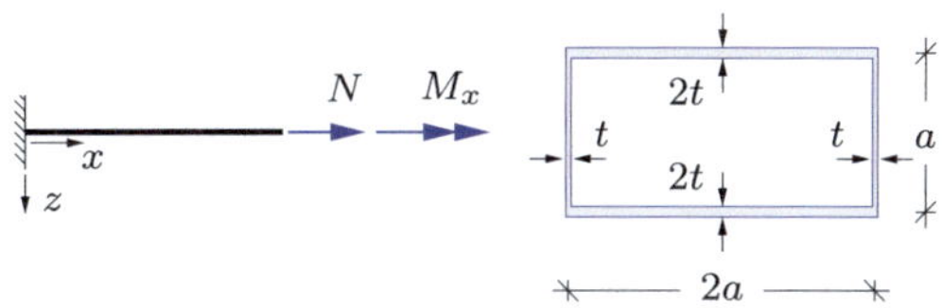

d) Determine the angle between the largest principal stress component σ_1 and the longitudinal axis x, and show the orientation of the principal stresses in a figure.

References

Z. P. Bažant and L. Cedolin. *Stability of Structures.* World Scientific Publishing, Singapore, 2010.

L. S. Beedle and L. Tall. Basic column strength. *Journal of the Structural Division, ASCE*, 86, 1960.

A. P. Boresi, K. P. Chong, and J. D. Lee. *Elasticity in Engineering Mechanics.* Wiley, New Jersey, 2011.

J. Chakrabarty. *Applied Plasticity.* Springer, New York, 2009.

W. F. Chen and D. J. Han. *Tubular Members in Offshore Structures.* Pitman, London, 1985.

R. D. Cook, D. S. Malkus, M. E. Plesha, and R. J. Witt. *Concepts and Applications of Finite Element Analysis.* Wiley, New York, 4th edition, 2002.

P. J. Dowling, P. Knowles, and G. W. Owens. *Structural Steel Design.* Butterworths, London, 1988.

C. L. Dym. *Stability Theory and it Application to Structural Mechanics.* Noordhoff, Leyden, 1974.

Y. C. Fung. *Foundations of Solid Mechanics.* Noordhoff, Prentice-Hall, Englewood Cliffs, N.J., 1965.

J. E. Gordon. *Structures or Why Things Don't Fall Down.* Da Capo Press, Cambridge, Mass., 2nd edition, 2003.

J. Heyman. *Coulombs Memoir on Statics.* Imperial College Press, London, 2nd edition, 1997.

J. Heyman. *Structural Analysis: A Historical Approach.* Cambridge University Press, Cambridge, 2nd edition, 1998.

R. C. Hibbeler. *Mechanics of Materials.* Prentice Hall, Upper Saddle River, N.J., 2005.

R. M. Jones. *Mechanics of Composite Materials.* Taylor & Francis, Philadelphia, Pa., 2nd edition, 1999.

S. Krenk, J. Høgsberg, *Statics and Mechanics of Structures*,
DOI 10.1007/978-94-007-6113-1,

S. Krenk. A general format for curved and nonhomogeneous beam elements. *Computers & Structures*, 50, 1994.

S. Krenk. Characteristic state plasticity for granular materials. Part 1: Basic theory. *International Journal of Solids and Structures*, 37, 2000.

S. Krenk. *Non-linear Modeling and Analysis of Solids and Structures.* Cambridge University Press, Cambridge, 2009.

P. Lade and J. M. Duncan. Cubical triaxial tests on cohesionless soil. *Journal of the Soil Mechanics and Foundations Division, ASCE*, 99, 1973.

S. G. Lekhnitskii. *Theory of Elasticity of an Anisotropic Elastic Body.* Holden-Day, San Francisco, 1963.

J. Lemaitre and J. L. Chaboche. *Mechanics of Solid Materials.* Cambridge University Press, Cambridge, 1990.

G. T. Mase and G. E. Mase. *Continuum Mechanics for Engineers.* CRC Press, Boca Raton, Florida, 2nd edition, 1999.

N. S. Ottosen and M. Ristinmaa. *The Mechanics of Constitutive Modeling.* Elsevier, Amsterdam, 2005.

J. N. Reddy. *Mechanics of Laminated Composite Plates and Shells: Theory and Analysis.* CRC Press, Boca Raton, Florida, 2nd edition, 2004.

G. Strang. *Linear Algebra and Its Applications.* Wellesley Cambridge Press, Wellesley, Mass., 3rd edition, 2001.

S. P. Timoshenko. *History of Strength of Materials.* Dover, Mineola, New York, 1983.

S. P. Timoshenko and J. M. Gere. *Theory of Elastic Stability.* Dover, Mineola, New York, 2nd edition, 2009.

S. P. Timoshenko and J. N. Goodier. *Theory of Elasticity.* McGraw-Hill, Singapore, 3rd edition, 1970.

O. C. Zienkiewicz and R. L. Taylor. *The Finite Element Method.* Butterworth-Heinemann, Oxford, 5th edition, 2000.

Index

S. Krenk, J. Høgsberg, *Statics and Mechanics of Structures*,
DOI 10.1007/978-94-007-6113-1,

MIX
Papier aus verantwortungsvollen Quellen
Paper from responsible sources
FSC® C105338

If you have any concerns about our products,
you can contact us on
ProductSafety@springernature.com

In case Publisher is established outside the EU,
the EU authorized representative is:
Springer Nature Customer Service Center GmbH
Europaplatz 3, 69115 Heidelberg, Germany

Printed by Libri Plureos GmbH
in Hamburg, Germany